Evolutionary biology of the fungi

Evolutionary biology of the fungi

SYMPOSIUM OF
THE BRITISH MYCOLOGICAL SOCIETY
HELD AT THE UNIVERSITY OF BRISTOL
APRIL 1986

EDITED BY

A.D.M. RAYNER, C.M. BRASIER &
DAVID MOORE

The right of the
University of Cambridge
to print and sell
all manner of books
was granted by
Henry VIII in 1534.
The University has printed
and published continuously
since 1584.

CAMBRIDGE UNIVERSITY PRESS

CAMBRIDGE
NEW YORK   PORT CHESTER
MELBOURNE   SYDNEY

CAMBRIDGE UNIVERSITY PRESS
Cambridge, New York, Melbourne, Madrid, Cape Town,
Singapore, São Paulo, Delhi, Tokyo, Mexico City

Cambridge University Press
The Edinburgh Building, Cambridge CB2 8RU, UK

Published in the United States of America by Cambridge University Press, New York

www.cambridge.org
Information on this title: www.cambridge.org/9780521279253

First published 1987
Reprinted 1989
First paperback edition 2011

A catalogue record for this publication is available from the British Library

Library of Congress Cataloguing in Publication data

British Mycological Society.   Symposium (1986 :
  University of Bristol)
  Evolutionary biology of the fungi.

  (British Mycological Society symposium series; 12)
  1. Fungi – Evolution – Congresses. I. Rayner,
  A. D. M. (Alan D. M.), 1950-   .   II. Brasier,
  C. M. (Clive Michael), 1942-   .   III. Moore,
  D. (David), 1942-   .   IV. Title.   V. Series.
  QK602.B75   1986   589.2´0438   86-33415

ISBN 978-0-521-33050-3 Hardback
ISBN 978-0-521-27925-3 Paperback

Contents

Contributors

A. M. Ainsworth, *School of Biological Sciences, University of Bath, Claverton Down, Bath BA2 7AY, UK.*

D. James Ballance, *Department of Microbiology, Medical School, University Walk, Bristol BS8 1TD, UK.*

John A. Barrett, *Department of Genetics, Downing Street, Cambridge CB2 3EA, UK.*

S. Bartnicki-Garcia, *Department of Plant Pathology, College of Natural & Agricultural Science, University of California at Riverside, Riverside, California 92521, USA.*

G. W. Beakes, *Department of Plant Biology, The University, Newcastle-upon-Tyne NE1 7RU, UK.*

C. M. Brasier, *Pathology Branch, Forest Research Station, Alice Holt Lodge, Wrecclesham, Farnham, Surrey GU10 4LH, UK.*

Sir John H. Burnett, *Office of the Principal, Old College, University of Edinburgh, South Bridge, Edinburgh EH8 9YL, UK*

Michael J. Carlile, *Department of Pure & Applied Biology, Imperial College at Silwood Park, Ascot, Berkshire SL5 7PY, UK.*

Jeane R. Cassidy, *Department of Biochemistry, Bowman Gray School of Medicine, 300 South Hawthorne Road, Winston-Salem, North Carolina 27103, USA.*

C. E. Caten, *Department of Genetics, University of Birmingham, Birmingham B15 2TT, UK.*

Tom Cavalier-Smith, *Department of Biophysics, Cell and Molecular Biology, King's College, University of London, 26-29 Drury Lane, London WC2B 5RL, UK.*

D. Coates, *School of Biological Sciences, University of Bath, Claverton Down, Bath BA2 7AY, UK.*

O'N. Ray Collins, *Department of Botany, University of California, Berkeley, California 94720, USA.*

R. C. Cooke, *Department of Botany, The University, Sheffield S10 2TN, UK.*

J. H. Croft, *Department of Genetics, University of Birmingham, PO Box 363, Birmingham B15 2TT, UK.*

R. L. Edwards, *Department of Organic Chemistry, The University, Bradford, UK.*

Everett M. Hansen, *Department of Botany and Plant Pathology, Oregon State University, Corvallis, Oregon 97331-2902, USA.*

D. L. Hawksworth, *CAB International Mycological Institute, Ferry Lane, Kew, Richmond, Surrey TW9 3AF, UK.*

Michèle C. Heath, *Department of Botany, University of Toronto, Toronto, Ontario M5S 1A1, Canada.*

Kari Korhonen, *Finnish Forest Research Institute, PL 18, SF-01301 Vantaa, Finland.*

David H. Lewis, *Department of Botany, The University, Sheffield S10 2TN, UK.*

O.K. Miller, *Department of Biological Sciences, Virginia Polytechnic Institute & State University, Blacksburg, Virginia 24061, USA.*

S. G. Oliver, *Department of Biochemistry and Applied Molecular Biology, University of Manchester Institute of Science and Technology, Sackville Street, Manchester M60 1QD, UK.*

H. Prillinger, *Institut fur Botanik II, Universitat Regensburg, Universitatstrasse 31, D-8400 Regensburg, Federal Republic of Germany*

Patricia J. Pukkila, *Department of Biology, Coker Hall 010A, University of North Carolina, Chapel Hill, North Carolina 27514, USA.*

A. D. M. Rayner, *School of Biological Sciences, University of Bath, Claverton Down, Bath BA2 7AY, UK.*

Eva R. Sansome, *'Stonecroft', Post Office Lane, Lighthorne, Warwickshire CV35 0AP, UK.*

Claudio Scazzocchio, *Institut de Microbiologie, Bat. 409, Université Paris-Sud, 91405 Orsay, France.*

M. A. A. Schipper, *Centraalbureau voor Schimmelcultures, Oosterstraat 1, PO Box 273, 3740 AG Baarn, The Netherlands.*

Geoffrey Turner, *Department of Microbiology, Medical School, University Walk, Bristol BS8 1TD, UK.*

Roy Watling, *Royal Botanic Garden, Edinburgh EH3 5LR, UK.*

John Webster, *Department of Biological Sciences, Hatherly Laboratories, University of Exeter, Prince of Wales Road, Exeter EX4 4PS, UK.*

A. J. S. Whalley, *Department of Biology, Liverpool Polytechnic, Byrom Street, Liverpool L3 3AF, UK.*

J. M. Whipps, *Department of Plant Pathology & Microbiology, Glasshouse Crops Research Institute, Littlehampton, West Sussex BN17 6LP, UK.*

Preface

One has sometimes been left with the impression that in some mycological circles evolution is deemed a subject unworthy of academic treatment. That it is, perhaps, a little too speculative to be scientifically respectable. Certainly, apart from some notable contributions from scattered activists, there seems to have been a failure to develop a substantial body of ideas concerning fungal evolutionary biology, resulting in an ever-growing and now strangely anachronistic gap in mycological philosophy. Yet probably no other field of mycology is potentially so unifying, encompassing the significance of every observable attribute – morphological, physiological or genetical – of every fungal molecule, cell, individual, population or tribe. And in probably no other field of mycology is the development of a fundamental framework of information and ideas so urgently needed to meet the challenge of a burgeoning era of molecular manipulation and gross environmental disturbance, with all the attendant evolutionary implications.

Fortunately, the advent of the strict methodological disciplines of molecular biology seems set to bring new impetus to the subject; hopefully, to the point where it will make its rightful contribution to mycological thought. Feeling that the time had come for an initiative in this field, we invited contributions to a symposium on 'Evolutionary Biology of Fungi' from mycologists with interests in fungal evolutionary biology in its broadest sense; ranging from genome organisation to the evolution of life styles, and from breeding units to phyletic origins. These were presented at the British Mycological Society's 4th General Meeting, held at the University of Bristol. This symposium volume represents a series of essays across the field rather than a comprehensive account, and we hope it will stimulate interest and provide a useful source of information. Taken as a whole, the chapters illustrate the variety of often unique fungal attributes which need to be taken into account in any consideration of fungal evolution. Many chapters demonstrate the extent to which the topic is becoming a fascinating blend of exact science and conceptual synthesis; one or two are unrepentantly and to some, perhaps, provocatively speculative. This seems appropriate enough in a field which, historically, has always excited speculation and expression of often vigorously held opinions. As editors, our standpoint has been that authors are entitled to, and responsible for, their own opinions. We have, therefore, made no attempt to impose uniformity or consensus even when our own views were at variance with those expressed.

It is a pleasure to thank Dr M. F. Madelin and his colleagues, Drs R. Campbell and A. Beckett, for their skilful organisation of the domestic arrangements which contributed so much to the success of the meeting. The Society also wishes to thank Harveys of Bristol for their refreshing contributions, and especially the British Council for donations in support of European visitors.

In the months preceding the Symposium we were saddened by the deaths of two colleagues whose work has contributed significantly to current ideas about fungal evolution, Drs Philip Gregory and Norman Todd. We hope that this volume will be a fitting tribute to their memory.

Clive Brasier
Alan Rayner
David Moore

1

Aspects of the macro- and micro-evolution of the fungi

SIR JOHN H. BURNETT

Office of the Principal, Old College, University of Edinburgh, Edinburgh EH8 9YL

Introduction

The limits of the fungi are not yet adequately defined, the origins and affinities of the groups conventionally classified as fungi are not known, their phylogenetic trends are still largely conjectural and their characteristic modes of speciation remain to be adequately assessed. Nevertheless, developments in biology and mycology over the last fifty years have provided an opportunity for major re-assessments of late nineteenth century phylogenetic ideas, as well as providing a platform from which population genetics and studies on speciation can be launched.

Phylogenetic ideas concerning fungi fell into disrepute earlier this century for it was evident they were based on little evidence. It must be remembered that the climate of ideas at the end of the last century was such that, for example, de Bary devoted an appreciable part of the preface of his book (1884) to refuting the notions that fungi arise spontaneously from either organic matter or the 'granulations moléculaire' of Béchamp (1883); and this was despite Micheli's (1729) demonstration both for moulds and higher fungi that, as he put it, ' . . . I have grown the seeds and have seen the fungi arise from them . . . '! It needs to be remembered, too, that although meiosis was first demonstrated in toadstools in 1884 by Strasburger, knowledge of cytological status and chromosomal cytology was inadequate or misleading; that sexuality in fungi was understood in terms of only the most obvious sexual differentiation; and that, especially in the absence of a fossil record, the fungi were studied and interpreted, in essence, by comparison with what was then known of green plants. Features now seen to be unique or distinctive were then regarded as aberrant or indeed, degenerate. Moreover, the situation was confused by the phenotypic plasticity and pleomorphism of the fungi, which often led to the false association of different species in a supposed life cycle if they happened to arise successively from the same substrate.

Early phylogenetic hypotheses

Three hypotheses for the origin of fungi were propounded in the late nineteenth century and have persisted until quite recently. (1) de Bary (1881, 1884) suggested a monophyletic derivation, from the green algae, of the Oomycetes through the Zygomycetes to the Ascomycetes, which he regarded as a central group, a point to which I shall return. From there he considered that evolution proceeded through the tremelloid rusts to the tremelloid Basidiomycetes from which the rest of the group were derived. Note that he was firmly of the opinion that the rusts and smuts were widely separated and distinct. (2) Sachs (1874) suggested a polyphyletic derivation from various algal groups. The Phycomycetes came largely from the Siphonales, the Ascomycetes from common ancestors of the Floridean red algae and, in due course, Ascomycetes gave rise to Basidiomycetes. The red algal ancestry subsequently received great, if uncritical, support. (3) A few others, notably Dangeard (1886), were impressed by the motile phases of the lower fungi and the free-living and parasitic slime moulds and suggested an ancestry in the Protozoa or from that heterogeneous agglomeration of microscopic forms, the Protista of Haeckel (1894).

These hypotheses rested largely on comparative morphology alone, although de Bary was exceptional in that he pioneered and included developmental and physiological studies as well. One such example is his clear separation of rusts from smuts. Another, which illustrates his biochemical prescience, is the distinction, albeit on staining properties only, of 'cellulose' from 'fungal cellulose'. Even so it was apparent then and for many years that fungal taxonomic groups were not natural and that the relative simplicity of organisation of the fungal thallus made it well-nigh impossible to separate truly primitive features from those reflecting secondary simplifications which had arisen from complex forms. Moreover, adaptive convergence of form and function were frequently not distinguished from true homology.

The predominantly simple microscopic morphology of fungi, the paucity of knowledge concerning their biology and the attitude of botanists towards them delayed the growth of knowledge for several decades. Dicta such as those of Gardner (1886) that ' . . . The fungi are destitute of chlorophyll and hence, or owing to their parasitic and saprophytic habits, any further development in them seems to have been arrested . . . ', although applied originally to fossil fungi, relegated mycology to a minor study of pathogenically tiresome organisms which, biologically speaking, were an evolutionary bye-line!

The expansion of mycology

In the period 1925 − 1955, however, the outlook was transformed. The intensive study of aquatic Phycomycetes and of morphogenetic studies of Ascomycetes and Basidiomycetes provided a sounder basis for comparative morphology. Nutritional studies led to enhanced understanding of metabolism and realisation of fungal biochemical diversity, an aspect furthered by the recognition of the significance of penicillin and the development of industrial fermentations in the 1940s. This, more than the impact of fungal diseases, changed the general attitude of the biological community to the fungi. So, when fungal genetics, developed in yeasts (Winge, 1935) and *Neurospora* (Lindegren, 1936), was matched with gene function in Beadle & Tatum's (1941) classic experiments with the latter fungus, widespread acceptance of the fungi as significant organisms followed.

Towards new phylogenies

The new knowledge which became available for phylogenetic speculation can be summarised under four heads. (a) *Structural features*, in particular the ultrastructure of walls, especially septa; cytoplasmic organelles, the mitotic and meiotic apparatus and the fine structure of motile cells. (b) *Chemical features*, in particular the composition of hyphal walls, mitochondrial and nuclear DNA, ribosomal RNA, and sundry metabolites. (c) *Nutritional and metabolic features*, especially a deeper understanding of the nature of saprotrophism, parasitism and symbiosis, and of metabolic pathways. (d) *Cytological states and sexuality*, notably the haploid and diploid phases, monokaryotic and dikaryotic states, and the variety of fungal mating systems.

The only major deficiencies which still remain in knowledge concerning fungi, compared with most other organisms, are in their fossil record and chromosomal cytology, but progress is being made with both of these.

Phycomycetes and fungal origins

This now discarded taxonomic group illustrates admirably the impact of the new knowledge and its consequential advantages and disadvantages for phylogenetic speculation.

Ivimey Cook's (1928) review of the 'Archimycetes' was heavily criticised (e.g. Bartlett, 1931) but did draw attention to the range of organisms with motile phases and uncertain affinities. Much of this confusion was clarified by the work of Sparrow (1943, 1958) and his students who gave detailed support to Lotsy's (1907) distinction of the importance of flagellation in the motile phase. So, and I shall here use somewhat outmoded terminology, the Chytridiomycetes with a single, usually posterior, whiplash flagellum were recognised together with the Hyphochytriomycetes with a single, anterior, tinsel flagellum in contrast to the biflagellate Plasmodiophoromycetes with two unequal whiplash flagella and the Oomycetes with a whiplash and a tinsel flagellum of nearly equal length. These groups of 'aquatic' fungi, together with the predominantly saprotrophic terrestrial Zygomycetes lacking a motile phase, and the somewhat ill-defined wholly parasitic Trichomycetes (Duboscq, Léger & Tuzet, 1948; Manier & Lichtwardt, 1968) as they came to be known, constituted the old Phycomycetes. What evidence is there of their origins and affinities?

Ultrastructural studies of flagella, their kinetosomes and centrioles, their rootlet complexes and their transition to the flagellum proper demonstrated that all motile phases possessed two centrioles (Fuller, 1966). In uniflagellate fungi only one was functional. Is this evidence of an original biflagellate ancestry or a step from the uniflagellate towards the biflagellate condition? Great diversity was demonstrated in the organisation of the motile cells. Within the Chytridiomycetes Lange & Olsen (1979, 1983) recognised four types plus the '*Physoderma*' type but it was not clear whether these differences were phylogenetically more significant than those between any one of these types and the zoospores of the Hypochytriomycetes of the biflagellate group. It was equally plausible that the differences reflected functional adaptations to a motile role. The Oomycetes, Hyphochytriomycetes, Plasmodiophoromycetes, and even some families of Chytridiomycetes possessed a Golgi dictyosome and mitochondria with inflated cristae; features found in some algae and green plants but not in the rest of the fungi. Does this suggest a common origin for the first group and a divergent phylogeny for the rest? These discoveries alone can give no clear answer but additional information was available. Examination of the fine structure of the mitotic apparatus (Heath, 1980a & b), for instance, has demonstrated that one pattern is highly uniform and virtually unique to the Zygomycetes. It differs greatly from that of the Oomycetes where it is also uniform save for *Thraustochytrium*, and from all other groups with flagellated phases save for the Plasmodiophoromycetes with which there are some similarities. Moreover, it became clear that the Oomycetes possessed a predominantly diploid life cycle, the Plasmodiophoromycetes were probably haploid – dikaryotic and the others, despite some uncertainties, predominantly haploid. Such evidence, in addition to that on the flagella, simply blurs the separation of the groups.

The introduction of chemical features added to the confusion (Klein & Cronquist, 1967). A notable division was created when it was shown that the Oomycetes and Hyphochytriomycetes shared with most algae and green plants the diaminopimelic acid pathway for lysine biosynthesis, while the rest of the fungi (the situation in the Plasmodiophoromycetes and Trichomycetes is unknown) shared the aminoadipic pathway with *Euglena* and the animal kingdom (Vogel, 1964, 1965). On the other hand, a comparison of the molecular weights of 25S and 18S components of ribosomal RNA distinguishes the Trichomycetes and Oomycetes from the rest of the Phycomycetes which, in turn, differ from the Ascomycetes and Basidiomycetes (Lovett & Haselby, 1971; Fraser & Buczacki, 1983). The dominant chemical character in phylogenetic speculation, however, has been that of the principal structural components of the cell wall (Bartnicki-Garcia, 1968, 1970; Chapter 26). Once again the Zygomycetes stand out clearly with chitin and chitosan as principal components. Chytridiomycetes resemble all higher fungi where chitin and β1-3, 1-6 glucans are equally characteristic. For some time it was thought that the Oomycetes were equally clearly defined with walls of a poorly crystalline, celluloselike, hexose polymer and β1-3, 1-6 glucans and this contrasted with the Hyphochytriomycetes where 'cellulose' and chitin co-existed, or the Plasmodiophoromycetes where chitin is associated with high levels of protein and lipid in the walls of the resting cells (Moxham & Buczacki, 1983). However, chitin has been demonstrated in the walls of *Leptomitus* and it is claimed to be present in small quantities in several other Oomycetes (Aronson & Lin, 1978; Vazira-Tehrani & Dick, 1980). The Thraustochytriaceae, in the Oomycetes, appear to have largely galactose polymers. Other variations are known. For example, in *Mucor rouxii*, sporangiospores, in contrast to hyphal and sporangial walls or walls of the yeast phase, have glucans and polyglucosamines as principal components. The Trichomycetes exhibit chitin in *Smittium* species, cellulose in the Eccrinales, and galactosamine or galactose polymers in the Amoebidiales (Lichtwardt, 1973). Do these variations reflect different ecological ways of life, phylogenetic differences or, in part, both?

Septal structure is also a characteristic. There are none in hyphal Oomycetes, although protruberances of cellulose do develop from time to time on the inner wall. Zygomycete septa are generally simple but in some merosporangiferous Mucorales and many Trichomycetes there is a central pore characteristically surrounded by a flared, ventricular inner margin. The pore itself is blocked by an electron dense plug, while in the Dimargaritaceae two large, amorphous global, electron-dense bodies lie on each side of the pore plug (Brain, Jeffries & Young, 1982). Does this signify some phylogenetic connection between these two groups or is it due to convergent evolution towards a functional end, such as a sealing device if rupture occurs near the septum?

It needs to be said that much of this new information often rests upon small sample sizes and this is especially true for the analyses of wall composition. In a sense, therefore, the position is not so very different from that of a century ago. We have more information, of course, but it is not clear how much of it is particular to one or a few species, and how much is of general validity.

Nevertheless, taken together available knowledge suggests that amongst the Phycomycetes: (a) groups with motile zoospores are probably all derived from biflagellate progenitors; (b) the Oomycetes have more in common with the Hyphochytriomycetes than with any other group; (c) the Zygomycetes seem to be remarkably isolated, having most in common with Trichomycetes but little in common with any group that has a motile phase; (d) virtually every permutation and combination of structure and ultrastructure, flagellation and chemical features and in fact, although not documented here, sexuality and life style and cycle, can be found in the group.

These propositions go little further in determining the origins and limits of the fungi than in the nineteenth century and merely suggest that a monophyletic origin is less likely than was thought by de Bary.

There are no discrepancies with the fossil record as it is known at present (Pirozynski, 1976). Although there is doubt about the identification of the earliest fungus-like filaments in the Middle and Upper Precambrian, about 850 million years ago, they are very comparable with an undoubted saprolegniacean-like fossil from the late Precambrian of South Australia. Endozoic Oomycete and Chytridiomycete-like fossils have been identified from the end of the Cambrian Period and, as is well known, a range of fungi existed in the Devonian. The commonest are branched, non-septate hyphae bearing terminal or intercalary vesicles or resting spores. They have been equated with Endogonaceae (Zygomycetes) and indeed, in somewhat later formations, chlamydospores comparable with those of the modern *Glomus* have been found (Wagner & Taylor, 1982). Nevertheless, Stubblefield, Taylor & Miller (1985) have demonstrated in a recent paper the uncertainties that still exist and, in particular, the uncertainty that such fungi represent early vesicular — arbuscular mycorrhizal forms. Other fungi in comparable deposits in Rhynie chert include forms closely resembling modern *Apodachlya* (Harvey, Lyon & Lewis, 1969) and within the spores of *Horneophyton*, sporangia extraordinarily reminiscent of those of the Hyphochytriomycetes (Illman, 1984). Thus, four of the main groups of the Phycomycetes were already present in the earliest fossil assemblages and, so far, Ascomycetes and Basidiomycetes have only been found in material dated at least 50 million years later. Thus, the fossil evidence supports the view that the most primitive fungi were Phycomycetes but gives no real clue to their origins or ancestry.

Neither the fossil record nor the vastly increased knowledge of contemporary fungi throws much light on the origin of the evolutionary features of fungi, such as nutritional status, apical growth, mitosis or sexuality. Although, as Cooke & Whipps (1980) have pointed out, the vast majority of living aquatic fungal groups are parasitic, this provides no evidence for their original nutritional status nor, indeed, as to whether the modes evolved successively or whether there were numerous transformations or reversions. Nor is there really convincing evidence for the attractive hypothesis of Pirozynski & Malloch (1975) that fungi colonised the land in biotrophic associations, although such a condition is probably very ancient (Lewis, Chapter 11).

The selective forces which led to apical, rather than generalised growth are not clear nor is the way in which the apical habit became established. Mitosis and sexuality are equally obscure in their origins and evolution. Heath (1980a) has attempted to assess those features of mitosis which are primitive, and amongst the great mitotic diversity in the Phycomycetes the Zygomycetes exhibit almost all of them, whereas in the Oomycetes mitotic features are relatively advanced. Presumably haploidy preceded diploidy, but the respective advantages of haploidy, diploidy and the dikaryotic condition are not evident in these fungi. Similarly, selfing presumably preceded outcrossing and although dimixis appears to be superimposed on basically self-fertilising bisexual forms in the Oomycetes, there is no such evidence in other groups, notably the Zygomycetes, although a comparative study of trisporic acid metabolism in homomictic and dimictic forms might be enlightening.

Evolution in the Eumycota

It is evident that the great increase in knowledge of lower fungi has failed to resolve their origins and affinities. This is equally true of the

higher fungi or Eumycota; the origin of the Ascomycetes and Basidiomycetes are at least as obscure as that of the Phycomycetes and clear
evolutionary lines within or between them have not been resolved. Here,
therefore, I shall only draw attention to a few intriguing topics as more
detailed proposals appear elsewhere (see Chapters 23 to 29).

Ascomycetes, as judged by species number, are the most successful
fungi, comprising some 45% of known species; the figure would increase
to 70% if, as seems reasonable, Deuteromycetes are included (Hawksworth, Sutton & Ainsworth, 1983). Both morphologically and ultrastructurally they are remarkably uniform, exhibiting a virtually invariant
haplophase with predominantly homomictic mating. There is, of course,
immense variation in conidiospore morphology and ontogeny, in
secondary metabolites, and a vast adaptive radiation into a variety of
ecological niches. The origin of this variation is a problem. The true
natural incidence of heterokaryosis is still unknown in these fungi
(Burnett, 1975). If heterokaryosis does occur, is that remarkable phenomenon of 'relative heterothallism' described in self-fertile *Aspergillus
nidulans* (Pontecorvo *et al.*, 1953), in which heterozygous asci appeared
to be favoured in heterokaryons (16 out of 27 tested), more widespread
than usually supposed? Natural mutation rates are an equally unresearched
area. Could variation be achieved through such a means over a vast period
of time by mutant selection, whether genic or cytological? Much more
evidence is needed from investigations in natural situations. However,
other alternatives need to be examined, such as variation in mitochondrial
DNA and indeed, of mitochondrial plasmids and their roles, since far
more have been identified in Ascomycetes than any other fungi (Grossman
& Hudspeth, 1985). Is this only a combination of chance and a small
sample size? Moreover, this group of fungi should be excellent for the
experimental study of general evolutionary problems, such as the nature
and genetic architecture of biotrophy, parasitism and saprotrophy since
groups of species, or in some cases apparently different strains, vary in
their nutritional behaviour. Present hypotheses differ and no clear
experimental tests are available (e.g. Lewis, 1973; Cooke & Whipps,
1980; and see Chapters 9 & 11).

However, one group of Ascomycetes is remarkably heterogeneous and
morphologically and biologically diverse, namely the yeasts. Their haplo-
diploid, haploid, or diploid life cycles; their unusual cell wall composition
– mannan-glucans or mannan-chitin reflecting their supposed ascomycetous or basidiomycetous affinities; their as yet unique mating system;
and the unusual genetic code of their mitochondrial genomes make them

distinct. Ultrastructurally, they exhibit a range of mitotic and meiotic structures, divide by fusion or budding, exhibit generalised or more polar growth, and have complete, simply poroid, or pseudo-dolipore-like septa. They are saprotrophic or biotrophic, including parasitic, and occur in virtually every terrestrial habitat from the most harsh and demanding to the most luxurious, as well as in fresh-water and marine environments. It is usual to regard these organisms as a 'rag-bag' of aberrant Ascomycetes and Basidiomycetes, but might they not equally well represent an ancestral group?

Even Basidiomycetes could develop from yeasts and the Heterobasideae certainly readily exhibit yeast-like phases. That having been said, the relationships between any of the diverse groups of yeast is as great as between any of them and the rest of the fungi. Savile (1954, 1955, 1968) has provided a plausible argument for a *Taphrina*-like ancestor for Basidiomycetes and his papers provide an excellent introduction to fungal phylogenetic problems in general. He has argued that the apparent similarities of the rusts and smuts represent convergence, rather than common ancestry, an argument borne out by further studies. Rusts exhibit a long haploid phase, relatively elaborate sexual differentiation, a dimictic mating system, and ultrastructural features, both cytoplasmic and in their mitotic and meiotic apparatus, which differentiate them from the Ustilaginales. The latter being more reminiscent of other Basidiomycetes with clamp connections (Heath & Heath, 1976; O'Donnell & McLaughlin, 1981, 1984). The relationship of the Heterobasideae to Homobasideae, if it exists, is unclear and supposed phylogenies within the latter are as diverse as those who speculate!

Microevolution in fungi

Since I have recently reviewed speciation in fungi (Burnett, 1983), and since much of this book deals with particular experimental situations (Chapters 16 − 22), I shall only draw attention here to the most neglected areas and techniques.

Some technical developments

Because hybridisation is apparently difficult between many fungi it is desirable to overcome the difficulties or to utilise techniques which provide information on genetic variation and relatedness. A method worthy of further exploration and application is the hybridisation of protoplasts, their reversion to the hyphal form and subsequent reproduction, a technique pioneered by Anné and Peberdy (1981) between

Penicillium species and used by Croft for *Aspergillus* species (Chapter 21).

Much genetic variation has been demonstrated in many diploid organisms by the study of electrophoretically separated isozymes and other proteins (e.g. Ferguson, 1980). To date this method has not been applied extensively to fungi and the limited applications have usually shown a remarkable degree of homogeneity, rather than the reverse (Erselius & Shaw, 1982). Nevertheless, the method is potentially applicable and might be of especial value with formally diploid or dikaryotic fungi.

The most promising techniques, however, are those which measure true genetic homology, i.e. those employing comparisons of nucleic acids. Nuclear and mitochondrial DNA, either complete single copy sequences or restriction endonuclease fragments, together with molecular weight or sequence studies of ribosomal RNA, are all potential materials for study (Chapters 2 − 6). A particularly instructive series of such studies are those made with species of *Neurospora* since several of these are capable of at least partial hybridisation, so molecular studies can be accompanied by genetic and cytological analysis (Perkins, Turner & Barry, 1976). The studies made by Dutta and his colleagues using DNA − DNA hybrid-isation have broadly borne out the results of direct hybridisation (Dutta, 1976; Dutta *et al.*, 1976; Williams, Mukhopadyha & Dutta, 1981). Greater resolution has been obtained through a comparison of cloned random fragments of nuclear DNA by D. O. Natvig & J. W. Taylor (per-sonal communication). By employing several strains of each species they were able to demonstrate intra-specific polymorphisms and, for example, showed that there was a considerable range of genetic diversity within *N. intermedia* comparable with that between any other species, although it was a species regarded by Perkins and Dutta as closer to *N. crassa* and *N. sitophila* than any others. Mitochondrial DNA and mitochondrial plasmid DNA has also been studied and compared in *Neurospora* (Grossman & Hudspeth, 1985; Kinsey, 1985; Natvig, May & Taylor, 1984; Taylor & Smolich, 1985; Taylor, Smolich & May, 1985). It is evident that distinct populations of mitochondrial DNA exist in *N. crassa*, some of which show a distinctive local geographical distribution, e.g. the Louisiana isolates. Both nonrandom length mutations and fewer nucleotide substitutions, determined from restriction site changes, were identified. Using plasmid DNA it was shown that there was more homology between plasmids from *N. crassa* and *N. intermedia* than between those from either of these species and *N. tetrasperma*. No plasmid DNA showed homology with mitochondrial DNA nor did presence or absence of plasmids appear to be affected by the nature of the mitochondrial DNA. Such studies are clearly in their infancy but they should be able, eventually, to throw light not only on relatedness and patterns of local divergence and isolation, but possibly on the history of genetic exchanges of all kinds.

Ribosomal RNA sequences are thought to be highly conserved and are, therefore, more likely to be of value in studies of the relationships of categories higher than species. Three kinds of comparisons have been made; the molecular weights of different fragments, complementarity with DNA, or nucleotide sequences of 5S RNA fragments. Reference has already been made to such comparisons but results so far are few (see Kurtzman, 1985; Pukkila, Chapter 5).

Some microevolutionary problems

On a number of topics there is almost complete ignorance. For instance, although the effects of natural selection may often be inferred, there is virtually nothing known about it by direct observation or experiment. This is also true for spontaneous natural mutation rates, gene dispersal, isolating mechanisms and, with the exception of the Oomycetes, chromosomal changes, whether structural or numerical. All are important if true pictures of the modes of fungal speciation are to be determined. The last is of particular interest, however, since the few chromosome numbers now emerging suggest that polyploidy does occur in Eumycota. This at times implies allopolyploidy and hence hybridisation although both are generally supposed to be rare in fungi (see, for example, Rogers, 1973; Burnett, 1983).

Allopatry appears to be an established mode of speciation, although a much wider range of supporting evidence is badly needed. However, it does look as though sympatric speciation, accompanied by the occurrence of sibling species may not be uncommon (Burnett, 1975, 1983). As more is learned of somatic incompatibility, its origins and operation, and its relationship to genetic isolation, so it will become easier to define the nature of the problems to be studied. Meanwhile, with the situations already described a number of questions can be investigated. For example, do all apparent sibling species arise in the same way? Are the differences determined initially by single gene mutations? Are cytological changes involved? Once sibling species are separated, what is the probability of their persistence or reversion? Can they show partial compatibility with their congeners, and what circumstances, if any, lead to their further divergence and the development of other character differences? Finally, it

12   *John H. Burnett*

may be asked whether or not sympatric speciation is always associated with the development of sibling species or, indeed, whether the hypothesised connection in fungi is in fact characteristic of fungal evolution.

This cursory survey of some of the issues which fungal evolution raises at the macro- or micro-level is unlikely to satisfy anyone or everyone. However, if, as I wrote ten years ago in another context ' . . . it will promote the progress of mycology whether by precept or provocation' then I shall be more than content and evolutionary biology will be the richer.

References

Anné, J. & Peberdy, J. F. (1981). Characterisation of inter-specific hybrids between *Penicillium chrysogenum* and *P. roquefortii* by iso-enzyme analysis. *Transactions of the British Mycological Society*, **77**, 401 − 408.

Aronson, J. M. & Lin, C. E. (1978). Hyphal cell wall chemistry of *Leptomitus lacteus*. *Mycologia*, **70**, 363 − 369.

Bartlett, A. W. (1931). Review. *New Phytologist*, **30**, 136 − 141.

Bartnicki-Garcia, S. (1968). Cell wall chemistry, morphogenesis, and taxonomy of fungi. *Annual Review of Microbiology*, **22**, 87 − 108.

Bartnicki-Garcia, S. (1970). Cell wall composition and other biochemical markers in fungal phylogeny. In *Phytochemical Phylogeny*, ed. Harborne, J. B., pp. 81 − 103. London: Academic Press.

Beadle, G. W. & Tatum, E. L. (1941). Genetic control of biochemical reactions in *Neurospora*. *Proceedings of the National Academy of Sciences, USA*, **27**, 499 − 506.

Béchamp, A. (1883). *Les Microzymes dans leurs rapports avec l'hétérogenie, l'histogénie, la physiologie et la pathologie*. Paris.

Brain, A. P. R., Jeffries, P. & Young, T. W. K. (1982). Ultrastructure of septa in *Tieghemiomyces californicus*. *Mycologia*, **74**, 173 − 181.

Burnett, J. H. (1973). *Fundamentals of Mycology*. London: Edward Arnold.

Burnett, J. H. (1975). *Mycogenetics*. London: John Wiley & Sons.

Burnett, J. H. (1983). Speciation in fungi. *Transactions of the British Mycological Society*, **81**, 1 − 14.

Cooke, R. C. & Whipps, J. M. (1980). The evolution of modes of nutrition in fungi parasitic on terrestrial plants. *Biological Reviews*, **55**, 341 − 363.

Dangeard, P. A. (1886). Recherches sur les organismes inférieures. *Annales des Sciences Naturelles; Botanique*, **4**, 241 − 341.

de Bary, A. (1881). Zur Systematik der Thallophyten. *Botanisches Zeitung*, **39**, 1 − 17; 33 − 36.

de Bary, A. (1884). *Vergleichende Morphologie und Biologie der Pilze. Mycetozoen und Bakterien*. Leipzig.

Duboscq, O., Léger, L. & Tuzet, O. (1948). Contribution à la connaissance des Eccrinales: les Trichomycètes. *Archives de Zoologie Experimentalle et generale*, **86**, 29 − 144.

Dutta, S. K. (1976). DNA homologies among heterothallic species of *Neurospora*. *Mycologia*, **68**, 388 − 401.

Dutta, S. K., Sheikh, I., Choppala, J., Aulakh, G. S. & Nelson, W. H. (1976). DNA

homologies among homothallic, pseudo-homothallic and heterothallic species
  of *Neurospora. Molecular and General Genetics*, **147**, 325 – 330.
Erselius, L. J. & Shaw, D. S. (1982). Protein and enzyme differences between
  Phytophthora palmivora and *megakarya*: evidence for self-fertilisation in
  pairings of the two species. *Transactions of the British Mycological Society*,
  78, 227 – 238.
Ferguson, A. (1980). *Biochemical Systematics and Evolution.* Glasgow & London:
  Blackie.
Fraser, R. S. S. & Buczacki, S. T. (1983). Ribosomal RNA molecular weights and
  affinities of the Plasmodiophorales. *Transactions of the British Mycological
  Society*, **80**, 107 – 112.
Fuller, M. S. (1966). Structure of the uniflagellate zoospores of aquatic Phycomycetes.
  In *The Fungus Spore*, ed. Madelin, M. F., pp. 67 – 84. London:
  Butterworth.
Gardner, J. S. (1886). *A Monograph of the British Eocene Flora I.* London:
  Palaeontological Society.
Grossman, L. T. & Hudspeth, M. E. S. (1985). Fungal mitochondrial genomes. In
  Gene Manipulations in Fungi, ed. Bennett, J. W. & Lasure, L. L.,
  pp. 65 – 103. Orlando, Florida: Academic Press.
Haeckel, E. H. P. A. (1894). *Systematische Phylogenie der Protisten und Pflanzen.*
  Berlin.
Harvey, R., Lyon, A. G. & Lewis, P. N. (1969). A fossil fungus from the Rhynie
  Chert. *Transactions of the British Mycological Society*, **53**, 155 – 156.
Hawksworth, D. L., Sutton, B. C. & Ainsworth, G. C. (1983). *Dictionary of the Fungi.*
  Kew, Surrey: Commonwealth Mycological Institute.
Heath, I. B. (1980a). Variant mitoses in lower eukaryotes: indicators of the evolution of
  mitosis. *International Review of Cytology*, **64**, 1 – 80.
Heath, I. B. (1980b). Fungal mitoses, the significance of variations on a theme.
  Mycologia, **72**, 229 – 250.
Heath, M. C. & Heath, I. B. (1976). Ultrastructure of mitosis in the cowpea rust fungus
  Uromyces phaseoli var. *vignae. Journal of Cell Biology*, **70**, 592 – 607.
Illman, W. I. (1984). Zoosporic fungal bodies in the spores of the Devonian fossil
  vascular plant *Horneophyton. Mycologia*, **76**, 545 – 547.
Ivimey-Cook, W. R. (1928). The inter-relationships of the Archimycetes. *New
  Phytologist*, **27**, 298 – 320.
Kinsey, J. A. (1985). *Neurospora* plasmids. In *Gene Manipulations in Fungi*, ed.
  Bennett, J. W. & Lasure, L. L., pp. 245 – 258. Orlando, Florida: Academic
  Press.
Klein, R. M. & Cronquist, A. (1967). A consideration of the evolutionary and
  taxonomic significance of some biochemical, micromorphological and
  physiological characters in the thallophytes. *Quarterly Review of Biology*, **42**,
  105 – 296.
Kurtzmann, C. P. (1985). Molecular taxonomy of the fungi. In *Gene Manipulations in
  Fungi*, ed. Bennett, J. W. & Lasure, L. L., pp. 35 – 63. Orlando, Florida:
  Academic Press.
Lange, L. & Olson, L. W. (1979). The uniflagellate Phycomycete zoospore. *Dansk
  Botanisk Arkiv*, **33**, 7 – 95.
Lange, L. & Olson, L. W. (1983). The fungal zoospore – its structure and biological
  significance. In *Zoosporic Plant Pathogens, a Modern Perspective*, ed.
  Buczacki, S. T., pp. 1 – 42. London: Academic Press.
Lewis, D. H. (1973). Concepts in fungal nutrition and the origin of biotrophy.
  Biological Reviews, **48**, 261 – 278.

14   *John H. Burnett*

Lichtwardt, R. W. (1973). The Trichomycetes: what are their relationships?. *Mycologia*, **55**, 1 — 20.

Lindegren, C. C. (1936). A six-point map of the sex-chromosome of *Neurospora crassa*. *Journal of Genetics*, **32**, 243 — 256.

Lotsy, J. P. (1907). *Algen und Pilze: Vorträge über Botanische Stammesgesichte*. Jena.

Lovett, S. J. & Haselby, J. A. (1971). Molecular weights of the ribosomal ribonucleic acid of fungi. *Archiv für Mikrobiologie*, **80**, 191 — 204.

Manier, J. F. & Lichtwardt, R. W. (1968). Revision de la sytematique des Trichomycètes. *Annales des Sciences Naturelles: Botanique, Séries 12*, **9**, 519 — 532.

Micheli, P. A. (1792). *Nova plantarum genera juxta Tournefortii methodium disposita*. Florence.

Moxham, S. E. & Buczacki, S. T. (1983). Chemical composition of the resting spore wall of *Plasmodiophora brassicae*. *Transactions of the British Mycological Society*, **80**, 297 — 304.

Natvig, D. O., May, G. & Taylor, J. W. (1984). Distribution and evolutionary significance of mitochondrial plasmids in *Neurospora*. *Journal of Bacteriology*, **159**, 288 — 293.

O'Donnell, K. L. & McLaughlin, D. J. (1981). Ultrastructure of meiosis in the Hollyhock rust fungus *Puccinia malvacearum*. III. Interphase I — Interphase II. *Protoplasma*, **108**, 265 — 288.

O'Donnell, K. L. & McLaughlin, D. J. (1984). Post-meiotic mitosis, basidiospore development, and septation in *Ustilago maydis*. *Mycologia*, **76**, 486 — 502.

Perkins, D. D., Turner, B. C. & Barry, E. G. (1976). Strains of *Neurospora* collected from nature. *Evolution*, **30**, 281 — 313.

Pirozynski, K. A. (1976). Fossil Fungi. *Annual Review of Phytopathology*, **14**, 337 — 346.

Pirozynski, K. A. & Malloch, D. W. (1975). The origin of land plants: a matter of mycotrophism. *BioSystems*, **6**, 153 — 164.

Pontecorvo, G., Roper, J. A., Hemmons, L. M., MacDonald, K. D. & Bufton, A. W. J. (1951). The genetics of *Aspergillus*. *Advances in Genetics*, **5**, 141 — 238.

Rogers, J. D. (1973). Polyploidy in fungi. *Evolution*, **27**, 253 — 260.

Sachs, J. (1874). *Lehrbuch der Botanik*. Leipzig.

Savile, D. B. O. (1954). Cellular mechanics, taxonomy and evolution in the Uredinales and Ustilaginales. *Mycologia*, **46**, 736 — 761.

Savile, D. B. O. (1955). A phylogeny of the Basidiomycetes. *Canadian Journal of Botany*, **33**, 60 — 104.

Savile, D. B. O. (1968). The possible interrelationships between fungal groups. In *The Fungi, An Advanced Treatise*, vol III, ed. Ainsworth, G. C. & Sussman, A. S., pp. 649 — 675. New York & London: Academic Press.

Sparrow, F. K. (1943). *The Aquatic Phycomycetes, exclusive of the Saprolegniaceae and Pythium*. Ann Arbor, Michigan: University of Michigan Press.

Sparrow, F. K. (1958). Interrelationships and phylogeny of the aquatic Phycomycetes. *Mycologia*, **50**, 797 — 813.

Strasburger, E. (1884). *Das botanische Practicum*. Jena

Stubblefield, S. P., Taylor, T. N. & Miller, C. E. (1985). Studies of Paleozoic fungi. IV. Wall ultrastructure of fossil Endogonaceous chlamydospores. *Mycologia*, **77**, 83 — 96.

Taylor, J. W. & Smolich, B. (1985). Molecular cloning and physical mapping of the *Neurospora crassa* 74-OR33-1A mitochondrial genome. *Current Genetics*, **9**, 597 — 603.

Taylor, J. W., Smolich, B. & May, G. (1985). Evolutionary comparisons of homologous mitochondrial plasmid DNA from three *Neurospora* species. *Molecular and General Genetics*, **201**, 161 − 167.

Vazira-Tehrani, B. & Dick, M. W. (1980). Neutral and amino sugars from the cell walls of Oomycetes. *Biochemical Systematics and Ecology*, **8**, 105 − 108.

Vogel, H. J. (1964). Distribution of lysine pathways among fungi: evolutionary implications. *American Naturalist*, **28**, 435 − 446.

Vogel, H. J. (1965). Lysine biosynthesis and evolution. In *Evolving Genes and Proteins*, ed. Bryson, V. & Vogel, H. J., pp. 25 − 40. New York: Academic Press.

Wagner, C. W. & Taylor, T. N. (1982). Fungal chlamydospores from the Pennsylvanian of North America. *Review of Palaeobotany and Palynology*, **37**, 317 − 328.

Williams, N. P., Mukhopadhyay, D. & Dutta, S. K. (1981). Homologies of *Neurospora* homothallic species using repeated and non-repeated DNA sequences. *Experientia*, **37**, 1157 − 1158.

Winge, O. (1935). On haplophase and diplophase in some Saccharomycetes. *Comptes rendus de Laboratoire de Carlesburg, séries Physiologique*, **21**, 77 − 112.

2

Structural variation and expression of fungal chromosomal genes

G . TURNER AND D . J . BALLANCE

Department of Microbiology, The Medical School, University of Bristol, Bristol BS8 1TD, UK

The molecular basis of evolution

Origins of variation

Natural selection, genetic drift, and more recently 'molecular drive' (Dover, 1982; Dover & Flavell, 1984), are processes which have been described to account for the origin of species. Recent advances in molecular biology have allowed us to study in detail the structure of genomes and of individual genes, and to make comparisons between different species. In addition, much more is being learned about the molecular mechanisms by which genomes and their genes evolve, and this information enables us to test the validity of the various hypotheses of speciation.

Random mutations in genes may be advantageous or deleterious to the organism and, as a consequence, become fixed in or lost from a population. Sometimes they may be neutral and become fixed or lost by the vagaries of genetic drift. Orgel & Crick (1980), and Doolittle & Sapienza (1980) put forward the notion of 'selfish' or 'parasitic' DNA to account for the relatively large amount of non-genic, apparently 'junk' DNA found in some eukaryotic organisms. Much of this DNA is repetitive, and duplicative transposition was suggested as a mechanism by which it might multiply and spread throughout the genome, thus counteracting the tendency for it to be deleted over a period of time. Since it has no function, quite large changes in its sequence or quantity may have little or no effect on the survival of the organism.

A number of eukaryotic genes, including those for ribosomal RNA species and sometimes histones, are also present in many copies in the eukaryotic genome. Repeated sequences are often referred to as families, whether they be functionally important (genic) or apparently non-essential (non-genic). Such sequences can be repeated in tandem, or be spread throughout the genome.

Ribosomal gene units are usually repeated in tandem, and include the genes for 26S, 18S and 5·8S rRNA species. In *Saccharomyces cerevisiae*, the unit also contains the 5S gene (Nath & Bollon, 1977), while in *Aspergillus nidulans* (Bartnik, Strugala & Stepien, 1981; Borsuk *et al.*, 1982), *Neurospora crassa* and *Schizosaccharomyces pombe* (Metzenberg *et al.*, 1985) the multicopy genes for 5S rRNA are dispersed elsewhere in the genome. The amount of repetition of rRNA genes varies with the species, being about 100 copies in the four species mentioned above.

Between the highly conserved coding regions of the rRNA gene repeats are 'spacers' in which the only critical sequences are, for instance, those acting as promoters or terminators of transcription. Sequence changes can occur relatively rapidly in such spacer regions compared with coding regions, and the way in which they change within a species and between species has been studied in a number of organisms. They tend to be uniform in sequence within a particular species, but differ between species (Dover, 1982; Pukkila & Cassidy, Chapter 5). This suggests that there are mechanisms which maintain homogeneity of repeated sequences within a species and that when a sequence alteration occurs in one of the members of a family it is either corrected, or may eventually spread throughout the family.

This kind of change in genome structure, which has little or no immediate consequence on the survival of the individual organism, has been termed 'molecular drive' by Dover (1982) who also considered possible implications for speciation. If such homogenisation of sequence families occurs much more slowly than the mixing of genomes by sexual reproduction within a population, the genome structure of an entire breeding population can change gradually without any individual being very different from another individual of that population. Eventually, this process can lead to quite large differences in genome structure between two isolated populations, and lead to reproductive incompatibility between them.

Recombination and genetic variation

What are the mechanisms which can lead to the spread of a variant through a family? They include transposition, unequal exchange during recombination, and gene conversion (Dover & Flavell, 1984), and all of these processes have been detected in fungi.

Duplicative transposition Transposable elements or transposons have been found in a wide range of prokaryotes and eukaryotes. Although the mechanisms of transposition differ, transposition is essentially a duplicative process, in which a transposable element present in one part of the genome is duplicated and inserted elsewhere in the genome. The Ty transposon of yeast is actually a kind of retrovirus (Boeke *et al.*, 1985) and is first transcribed into RNA which can be encapsulated in a virus particle. The transposon encodes its own reverse transcriptase which makes a DNA copy of the RNA that can then insert elsewhere into the yeast genome. The insertion event is sometimes mutagenic. Yeast carries about 35 copies of Ty-1 in its genome and homologous recombination between these repeated sequences can lead to large-scale chromosome changes and aberrations, such as deletions, inversions and translocations (Roeder & Fink, 1983; Oliver, Chapter 3). To date there are no reports of transposons in filamentous fungi.

Unequal exchange in recombination When homologous chromosomes carry tandem repeats, such as the ribosomal units, it is possible for recombination within these regions, between homologous chromosomes, to increase or decrease the number of repeats as a result of misalignment (Fig. 2.1). Successive rounds of increase and decrease can lead to the loss or spread of a variant through the family. While such exchanges might be expected to occur most frequently at meiosis, when recombination is most frequent, it could also occur during mitosis, either between sequences on homologous chromosomes or between sequences on the same or different chromosomes. Unequal meiotic recombination within ribosomal gene arrays has been demonstrated in *Saccharomyces cerevisiae* (Petes, 1980).

In *Aspergillus nidulans*, tandemly repeated sequences have been

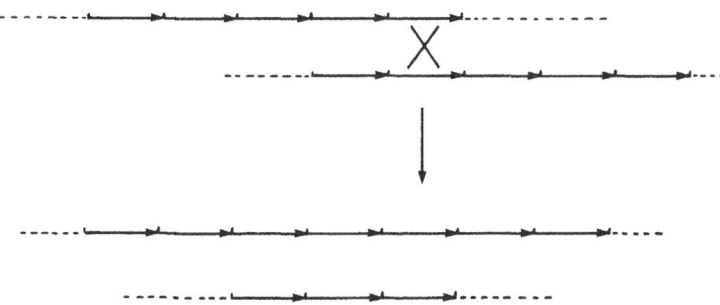

Fig. 2.1. Unequal crossover between tandemly repeated sequences.

generated in genetically manipulated strains. Transformation of the fungus with the structural gene for acetamidase, *amdS*, often leads to multiple integration of the transforming DNA, giving tandemly repeated *amdS* of up to around 100 copies (Tilburn *et al.*, 1983; Kelly & Hynes, 1985). If such strains are selfed, rearrangements of the integrated sequences are observed in the progeny (Wernars *et al.*, 1985). *Aspergillus* thus offers a useful model system in which the behaviour of tandemly repeated sequences can be studied during meiosis and mitosis. Similar behaviour has also been observed following transformation of *Neurospora crassa* with the cloned *am* (NADP-specific glutamate dehydrogenase) gene (Bull & Wootton, 1984).

Gene conversion If two variants of a family undergo a recombinational interaction such that one of the variants is made identical to the other, the process is called gene conversion. Gene conversion has been known in fungi for many years and has been incorporated into models of genetic recombination (e.g. Meselson & Radding, 1975). It can occur during meiosis and mitosis and we have recently carried out some studies on *Aspergillus nidulans* transformants which clearly demonstrate the occurrence of gene conversion even during vegetative growth (Ward, Wilkinson & Turner, 1986).

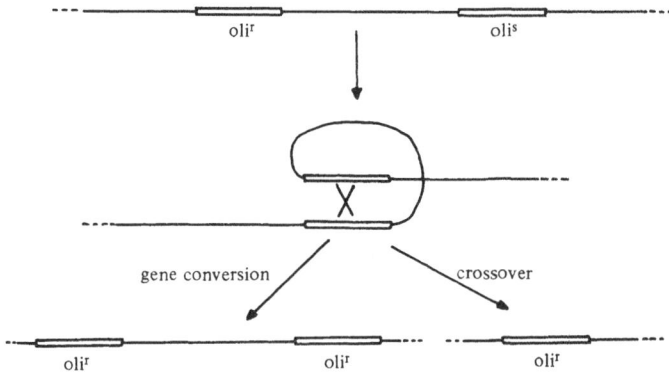

Fig. 2.2. Recombination between duplicated *oliC* genes in *Aspergillus nidulans*, the gene for subunit 9 of the mitochondrial ATP synthase. The duplication was generated by transformation of a sensitive strain with the cloned resistance gene. The wild type gene, *oliC$^+$*, encodes a drug-sensitive subunit (*oli^s*), and the mutant gene, *oliC31*, a resistant subunit (*oli^r*). The events shown are detected by plating conidia on drug-containing medium (Ward *et al.*, 1986).

Transforming *A. nidulans* with the mutant *oliC31* gene, which confers resistance to the drug oligomycin (Oli^r), leads to a duplication of the gene, the transforming gene conferring resistance and the native gene sensitivity (Fig. 2.2); the resulting phenotype is semi-resistant to the drug. If such a transformed strain is maintained on medium containing oligomycin, rapidly growing sectors of fully resistant derivatives arise. Analysis of these shows that they can arise by either gene conversion, giving a duplicated *oliC31*, or a crossover event which results in loss of the sensitivity gene together with the sequence between the duplicated genes. The relative ratio of conversion:crossover is about 3:2. The frequency with which either event occurs is quite low, in less than 1 in 10^3 nuclei during vegetative growth, though this would increase greatly during meiosis (Miller, Miller & Timberlake, 1985). If the second copy of the gene is inserted on a different chromosome then the recombinational interaction between them is barely detectable above the spontaneous mutation rate (10^{-8}). Such results suggest that homogenisation of tandem arrays would occur more quickly than homogenisation of repeated sequences located on different chromosomes.

Gene conversion can be asymmetric, that is show a bias in favour of one variant over another, even in the absence of selection at the level of the whole organism. Any variant which shows a positive bias would naturally tend to spread through a family and through a population. In the *oliC31* example given above, selection pressure was applied in favour of the *oliC31* allele, so we cannot judge from those data whether the conversion event was really asymmetric.

We have observed what appears to be asymmetric gene conversion with certain mitochondrial introns in the *Aspergillus nidulans* species group (Earl *et al.*, 1981). Closely related members of this group have mitochondrial genomes which vary in the number of introns present in certain mitochondrial genes. If two different mitochondrial types, with and without these introns, are brought together in the same cytoplasm recombination occurs between the mitochondrial genomes and all the hybrid mitochondria contain certain of these optional introns. A similar phenomenon had already been observed in yeast (Heyting & Menke, 1979), where an intron (*omega*$^+$) in the mitochondrial large ribosomal RNA gene converts intron-less genes (*omega*$^-$) with a high degree of asymmetry. Recent studies on the *omega* conversion have demonstrated that the intron encodes a double strand endonuclease which initiates the conversion event by cleaving the mitochondrial rRNA gene of *omega*$^-$ strains at a specific site (Colleaux *et al.*, 1986).

Repetitive DNA and recombination in fungi

The molecular drive hypothesis put forward by Dover (1982) concentrates on the behaviour of multicopy families of sequences, and their potential role in speciation. In general, fungi do not seem to have anything like as much high- and middle-repetitive DNA as higher eukaryotes. Timberlake (1978) showed that *A. nidulans* has only 2 − 3% repetitive sequences, and this must include the rRNA repeats. We have detected an AT-rich, non-genic sequence of a few kilobases which is partly repeated about twelve times in the genome (Ballance & Turner, 1985). It does not appear to be a highly variable region as judged by its analysis in closely related strains of *A. nidulans* (D. J. Ballance, unpublished).

Repetitive DNA is a somewhat ambiguous term, since there is no sharp dividing line between 'unique' and repetitive sequences. Just how much homology is required between DNA sequences before recombination events can easily occur between them? In *Saccharomyces cerevisiae*, recombination between two sequences shows strong dependence on the length of homology which they share (Hicks, Hinnen & Fink, 1979). However, studies on the transformation of filamentous Ascomycetes have clearly shown that these organisms are much less demanding of a high degree of homology for recombination to occur (Ballance & Turner, 1985). This implies that sequences not defined as identical or even very similar on the basis of hybridisation analysis can still participate in recombinational interactions. Just what the minimal level requirement is for recombination to occur is as yet undefined for these fungi. 'Non-homologous' recombination could be fatal if it involved genic sequences, but such recombination in 'low-repetitive' inter-gene sequences might be harmless.

Since rRNA repeat units are tandemly arranged in a number of fungi, their homogeneity may be maintained by the recombinational mechanisms outlined above. A problem arises with the 5S rRNA genes of *Neurospora crassa*, which have been analysed in some detail recently (Metzenberg *et al.*, 1985). Although they exist in more than one form, known as alpha, beta, etc., isotypes, it seems that these isotypes may have some functional significance, since related isotypes are found in other fungi. Any particular isotype, however (e.g. alpha, the major class), shows intra-species homogeneity. The puzzle is just how this homogeneity is maintained, since the 50 or so copies of alpha 5S rRNA are dispersed, never tandemly repeated. This rules out unequal crossing over as a homogenising mechanism. Gene conversion could be the favoured alternative, but it seems that 5′ and 3′ flanking sequences immediately outside the coding region differ drastically from gene to gene. When one sequence 'corrects' another during gene conversion the correction would not necessarily be confined to the coding region and one would expect to find flanking regions of genes (equivalent to the inter-gene 'spacers' previously mentioned) showing similarities in different genes. Does this mean that the 5Sr RNA itself is somehow involved in the 'correction' process, rather than a direct DNA/DNA interaction?

Gene structure and comparisons
Genome relatedness

Attempts at judging evolutionary relatedness by comparison of genome composition of different fungi has been reviewed recently (Kurtzman, 1985). One of the earliest methods was comparison of GC content, but this is rather insensitive and fairly limited in application. An improvement on this is measurement of hybridisation between DNAs of different organisms. When this is carried out with total DNA the presence of large amounts of repetitive DNA in one or both of the organisms can give misleading results. Attempts have been made to correlate DNA relatedness measured in this way with genetic relatedness as judged by hybrid formation. Sometimes DNA relatedness as low as 25% still permitted genetic exchange and viable offspring. Thus, the total DNA approach is limited to detection of sibling species.

When working with such closely related species the mitochondrial genome can provide useful taxonomic data. Since it is much smaller than the nuclear genome it is quite easy to generate recognisable restriction endonuclease digest fragment patterns with a number of different enzymes, and compare these patterns (i.e. fragment sizes) for different species of a genus. Closely related members of the *A. nidulans* species group have been compared in this way (Earl *et al.*, 1981), as have rather more distantly related species of the genus *Aspergillus* (Kozlowski & Stepien, 1982).

Analogous restriction fragment length polymorphism data can be obtained for nuclear DNA if a single nuclear gene or chromosomal fragment from one species is used as a radiolabelled probe against restriction digests of other strains.

For relatedness above the species level, rRNA homology has been used, since these gene sequences are highly conserved. In particular, the 5S rRNAs are small enough to be sequenced quickly and have been used extensively in the construction of phylogenetic trees, providing one of the best molecular approaches to fungal taxonomy which is especially useful for imperfect fungi (Chen *et al.*, 1984).

Genes coding for proteins

Recent years have seen an increase in the isolation of fungal genes coding for proteins and their eventual sequencing (Bennett & Lasure, 1985; Timberlake, 1985). Studies on gene expression and regulation are often the main aims behind this type of work, rather than sequence comparison, but there is some taxonomic spin-off as increasing numbers of similar genes are sequenced from a range of organisms. Naturally, some genes show greater sequence conservation than others throughout nature and it is not especially easy to judge relatedness by looking at genes for a single protein. Furthermore, even within a gene some regions are conserved more strongly than others, reflecting differential biological importance. We can illustrate these points by comparing the derived amino acid sequences of two genes, one coding for subunit 9 of the mitochondrial ATP synthetase, and the other for phosphoribosylanthranilate isomerase, a step in tryptophan synthesis. The former has been sequenced in *Neurospora crassa* (Viebrock, Perz & Sebald, 1982) and *Aspergillus nidulans* (Ward & Turner, 1986), the latter in *N. crassa* (Schechtman & Yanofsky, 1983), *A. nidulans* (Mullaney *et al.*, 1985) and *A. niger* (C. A. M. J. J. van den Hondel, personal communication).

The subunit 9 protein of *N. crassa* shows amino acid homologies as follows: with *A. nidulans* 80%, *Saccharomyces cerevisiae* 50%, bovine 50%, *Escherichia coli* 22% (Hoppe & Sebald, 1984). Although we have no sequence data for the *A. niger* gene, hybridisation of the *N. crassa* cDNA against *A. niger* and *A. nidulans* shows clearly that there is greater DNA homology between *Neurospora* and *A. niger* than between *Neurospora* and *A. nidulans*. Examination of the amino acid homologies for the phosphoribosylanthranilate isomerase gene shows: *N. crassa*: *A. nidulans* 60%; *N. crassa*: *A. niger* 58%; *A. nidulans*: *A. niger* 78%.

These data are broadly in agreement with those obtained by 5S rRNA analysis, but provide little extra taxonomic insight. It would be difficult to justify the sequencing of protein coding genes on taxonomic grounds alone, but such sequences will accumulate rapidly in the literature and in computer banks in future years and will thus be available to the taxonomist.

Expression of fungal genes

Transcription

In higher eukaryotes, as in prokaryotes, characteristic DNA sequences (promoters) are present in the 5' region of the gene, i.e. 'upstream' of the coding region. Higher eukaryote promoters appear to contain at least two types of signal; one involved in control of transcription, and one responsible for determining the site of initiation of transcription by RNA polymerase. The latter is thought to consist of a TATA box 20 − 40 bases upstream of the RNA start site (Gannon *et al.*, 1979; Nussinov, 1986) and, to a lesser extent, a CAAT box ($GG^C/_GCAATCT$) 70 − 90 bases upstream of the start point (Benoist *et al.*, 1980).

As gene sequences of yeast and later, filamentous fungi, began to appear, the search started for equivalent promoter sequences. Initially, this consisted of examination of sequences for similar motifs to those seen in higher eukaryotes and recognition of patterns common to all genes or similar kinds of genes. This approach has to be backed up with identification of transcription start sites by hybridisation of mRNA to the gene, and by deletion of putative promoter sequences.

Sequences related to the consensus sequence $TAT^A/_TA^A/_T$ (Corden *et al.*, 1980) have been found upstream of the major mRNA start point in many yeast genes, though at a more variable distance from this point than in higher eukaryotic genes (Hahn, Hoar & Guarente, 1985). This calls into question the proposed central role of this sequence in determining initiation of transcription a set distance downstream (Grosschedl & Birnstiel, 1980). Recent work has suggested that, although such sequences may be important in initiation of transcription, the sequences between the TATA box and the initiation point probably have a greater role than in higher eukaryotes (Chen & Struhl, 1985).

Examination of the 5′ non-coding regions of the genes of filamentous fungi sequenced to date reveals an equivocal situation. Some of these genes do possess sequences reminiscent of the canonical TATA sequence at approximately the expected position (Boel *et al.*, 1984; Kinnaird & Fincham, 1983; McKnight *et al.*, 1985; Munger, Germann & Lerch, 1985; Clements & Roberts, 1986) whereas others do not have such a recognisable motif (Alton *et al.*, 1982; Rutledge, 1984; Sebald & Kruse, 1984; Legerton & Yanofsky, 1985; Mullaney *et al.*, 1985). Some of the latter do, however, contain AT-rich stretches in this region. The CAAT box is even harder to find (or easier, depending on one's point of view and imagination). Consequently, the importance of the TATA and CAAT boxes in initiation of transcription in fungi, and particularly in filamentous species, is unresolved. In most instances, however, the mRNA start site (or sites) is preceded by a run of pyrimidines (C and T) (Arends & Sebald, 1984; Boel *et al.*, 1984; Sebald & Kruse, 1984; Kinnaird & Fincham, 1983; Munger *et al.*, 1985; Clements & Roberts, 1986), this being particularly marked in the *oliC* gene of *Aspergillus nidulans* (Ward & Turner, 1986). Such a feature is also present in some yeast genes (Dobson *et al.*, 1982), especially highly expressed genes, and has lead to the speculation that the extent of the 'pyrimidine block' influences the level of transcription, perhaps due to stabilisation of the RNA polymerase molecule. Preliminary functional analysis of the *A. nidulans trpC* promoter has confirmed the importance of the pyrimidine block in transcription initiation; deletion leading to failure of transcription (W. E. Timberlake, personal communication).

Intervening sequences

Many protein encoding genes in eukaryotes are interrupted by non-translated intervening sequences or introns. In higher eukaryotes the majority of genes contain introns and in most cases these are relatively large (often 10 kb or more). This contrasts sharply with *Saccharomyces cerevisiae* where introns are very much the exception rather than the rule and tend to be relatively short (a few hundred bases or less). Introns in the genes of filamentous Ascomycetes appear to be somewhere in between, being common (present in about two-thirds of the genes analysed so far) but very short (55 − 192 base pairs). Interestingly, the gene encoding the proteolipid of the mitochondrial ATPase lacks introns in *Aspergillus nidulans* (Ward & Turner, 1986), but contains two introns in *Neurospora crassa* (Sebald & Kruse, 1984). The glutamate dehydrogenase genes of these two organisms appear to have two introns in exactly the same positions (Kinnaird & Fincham, 1983; S. Gurr and J. R. Kinghorn, personal communication).

Recent studies have centred on the importance of the sequence at the sites at which splicing of the immature RNA occurs. It is possible to compose a consensus for these junctions, the rationale being that conservation of particular bases at particular positions reflects the importance of such sequences in the splicing process. This has been done for higher eukaryotes (Mount, 1982) and *S. cerevisiae* (Langford & Gallwitz, 1983), and is shown here for filamentous fungi (Table 2.1). The element upstream of the 3′ splice site interacts with the 5′ terminal G of the intron to form an intermediate (called a lariat) in the splicing process (Ruskin *et al.*, 1984). As can be seen there are certain similarities in these consensus sequences, though *S. cerevisiae* and filamentous ascomycete species lack the pyrimidine stretch which precedes the 3′ splice site in higher eukaryotes. It is noteworthy that the invariant TACTAAC seen in *S. cerevisiae* is rarely seen in filamentous ascomycetes.

Table 2.1. *Conservation of intron/exon junction sequences.*
The intron/exon boundary (= splice site) is denoted by |. The nucleotide bases are indicated by their initial letters, N represents any nucleotide, Y represents a pyrimidine (i.e. C or T).

5′ splice site	
Higher eukaryotes	CAG|GTAAGT
	A G
Yeast	N|GTATGT
	C
Filamentous fungi	NNN|GTAAGTTANC
	C T
	T
Splicing signal (lariat formation)	
Higher eukaryotes	CTAAT
	C C
Yeast	TACTAAC
Filamentous fungi	AGCTGAC
	T A
3′ splice site	
Higher eukaryotes	$(T/C)_n$NCAG|G
	T
Yeast	YYNYAG|N
Filamentous fungi	NNNNNNNNTATAG|N
	CCC
	(46−65% CT)

Heterologous gene expression

A number of genes of *Saccharomyces cerevisiae* were isolated by virtue of their ability to complement the equivalent mutation in *Escherichia coli* (Struhl, Cameron & Davis, 1976; Walz, Ratzkin & Carbon, 1978; Ratzkin & Carbon, 1977). More recently, several genes from filamentous fungi have been isolated in a similar manner, and one (*argB* of *Aspergillus nidulans*) by complementation of a yeast mutant (Berse *et al.*, 1983). However, this approach has failed on many other occasions, suggesting barriers to heterologous gene expression. Other work has highlighted failure of *A. nidulans* and *A. niger* genes to be expressed in *S. cerevisiae* (Pentilla *et al.*, 1984) and failure of *S. cerevisiae* genes to be expressed in *Neurospora crassa* (Case, 1982) and *A. nidulans* (D. J. Ballance, unpublished).

Conversely, the *N. crassa pyr4* gene is able to complement the equivalent mutation in *A. nidulans* and, poorly, in *S. cerevisiae* (Ballance,

Buxton & Turner, 1983). The *A. nidulans argB* gene is expressed in *N. crassa* (Weiss, Puetz & Cybis, 1985), *S. cerevisiae* (Berse *et al.*, 1983) and *A. niger* (Buxton, Gwynne & Davies, 1985) and the *trpC* genes of *N. crassa*, *A. nidulans* and *A. niger* are probably interchangeable (Yelton, Timberlake & van den Hondel, 1985; Kos *et al.*, 1985). The *A. nidulans amdS* gene functions in *A. niger* (Kelly & Hynes, 1985) and *Cochliobolus heterostrophus* (Turgeon, Garber & Yoder, 1985). A number of other, as yet unpublished, observations support the supposition that the genes of filamentous ascomycetes and related deuteromycetes are to a large extent interchangeable.

The data suggest that where only a small subset of genes are expressed in another organism, the reason for this expression is the fortuitous occurrence of sequences similar to those present in the resident genes, rather than to similarities in the normal promoters of the two organisms. For instance, the *N. crassa qa-2* gene is able to complement an *Escherichia coli* mutant because a perfect *E. coli* ribosome-binding site is present, though this obviously does not have the same function in *N. crassa* (Alton *et al.*, 1982).

However, the fact that filamentous fungal genes are interchangeable but rarely function in *S. cerevisiae* suggests that filamentous fungal promoters, intron splice signals and terminators are basically similar but differ significantly from those of *S. cerevisiae*, supporting the taxonomic distancing of *S. cerevisiae* from filamentous Ascomycetes. There are examples in the literature of failure of *S. cerevisiae* to splice filamentous fungal introns and of failure to transcribe some of them unless yeast transcription signals are provided (Innis *et al.*, 1985).

References

Alton, N. K., Buxton, F., Patel, V., Giles, N. H. & Vapnek, D. (1982). 5'-untranslated sequences of two structural genes in the *qa* gene cluster of *Neurospora crassa*. *Proceedings of the National Academy of Sciences, USA*, **79**, 1955 – 1959.

Arends, H. & Sebald, W. (1984). Nucleotide sequence of the cloned mRNA and gene of the ADP/ATP carrier from *Neurospora crassa*. *EMBO Journal*, **3**, 377 – 382.

Ballance, D. J., Buxton, F. P. & Turner, G. (1983). Transformation of *Aspergillus nidulans* by the orotidine-5'-phosphate decarboxylase gene of *Neurospora crassa*. *Biochemical and Biophysical Research Communications*, **112**, 284 – 289.

Ballance, D. J. & Turner, G. (1985). Development of a high-frequency transforming vector for *Aspergillus nidulans*. *Gene*, **36**, 321 – 331.

Bartnik, E., Strugala, K. & Stepien, P. P. (1981). Cloning and analysis of recombinant plasmids containing genes for *Aspergillus nidulans* 5S rRNA. *Current Genetics*, **4**, 173 − 176.

Bennett, J. W. & Lasure, L. L. (1985). *Gene Manipulations in Fungi*. Orlando, Florida: Academic Press.

Benoist, C., O'Hare, K., Breathnach, R. & Chambon, P. (1980). The ovalbumin gene − sequence of putative control regions. *Nucleic Acids Research*, **8**, 127 − 142.

Berse, B., Dmochowska, A., Skrzypek, M., Weglenski, P., Bates, M. A. & Weiss, R. L. (1983). Cloning and characterization of the ornithine carbamoyltransferase gene from *Aspergillus nidulans*. *Gene*, **25**, 109 − 117.

Boeke, J. D., Garfinkel, D. J., Styles, C. A. & Fink, G. R. (1985). Ty elements transpose through an RNA intermediate. *Cell*, **40**, 491 − 500.

Boel, E., Hansen, M. T., Hjort, I., Hoegh, I. & Fiil, N. P. (1984). Two different types of intervening sequences in the glucoamylase gene from *Aspergillus niger*. *EMBO Journal*, **3**, 1581 − 1585.

Borsuk, P. A., Nagiec, M. M., Stepien, P. P. & Bartnik, E. (1982). Organization of the ribosomal RNA gene cluster in *Aspergillus nidulans*. *Gene*, **17**, 147 − 152.

Bull, J. H. & Wootton, J. C. (1984). Heavily methylated amplified DNA in transformants of *Neurospora crassa*. *Nature*, **310**, 701 − 704.

Buxton, F. P., Gwynne, D. I. & Davies, R. W. (1985). Transformation of *Aspergillus niger* using the *argB* gene of *Aspergillus nidulans*. *Gene*, **37**, 207 − 214.

Case, M. E. (1982). Transformation of *Neurospora crassa*. In *Engineering of Microorganisms for Chemicals*, Basic Life Sciences, vol. 19, ed. Hollaender, A., De Moss, R. D., Koninsky, J., Savage, D. & Wolfe, R. S., pp. 87 − 100. New York: Plenum Press.

Chen, M. W., Anne, J., Volckaert, G., Huysmans, E., Vandenberghe, A. & De Wachter, R. (1984). The nucleotide sequences of the 5S RNAs of seven molds and a yeast and their use in studying ascomycete phylogeny. *Nucleic Acids Research*, **12**, 4881 − 4892.

Chen, W. & Struhl, K. (1985). Yeast mRNA initiation sites are determined primarily by specific sequences, not by the distance from the TATA element. *EMBO Journal*, **4**, 3273 − 3280.

Clements, J. M. & Roberts, C. R. (1986). Transcription and processing signals in the 3-phosphoglycerate kinase (PGK) gene from *Aspergillus nidulans*. *Gene*, in press,

Colleaux, L., d'Auriol, L., Betermier, M., Cottarel, G., Jacquier, A., Galibert, F. & Dujon, B. (1986). Universal code equivalent of a yeast mitochondrial intron reading frame is expressed into *E. coli* as a specific double strand endonuclease. *Cell*, **44**, 521 − 533.

Corden, J., Wasylyk, B., Buchwalder, A., Sassone-Corsi, P., Kedinger, C. & Chambon, P. (1980). Promoter sequences of eukaryotic protein-coding genes. *Science*, **209**, 1406 − 1414.

Dobson, M. J., Tuite, M. F., Roberts, N. A., Kingsman, A. J. & Kingsman, S. M. (1982). Conservation of high efficiency promoter sequences in *Saccharomyces cerevisiae*. *Nucleic Acids Research*, **10**, 2625 − 2637.

Doolittle, W. F. & Sapienza, C. (1980). Selfish genes, the phenotype paradigm and genome evolution. *Nature*, **284**, 601 − 603.

Dover, G. A. (1982). Molecular drive: a cohesive model of species evolution. *Nature*, **299**, 111 − 117.

Dover, G. A. & Flavell, R. B. (1984). Molecular coevolution: DNA divergence and the maintenance of function. *Cell*, **38**, 622 − 623.

Earl, A. J., Turner, G., Croft, J. H., Dales, R. B. G., Lazarus, C. M., Lunsdorf, H. & Kuntzel, H. (1981). High frequency transfer of species specific mitochondrial DNA sequences between members of the Aspergillaceae. *Current Genetics*, 3, 221 − 228.

Gannon, F., O'Hare, K., Perrin, F., Le Pennec, J. P., Benoist, C., Cochet, M., Breathnach, R., Royal, A., Garapin, A., Cami, B. & Chambon, P. (1979). Organisation and sequences at the 5′ end of a cloned complete ovalbumin gene. *Nature*, 278, 428 − 434.

Grosschedl, R. & Birnstiel, M. L. (1980). Identification of regulatory sequences in the prelude sequence of an H2A histone gene by the study of specific deletion mutants *in vivo*. *Proceedings of the National Academy of Sciences, USA*, 77, 1432 − 1436.

Hahn, S., Hoar, E. T. & Guarente, L. (1985). Each of the 'TATA elements' specifies a subset of the transcription initiation sites at the CYC-1 promoter of *Saccharomyces cerevisiae*. *Proceedings of the National Academy of Sciences, USA*, 82, 8562 − 8566.

Heyting, C. & Menke, H. (1979). Fine structure of the 21S ribosomal RNA region on yeast mitochondrial DNA. III. Physical location of mitochondrial genetic markers and the molecular nature of *omega*. *Molecular and General Genetics*, 168, 279 − 291.

Hicks, J. B., Hinnen, A. & Fink, G. R. (1979). Properties of yeast transformation. *Cold Spring Harbor Symposia on Quantitative Biology*, 43, 1305 − 1313.

Hoppe, J. & Sebald, W. (1984). The proton conducting Fo-part of bacterial ATP synthases. *Biochimica et Biophysica Acta*, 768, 1 − 27.

Innis, M. A., Holland, M. J., McCabe, P. C., Cole, G. F., Wittman, V. P., Tal, R., Watt, K. W. K., Gelfand, D. H., Holland, J. P. & Meade, J. H. (1985). Expression, glycosylation and secretion of an *Aspergillus* glucoamylase by *Saccharomyces cerevisiae*. *Science*, 228, 21 − 26.

Kelly, J. M. & Hynes, M. J. (1985). Transformation of *Aspergillus niger* by the *amdS* gene of *Aspergillus nidulans*. *EMBO Journal*, 4, 475 − 479.

Kinnaird, J. H. & Fincham, J. R. S. (1985). The complete nucleotide sequence of the *Neurospora crassa am* (NADP-specific glutamate dehydrogenase) gene. *Gene*, 26, 253 − 260.

Kos, A., Kuijvenhoven, J., Wernars, K., Bos, C. J., van den Broek, H. W. J., Pouwels, P. H. & van den Hondel, C. A. M. J. J. (1985). Isolation and characterization of the *Aspergillus niger trpC* gene. *Gene*, 39, 231 − 238.

Kozlowski, M. & Stepien, P. P. (1982). Restriction enzyme analysis of mitochondrial DNA of members of the genus *Aspergillus* as an aid in taxonomy. *Journal of General Microbiology*, 128, 471 − 476.

Kurtzman, C. P. (1985). Molecular taxonomy of the fungi. In *Gene Manipulations in Fungi*, ed. Bennett, J. W. & Lasure, L. L., pp. 35 − 63. Orlando, Florida: Academic Press.

Langford, C. J. & Gallwitz, D. (1983). Evidence for an intron-contained sequence required for the splicing of yeast RNA polymerase II transcripts. *Cell*, 33, 519 − 527.

Legerton, T. L. & Yanofsky, C. (1985). Cloning and characterization of the multifunctional *his-3* gene of *Neurospora crassa*. *Molecular and General Genetics*, 201, 450 − 453.

McKnight, G. L., Kato, H., Upshall, A., Parker, M. D., Saari, G. & O'Hara, P. J. (1985). Identification and molecular analysis of a third *Aspergillus nidulans* alcohol dehydrogenase gene. *EMBO Journal*, 4, 2093 − 2099.

Meselson, M. & Radding, C. (1975). A general model for genetic recombination. *Proceedings of the National Academy of Sciences, USA*, **72**, 358 − 361.

Metzenberg, R. L., Selker, E. U., Morzycka-Wroblewska, E. & Stevens, J. N. (1985). Dispersed multiple copy genes for 5S RNA: what keeps them honest? In *The Molecular Genetics of Filamentous Fungi*, UCLA Symposia on Molecular and Cellular Biology, New Series, volume 34, ed. Timberlake, W. E., pp. 295 − 307. New York: Alan R. Liss.

Miller, B. L., Miller, K. Y. & Timberlake, W. E. (1985). Direct and indirect gene replacements in *Aspergillus nidulans*. *Molecular and Cellular Biology*, **5**, 1714 − 1721.

Mount, S. M. (1982). A catalogue of splice junction sequences. *Nucleic Acids Research*, **10**, 459 − 472.

Mullaney, E. J., Hamer, J. E., Roberti, K. A., Yelton, M. M. & Timberlake, W. E. (1985). Primary structure of the *trpC* gene from *Aspergillus nidulans*. *Molecular and General Genetics*, **199**, 37 − 45.

Munger, K., Germann, U. A. & Lerch, K. (1985). Isolation and structural organization of the *Neurospora crassa* copper metallothionein gene. *EMBO Journal*, **4**, 2665 − 2668.

Nath, K. & Bollon, A. P. (1977). Organization of the yeast ribosomal gene cluster via cloning and restriction analysis. *Journal of Biological Chemistry*, **252**, 6562 − 6571.

Nussinov, R. (1986). Some guidelines for identification of recognition sequences; regulatory sequences frequently contain (T)GTG/CAC(A), TGA/TCA and (T)CTC/GAG(A). *Biochimica et Biophysica Acta*, **866**, 93 − 108.

Orgel, L. E. & Crick, F. H. C. (1980). Selfish DNA: the ultimate parasite. *Nature*, **284**, 604 − 607.

Pentilla, M. E., Nevalainen, K. M. H., Raynal, A. & Knowles, J. K. C. (1984). Cloning of *Aspergillus niger* genes in yeast. Expression of the gene coding *Aspergillus* beta-glucosidase. *Molecular and General Genetics*, **194**, 494 − 499.

Petes, T. D. (1980). Unequal meiotic recombination within tandem arrays of yeast ribosomal DNA genes. *Cell*, **19**, 765 − 774.

Ratzkin, B. & Carbon, J. (1977). Functional expression of cloned yeast DNA in *Escherichia coli*. *Proceedings of the National Academy of Sciences, USA*, **74**, 487 − 491.

Roeder, G. S. & Fink, G. R. (1983). Transposable elements in yeast. In *Mobile Genetic Elements*, ed. Shapiro, J. A., pp. 299 − 328. Orlando, Florida: Academic Press.

Ruskin, B., Krainer, A. R., Maniatis, T. & Green, M. R. (1984). Excision of an intact intron as a novel lariat structure during pre-mRNA splicing *in vitro*. *Cell*, **38**, 317 − 331.

Rutledge, B. J. (1984). Molecular characterization of the *qa-4* gene of *Neurospora crassa*. *Gene*, **32**, 275 − 287.

Schechtman, M. G. & Yanofsky, C. (1983). Structure of the trifunctional *trp1* gene from *Neurospora crassa* and its aberrant expression in *Escherichia coli*. *Journal of Molecular and Applied Genetics*, **2**, 83 − 99.

Sebald, W. & Kruse, B. (1984). Nucleotide sequence of the nuclear genes for the proteolipid and delta subunit of the mitochondrial ATP synthase from *Neurospora crassa*. In *H+-ATPase (ATP synthase): Structure, Function, Biogenesis. The $F_0 F_1$ Complex of Coupling Membranes*, ed. Papa, S., Altendorf, K., Ernster, L. & Packer, L., pp. 67 − 75. Bari: Adritica Editrice.

Struhl, K., Cameron, J. R. & Davis, R. W. (1976). Functional genetic expression of eukaryotic DNA in *Escherichia coli*. *Proceedings of the National Academy of Sciences, USA*, **73**, 1471 − 1475.

Tilburn, J., Scazzocchio, C., Taylor, G. G., Zabicky-Zissman, J. H., Lockington, R. A. & Davies, R. W. (1983). Transformation by integration in *Aspergillus nidulans*. *Gene*, **26**, 205 − 221.

Timberlake, W. E. (1978). Low repetitive DNA content in *Aspergillus nidulans*. *Science*, **202**, 973 − 975.

Timberlake, W. E. (1985). *Molecular Genetics of Filamentous Fungi*. UCLA Symposia on Molecular and Cellular Biology, New Series, volume 34. New York: Alan R. Liss.

Turgeon, B. G., Garber, R. C. & Yoder, O. C. (1985). Transformation of the fungal maize pathogen *Cochliobolus heterostrophus* using the *Aspergillus nidulans amdS* gene. *Molecular and General Genetics*, **201**, 450 − 453.

Viebrock, A., Perz, A. & Sebald, W. (1982). The imported pre-protein of the proteolipid subunit of the mitochondrial ATP synthase from *Neurospora crassa*. Molecular cloning and sequencing of the mRNA. *EMBO Journal*, **1**, 565 − 571.

Walz, A., Ratzkin, B. & Carbon, J. (1978). Expression of a cloned yeast (*Saccharomyces cerevisiae*) gene (*trp5*) by a bacterial insertion element (IS2). *Proceedings of the National Academy of Sciences, USA*, **75**, 6172 − 6176.

Ward, M., & Turner, G. (1986). The ATP synthase subunit 9 gene of *Aspergillus nidulans: sequence and transcription. Molecular and General Genetics*, **205**, 331 − 338.

Ward, M., Wilkinson, B. & Turner, G. (1986). Transformation of *Aspergillus nidulans* with a cloned, oligomycin-resistant ATP synthase subunit 9 gene. *Molecular and General Genetics*, **202**, 265 − 270.

Weiss, R. L., Puetz, D. & Cybis, J. (1985). Expression of *Aspergillus* genes in *Neurospora*. In *Gene Manipulations in Fungi*, ed. Bennett, J. W. & Lasure, L. L., pp. 280 − 292. Orlando, Florida: Academic Press.

Wernars, K., Goosen, T., Wennkes, L. M. J., Visser, J., Bos, C. J., van den Broek, H. W. J., van Gorcom, R. F. M., van den Hondel, C. A. M. J. J. & Pouwels, P. H. (1985). Gene amplification in *Aspergillus nidulans* by transformation with vectors containing the *amdS* gene. *Current Genetics*, **5**, 361 − 368.

Yelton, M. M., Timberlake, W. E. & van den Hondel, C. A. M. J. J. (1985). A cosmid for selecting gene by complementation in *Aspergillus nidulans*: selection of the developmentally regulated *yA* locus. *Proceedings of the National Academy of Sciences, USA*, **82**, 834 − 838.

3

Chromosome organisation and genome evolution in yeast

STEPHEN G. OLIVER

Department of Biochemistry and Applied Molecular Biology, University of Manchester Institute of Science and Technology, Manchester M60 1QD

Introduction

In the preface to a previous Symposium volume of this Society, Keith Gull and I wrote that ' . . . the small size of the fungal genome is a great attraction to fungal geneticists, but the correspondingly small size of the nucleus has posed problems for classical cytologists . . . ' (Gull & Oliver, 1981). The intervening five years has provided ample evidence for the advantages of the fungi to the geneticist and molecular biologist. Most progress has been made with the budding yeast, *Saccharomyces cerevisiae*, and in this chapter I wish to explain how studies on the nuclear genome of this organism illuminate the subject of fungal evolution.

S. cerevisiae has a haploid genome of about $1 \cdot 4 \times 10^4$ kb (kilobase pairs), less than four times the size of that of the bacterium *Escherichia coli*, and is organised in a characteristically eukaryotic fashion. Classical genetic studies have provided evidence for the division of the yeast genome into 17 linkage groups which are now known to be equivalent to chromosomes (Mortimer & Schild, 1985). Each of these chromosomes appears to be a single, linear DNA molecule about the same size as the DNA molecule from a 'T' bacteriophage. Their small size enables separation by two novel electrophoretic techniques, orthogonal field agarose gel electrophoresis (OFAGE; Schwartz & Cantor, 1984; Carle & Olson, 1984) and field inversion agarose gel electrophoresis (FIGE; Carle, Frank & Olson, 1986). Fig. 3.1 shows the separation of the 17 chromosomes of *S. cerevisiae* by OFAGE and this technique is now being applied to a range of other yeasts (de Jonge, *et al.*, 1986) to provide the karyotypes which were unobtainable by classical cytology.

A striking thing about the *Saccharomyces* genetic map is its relative

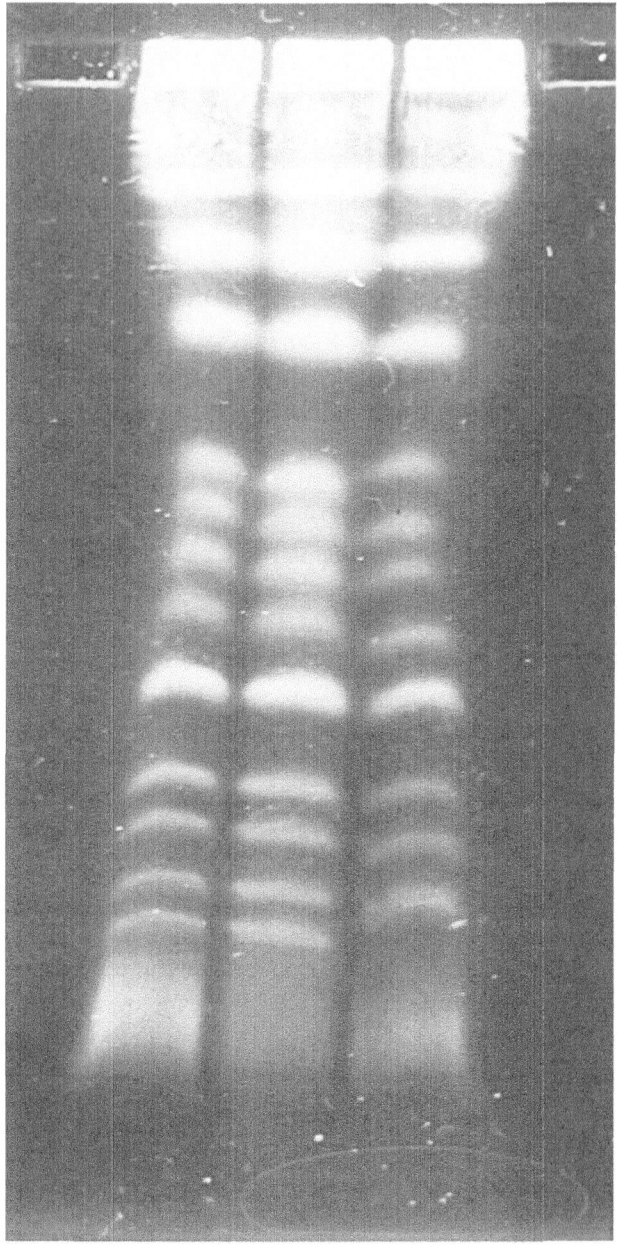

Fig. 3.1. Separation of yeast chromosomes by orthogonal field agarose gel electrophoresis. The two tracks on the left are from strain S1795a, the right hand track is from strain D6B. The two strains differ at two positions. (Photograph kindly supplied by D. H. Williamson.)

constancy between strains. This constancy is important to the evolutionary process since it means that natural selection can act in a very specific way on variants that arise, it also permits favourable mutations or gene arrangements to be spread through the population via the mating process. Mechanisms to maintain the genetic integrity of a species are built into the structure of the chromosomes and, in yeast, this structure is being analysed in detailed molecular terms.

Chromosome structure and function

Three different structures are essential to the function of eukaryotic chromosomes. Replication origins are required to initiate faithful duplication of chromosomes during the S-phase of the cell cycle. Centromeres are essential for correct segregation of newly replicated chromosomes into the two daughter nuclei. Finally, specialised structures called telomeres are required on the ends of the linear chromosomes to permit their complete replication, protect them from degradation and prevent them from acting as foci for uncontrolled recombination.

Replication origins

Yeast, like all eukaryotes, replicates its DNA bidirectionally from multiple internal origins (Newlon *et al.*, 1974; Petes *et al.*, 1973; Petes & Newlon, 1974; Petes & Williamson, 1975; Newlon & Burke, 1980; Rivin & Fangman, 1980). This is surprising since, given a replication fork propagation rate of $2 - 6$ kb min^{-1} (Petes & Williamson, 1975; Johnston & Williamson, 1978; Rivin & Fangman, 1980), most yeast chromosomes could be completely replicated from just two replication origins during the 30 min S-phase of the organism (Williamson, 1965). Instead, there are about 20 replication origins per chromosome on average, the modal replicon size being about 35 kb.

An attempt has been made to analyse these origins by cloning fragments of yeast chromosomal DNA which confer replicative ability on recombinant plasmids in yeast (Struhl *et al.*, 1979). Such fragments are known as *ARS* elements (*A*utonomously *R*eplicating *S*egment; Stinchcomb, Struhl & Davis, 1979). A large number have been cloned and this has permitted the definition of a consensus sequence for *ARS* activity — $5'$-$^A/_T$TTTAT$^A/_G$TTT$^A/_T$-$3'$ (Kearsey, 1984). While this sequence (or a near match to it) is essential to *ARS* activity in *S. cerevisiae*, it is not sufficient. The sequence context in which the consensus is found has a large influence on whether it has *ARS* activity or not, and how efficient an *ARS* it is (Celnicker *et al.*, 1984). We have recently defined a second consensus sequence, 3' to the one given above, which is required in fully functional *ARS* elements (Palzkill, Oliver & Newlon, 1986).

ARS elements may be cloned from the DNA of virtually any eukaryotic organism, but it seems unlikely that the fragments isolated represent functional origins of replication in their donor organisms. Maundrell *et al.* (1985) have shown that a significant proportion of DNA fragments which showed *ARS* activity in *Schizosaccharomyces pombe* failed to show such activity in *Saccharomyces cerevisiae*, while a number of fragments showed activity in both. We have obtained similar data with the lactose-fermenting yeast *Kluyveromyces lactis* which is much more closely related to *Saccharomyces cerevisiae* than is *Schizosaccharomyces pombe*. Thompson & Oliver (1986) cloned fragments from the linear killer plasmid, k1 (Gunge *et al.*, 1981), of *K. lactis* which exhibited *ARS* activity in both *K. lactis* and *Saccharomyces cerevisiae*. On subcloning we were able to separate the activities for the two organisms and to define an *ARS* consensus for *K. lactis*, 5'-TCATAATATA-3', which is quite distinct from that in *S. cerevisiae*. The sequence requirements for replicative ability are therefore species specific.

The correspondence of *ARS* elements to chromosomal origins of replication is not proven (Williamson, 1985) and it should be remembered that the assay system imposes a very strong selection pressure for replicative ability. It is perhaps safest to regard *ARS* elements as *potential* origins of replication and not to infer that they necessarily or normally act as origins within the chromosomes from which they were derived. This point was emphasised by studies on the replication of the array of about 120 genes which encode rRNA in yeast (Walmsley, Johnston *et al.*, 1984). All of the genes in the array appear to be identical (Petes, Hereford & Skryabin, 1978; Szostak & Wu, 1979; Zamb & Petes, 1982) and all of them contain an *ARS* element (Szostak & Wu, 1979). Our measures of replicon size for rDNA showed that a functional origin occurred once in every five rRNA genes (i.e. at 45 kb intervals; Walmsley, Johnston *et al.*, 1984). Thus, every *ARS* element need not be functional as an origin and they may be activated by a mechanism which involves factors other than the recognition of specific DNA sequences. For instance, the higher order structure of chromatin may play an important role in selecting potential origins for use.

Centromeres

The centromere is the site on the chromosome which is responsible for spindle attachment and which enables the correct segregation of homologous chromosomes at mitosis and meiosis. Recombinant DNA technology has enabled the isolation of chromosome fragments containing centromeres. This was first done by 'chromosome walking' from clones of genes which were known, from classical genetic analysis, to be closely linked to their centromere (Clarke & Carbon, 1980; Fitzgerald-Hayes *et al.*, 1982; Stinchcomb, Mann & Davis, 1982; Panzeri & Philippsen, 1982). The insertion of a centromere into a YRp plasmid, which is replicated from an *ARS* fragment, has two consequences. It stabilises the plasmid, causing it to segregate in a manner analogous to that of true chromosomes at both mitosis and meiosis (Clarke & Carbon, 1980; Hsiao & Carbon, 1981), and it lowers the copy number of the plasmid from 10 − 20 to 1 − 2 molecules per cell. This last feature has been exploited by Hieter *et al.* (1985) to clone ten different yeast centromeres by counter-selection against a suppressor gene which is lethal in high copy number. These ten centromere-containing fragments have been sequenced and the consensus shown in Fig. 3.2 derived. There is a remarkable conservation in their primary structure and it seems unlikely that the centromere identifies a chromosome and ensures that homologues disjoin correctly during nuclear division. Indeed, Clarke & Carbon (1983) showed that if the centromere of one copy of chromosome III in a diploid strain was replaced with that from chromosome XI, the altered chromosome still segregated correctly from its homologue during meiosis. It seems likely that homology between chromosomes is the major determinant of correct segregation and it remains to be seen whether homology over the entire length of the two chromosomes is required, or whether there are dis-tributed sites which must exhibit homology in order that disjunction be effected.

Telomeres

It was Watson (1972) who first noted that the inability of DNA

CDE I	CDE II	CDE III
uTCACuTG	(78–86 bp) > 90% AT	TGTTT^T_ATG.TTTCCGAAA....AAA

9 979.9   .7 9   6....898

Fig. 3.2. Consensus sequence for yeast centromeres (Hieter *et al.*, 1985). This consensus is derived from the sequences of ten cloned centromeres and shows three conserved centromere DNA sequence elements (CDE I, II, III). Dyad symmetry in CDE III is indicated by the arrows. The numbers on the bottom line show the number of elements which contain the indicated nucleotide when it is not found in all centromeres analysed (u = purine.)

polymerases to extend polynucleotide chains in the 3' to 5' direction meant that it was impossible to replicate completely the lagging strand in a linear DNA duplex. Special structures are required on the ends of linear chromosomes to enable complete duplication of the molecule and prevent its becoming progressively shorter in succeeding generations. These special structures, telomeres, must also protect the chromosome from degradation by exonucleases and protect the genome from the recombinogenic properties of naked ends. Yeast telomeres have been cloned using linear recombinant plasmids having the terminal portion of the *Tetrahymena* rDNA extrachromosome on their ends (Szostak & Blackburn, 1982). Removal of the *Tetrahymena* terminus from one end of the linear molecule permitted its replacement with a yeast fragment containing a functional telomere. The structure of yeast telomeres has been analysed in detail (Fig. 3.3). They invariably contain an *ARS* element called X and may contain a second, known as Y' (Chan & Tye, 1983; Walmsley, Chan *et al.*, 1984). These elements are presumably required for the efficient replication of the termini. The most striking feature of the telomeres is that they all contain the sequence motif $C_{1-3}A$ repeated many times (Shampay, Szostak & Blackburn, 1984; Walmsley, Chan *et al.*, 1984). Walmsley, Chan *et al.* (1984) have suggested that this terminal redundancy may be important in preventing chromosome shortening during replication. They suggest that telomere resolution involves the incomplete strand making up its full length by 'borrowing' a $(C_{1-3}A)_n$ stretch from the subterminal region of another chromosome, using a recombinational process (Fig. 3.4).

This mechanism may have important evolutionary consequences. It is noteworthy that essential genes do not occur near the ends of yeast chromosomes. This is exemplified by the fact that cells containing a

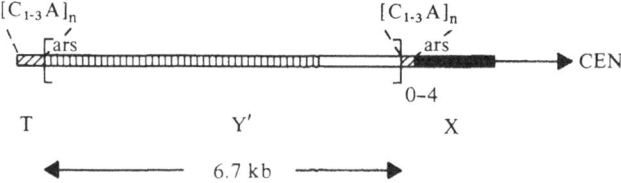

Fig. 3.3. Organisation of yeast telomeres. Repeated sequences of $[C_{1-3}A]_n$ (diagonal hatching) are found at the terminus (T) itself and are immediately followed by the X element (solid black) which has *ARS* activity. Some telomeres consist of just T and X but others have between one and four copies of the element Y' (vertical hatching) inserted between them. Y' also has *ARS* activity and most yeast telomeres have a single copy of this element. (Figure kindly provided by R. M. Walmsley.)

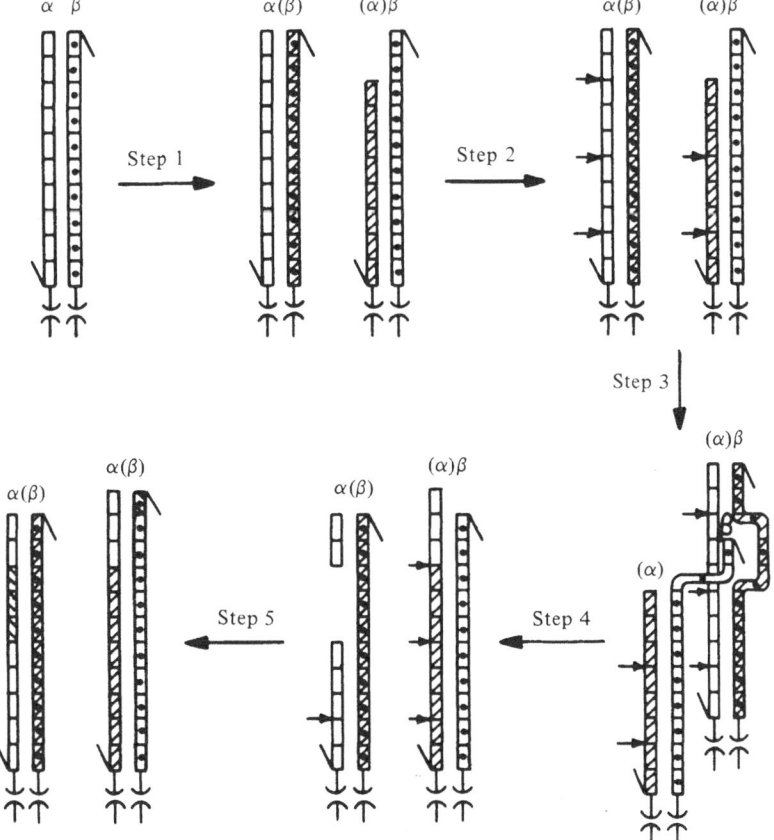

Fig. 3.4. Recombinational model of telomere resolution (Walmsley *et al.*, 1984). The two DNA strands in the telomere are represented by a series of rectangles, each of which indicates a short repeated DNA sequence. Thus open rectangle ([]) represents the G_xT_y repeat on the alpha strand and dotted rectangle ([:]) the complementary C_xA_y repeat on the beta strand. Crosshatched rectangles represent DNA synthesised by semi-conservative replication or, in the case of step 5, repair synthesis. Horizontal arrows indicate the position of single-strand nicks and vertical arrows point in the 3' direction on the DNA strand. *Step 1*: semi-conservative replication followed by primer excision at the 5' end of the lagging strand. *Step 2*: the alpha strand is nicked in both daughter DNA molecules. *Step 3*: the two daughter molecules align unequally and the 3' overhanging end of the $(\alpha)\beta$ molecule invades the $\alpha(\beta)$ molecule, displacing the (β) strand. *Step 4*: the nicks on the (α) strand permit a recombination event which transfers G_xT_y repeats to the terminus of the (α) strand. This leaves the previously truncated (α) strand with an overhanging end and the α with an internal gap. *Step 5*: the gap is filled by repair synthesis and nicks are sealed by DNA ligase. The mechanism as shown would result in the growth of the telomere at each generation. Since this is not observed, it is postulated that the 3' overhang on the (α) strand is removed by an exonuclease. (Figure kindly supplied by R. M. Walmsley.)

circular derivative of chromosome III, generated by a recombination event between the two 'silent' mating type cassettes which occur within 10 − 12 kb of the termini, are perfectly viable in the haploid state (Klar *et al.*, 1983). It may be that mistakes sometimes occur in telomere resolution and that gene arrangements which place essential genes near the ends of chromosomes have been selected against. Those genes which are found close to the telomeres, such as the fermentation markers *SUC* (Carlson & Botstein, 1983) and *MAL* (Michels & Needleman, 1983) are repeated 5 − 6 times in the genome. Many of the *SUC* gene copies are found between the X and Y′ (Fig. 3.3) elements of the telomeres (Carlson, Celenza & Eng, 1985); *SUC* duplications may thus be a direct consequence of the recombinatorial mechanism of telomere resolution. The reiteration of telomere-associated genes also provides a protection against loss of genetic information due to chromosome shortening as a result of defective resolution events.

Mobile genetic elements and genome plasticity

The ends of chromosomes are not the only regions of the yeast genome where a high degree of mobility is observed. Yeast chromosomes contain a number of genetic elements which are capable of transposition; these include the Ty transposons (Williamson, 1983; Mellor, Kingsman & Kingsman, 1986), the related sequence tau (Genbauffe, Chisholm & Cooper, 1984) and the sigma elements (del Rey, Donahue & Fink, 1982; Sandmeyer & Olson, 1982). The mating type genes should probably be added to this list.

There are three loci on chromosome III which contain mating type information. The *MAT* locus, close to the centromere on the right arm of the chromosome, is the expression locus and confers either the *a* or *α* mating type on a haploid yeast cell according to the identity of the information it contains. There are two 'silent cassettes' of mating type information, *HML* and *HMR,* which are found close to the telomeres on each arm of chromosome III. In homothallic yeast strains, cells change their mating type at successive cell divisions by transposition of mating type information from the silent cassettes into the *MAT* locus (for review, see Herskowitz & Oshima, 1981).

The Ty element

The small genome size of yeast has already been noted and it is evident that there are selection pressures which prevent the accumulation of repetitive DNA. There are about 30 copies of the Ty element in most strains of *Saccharomyces cerevisiae* (Cameron, Loh & Davis, 1979). These elements consist of a 5·2 kb unique 'epsilon' region flanked by direct repeats of about 330 base pairs called 'delta' sequences (Fig. 3.5). Homologous recombination events between the two delta sequences of a single Ty element result in the excision of the epsilon region together with one copy of delta, a 'solo delta' being left behind on the chromosome (see below). Most laboratory strains of *S. cerevisiae* contain about 100 solo delta elements (Gafner & Philippsen, 1980). The limited number of these repetitive elements in the yeast genome contrasts with the situation in metazoa and even the slime moulds − *Physarum polycephalum* has been found to contain 2000 − 5000 copies of a transposon called Tp1 (Pearston, Gordon & Hardman, 1985). There may be lesser selection pressures against accumulation of such elements in this organism, which has an overall genome size of 2×10^6 kb, than there are in yeast.

The structure of the yeast Ty elements (Fig. 3.5) is reminiscent of that of bacterial transposons and of retroviral proviruses (Heffron, 1983; Baltimore, 1985). The designation of the Ty element as a retrotransposon which is able to spread through an RNA intermediate has been elegantly confirmed by Boeke *et al.* (1985). Ty is transcribed from a site within the left-hand delta sequence to produce a 5·7 kb RNA species which terminates within the right hand delta sequence (Elder, Loh & Davis, 1983). It is this RNA molecule which acts as a template for the synthesis of a DNA copy by the action of reverse transcriptase. However, the transposed DNA copy (unlike its RNA template) represents a complete Ty element (Boeke *et al.*, 1985). A strand switching model has been proposed to explain how the missing portions of the two delta sequences are replaced in the DNA copy (Warmington *et al.*, 1985). It is likely that the primary RNA transcript is encapsidated in a virus-like particle (VLP). These VLPs or 'transposisomes' have been observed in very large numbers in yeast

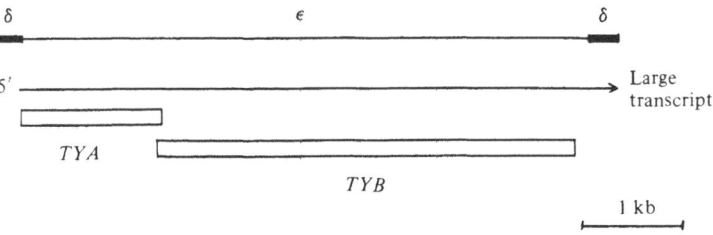

Fig. 3.5. Generalised structure of a yeast Ty element. The large transcript, which is an intermediate in the transposition process, is shown, as are the two overlapping reading frames, *TYA* and *TYB*.

cells over-expressing Ty genes (Garfinkel, Boeke & Fink, 1985; Mellor *et al.*, 1985).

Determination of the DNA sequence of Ty has revealed coding regions showing some homology to the *gag* and *pol* regions of retroviruses (Clare & Farabaugh, 1985; Hauber, Nelbock-Hochstetter & Feldmann, 1985; Warmington *et al.*, 1985). In retroviruses, *gag* is an important structural component which binds the RNA intermediate and *pol* contains three enzymic functions involved in transposition – protease, integrase and reverse transcriptase. Retroviruses contain, in addition, a third coding region, called *env*, which specifies the viral envelope and thus mediates cell to cell transmission. The Ty element lacks such a region, which may explain why it can only be transmitted by cell division and mating. Retention of the other structural components of the viral particle in yeast may indicate that VLPs are essential to transposition. Alternatively, selection pressure may have favoured retaining the VLP, so that reverse transcriptase is sequestered and prevented from producing DNA copies of yeast mRNA molecules. A more radical idea is that such cDNA molecules are (or have been) produced and integrated into the genome; perhaps accounting for the low incidence of intron-containing genes in yeast chromosomes (Baltimore, 1985).

Genetic consequences of Ty transposition

Yeast Ty elements appear to be classic examples of so-called 'selfish' DNA molecules (Dawkins, 1976; Doolittle & Sapienza, 1980; Orgel & Crick, 1980) which have an elaborate and efficient mechanism of spreading themselves through the yeast genome. In addition, though, Ty transposition can have profound effects on the activity and regulation of chromosomal genes and such effects may have important evolutionary consequences. There is no obvious conservation of sequence among the chromosomal target sites for Ty transposition, but Ty elements are frequently found associated with tRNA genes (Eigerl & Feldmann, 1981) and sigma elements are always associated with such genes (Sandmeyer & Olson, 1982). Ty elements also show a propensity to insert into promoter regions of structural genes (Eibel & Philippsen, 1984), perhaps because of the AT-rich nature of such regions. These insertions are very likely to affect expression or regulation of the structural gene involved.

An early observation was that two histidine auxotrophic mutations, *his4*-917 and *his4*-912, resulted from the insertion of Ty elements upstream of the *HIS4* gene (Roeder *et al.*, 1980). In *his4*-912 the insertion oriented the element such that transcription of Ty occurred in the same direction as that of *HIS4*, while in *his4*-917 Ty was in the opposite orientation. Other Ty insertions result in a reduction of transcription of the adjacent gene, rather than a complete block (e.g. *LYS2*; Eibel & Philippsen, 1984) and can even cause reversion of an auxotrophic mutation. Scherer, Mann & Davis (1982) found that 6/14 revertants of a deletion mutation at *LYS2* were caused by Ty insertions, the promoter region within delta presumably substituting for the deleted function.

One of the most striking effects of Ty insertion is to place expression of an adjacent gene under mating type control. This is the ROAM mutation (*R*egulated *O*verproducing *A*lleles responding to *M*ating type; Errede *et al.*, 1980), in which Ty insertion leads to an elevated level of expression of the adjacent gene in cells capable of mating (e.g. *MATa* or *MAT*α haploids and *MATa/MATa* diploids) but a normal level of expression in non-mating strains (e.g. *MATa/MAT*α diploids and certain sterile mutants). In all ROAM mutants so far characterised, transcription of the Ty element and that of the adjacent gene are divergent, so the effect cannot be mediated by transcription of the gene from the promoter within delta. ROAM mutations are *cis*-dominant and *trans*-recessive (Sherman *et al.*, 1978; Rothstein & Sherman, 1980; Dubois *et al.*, 1978; Deschamps & Wiame, 1979; Lemoine, Dubois & Wiame, 1978) so it seems likely that an enhancer-like mechanism is operating. Errede *et al.* (1984, 1985) have localised, by deletion analysis, the region of Ty responsible for the ROAM mutation of *CYC7*; it contained two segments with homology to the SV40 enhancer consensus (Khoury & Gruss, 1983). It also contained one (in class II Ty elements) or two (in class I elements) sequences which are a 16/18 base pair match with the *MAT*α 'diploid control site'. This is a sequence which lies between the two transcribed portions of the *MAT* gene and regulates their expression (Siliciano & Tatchell, 1984). The homologous sequence which is conserved in both classes of Ty element is immediately downstream of an enhancer element (Warmington *et al.*, 1985). Our current picture of the ROAM mutation is that it results from the *cis*-directed effect of an enhancer region within Ty that may be negated by binding of regulatory protein(s) at the diploid control site between the enhancer and the affected gene. Transcription of Ty is under mating type control (Dubois, Jacobs & Jauniaux, 1982) and the same sequence elements may be involved as in the ROAM effect.

Ty interactions and genome rearrangements

The effects of Ty elements on gene expression outlined above may be an important source of variation upon which natural selection can work.

The other major contribution these transposable elements might make to genome evolution, particularly to speciation, is to promote chromosomal rearrangements. These could result from recombination events either between the two repeated delta sequences within a single Ty or between separate Ty elements on the same or different chromosomes. Such interactions between the 30 or so complete Ty copies and the 100 solo delta elements in the yeast genome have the potential to generate major sequence rearrangements.

Recombination events between Ty elements have been found to involve both reciprocal and non-reciprocal exchanges. Roeder & Fink (1982), using a Ty element marked with *URA3*, found that conversion always involved the entire Ty; so that, for instance, no novel transposons were created which represented recombinants of class I and class II elements. Ty replacements detected never involved rearrangement of flanking sequences – an important point since exchanges outside Ty would result in major chromosome rearrangements. Jinks-Robertson & Petes (1985) studied the analogous situation of conversion between genes duplicated on non-homologous chromosomes. They found that the frequency of conversions in this situation was the same as that observed between the genes when they occupied allelic loci on homologous chromosomes and again, no recombination of flanking markers was observed. One of a number of explanations offered was that conversion between non-homologous chromosomes might involve a small, diffusible, single-stranded fragment, such as an Okazaki fragment (Jinks-Robertson & Petes, 1985). In the case of Ty conversion, one might speculate that the diffusible fragment is a single-stranded cDNA copy of the element produced by reverse transcription.

While reciprocal exchanges leading to translocations, inversions or deletions may not occur in the chromosomal regions adjoining Ty conversion events, crossing over between Ty elements can produce such aberrations. The most common event in this class is a crossover between the identical delta sequences at the ends of the same Ty. As shown in Fig. 3.6, this results in excision of the epsilon sequence together with one copy of delta and leaves behind a solo delta element. Such delta – delta interactions must be a major factor limiting the number of intact Ty in the yeast genome and may be balanced by the fact that delta sequences act as 'hot-spots' for subsequent transposition events (Warmington *et al.*, 1986).

Reciprocal exchanges between two different Ty elements, two solo delta sequences, or a Ty and a delta will lead to deletions if they occur within the same chromosome, or translocations if the elements are located on different chromosomes (Fig. 3.6). Ty-associated deletions have been recorded by a number of workers (Liebman, Singh & Sherman, 1979; Rothstein, 1979; Roeder & Fink, 1982). It has been suggested (Roeder &

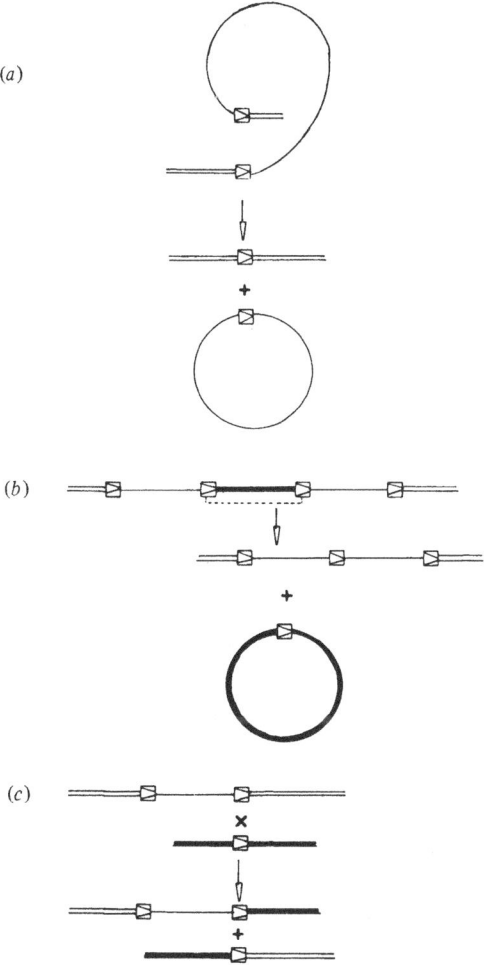

Fig. 3.6. Chromosomal rearrangements resulting from delta−delta recombination events. (a) Cross-over between two delta sequences on the same element, leading to excision of Ty with a 'solo delta' sequence left in the chromosome. (b) Cross-over between two delta sequences on different Ty elements on the same chromosome. This leads to fusion of the two Ty elements and the deletion of a portion of unique chromosomal DNA. The latter is incorporated into a novel delta-containing element which may itself be mobile. (c) Cross-over between two delta sequences on different chromosomes resulting in a reciprocal translocation. [>], delta; _____, epsilon; ▭ & ▬▬▬ , chromosomal sequences.

Fink, 1982) that some of these deletions result from conversion events where a single Ty replaces two such elements and the region between them. However, a deletion on the right arm of chromosome III which we have characterised in detail (J. R. Warmington, C. S. Newlon & S. G. Oliver, unpublished) is not the product of a conversion. It arose from a reciprocal exchange between two delta sequences on tandemly arranged Ty elements. The products of this exchange are a fused double transposon (Fig. 3.6) and, we postulate, a circular DNA molecule containing a solo delta and a region of chromosome III DNA. Such a molecule would represent a novel type of mobile element since it could reinsert into the genome at many sites via a second delta − delta recombination. The chances of reinsertion would be improved if the chromosomal segment contained an *ARS* sequence. A similar exchange between a Ty and a solo delta would be less readily detected since a complete Ty would be left in the chromosome, indistinguishable from other Ty elements, except that it would not be flanked by a target site duplication. Such an event might explain the Ty-associated deletion within the rRNA gene array recently discovered by Vincent & Petes (1986).

The ability of tandemly arranged Ty elements to mobilise the chromo-somal DNA between them is exemplified by the work of Stiles *et al.* (1981) on the COR region of chromosome X. COR contains three genes (*CYC1-OSM1-RAD7*) and comprises about 12 kb of DNA. Mutants were found which had an enhanced level of the *CYC1* gene product, *iso*-1cytochrome *c*, as a result of translocation of COR from chromosome X to chromosome VII and its duplication at the new site. Since COR was retained at its original location, this particular translocation appears to be the result of transposition or conversion.

Gene duplication, followed by divergence, is a major force for innovation in evolution (Ohno, 1970) and McKnight, Cardillo & Sherman (1981) have presented evidence for such an evolutionary path involving the COR region. They found that the *CYC7*-43 mutation on chromosome V, which leads to over-production of *iso*-2-cytochrome *c*, also conferred UV sensitivity and osmotic sensitivity. This discovery defined a new gene cluster *ANP1-RAD23-CYC7* which is functionally and structurally analogous to the COR cluster (see above) on chromosome X. Since the gene orders of the two are related by a circular permutation it was postulated that a translocation event involving a circular intermediate was responsible for the duplication of the primordial gene cluster.

The *his4*-912 mutation is the result of a Ty insertion upstream of *HIS4* on chromosome III. A series of his[+] revertants were isolated which exhibited a range of chromosomal rearrangements. Some were limited in extent; e.g. deletion of Ty via delta – delta recombination, small deletions extending from Ty into the 5' non-coding portion of the *HIS4* structural gene and transposition of *HIS4* from chromosome III to chromosome VIII (Roeder & Fink, 1980). However, some of the revertants involved major chromosomal aberrations, such as the paracentric inversion of most of the left arm of chromosome III, and reciprocal translocations between the left arm of III and the right arm of either chromosome I or VIII (Chaleff & Fink, 1980). Such major rearrangements can result in lethal segregation events at meiosis, but once established in a homozygous diploid, are a step towards speciation.

Viewed as a retrovirus which has lost the capacity for horizontal transmission, the Ty element is an interesting evolutionary situation. Variants which increase efficiency of Ty propagation to the extent that its activities become detrimental to the host yeast cell will be selected against. Correspondingly, it might be expected that mutations of the host genome which act to limit the activities of Ty would be favoured. Rothstein (1986) has discovered a gene, *EDR* (*E*levated *D*elta *R*ecombination), recessive mutations of which enhance the frequency of delta – delta recombination. The effect of the wild type allele of this gene may be to limit the frequency with which major chromosomal rearrangements occur as a result of recombination between delta sequences.

Conclusions

The genome of the yeast *Saccharomyces cerevisiae* is very small, yet its 17 chromosomes are organised and replicated in a typically eukaryotic manner. There are a number of sequence elements which act to ensure the efficiency and fidelity of chromosome replication and segregation. Other elements, particularly the Ty transposons, may be a major source of the variations involved in genome evolution. The activity of these elements, which may only be transmitted between cells by mating and division, is probably limited by functions encoded by the host cell. While our understanding of this role of Ty elements is increasing, the part played by other repeated sequence elements, such as tracts of poly(dA), poly(dGdT) and poly(dGdC), is completely unknown. Since these elements are found throughout the eukaryotes, an understanding of their role in genome evolution will have a significance beyond the yeasts and fungi.

Acknowledgements I am grateful to Richard Walmsley and John Warmington for their critical reading of the manuscript and to Don

Williamson for providing Fig. 3.1. Work in my own laboratory has been supported by the Science and Engineering Research Council and by a NATO Collaborative Award held with Carol S. Newlon.

References

Baltimore, D. (1985). Retroviruses and retrotransposons: the role of reverse transcriptases in shaping the eukaryotic genome. *Cell*, **40**, 481 − 482.

Boeke, J. D., Garfinkel, D. J., Styles, C. A. & Fink, G. R. (1985). Ty elements transpose through an RNA intermediate. *Cell*, **40**, 491 − 500.

Cameron, J. R., Loh, E. Y. & Davis, R. W. (1979). Evidence for transposition of dispersed repetitive DNA families in yeast. *Cell*, **16**, 739 − 751.

Carle, G. F., Frank, M. & Olson, M. V. (1986). Electrophoretic separations of large DNA molecules by periodic inversion of the electric field. *Science*, **232**, 685 − 689.

Carle, G. F. & Olson, M. V. (1984). Separation of chromosomal DNA molecules from yeast by orthogonal-field-alternation gel electrophoresis. *Nucleic Acids Research*, **12**, 5647 − 5664.

Carlson, M. & Botstein, D. (1983). Organization of the *SUC* gene family in *Saccharomyces*. *Molecular and Cellular Biology*, **3**, 351 − 359.

Carlson, M., Celenza, J. L. & Eng, F. J. (1985). Evolution of the dispersed *SUC* gene family of *Saccharomyces* by rearrangements with chromosomal telomeres. *Molecular and Cellular Biology*, **5**, 2894 − 2902.

Celnicker, S. E., Sweder, K., Srience, F., Bailey, J. E. & Campbell, J. L. (1984). Deletion mutations affecting autonomously replicating sequence *ars1* of *Saccharomyces cerevisiae*. *Molecular and Cellular Biology*, **4**, 2455 − 2466.

Chaleff, D. T. & Fink, G. R. (1980). Genetic events associated with an insertion mutation in yeast. *Cell*, **21**, 227 − 237.

Chan, C. S. M. & Tye, B.-K. (1983). Organization of DNA sequences and replication origins at yeast telomeres. *Cell*, **33**, 563 − 573.

Clare, J. J. & Farabaugh, P. (1985). Nucleotide sequence of a yeast Ty element: evidence for an unusual mechanism of gene expression. *Proceedings of the National Academy of Sciences, USA*, **82**, 2829 − 2833.

Clarke, L. & Carbon, J. (1980). Isolation of a yeast centromere and construction of functional small circular chromosomes. *Nature*, **287**, 504 − 509.

Clarke, L. & Carbon, J. (1983). Genomic substitutions of centromeres in *S. cerevisiae*. *Nature*, **305**, 23 − 28.

Dawkins, R. (1976). *The Selfish Gene*. Oxford University Press.

de Jonge, P., de Jongh, F. C. M., Meijers, R., Steensma, H. Y. & Scheffers, W. A. (1986). Orthogonal-field-alternation gel electrophoresis banding patterns from yeasts. *Yeast*, **2**, 193 − 204.

del Rey, F. J., Donahue, T. F. & Fink, G. R. (1982). Sigma, a repetitive element found adjacent to tRNA genes of yeast. *Proceedings of the National Academy of Sciences, USA*, **79**, 4138 − 4142.

Deschamps, J. & Wiame, J.-M. (1979). Mating-type effect on *cis* mutants leading to constitutivity of ornithine transaminase in diploid cells of *Saccharomyces cerevisiae*. *Genetics*, **92**, 749 − 758.

Doolittle, W. F. & Sapienza, C. (1980). Selfish genes, the phenotype paradigm and genome evolution. *Nature*, **284**, 601 − 603.

Dubois, E., Hiernaux, D., Grenson, M. & Wiame, J.-M. (1978). Specific induction of catabolism and its relation to repression of biosynthesis in arginine metabolism of *Saccharomyces cerevisiae*. *Journal of Molecular Biology*, **122**, 383 − 406.

Dubois, E., Jacobs, E. & Jauniaux, J. C. (1982). Expression of the ROAM mutations in *Saccharomyces cerevisiae*: involvement of trans-acting regulatory elements and relation with Ty1 transcription. *EMBO Journal*, **1**, 1133 − 1139.

Eibel, H. & Philippsen, P. (1984). Preferential integration of yeast transposable element Ty into a promoter region. *Nature*, **307**, 386 − 388.

Eigerl, A. & Feldmann, H. (1982). Ty1 and delta elements occur adjacent to several tRNA genes in yeast. *EMBO Journal*, **1**, 1245 − 1250.

Elder, R. T., Loh, E. Y. & Davis, R. W. (1983). RNA from the yeast transposable element Ty1 has both ends in direct repeats, a structure similar to retroviruses. *Proceedings of the National Academy of Sciences, USA*, **80**, 2432 − 2436.

Errede, B., Cardillo, T. S., Teague, M. A. & Sherman, F. (1984). Identification of regulatory regions within the Ty1 transposable element that regulate iso-2-cytochrome c production in the *CYC7-H2* mutant. *Molecular and Cellular Biology*, **4**, 1393 − 1401.

Errede, B., Cardillo, T. S., Wever, G. & Sherman, F. (1980). ROAM mutations causing increased expression of yeast genes: their activation by signals directed toward conjugation functions and their formation by insertion of Ty 1 repetitive elements. *Cold Spring Harbor Symposia on Quantitative Biology*, **45**, 593 − 602.

Errede, B., Company, M., Ferchak, J. D., Hutchinson, C. A. & Yarnell, W. S. (1985). Activation regions in a yeast transposon have homology to mating type control and to mammalian enhancers. *Proceedings of the National Academy of Sciences, USA*, **82**, 5423 − 5427.

Fitzgerald-Hayes, M., Buhler, J.-M., Cooper, T. G. & Carbon, J. (1982). Isolation and subcloning analysis of functional centromere DNA (*CEN11*) from *S. cerevisiae* chromosome XI. *Molecular and Cellular Biology*, **2**, 82 − 87.

Gafner, G. & Philippsen, P. (1980). The yeast transposon Ty 1 generates duplications of target site DNA on insertion. *Nature*, **286**, 414 − 418.

Garfinkel, D. J., Boeke, J. D. & Fink, G. R. (1985). Ty element transposition: reverse transcriptase and virus-like particles. *Cell*, **42**, 507 − 517.

Genbauffe, D. J., Chisholm, G. E. & Cooper, T. G. (1984). Tau, sigma and delta. A family of repeated elements in yeast. *Journal of Biological Chemistry*, **259**, 10518 − 10525.

Gull, K. & Oliver, S. G. (1981). *The Fungal Nucleus*. British Mycological Society Symposium volume 5. Cambridge University Press.

Gunge, N., Tamura, A., Ozawa, F. & Sakaguchi, L. (1981). Isolation and characterization of linear deoxyribonucleic acid plasmids from *Kluyveromyces lactis* and the plasmid-associated killer character. *Journal of Bacteriology*, **145**, 382 390.

Hauber, J., Nelbock-Hochstetter, P. & Feldmann, H. (1985). Nucleotide sequence and characteristics of a Ty element from yeast. *Nucleic Acids Research*, **13**, 2745 − 2758.

Heffron, F. (1983). Tn3 and its relatives. In *Mobile Genetic Elements*, ed. Shapiro, J. A., pp. 223 − 260. New York: Academic Press.

Herskowitz, I. & Oshima, Y. (1981). Control of cell type and mating type interconversion. In *The Molecular Biology of the Yeast Saccharomyces: Life Cycle and Inheritance*, ed. Strathern, J. N., Jones, E. W. & Broach, J. R., pp. 181 − 207. New York: Cold Spring Harbor Laboratory.

Hieter, P., Pridmore, D., Hegemann, J. H., Thomas, M., Davis, R. W. & Philippsen, P. (1985). Functional selection and analysis of yeast centromeric DNA. *Cell*, **42**, 913 – 921.

Hsiao, C.-L. & Carbon, J. (1981). Direct selection procedure for the isolation of functional centromeric DNA. *Proceedings of the National Academy of Sciences, USA*, **78**, 3760 – 3764.

Jinks-Robertson, S. & Petes, T. D. (1985). High frequency meiotic gene conversion between repeated genes on non-homologous chromosomes. *Proceedings of the National Academy of Sciences, USA*, **82**, 3350 – 3354.

Johnston, L. H. & Williamson, D. H. (1978). An alkaline sucrose gradient analysis of the mechanism of nuclear DNA synthesis in the yeast *Saccharomyces cerevisiae*. *Current Genetics*, **2**, 175 – 180.

Kearsey, S. E. (1984). Structural requirements for the function of a yeast chromosomal replicator. *Cell*, **37**, 299 – 307.

Khoury, G. & Gruss, P. (1983). Enhancer elements. *Cell*, **33**, 313 – 314.

Klar, A. J. S., Strathern, J. N., Hicks, J. B. & Prudente, D. (1983). Efficient production of a ring derivative of chromosome III by mating-type switching mechanism in *Saccharomyces cerevisiae*. *Molecular and Cellular Biology*, **3**, 803 – 810.

Lemoine, Y., Dubois, E. & Wiame, J.-M. (1978). The regulation of urea amidolyase of *Saccharomyces cerevisiae*. Mating type influence on a constitutivity mutation acting in *cis*. *Molecular and General Genetics*, **166**, 251 – 258.

Liebman, S. W., Singh, A. & Sherman, F. (1979). A mutator affecting the region of the iso-1-cytochrome c gene in yeast. *Genetics*, **92**, 783 – 802.

Maundrel, K., Wright, A. P. H., Piper, M. & Shall, S. (1985). Evaluation of heterologous *ARS* activity in *S. cerevisiae* using cloned DNA from *S. pombe*. *Nucleic Acids Research*, **13**, 3711 – 3723.

McKnight, G. L., Cardillo, T. S. & Sherman, F. (1981). An extensive deletion causing overproduction of yeast iso-1-cytochrome c. *Cell*, **25**, 409 – 419.

Mellor, J., Kingsman, A. J. & Kingsman, S. M. (1986) Ty, an endogenous retrovirus of yeast? *Yeast*, **2**, 145 – 152.

Mellor, J., Malim, M. H., Gull, K., Tuite, M. F., McCready, S., Dibbayawan, T., Kingsman, S. M. & Kingsman, A. J. (1985). Reverse transcriptase activity and Ty RNA are associated with virus-like particles in yeast. *Nature*, **318**, 583 – 586.

Michels, C. A. & Needleman, R. B. (1983). Repeated family of genes controlling maltose fermentation in *Saccharomyces carlsbergensis*. *Molecular and Cellular Biology*, **3**, 796 – 802.

Mortimer, R. K. & Schild, D. (1985). Genetic map of *Saccharomyces cerevisiae*, edition 9. *Microbiological Reviews*, **49**, 181 – 212.

Newlon, C. S. & Burke, W. (1980). Replication of small chromosomal DNAs in yeast. In *Mechanistic Studies of DNA Replication and Genetic Recombination* (ICN-UCLA Symposia on Molecular and Cellular Biology, vol 19), ed. Alberts, B. & Fox, C. N., pp. 339 – 409. New York: Academic Press.

Newlon, C. S., Petes, T. D., Hereford, L. M. & Fangman, W. L. (1974). Replication of yeast chromosomal DNA. *Nature*, **247**, 32 – 35.

Ohno, S. (1970). *Evolution by Gene Duplication*. London: Allen & Unwin.

Orgel, L. E. & Crick, F. H. C. (1980). Selfish DNA: the ultimate parasite. *Nature*, **284**, 604 – 607.

Palzkill, T. G., Oliver, S. G. & Newlon, C. S. (1986). DNA sequence analysis of *ARS* elements from chromosome III of *Saccharomyces cerevisiae*. *Nucleic Acids Research*, **14**, 6247 – 6264.

Panzeri, L. & Philippsen, P. (1982). Centromeric DNA from chromosome VI in *S. cerevisiae* strains. *EMBO Journal,* **1**, 1605 − 1611.

Pearston, D. H., Gordon, M. & Hardman, N. (1985). Transposon-like properties of the major, long repetitive sequence family in the genome of *Physarum polycephalum*. *EMBO Journal*, **4**, 3557 − 3562.

Petes, T. D., Hereford, L. M. & Skryabin, K. G. (1978). Characterization of two types of yeast ribosomal RNA genes. *Journal of Bacteriology*, **134**, 295 − 305.

Petes, T. D. & Newlon, C. S. (1974). Structure of DNA in DNA replication mutants of yeast. *Nature*, **251**, 637 − 639.

Petes, T. D., Newlon, C. S., Byers, B. & Fangman, W. L. (1973). Yeast chromosomal DNA: size, structure and replication. *Cold Spring Harbor Symposia on Quantitative Biology*, **38**, 9 − 16.

Petes, T. D. & Williamson, D. H. (1975). Fiber autoradiography of replicating yeast DNA. *Experimental Cell Research*, **95**, 103 − 110.

Rivin, C. J. & Fangman, W. L. (1980). Replication fork rate and origin activation during the S-phase of *Saccharomyces cerevisiae*. *Journal of Cell Biology*, **85**, 108 − 115.

Roeder, G. S., Farabaugh, P. J., Chaleff, D. T. & Fink, G. R. (1980). The origins of gene instability in yeast. *Science*, **209**, 1375 − 1380.

Roeder, G. S. & Fink, G. R. (1980). DNA rearrangements associated with a transposable element in yeast. *Cell*, **21**, 239 − 249.

Roeder, G. S. & Fink, G. R. (1982). Movement of transposable elements by gene conversion. *Proceedings of the National Academy of Sciences, USA*, **79**, 5621 − 5625.

Rothstein, R. (1979). Deletions of a tyrosine tRNA gene in *S. cerevisiae*. *Cell*, **17**, 185 − 190.

Rothstein, R. (1986). Genetic evidence for heteroduplex DNA during recombination. In *Biochemistry and Molecular Biology of Industrial Yeasts*, ed. Stewart, G. G., Russell, I., Klein, R. D., & Hiebsch, R.R., in press. Boca Raton, Florida: CRC Press.

Rothstein, R. J. & Sherman, F. (1980). Dependence on mating-type for the over-production of iso-2-cytochrome c in the yeast mutant *CYC7-H2*. *Genetics*, **94**, 891 − 898.

Sandmeyer, S. B. & Olson, M. V. (1982). Insertion of a repetitive element at the same position in the 5′-flanking regions of two dissimilar yeast tRNA genes. *Proceedings of the National Academy of Sciences, USA*, **79**, 7674 − 7678.

Scherer, S., Mann, C. & Davis, R W. (1982). Reversion of a promoter deletion in yeast. *Nature*, **298**, 815 − 819.

Schwartz, D. C. & Cantor, C. R. (1984). Separation of yeast chromosome-sized DNAs by pulsed field gradient gel electrophoresis. *Cell*, **37**, 67 − 75.

Shampay, J., Szostak, J. W. & Blackburn, E. H. (1984). DNA sequences of telomeres maintained in yeast. *Nature*, **310**, 154 − 157.

Sherman, F., Stewart, J. W., Helms, C. & Downie, J. A. (1978). Chromosome mapping of the *CYC7* gene determining yeast iso-2-cytochrome c: structural and regulatory regions. *Proceedings of the National Academy of Sciences, USA*, **75**, 1437 − 1441.

Siliciano, P. G. & Tatchell, K. (1984). Transcription and regulatory signals at the mating-type locus in yeast. *Cell*, **37**, 969 − 978.

Stiles, J. I., Friedman, L. R., Helms, C., Consaul, S. & Sherman, F. (1981). Transposition of the gene cluster *CYC1-OSM1-RAD7* in yeast. *Journal of Molecular Biology*, **148**, 331 − 336.

Stinchcomb, D. T., Mann, C. & Davis, R. W. (1982). Centromeric DNA from

Saccharomyces cerevisiae. Journal of Molecular Biology, **158**, 157 — 179.

Stinchcomb, D. T., Struhl, K. & Davis, R. W. (1979). Isolation and characterization of a yeast chromosomal replicator. *Nature*, **282**, 39 — 43.

Struhl, K., Stinchcomb, D. T., Scherer, S. & Davis, R. W. (1979). High frequency transformation of yeast: autonomous replication of hybrid DNA molecules. *Proceedings of the National Academy of Sciences, USA*, **76**, 1035 — 1039.

Szostak, J. W. & Blackburn, E. H. (1982). Cloning of yeast telomeres on linear plasmid vectors. *Cell*, **29**, 245 — 255.

Szostak, J. W. & Wu, R. (1979). Insertion of a genetic marker into the ribosomal DNA of yeast. *Plasmid*, **2**, 536 — 554.

Thompson, A. & Oliver, S. G. (1986). Physical separation and functional interaction of *Kluyveromyces lactis* and *Saccharomyces cerevisiae* ARS elements derived from killer plasmid DNA. *Yeast*, **2**, 179 — 192.

Vincent, A. & Petes, T. D. (1986). Isolation and characterization of a Ty element inserted into the ribosomal DNA of the yeast *Saccharomyces cerevisiae*. *Nucleic Acids Research*, **14**, 2939 — 2949.

Walmsley, R. M., Chan, C. S. M., Tye, B.-K. & Petes, T. D. (1984). Unusual DNA sequences associated with the ends of yeast chromosomes. *Nature*, **310**, 157 — 160.

Walmsley, R. M., Johnson, L. H., Williamson, D. H. & Oliver, S. G. (1984). Replicon size of yeast ribosomal DNA. *Molecular and General Genetics*, **195**, 260 — 266.

Warmington, J. R., Anwar, R., Newlon, C. S., Davies, R. W., Indge, K. J. & Oliver, S. G. (1986). A 'hot-spot' for Ty transposition on the left arm of chromosome III. *Nucleic Acids Research*, **14**, 3475 — 3485.

Warmington, J. R., Waring, R. B., Newlon, C. S., Indge, K. J. & Oliver, S. G. (1985). Nucleotide sequence characterization of Ty 1-17, a class II transposon from yeast. *Nucleic Acids Research*, **13**, 6679 — 6693.

Watson, J. D. (1972). Origin of concatameric T7 DNA. *Nature New Biology*, **239**, 197 — 200.

Williamson, D. H. (1965). The timing of deoxyribonucleic acid synthesis in the cell cycle of *Saccharomyces cerevisiae*. *Journal of Cell Biology*, **25**, 517 — 528.

Williamson, D. H. (1985). The yeast ARS element six years on: a progress report. *Yeast*, **1**, 1 — 14.

Williamson, V. M. (1983). Transposable elements in yeast. *International Review of Cytology*, **83**, 1 — 25.

Zamb, T. J. & Petes, T. D. (1982). Analysis of the junction between ribosomal RNA genes and single copy sequences in the yeast *Saccharomyces cerevisiae*. *Cell*, **28**, 355 — 364.

4

The natural history of fungal mitochondrial genomes

CLAUDIO SCAZZOCCHIO

Institut de Microbiologie, Université de Paris-Sud, Centre d'Orsay, 91405 Orsay, France

Introduction

The title of this chapter derives both from the vagaries of fungal taxonomy and from the fragmentary nature of the information available to us. While we have no doubt that ascomycetes and basidiomycetes are quite closely related organisms, the relationships of other groups, especially oomycetes and zygomycetes, are uncertain. We have some fragmentary molecular information from oomycetes and basidiomycetes, but all the 'hard' molecular data come from a restricted number of ascomycete species. With such limitations, it is impossible to speculate on evolutionary relationships of fungal mitochondrial genomes, or to use such data as the basis for wider speculation. All that can be done is to identify some features that might have evolutionary implications.

General organisation of fungal mitochondrial genomes

Mitochondrial genomes of metazoans have a fairly constant size, of around 16 kilobases (kb); and those of plants are usually large (and variable). Mitochondrial genomes of fungi vary from a metazoan-like size, 18·9 kb in *Torulopsis glabrata* (Clark-Walker *et al.*, 1980), to almost plant-like size, 176·3 kb in *Agaricus bitorquis* (Hintz *et al.*, 1985). It seems clear that the variations in genome size are due to variation in the size of intergenic regions and on the number of introns that genes contain (for reviews see Dujon, 1983; Grossman & Hudspeth, 1985). The mitochondrial genomes of *Aspergillus nidulans* (32 kb) and of *Neurospora crassa* (75 kb) have almost exactly the same genes (Nelson & Macino, 1985; Breitenberger & RajBhandary, 1985; Brown *et al.*, 1985) but intron content differs drastically (see below) and *N. crassa* also possesses an undetermined number of GC-rich repetitive sequences. Weber *et al.*

(1986) have analysed the mitochondrial genomes of *Coprinus cinereus* and
C. stercorarius, which were respectively 43·3 kb and 91·1 kb long. A
number of genes were recognised by hybridisation with probes from
Saccharomyces cerevisiae and, again, it is·obvious that most of the size
variation is due to intergenic regions.

Most fungal mitochondrial genomes are circular. One linear mitochondrial DNA has been reported, from the yeast *Hansenula mrakii*
(Weslowski & Fukuhara, 1981). *Achlya ambisexualis* and other *Achlya*
species have an inverted repeat containing the genes specifying the rRNAs
(Shumard, Grossman & Hudspeth, 1986). This peculiarity, shared by
most (but not all) oomycetes, is a characteristic of the chloroplast genome
of higher plants. The yeasts *Kloechera africana* and *Hanseniospora vineae*
also have an inverted repeat, but this, at least in the former species, is
confined to part of the LrRNA gene and some tRNA genes (Clark-Walker,
McArthur & Sriprakash, 1981). It may be that inverted repeats can be
generated independently and relatively frequently; the event may have
occurred early in chloroplast evolution (perhaps in the original endo-
symbiont) and at many other times at different evolutionary branch points.
The only constant feature is that it seems to involve rRNA genes, which
may, therefore, contain some 'duplication-prone' sequence.

Gene order is not a useful basis for attempts to establish evolutionary
relationships. The rate of recombination in mitochondrial genomes is
probably quite high, a point discussed by Dujon *et al.* (1974) and Waring
& Scazzocchio (1983). In the absence of selective constraints, any order
could be generated from any other on an evolutionary time-scale. The
positions of tRNA genes seem to differ quite considerably between the
ascomycete species so far examined (Dujon, 1983; Lang *et al.*, 1983).
However, there are some striking similarities. For example, in both *N.
crassa* and *A. nidulans* between the SrRNA and LrRNA genes there is a
cluster of tRNA genes, interrupted by the gene coding for subunit III of
cytochrome oxidase (*cox*III) and an open reading frame (ORF) of
unknown function (URFA1), though the relative order of these is inverted
(Nelson & Macino, 1985).

Genes present in fungal mitochondrial genomes

All fungal mitochondrial genomes studied to date contain tRNA
and rRNA genes, the genes coding for subunits I, II, and III of cytochrome
oxidase, the gene coding for apocytochrome *b* and the genes coding for
subunits 6 and 8 of the mitochondrial ATPase. The gene coding for subunit
9 of ATPase is present (and active) in the mitochondrial genome of

Saccharomyces cerevisiae and *Schizosaccharomyces pombe*. The ATPase subunit 9 expressed during vegetative growth in both *Neurospora crassa* and *Aspergillus nidulans* is encoded in the nucleus, although an intact gene exists in the mitochondrial genomes of both organisms (Van den Boogaart, Samall & Agsteribbe, 1982; Brown *et al.*, 1985). Using a mitochondrial *S. cerevisiae* probe, the subunit 9 gene has been detected in the mitochondrial genomes of two species of *Coprinus* (Weber *et al.*, 1986). This probe does not detect a nuclear copy of the gene, but the crucial experiment of using a nuclear gene from *A. nidulans* or *N. crassa* as a probe has not been reported.

The *var* gene codes for a ribosomal protein in *Saccharomyces cerevisiae*, but has not been found in *Schizosaccharomyces pombe*, *A. nidulans* or *N. crassa*. Sequences hybridising with a *var* probe have been found in the mitochondrial genome of *Coprinus stercorarius*, but in the nuclear genome of *C. cinereus* (Weber *et al.*, 1986); the significance of this result is not yet clear.

Complete sequencing of a number of metazoan mitochondrial genomes revealed eight ORFs coding for unidentified proteins; these were called URFs (e.g. Anderson *et al.*, 1981; Clary & Wolstenholme, 1985). One, URFA6L, has been shown to code for ATPase subunit 8 and URFs 1, 2, 3, 4, 4L, 5 and 6 code for components of the mitochondrial NADH dehydrogenase (Chomyn *et al.*, 1985). To date, URFs 1, 2, 3, 4, and 5 have been identified in *A. nidulans*, whilst URFs 1, 2, 3 and 5 have also been identified in *N. crassa* and URF1 in *Podospora anserina* and *Cephalosporium acremonium* (Brown *et al.*, 1983a, 1985; Nelson & Macino, 1985; de Vries *et al.*, 1986; Cummings *et al.*, 1985; Peñalva & Garcia, 1986). Although the mitochondrial DNA of *A. nidulans* has been sequenced almost completely, enough gaps exist in the known *N. crassa* sequence to accommodate some additional URFs. Recently, Macino and co-workers (personal communication) have identified URF4L in *N. crassa* and have observed its presence in *A. nidulans* sequences supplied by our group. Interestingly, genes homologous to these URFs have not been found in either *Saccharomyces cerevisiae*, or in the completely sequenced mitochondrial genome of *Schizosaccharomyces pombe*. As, obviously, these yeasts have NADH dehydrogenase in their mitochondria, the corresponding genes must occur in their nuclear genomes.

No direct experimental data yet exist on 'movement' of DNA sequences between cell compartments. The only evidence is that sequences found in the nuclear genome of one organism may be in the mitochondrial genome of another or, as in the case of the ATPase subunit 9 discussed above, may be present in the genomes of both compartments. Farrelly & Butow (1983) have reported the presence of non-functional mitochondrial sequences in the nuclear genome of *Saccharomyces cerevisiae*.

In both *A. nidulans* and *N. crassa*, duplications of portions of the mitochondrial genome involving tRNA genes have been detected and, in both, one duplication also involves the N-terminal region of an ORF. In *A. nidulans*, a 300 base-pair (bp) duplication contains a tRNA$_{asn}$, an intergenic region and 108 bp of ORF. In one position this duplication is immediately upstream of the ATPase subunit 8 gene and includes the N-terminal end of this; in the other position, it includes the N-terminal end of URFA3, an ORF not yet found in any other species (Brown *et al.*, 1983b). Thus, there is a potential for gene duplications to generate new hybrid genes.

Mitochondrial plasmids, petites and related phenomena

Obviously related to the subject of movement and transfer of sequences within the mitochondrial genome and between the nuclear and mitochondrial compartments, is the widespread presence of mitochondrial plasmids in fungi. Some of these plasmids are unrelated to mitochondrial genomic sequences and, in the few cases where the test has been performed, no relation to nuclear sequences has been found. A systematic search has not been done, but plasmids have been reported in fungi as diverse as *Claviceps*, *Cochliobolus*, *Neurospora* (Grossman & Hudspeth, 1985; Kinsey, 1985) and *Fusarium* (Kistler & Leong, personal communication). It can be speculated that such plasmids represent means by which sequences may transfer from one compartment to another, or even between species. Particularly interesting, among *Neurospora crassa* plasmids, is the plasmid present in the Mauriceville strain; for, while being different from the mitochondrial genome of this species, it does have some of the characteristics of mitochondrial sequences, especially intronic sequences (Nargang *et al.*, 1984). Also of special interest is a 9 kb element in *kalilo* strains of *N. intermedia* which exists as a free plasmid in high copy number in the nucleus, but which can transpose into the mitochondrial genome where it does not replicate autonomously. Insertion of this element is directly related to the onset of senescence in these strains (Bertrand, Chan & Griffiths, 1985, and personal communication).

Classical mitochondrial genetics in *Saccharomyces cerevisiae* started with the isolation of cytoplasmically inherited *petite* mutations. A class of these strains have mitochondrial sequences repeated many times, either as tandem or reverse repeats, in circular plasmid-like structures while the rest of the mitochondrial genome is deleted (Dujon, 1981). Similar phenomena occur in a number of ascomycetes, with or without conservation of the mitochondrial genome. The *stopper* mutations of *N. crassa* (Gross, Hsieh & Levine, 1984; de Vries *et al.*, 1986) arise from deletion of genomic sequences with or without amplification of other sequences (Collins & Lambowitz, 1981). Deletions in *stopper* mutants can result from intramolecular recombination; Gross *et al.* (1984) established that the recombination frequently occurs at a direct repeat comprising the tRNA$_{met}$. While similar direct repeats exist in *A. nidulans*, *stopper*-like mutations have not been reported in this species. However, in *A. amstelodami*, a similar phenotype, *ragged*, originates from excision and amplification of specific regions of the mitochondrial genome (Lazarus *et al.*, 1980; Lazarus & Kuntzel, 1981).

Neither *stopper* nor *ragged* are lethal, possibly because the complete mitochondrial gene complement is conserved, though under-represented. The case is very different for the senescence phenomenon in *Podospora anserina*, however. This lethal event is correlated with the presence of a number of plasmids which obviously originate from a limited number of regions of the mitochondrial genome (Wright, Horrum & Cummings, 1982; for review see Benne & Tabak, 1986). Interestingly, one such plasmid, α, is a precise excision at the DNA level of a class II intron of the cytochrome oxidase subunit I gene which contains an ORF (Osiwecz & Esser, 1984), Belcour & Vierny (1986) showed that besides 'α-sequences', senescent cultures contain a residual DNA with imprecise excision of the same α-intron. A number of themes seem to be converging: the existence of mitochondrial plasmids which are intron-like and, in some cases, transcribed; the possibility that proteins coded by intron ORFs might catalyse intron excision at the DNA level (see below), and that imprecise excision might be involved in the onset of the senescence phenotype. The *kalilo* strains of *Neurospora intermedia* also show a senescence-like phenomenon; but in this case it is the result of insertional inactivation of mitochondrial DNA sequences. As pointed out by Bertrand *et al.* (1985), while the physiology of senescence might well be the same in *Neurospora* and *Podospora*, the molecular mechanism at the onset of the phenomenon cannot be more different. There are many roads to the same (senescent) end.

Mitochondrial genetic code and codon usage

The sequencing of a number of mitochondrial genomes revealed that in every case one or more codons had a meaning different from that

Table 4.1. *Departures from the universal genetic code*

	Codon					
	UGA	UAU	CUN	AGP	CGG	UAP
Universal code	Ter	Ile	Leu	Arg	Arg	Ter
Mitochondrial codes						
Saccharomyces cerevisiae	Trp	Met	Thr	Arg	Arg?	Ter
Neurospora crassa, Aspergillus nidulans and *Schizosaccharomyces*						
pombe	Trp	Ile	Leu	Arg	Arg?	Ter
Mammals	Trp	Met	Leu	Ter	Arg	Ter
Drosophila	Trp	Met	Leu	Ser	Arg	Ter
Trpanosomes	Trp	Ile	Leu	Arg	Arg	Ter
Plants	Ter	Ile	Leu	Arg	Trp	Ter
Other variations						
Mycoplasma capricolum (Eubacteriales)	Trp	Ile	Leu	Arg	Arg	Ter
Ciliate macronucleus	Ter	Ile	Leu	Arg	Arg	Gin

N means any of the four bases, P means a purine (A or G). References are in the text, but for trypanosomes see Simpson *et al.*, 1985; and for plants, Schuster & Brennicke, 1985 and references therein. Some slight simplifications have been made, e.g. AGA specifies serine in *Drosophila yakuba*, but AGG is never found (Clary & Wolstenholme, 1985).

of the universal code. Table 4.1 shows the situation in a number of ascomycetes compared with other groups of organisms. It is striking that even within ascomycetes the code is not identical. *Schizosaccharomyces pombe*, which is similar in gene content to *Saccharomyces cerevisiae*, nevertheless employs a genetic code that is probably identical to that of *A. nidulans* and *N. crassa*. A different 'wobble rule' (Heckman *et al.*, 1980), where a U in the first position of the anticodon enables a tRNA to read all four codons of a given family, limits the number of tRNAs necessary to read all codons to 24.

Some particular cases deserve discussion. Arginine is coded by two families in the universal code; the AGA − AGG family is used preferentially in all fungi, whilst the CGN family is used almost exclusively in intron ORFs in *Saccharomyces cerevisiae* and *N. crassa*, though CGA is found in exons of *Schizosaccharomyces pombe*. In *A. nidulans*, twelve instances of the use of CGT in exons have been recorded, while both CGT and CGG have been found in intron ORFs. In all animal mitochondria and in *Saccharomyces cerevisiae*, *N. crassa* and *A. nidulans*, UGA is read as tryptophan. In *Schizosaccharomyces pombe* this codon is found only in the *cob* gene (cytochrome *b*) intron ORFs and, moreover, the sequenced tRNA should not be able to recognise UGA. As none of the positions where UGA is found happen to be in an ORF coding for a known product, it is difficult to decide whether this codon codes for tryptophan or serves as a stop codon, as it does in the universal code. An indirect argument, developed by Lang (1984), suggests that UGA codons of *Schizosaccharomyces pombe* are, indeed, read as tryptophan, though inefficiently. While the start codon is, in nuclear and prokaryotic genomes, always AUG (methionine), there are clear instances where other initiator codons are used in mitochondrial genomes. The most striking case is the cytochrome oxidase subunit I gene (*cox*I, *oxi*A) which does not have an obvious start codon in either *N. crassa* or *A. nidulans*, or *Drosophila yakuba*. It has been suggested that the quadruplet AUAA or AUGA is used as start codon in these cases (Clary & Wolstenholme, 1983; Waring *et al.*, 1984).

What is of interest here is the variation of the mitochondrial coding system even between related organisms. Until very recently this could be contrasted with the fixity of the nuclear and prokaryotic code. But it is now known that in the eubacterium *Mycoplasma capricolum* UGA is read as tryptophan (Yamao *et al.*, 1985) and that in the macronucleus of ciliates, UAG and UAA are not stop codons (Caron & Meyer, 1985; Preer *et al.*, 1985) but are read as glutamine (Kucino *et al.*, 1985). Whether other small variations from the universal code are used in the nuclear genome of organisms of significant taxonomic position (e.g. choanociliates − see Cavalier-Smith, Chapter 23) is a matter of obvious interest. It remains the fact that mitochondrial genomes have been able to diverge almost without tRNA redundancy. The mechanism of this divergence is something which can tax the imagination of the evolutionary biologist.

Introns

Location and number

One of the amazing characteristics of fungal (and plant) mitochondrial genomes is that they contain large introns, whilst metazoan mitochondrial genomes are totally free of them. In *Saccharomyces cerevisiae* and *A. nidulans* the *cob* (apocytochrome *b*) gene, *cox*I and LrRNA genes have introns; no other mitochondrial gene has them in either species. Examination of *N. crassa* disturbs this apparent pattern: *cox*I has no introns, but the ATPase subunit 6 gene and URFs1, 3, 4 and 4L (all

URFs absent from *S. cerevisiae* but intron-less in *A. nidulans*) have introns; the URF1 of *Podospora anserina* also has introns (Dujon, 1983; Nelson & Macino, 1985; Cummings *et al.*, 1985).

The phylogenetic mobility of introns in ascomycetes is highlighted when related species or even different strains of the same species are compared. Strains of *S. cerevisiae* can differ in the number of introns in the LrRNA gene (0 or 1), in the *cob* gene (2 or 5) or in the *cox*I gene (5 or 7) (Dujon, 1980; Lazowska, Jacq & Slonimski, 1980; Nobrega & Tragaloff, 1980; Bonitz *et al.*, 1980). Strains with a variable number of introns can be generated by genetic recombination in *S. cerevisiae* (Labousse & Slonimski, 1983). *Aspergillus equinulatus* has additional introns in *cob* and *cox*I when compared with the closely related *A. nidulans*, whilst *A. quadrilineatus* has none (Earl *et al.*, 1981; Turner, Earl & Greaves, 1982). Some of these data are summarised in Table 4.2.

Introns are rarely found in the same positions in homologous genes of different organisms. However, the intron in the LrRNA gene which is optional in *Saccharomyces cerevisiae* is, when present, in precisely the same position as in *N. crassa*, *A. nidulans* and *Kluyveromyces thermotolerans*, which also coincides with one of the two positions where introns are present in the *nuclear* LrRNA gene of *Physarum polycephalum* (Burke & RajBhandary, 1982; Michel, Jacquier & Dujon, 1982). The mitochondrial introns inserted in the LrRNA gene all belong to the same subclass of class I introns, though their ORFs code for proteins of different functions (see below). Two other exceptions are even more striking. Intron 3 of the *cob* gene of *S. cerevisiae* is present only in the so-called 'long' strains, not in 'short' strains (where 'long' and 'short' refer to the length of the *cob* and *cox*I genes). Surprisingly, it is the only intron present in the *cob* gene of *A. nidulans*. Not only are both introns in exactly the same position, but their sequences, at the nucleotide level and in their ORF, are clearly related (Waring *et al.*, 1981, 1982; Lazowska *et al.*, 1981; Holl, Schmidt & Schweyen, 1985). The similarity between the third intron of the *A. nidulans cox*I gene and the second intron of the homologous gene in *Schizosaccharomyces pombe* is even more striking. The positions of insertion are identical and homology is very high, reaching 70% of amino acid homology in the intron ORF (Waring *et al.*, 1984; Lang, 1984). Lang (1984) proposed that this situation arose by horizontal transmission between the mitochondria of the two fungi. In the region of highest homology, the intron of *A. nidulans* has an additional surprising feature: a 37 bp insert flanked by a 5 bp direct repeat. This insert is the most GC-rich stretch of the entire mitochondrial genome of *A. nidulans*.

It is obviously non-mitochondrial in origin, and the presence of the 5 bp repeat suggests the possibility that it is the 'fossil' of a nuclear transposable element. An element capable of doing precisely this is present in the *kalilo* strains of *Neurospora intermedia* (see above).

It is difficult to explain these two instances where homologous introns occur in exactly the same place in species that are not closely related, against the background of extensive variability in intron number and position. One hypothesis is that the mitochondrial genome of the 'primeval' fungus had all the introns we have found or can find and that these were lost differentially during evolution. The second is that introns actually move around. Loss of introns by exact excision has been observed in *Saccharomyces cerevisiae* and one of the senescence plasmids of *Podospora anserina* is a precisely excised type II intron. Gain of introns, though, has only been observed in the case of the highly polarised gene conversion involving the LrRNA intron of *S. cerevisiae*, where all recombinational events in the mitochondrial genome between an intron-carrying and an intron-less strain give intron-carrying progeny (Jacquier & Dujon, 1985). If introns are, or derive from, mobile genetic elements they might land, in some cases, exactly in the same position due to the requirement for formation of a correct internal guide sequence (see below).

Intron open reading frames

Both classes of intron (type I and type II — see below) might have sizeable ORFs; Kotylak *et al.* (1985) have summarised the evolutionary significance of these. With the exception of the protein coded by the intron in the LrRNA gene of *Neurospora crassa* and *Aspergillus nidulans*, which is possibly a ribosomal protein (Burke & RajBhandary, 1982), all intron coded proteins seem to be involved in reactions affecting DNA, RNA or both.

In *Saccharomyces cerevisiae*, some ORFs of class I introns have been shown to be necessary for correct excision of the intron itself; in one case, for excision of an intron in another gene as well (Lazowska *et al.*, 1980; Schemelzer, Schmidt & Schweyen,1981; Weiss-Brummer *et al.*, 1982; Anziano *et al.*, 1982; Dujardin, Jacq & Slonimski, 1982). The protein coded by these ORFs has been called maturase (Lazowska *et al.*, 1980); a diagnostic feature of maturase-like proteins is the presence of two repeated decapeptides separated by approximately 100 amino acids (Waring *et al.*, 1982, 1984; Waring & Davies, 1984; Hernsgens *et al.*, 1983). Such decapeptides have been found in ORFs of class I introns of *A. nidulans*, *N.*

Table 4.2. *Introns in fungal mitochondrial genomes*

Gene	Species			
	Sacharomyces cerevisiae	*Schizosaccharomyces pombe*	*Aspergillus nidulans*	*Neurospora crassa*
cob (apocytochrome b)	2 class I (obligatory) 2 class I (optional) 1 class II (optional)	1 class II	1 class I	2 class I
cox I, *oxi* A (cytochrome oxidase subunit I)	2 class I, 3 class II (obligatory) 2 class I (optional)	2 class I	3 class I	none

oli 2, *oli* A (ATPase subunit 6)	none	none	none	1 class I 1 unique (see text)
URF1 subunit of NADH dehydrogenase	gene absent	gene absent	none	1 class I
URF5 (subunit of NADH dehydrogenase)	gene absent	gene absent	none	3 class I
URF4L (subunit of NADH dehydrogenase)	gene absent	gene absent	possibly none	1 class I
LrRNA (large ribosomal RNA)	1 class I (optional)	none	1 class I	1 class I

For references see text and Helmer-Citterich *et al.* (1983).

crassa, Schizosaccharomyces pombe and *Podospora anserina*. The two decapeptides also exist in a non-intronic ORF of the mitochondrial DNA of *Saccharomyces cerevisiae* (Waring *et al.*, 1982). A degree of homology of uncertain significance has been found between the inter-decapeptide sequence of maturases and the p30 protein of tobacco mosaic virus (Zimmern, 1983), whilst the LrRNA of the archaebacterium *Desulfurococcus mobilis* contains a 622 bp intron with features in common with class I mitochondrial introns (Kjems & Garett, 1985). This intron shows an ORF with one clear decapeptide (P. P. Slonimski, personal communication).

The ORFs in class I introns can code for proteins with functions other than RNA splicing. Indeed, the maturase protein may itself be involved in (DNA-level) intron deletion (Gargouri, Lazowska & Slonimski, 1983) and the ORF of the LrRNA intron of *Saccharomyces cerevisiae* is essential for the duplicative transposition of that intron. Interestingly, the sequence recognised by this protein is similar to the sequence recognised by the *HO* gene product, which is essential for mating type switching in *S. cerevisiae* (Jacquier & Dujon, 1985; Macreadie *et al.*, 1985). Kotylak *et al.* (1985) have shown that the ORF of intron *cob*I4 can induce homologous recombination of other parts of the genome, and that this process is not site-specific. Finally, Michel & Lang (1986) have observed that ORFs of class II introns have homologies with viral reverse transcriptases.

Intron secondary structure and the mechanism of intron excision

With the possible exception of a small intron in the gene coding for subunit 6 of ATPase in *Neurospora crassa* (Morelli & Macino, 1984), all mitochondrial introns can be divided into two classes according to their secondary structure, and two independent methods have given secondary structure models which are strikingly similar (Davies *et al.*, 1982; Michel *et al.*, 1982; Michel & Dujon, 1983). A considerable number of papers and reviews have recently dealt with both the structure and excision mechanism of both classes of intron (Waring & Davies, 1984; Cech, 1986; Tabak & Grivell, 1986).

The secondary structure of class I introns is characterised by pairings between four highly conserved sequences (P & Q and R & S) and two other sequences, E and E', which are exactly complementary and in conserved positions even if their sequences differ between different introns. When a secondary structure is constructed such that these pairings are respected, a number of other pairings appear (Fig. 4.1). The most striking of these is between a sequence (internal guide sequence, IGS) near the junction of the 5′ extreme of the intron and the 3′ extreme of the upstream exon and the 5′ extreme of the downstream exon. The IGS determines the splicing point precisely in a number of class I introns. The evidence that the proposed pairings have physiological importance came from genetic studies in *Saccharomyces cerevisiae*, where *cis*-dominant and *trans*-recessive mutations mapping in R, S, E or E′ have been identified, while other mutations, both *cis*- and *trans*-recessive, map in the putative maturase ORF implying that both the secondary structure and at least the maturase protein were necessary for correct splicing (Weiss-Brummer *et al.*, 1982; Anziano *et al.*, 1982; De La Salle, Jacq & Slonimski, 1982; Jacq *et al.*, 1982; Netter *et al.*, 1982).

It was soon realised that plant mitochondria and chloroplasts have both class I and class II introns. Two 'almost type I' introns are present in the *nuclear* LrRNA genes of *Physarum polycephalum*, but, most surprisingly, the intron of the LrRNA nuclear genes of *Tetrahymena pigmentosa* and *T. thermophila* are perfect examples of class I introns (Waring *et al.*, 1983; Michel & Dujon, 1983; Cech *et al.*, 1983). This was surprising because this intron has been shown to self-splice *in vitro* in the absence of any proteins and requiring only a guanosine co-factor (Cech, Zaug & Grabowski, 1981; Grabowski, Zaug & Cech, 1981; Kruger *et al.*, 1982).

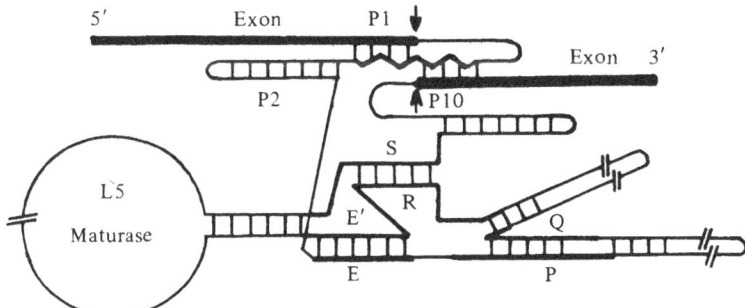

Fig. 4.1. Generalised diagrammatic representation of the secondary structure of class I introns. The 'central core' is formed by the pairing of sequences labelled P and Q, of E and E′, R and S and adjacent associated pairings. A maturase open reading frame (ORF) is usually contained in the L5 loop, but this loop can be very small in some class I introns which have no ORF. The internal guide sequence (IGS) is indicated with a wavy line and is shown paired with the upstream and downstream exons (P1 and P10). The distance between P2 and E is only about four nucleotides so the line joining these sequences is drawn out of scale for graphical convenience. For more details see Davies *et al.* (1982) from which this figure is adapted.

The interest shifted then to assess whether the model proposed by Davies *et al.* (1982) could account for self-splicing. A number of *in vitro* and *in vivo* assays (Waring *et al.*, 1985; Price & Cech, 1985) have confirmed that almost all the postulated pairings are necessary for self-splicing of the *Tetrahymena* intron. Of particular interest is the confirmation of at least part of the role of the IGS in determining splicing specificity (Waring *et al.*, 1985, 1986). In addition, a number of class I introns from fungal mitochondria have been found to be self-splicing, i.e. not to require a maturase (Garriga & Lambowitz, 1984; Van der Horst & Tabak, 1985).

The secondary structure model of class II introns had less heuristic value than the corresponding model for class I. The secondary structure is conserved, but conserved sequences are limited to the 3' terminal fragment of the intron. Recently, however, evidence has been found both for the role of RNA pairings and of maturase-like proteins in class II intron splicing (Carignani *et al.*, 1983; Schmelzer *et al.*, 1983). Here again, one member of this class of intron has been found to be self-splicing. The excision involves formation of a 'lariat' structure which suggests a mechanism not dissimilar from that used in excision of introns from nuclear mRNA precursors (Peebles *et al.*, 1986; van den Veen *et al.*, 1986).

There is an apparent contradiction between the demonstrated need for maturases and the newly discovered self-splicing reactions which might reflect the influence of maturase (and other) proteins on the kinetics of the reaction and/or the stability of the RNA secondary structures.

These two classes of intron seem to represent two types of RNA which can take secondary structures that enable them to self-catalyse reactions involving the RNA molecule itself. It might be a semantic point whether such molecules can be called 'enzymes' (or ribozymes) as they are self-destroyed in the process of catalysis. A third structure, the RNA component of RNAaseP from *Escherichia coli* and *Bacillus subtilis* is, indeed, an enzyme (Guerrier-Takada *et al.*, 1983). Moreover, Zaug & Cech (1986) have shown that a modified form of the *Tetrahymena thermophila* intron can elongate pentacytidilic acid, working effectively as an RNA polymerase; while Waring *et al.* (1986) have shown that a mutant of the intron affected in 5' excision can catalyse a *trans*-circularisation reaction. The presence of these RNA structures in places as diverse as the nucleus of *Tetrahymena*, fungal mitchondria, chloroplast genomes, or even the DNA of archeobacteria bears witness to their antiquity. In addition, an intron which might be self-splicing has been identified in the thymidilate synthase gene of bacteriophage T4 (Belfort *et al.*, 1985).

Thus, to the evolutionary biologist, the most interesting result of this analysis of the natural history and comparative molecular anatomy of fungal mitochondrial genomes might well be the discovery of these peculiar RNA objects which bridge between plastids and nuclei, eubacteria, archeobacteria and eukaryotes and might point towards the origin of catalysis and, perhaps, of life itself.

Acknowledgements I thank all the members of the Essex — Manchester mitochondrial team for fruitful discussions through the years we worked together. G. Macino, T. Brown, S. Leong and P. Slonimski are thanked for communicating unpublished data.

References

Anderson, S., Bankier, A. T., Barrell, B. G., de Brujin, M. H. L., Coulson, A. R., Drouin, J., Eperon, I. C., Nierlich, D. P., Roe, B. A., Sanger, F., Schereier, P. H., Smith, A. J. H., Staden, R. & Young, I. G. (1981). Sequence organisation of the human mitochondrial genome. *Nature*, **290**, 457 − 465.

Anziano, P. Q., Hanson, D. K., Mahler, H. R. & Perlman, P. S. (1982). Functional domains in introns: identification of the probable *trans*-acting element (maturase) and a *cis*-acting splicing signal in intron 4 of the cytochrome *b* gene of yeast mitochondrial DNA. *Cell*, **30**, 925 − 932.

Belcour, L. & Vierny, C. (1986). Variable DNA splicing sites of a mitochondrial intron: relationship to the senescence process in *Podospora*. *EMBO Journal*, **5**, 609 − 614.

Belfort, M., Pederson-Lane, J., West, D., Ehrenman, K., Maley, G., Chu, F. & Maley, F. (1985). Processing of intron containing thymidilate synthase (td) gene of phage T4 is at the RNA level. *Cell*, **41**, 275 − 282.

Benne, R. & Tabak, H. F. (1986). Senescence comes of age. *Trends in Genetics*, **2**, 147 − 148.

Bertrand, H., Chan, B. S. S. & Griffiths, A. J. F. (1985). Insertion of a foreign nucleotide sequence into mitochondrial DNA causes senescence in *Neurospora intermedia*. *Cell*, **41**, 877 − 884.

Bonitz, S. G., Homison, G., Thalenfeld, B. E., Tzagoloff, A. & Macino, G. (1980). Assembly of the mitochondrial membrane system: structural nucleotide sequence of the gene coding for subunit 1 of yeast cytochrome oxidase. *Journal of Biological Chemistry*, **255**, 11927 − 11941.

Breitenberger, C. A. & RajBhandary, U. L. (1985). Some highlights of mitochondrial research based on analysis of *Neurospora crassa* mitochondrial DNA. *Trends in Genetics*, **1**, 478 − 483.

Brown, T. A., Davies, R. W., Waring, R. R. B. & Scazzocchio, C. (1983a). The mitochondrial genome of *Aspergillus nidulans* contains reading frames homologous to human URFs1 and 4. *EMBO Journal*, **2**, 427 − 435.

Brown, T. A., Davies, R. W., Waring, R. R. B. & Scazzocchio, C. (1983b). DNA duplication has resulted in transfer of an amino terminal peptide between two mitochondrial frames. *Nature*, **302**, 721 − 723.

Brown, T. A., Waring, R. R. B., Scazzocchio, C. & Davies, R. W. (1985). The

Aspergillus nidulans mitochondrial genome. *Current Genetics*, **9**, 113 – 117.

Burke, J. M. & RajBhandary, U. L. (1982). Intron within the large rRNA gene of *N. crassa*: a long open reading frame and a consensus sequence possibly important in splicing. *Cell*, **31**, 509 – 520.

Carignani, G., Groudinsky, O., Frezza, D., Schiavon, E., Bergantino, E. & Slonimski, P. P. (1983). An mRNA maturase is encoded by the first intron of the mitochondrial gene for the subunit 1 of cytochrome oxidase in *S. cerevisiae*. *Cell*, **35**, 733 – 742.

Caron, F. & Mayer, E. (1985). Does *Paramecium primaurelia* use a different genetic code in its macronucleus? *Nature*, **314**, 185 – 188.

Cech, T. R. (1986). The generality of self-splicing RNA: relationship to nuclear mRNA splicing. *Cell*, **44**, 207 – 210.

Cech, T. R., Tanner, N. K., Pinoco, I., Wier Bruc, R., Zuker, M. & Perlman, P. (1983). *Tetrahymena* ribosomal RNA intervening sequence: structural homology with fungal mitochondrial intervening sequences. *Proceedings of the National Academy of Sciences, USA*, **80**, 3903 – 3908.

Cech, T. R., Zaug, A. J. & Grabowski, P. J. (1981). In vitro splicing of the ribosomal RNA precursor of *Tetrahymena*: involvement of a guanosine nucleotide in the excision of the intervening sequence. *Cell*, **27**, 487 – 496.

Chomyn, A., Mariottini, P., Cleeter, M. W. J., Ragan, I. C., Russel, F., Doolittle, F., Yagi, A. M., Hatefi, Y. & Attardi, G. (1985). Functional assignment of the products of the unidentified reading frames of human mitochondrial DNA. In *Achievements and Perspectives of Mitochondrial Research*, vol. II, Biogenesis, ed. Quagliariello, E., pp. 259 – 275. Elsevier Science Publishers, BV.

Clark-Walker, G. D., McArthur, C. R. & Sriprakash, K. S. (1981). Partial duplication of the large ribosomal RNA sequence in an inverted repeat in circular mitochondrial DNA from *Kloechera africana*. *Journal of Molecular Biology*, **147**, 399 – 415.

Clark-Walker, G. D., Sriprakash, K. S., McArthur, C. R. & Azad, A. A. (1980). Mapping of mitochondrial DNA from *Torulopsis glabrata*: location of ribosomal and transfer RNA genes. *Current Genetics*, **1**, 209 – 217.

Clary, D. O. & Wolstenholme, D. R. (1983). Genes for cytochrome *c* oxidase subunit 1, URF2 and three tRNAs in *Drosophila* mitochondrial DNA. *Nucleic Acids Research*, **11**, 6854 – 6872.

Clary, D. O. & Wolstenholme, D. R. (1985). The mitochondrial DNA molecule of *Drosophila yakuba*: nucleotide sequence, gene organization, and genetic code. *Journal of Molecular Evolution*, **22**, 252 – 271.

Collins, R. A. & Lambowitz, A. M. (1981). Characterisation of a variant *Neurospora crassa* mitochondrial DNA which contains tandem reiterations of a 1 – 9 kb sequence. *Current Genetics*, **4**, 131 – 133.

Cummings, D. J., MacNeil, I. A., Domenico, J. & Matsuura, E. T. (1985). Excision amplification of mitochondrial DNA during senescence in *Podospora anserina*. *Journal of Molecular Biology*, **185**, 659 – 680.

Davies, R. W., Waring, R. B., Ray, J. A., Brown, T. A. & Scazzocchio, C. (1982). Making ends meet: a model for RNA splicing in fungal mitochondria. *Nature*, **300**, 719 – 724.

De La Salle, H., Jacq, C. & Slonimski, P. P. (1982). Critical sequences within mitochondrial introns: pleiotropic mRNA maturase ad *cis*-dominant signals for the *box* intron controlling reductase and oxidase. *Cell*, **28**, 721 – 732.

de Vries, H., Alzner-Deweerd, B., Breitenberger, C. A., Chang, D. D., de Jonge, J. C. & RajBhandary, U. L. (1986). The E35 stopper mutant of *Neurospora crassa*: precise localisation of deletion endpoints in mitochondrial DNA and evidence that the deleted DNA codes for a subunit of the NADH dehydrogenase. *EMBO Journal*, **5**, 779 − 785.

Dujardin, G., Jacq, C. & Slonimski, P. P. (1982). Single base substitution in an intron of oxidase gene compensates splicing defects of the cytochrome *b* gene. *Nature*, **298**, 628 − 632.

Dujon, B. (1980). Sequence of the intron and flanking exons of the mitochondrial 21S rRNA gene of yeast strains having different alleles at the omega and *rib-1* loci. *Cell*, **20**, 185 − 197.

Dujon, B. (1981). Mitochondrial genetics and functions. In *The Molecular Biology of the Yeast Saccharomyces: Life Cycle and Inheritance*, ed. Strathern, J. N., Jones, E. W. & Broach, J. R., pp. 505 − 635. New York: Cold Spring Harbor Laboratory.

Dujon, B. (1983). Mitochondrial genes, mutants and maps: a review. In *Mitochondria 1983*, ed. Schweyen, R. J., Wolf, K. & Kaudewitz, F., pp. 1 − 24. Berlin & New York: Walter de Gruyter & Co.

Dujon, B., Slonimski, P. P. & Weill, L. (1974). Mitochondrial genetics IX. A model for recombination and segregation of mitochondrial genomes in *Saccharomyces cerevisiae*. *Genetics*, **78**, 415 − 437.

Earl, A. J., Turner, G., Croft, J. H., Dales, R. B. G., Lazarus, C. M., Lunsdorf, H. & Kuntzel, H. (1981). High frequency transfer of species specific mitochondrial DNA sequences between members of the Aspergillaceae. *Current Genetics*, **3**, 221 − 228.

Farrelly, F. & Butow, R. A. (1983). Rearranged mitochondrial genes in the yeast nuclear genome. *Nature*, **301**, 296 − 301.

Gargouri, A., Lazowska, J. & Slonimski, P. P. (1983). DNA − splicing of introns in the genes. A general way of reverting intron mutations. In *Mitochondria 1983*, ed. Schweyen, R. J., Wolf, K. & Kaudewitz, F., pp. 259 − 268. Berlin & New York: Walter de Gruyter & Co.

Garriga, G. & Lambowitz, A. M. (1984). RNA splicing in *Neurospora* mitochondria: self-splicing of a mitochondrial intron *in vitro*. *Cell*, **38**, 631 − 641.

Grabowski, P. J., Zaug, A. J. & Cech, T. R. (1981). The intervening sequence of the ribosomal RNA precursor is converted to a circular RNA in isolated nuclei of *Tetrahymena*. *Cell*, **23**, 467 − 476.

Gross, S. R., Hsieh, T. & Levine, P. H. (1984). Intermolecular recombination as a source of mitochondrial chromosome heteromorphism in *Neurospora*. *Cell*, **38**, 233 − 239.

Grossman, L. T. & Hudspeth, E. S. (1985). Fungal mitochondrial genomes. In *Gene Manipulations in Fungi*, ed. Bennett, J. W. & Lasure, L. L., pp. 66 − 91. Orlando, Florida: Academic Press.

Guerrier-Takada, C., Gardiner, C., Marsh, T., Pace, N. & Altman, S. (1983). The RNA moiety of ribonuclease P is the catalytic subunit of the enzyme. *Cell*, **35**, 849 − 857.

Heckman, J. E., Sarnoff, J., Alzner de Weerd, B., Yin, S. & RajBhandary, U. L. (1980). Novel features in the genetic code and codon reading patterns in *Neurospora crassa* mitochondria based on sequences of six mitochondrial tRNAs. *Proceedings of the National Academy of Sciences, USA*, **77**, 3159 − 3163.

Helmer-Citterich, M., Morelli, G., Nelson, M. A. & Macino, G. (1983). Expression of split genes of the *Neurospora crassa* mitochondrial genome. In *Mitochondria 1983*, ed. Schweyen, R. J., Wolf, K. & Kaudewitz, F., pp. 357 − 369. Berlin & New York: Walter de Gruyter & Co.

Hernsgens, L. A. M., Bonen, L., De Haan, J., Van der Horst, G. & Grivell, L. A.

(1983). The sequence of two optional introns in the mitochondrial gene for subunit 1 of cytochrome oxidase in yeast reveals homologies among URFs containing introns and strain dependent variation in flanking exons. *Cell*, **32**, 379 – 389.

Hintz, W. E., Mohan, M., Anderson, J. B. & Horgen, P. A. (1985). The mitochondrial DNA of *Agaricus*: heterogeneity in *A. bitorquis*and homogenity in *A. brunnescens*. *Current Genetics*, **9**, 127 – 132.

Holl, J., Schmidt, C. & Schweyen, R. J. (1985). Cob intron 3 in yeast mtDNA: nucleotide sequence and mutations in a novel RNA domain. In *Achievements and Perspectives of Mitochondrial Research*, vol. II, Biogenesis, ed. Quagliariello, E., pp. 227 – 236. Elsevier Science Publishers BV.

Jacq, C., Pajot, P., Lazowska, J., Dujardin, G., Claisse, M., Groudinsky, O. De La Salle, H., Grandchamp, C., Labouesse, M., Gargouri, A., Guiard, B., Spyridakis, A., Dreyfus, M. & Slonimski, P. P. (1982). Role of introns in the yeast cytochrome *b* gene: *cis*- and *trans*-acting signals, intron manipulation, expression and intergenic communications. In *Mitochondrial Genes*, ed. Slonimski, P. P., Borst, P. & Attardi, G., pp. 155 – 183. New York: Cold Spring Harbor Laboratory.

Jacquier, A. & Dujon, B. (1985). An intron-encoded protein is active in a gene conversion process that spreads an intron into a mitochondrial gene. *Cell*, **41**, 383 – 394.

Kinsey, J. A. (1985). *Neurospora* plasmids. In *Gene Manipulations in Fungi*, ed. Bennett, J. W. & Lasure, L. L., pp. 245 – 258. Orlando, Florida: Academic Press.

Kjems, J. & Garrett, R. A. (1985). An intron in the 23S ribosomal RNA gene of the Archaebacterium *Desulforococcus mobilis*. *Nature*, **318**, 675 – 677.

Kotylak, Z., Lazowska, J., Hawthorne, D. C. & Slonimski, P. P. (1985). Intron encoded proteins of mitochondria: key elements of gene expression and genomic evolution. In *Achievements and Perspectives of Mitochondrial Research*, vol. II, Biogenesis, ed. Quagliariello, E., pp. 1 – 20. Elsevier Science Publishers BV.

Kruger, K., Grabowski, P. J., Zaug, A. J., Sands, J., Gottschling, D. E. & Cech, T. R. (1982). Self splicing RNA: autoexcision and autocyclization of the ribosomal RNA intervening sequence of *Tetrahymena*. *Cell*, **31**, 147 – 157.

Kucino, Y., Hanyo, N., Tashiro, F. & Nishimura, S. (1985). *Tetrahymena thermophila* glutamine tRNA and its gene that corresponds to UAA termination codon. *Proceedings of the National Academy of Sciences, USA*, **82**, 4758 – 4762.

Labouesse, M. & Slonimski, P. P. (1983). Construction of novel cytochrome *b* genes in yeast mitochondria by subtraction or addition of introns. *EMBO Journal*, **2**, 269 – 276.

Lang, B. F. (1984). The mitochondrial genome of the fission yeast *Schizosaccharomyces pombe*: highly homologous introns are inserted at the same position of the otherwise less conserved *cox1* genes in *Schizosaccharomyces pombe* and *Aspergillus nidulans*. *EMBO Journal*, **3**, 2129 – 2136.

Lang, B. F., Ahne, F., Distler, S., Trinkl, H., Kaudewitz, F. & Wolf, K. (1983). Sequencing of the mitochondrial DNA, arrangement of genes and processing of their transcripts in *Schizosaccharomyces pombe*. In *Mitochondria 1983*, ed. Schweyen, R. J., Wolf, K. & Kaudewitz, F., pp. 313 – 329. Berlin & New York: Walter de Gruyter & Co.

Lazarus, C. M., Earl, A. J., Turner, G. & Kuntzel, H. (1980). Amplification of a mitochondrial DNA sequence in the cytoplasmically inherited ragged mutant of *Aspergillus amstelodami*. *European Journal of Biochemistry*, **106**,

633 − 641.

Lazarus, C. M. & Kuntzel, H. (1981). Anatomy of amplified DNA in 'ragged' mutants of *Aspergillus amstelodami*: excision points within protein genes and a common 215 bp segment containing a possible origin of replication. *Current Genetics*, **4**, 99 − 107.

Lazowska, J., Jacq, C. & Slonimski, P. P. (1980). Sequence of introns and flanking exons in wild type and *box3* mutants of cytochrome *b* reveals an interlaced splicing protein coded by an intron. *Cell*, **22**, 333 − 348.

Lazowska, J., Jacq, C. & Slonimski, P. P. (1981). Splice points of the third intron in the yeast mitochondrial cytochrome *b* gene. *Cell*, **27**, 12 − 14.

Macreadie, I. G., Scott, R. M., Zin, A. R. & Butow, R. A. (1985). Transposition of an intron in yeast mitochondria requires a protein encoded by that intron. *Cell*, **41**, 395 − 402.

Michel, F. & Dujon, B. (1983). Conservation of RNA secondary structures in two intron families including mitochondrial-, chloroplast and nuclear-encoded members. *EMBO Journal*, **2**, 33 − 38.

Michel, F., Jacquier, A. & Dujon, B. (1982). Comparison of fungal mitochondrial introns reveals extensive homologies in RNA secondary structure. *Biochimie*, **64**, 867 − 881.

Michel, F. & Lang, F. B. (1986). Mitochondrial class II introns encode proteins related to the reverse transcriptases of retroviruses. *Nature*, **316**, 641 − 643.

Morelli, G. & Macino, G. (1984). Two intervening sequences in the ATPase subunit 6 gene of *Neurospora crassa*. A short intron (93 base pairs) and a long intron that is stable after excision. *Journal of Molecular Biology*, **178**, 491 − 507.

Nargang, F. E., Bell, J. B., Stohl, L. L. & Lambowitz, A. M. (1984). The DNA sequences and genetic organisation of a *Neurospora* plasmid suggests a relationship to the mitochondrial introns and mobile genetic elements. *Cell*, **38**, 441 − 453.

Nelson, M. A. & Macino, G. (1985). Gene organisation and expression in *Neurospora crassa* mitochondria. In *Achievements and Perspectives of Mitochondrial Research*, vol. II, Biogenesis, ed. Quagliariello, E., pp. 293 − 304. Elsevier Science Publishers BV.

Netter, P., Jacq, C., Carignani, G. & Slonimski, P. P. (1982). Critical sequences within mitochondrial introns: *cis*-dominant mutations of cytochrome *b*-like intron of the oxidase gene. *Cell*, **28**, 733 − 738.

Nobrega, F. G. & Tzagoloff, A. (1980). Assembly of the mitochondrial membrane system. DNA sequence and organization of the cytochrome *b* genes in *Saccharomyces cerevisiae* D273-10B. *Journal of Biological Chemistry*, **255**, 9828 − 9837.

Osiewacz, H. D. & Esser, K. (1983). DNA sequence analysis of the mitochondrial plasmid of *Podospora anserina*. *Current Genetics*, **1**, 219 − 223.

Peebles, C. L., Perlman, P. S., Mecklenburg, K. L., Petrillo, M. L., Tabor, J. H., Jarell, K. A. & Cheng, H. L. (1986). A self splicing RNA excises an intron lariat. *Cell*, **44**, 213 − 223.

Penalva, M. A. & Garcia, J. L. (1986). The subunit I of mitochondrial NADH dehydrogenase from *Cephalosporium acremonium*: the evolution of a mitochondrial gene. *Current Genetics*, **in press**,

Preer, J. R., Preer, L. B., Rudman, B. M. & Barnett, J. A. (1985). Deviation from the universal code shown by the gene for a surface protein SIA in *Paramecium*. *Nature*, **314**, 188 − 190.

Price, J. V. & Cech, T. R. (1985). Coupling of *Tetrahymena* ribosomal RNA splicing to β-galactosidase expression in *Escherichia coli*. *Science*, **228**, 719 − 722.

Schmelzer, C., Schmidt, C., May, K. & Schweyen, R. J. (1983). Determination of functional domains of intron bII of yeast mitochondrial RNA by studies of mitochondrial mutations and a nuclear suppression. *EMBO Journal*, **2**, 2047 − 2052.

Schuster, W. & Brennicke, A. (1985) TGA-Termination codon in the apocytochrome *b* gene from *Oonothera* mitochondria. *Current Genetics*, **9**, 157 − 163.

Shumard, D. S., Grossman, L. I. & Hudspeth, M. E. S. (1986). *Achlya* mitochondrial DNA: gene localization and analysis of inverted repeats. *Molecular and General Genetics*, **202**, 16 − 23.

Simpson, L., Simpson, A. M., de la Cruz, V., Neckelmann, N., & Munich, M. (1985). Genomic organization and transcription of mitochondrial maxicircle DNA in trypanosomid protozoa. In *Achievements and Perspectives of Mitochondrial Research*, vol. II, Biogenesis, ed. Quagliariello, E., pp. 99 − 110. Elsevier Science Publishers BV.

Tabak, H. F. & Grivell, L. A. (1986). RNA catalysis in the excision of yeast mitochondrial introns. *Trends in Genetics*, **2**, 51 − 55.

Turner, G., Earl, A. J. & Geaves, D. R. (1982). Interspecies variation and recombination of mitochondrial DNA in the *Aspergillus nidulans* species group and the selection of species specific sequences by nuclear background. In *Mitochondrial Genes*, ed. Slonimski, P. P., Borst, P. & Attardi, G., pp. 411 − 414. New York: Cold Spring Harbor Laboratory.

van den Boogaart, P., Samall, J. & Agsteribbe, E. (1982). Similar genes for a mitochondrial ATPase subunit in the nuclear and mitochondrial genomes of *Neurospora crassa*. *Nature*, **298**, 187 − 189.

van den Veen, R., Arnberg, A. C., van der Horst, G., Bonen, L., Tabak, H. F. & Grivell, L. A. (1986). Excised group II introns in yeast mitochondria are lariats and can be formed by self-splicing *in vitro*. *Cell*, **44**, 225 − 234.

van der Horst, G. & Tabak, H. F. (1985). Self splicing of yeast mitochondrial ribosomal and messenger RNA precursors. *Cell*, **40**, 754 − 766.

Waring, R. B., Brown, T. A., Ray, I. A., Scazzocchio, C. & Davies, R. W. (1984). Three variant introns in the same general class in the mitochondrial gene for cytochrome oxidase subunit 1 in *Aspergillus nidulans*. *EMBO Journal*, **3**, 2121 − 2128.

Waring, R. B. & Davies, R. W. (1984). Assessment of a model of intron secondary structure relevant to RNA self-splicing − a review. *Gene*, **28**, 277 − 291.

Waring, R. B., Davies, R. W., Lee, S., Grisi, E., McPhails Berks, M. & Scazzocchio, C. (1981). The mosaic organization of the apocytochrome *b* gene of *Aspergillus nidulans* revealed by DNA sequencing. *Cell*, **27**, 4 − 11.

Waring, R. B., Davies, R. W., Scazzocchio, C. & Brown, T. (1982). Internal structure of a mitochondrial intron in *Aspergillus nidulans*. *Proceedings of the National Academy of Sciences, USA*, **79**, 6332 − 6336.

Waring, R. B., Ray, J. A., Edwards, S. W., Scazzocchio, C. & Davies, R. W. (1985). The *Tetrahymena* rRNA intron self-splices in *E. coli*: *in vivo* evidence for the importance of key base-paired regions of RNA for RNA enzyme function. *Cell*, **40**, 371 − 380.

Waring, R. B. & Scazzocchio, C. (1983). Mitochondrial four-point crosses in *Aspergillus nidulans*: mapping of a suppressor of a mitochondrially inherited cold-sensitive mutation. *Genetics*, **103**, 409 − 428.

Waring, R. B., Scazzocchio, C., Brown, T. A. & Davies, R. W. (1983). Close relationship between certain nuclear and mitochondrial introns. Implications for the mechanism of RNA splicing. *Journal of Molecular Biology*, **167**,

595 − 605.

Weber, C. A., Hudspeth, M. S., Moore, G. P. & Grossman, L. I. (1986). Analysis of the mitochondrial and nuclear genomes of two basidiomycetes, *Coprinus cinereus* and *Coprinus stercorarius*. *Current Genetics*, **10**, 515 − 525.

Weiss-Brummer, B., Holl, J., Schweyen, R. J., Rodel, G. & Kaudewitz, F. (1983). Processing of yeast mitochondrial RNA: involvement of intramolecular hybrids in splicing of the *cob* intron 4 by mutation and reversion. *Cell*, **33**, 195 − 202.

Weiss-Brummer, B., Rodel, G., Schweyen, R. J. & Kaudewitz, F. (1982). Expression of the split gene cob in yeast: evidence for a precursor of a 'maturase' protein translated from intron 4 and the preceding exons. *Cell*, **29**, 527 − 536.

Weslowski, M. & Fukuhara, H. (1981). Linear mitochondrial deoxyribonucleic acid from the yeast *Hansenula mrakii*. *Molecular and Cellular Biology*, **1**, 387 − 393.

Wright, R. M., Horrum, M. A. & Cummings, D. J. (1982). Are mitochondrial structural genes selectively amplified during senescence in *Podospora anserina*. *Cell*, **29**, 505 − 515.

Yamao, F., Muto, A., Kawauchi, Y., Ieami, M., Iwagami, S., Azumi, Y. & Osowa, S. (1985). UGA is read as tryptophan in *Mycoplasma capricolum*. *Proceedings of the National Academy of Sciences, USA*, **82**, 2306 − 2309.

Zaug, A. I. & Cech, T. R. (1986). The intervening sequence RNA of *Tetrahymena* is an enzyme. *Science*, **231**, 470 − 475.

Zimmern, D. (1983). Homologous proteins encoded by yeast mitochondrial introns and by a group of RNA viruses from plants. *Journal of Molecular Biology*, **171**, 345 − 352.

5

Varying patterns of ribosomal RNA gene organisation in basidiomycetes

PATRICIA J. PUKKILA
AND JEANE R. CASSIDY

Department of Biology and Curriculum in Genetics, University of North Carolina, Chapel Hill, NC 27514, USA

Introduction

Although the genomic organisation of the three largest ribosomal RNAs (26S, 18S and 5·8S) is relatively uniform among eukaryotes, the location of the gene encoding the fourth RNA component of the ribosome (the 5S RNA) is quite variable (for review see Gerbi, 1985). We have shown that in *Coprinus cinereus* the genes encoding the 5S RNA are part of a repeating unit which also encodes the other rRNAs. Furthermore, all four RNAs are transcribed in the same direction (Cassidy *et al.*, 1984). This genomic organisation had not been reported previously. To ascertain whether this organisation is unique to *Coprinus cinereus*, we determined the organisation in five other members of the order Agaricales (*Flammulina velutipes*, *Agaricus bisporus*, *Coprinus atramentarius*, *Coprinus comatus* and *Coprinus micaceus*). These studies made use of a method we developed to determine the direction of transcription of these genes using crude preparations of genomic DNA. Since the method eliminates the need to obtain cloned copies of the genes from each organism before studying their genomic arrangement, the technique could be applied to a large number of fungi to assist in establishing phylogenetic relationships between them.

Pattern of ribosomal RNA gene organisation in *Coprinus cinereus*

To facilitate molecular analysis of the ribosomal (rRNA) genes we isolated a cloned copy of these sequences (Wu, Cassidy & Pukkila, 1983). Genomic DNA was digested with the restriction enzyme *Bam* HI and fragments approximately 9 kilobases (kb) in size were isolated using sucrose gradient sedimentation, and ligated to the plasmid pBR322. Clone pCc1, which contains a 9·3 kb insert encoding all four rRNAs, was identified using colony hybridisation. The locations of the recognition sites for several restriction endonucleases were determined after gel electrophoresis (Fig. 5.1A). The positions of the coding regions for the major ribosomal RNAs, indicated in Fig. 5.1B, were determined as follows. The cloned DNA was first digested with restriction endonucleases whose cutting sites had been mapped and the resulting fragments were separated using gel electrophoresis. The fragments were transferred to nitrocellulose filters using the procedure of Southern (1975), and the filters were probed with purified ribosomal RNA species radioactively labelled with ^{32}P. The positions of the bands which hybridised to each probe were revealed by autoradiography. We also determined the number of copies of the 9·3 kb ribosomal DNA (rDNA) unit in the *Coprinus* genome in quantitative hybridisation experiments. The amount of hybridisation to 14 − 22 ng of cloned rDNA matched the amount of hybridisation in 1 μg of total genomic DNA. Since the genome size is 37,500 kb, each genome must contain 60 − 90 copies of the 9·3 kb repeat. These copies segregate as a single Mendelian gene, called *RDN1*, that shows centromere linkage (Cassidy *et al.*, 1984).

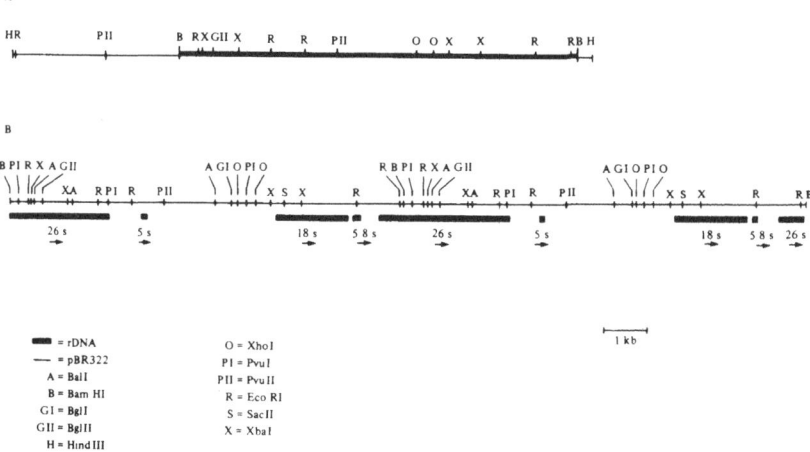

Fig. 5.1. Map of restriction enzyme recognition sites in *Coprinus cinereus* rDNA. In A a partial restriction map of the plasmid pCc1 is shown; this plasmid contains one complete repeat of rDNA. Note the location of the *Bam* HI (symbolised with the letter B) and *Bgl* II (GII) sites used to determine the direction of transcription of the rRNA genes. In B a more complete restriction map is presented, which includes two of the tandem rDNA repeats as found in genomic DNA. The locations of the coding regions are also indicated (Cassidy *et al.*, 1984).

To determine the direction of transcription of these RNAs, we digested clone pCc1 with *Bam* HI and loaded the DNA on the gel in the presence of 1M NaOH. Hayward (1972) showed that the two complementary strands of bacteriophage lambda migrated to different positions after such treatment, and the same proved to be true for the rDNA sequences (but not the plasmid vector). After Southern transfer, the more slowly migrating fragment hybridised to ^{32}P-labelled 26S and 18S RNA, while the more quickly moving fragment hybridised to complementary DNA (cDNA) prepared to 5·8S and 5S RNA. Thus, the more slowly moving fragment is the coding strand for all four ribosomal RNAs. By labelling the 5' ends of the *Bam* HI fragment and digesting with *Bgl* II, we showed that this slowly moving fragment had the *Bgl* II site near its 3' end (since the 5' end label remained after the *Bgl* II digestion). We were thus able to orient the direction of transcription relative to the restriction enzyme recognition sites as shown in Fig. 5.1B (Cassidy *et al.*, 1984). In most eukaryotes, the 5S RNA gene is not found as part of the repeating rDNA unit. Instead, repeated 5S genes are clustered at one or several sites in the genome, or widely dispersed (Gerbi, 1985). In some fungi the 5S gene is part of the repeating rDNA unit, but in the previously reported cases such as *Saccharomyces cerevisiae* and *Torulopsis utilis* (Bell *et al.*, 1977; Valenzuela *et al.*, 1977; Tabata, 1980), the gene was transcribed in the opposite direction, and thus, from the opposite strand.

Pattern of ribosomal RNA gene organisation in other agarics

It was of interest to determine if the novel gene arrangement seen in *Coprinus cinereus* was unique to this species or more widespread. We determined whether the strand separation procedure could be applied to total genomic DNA, since if the two strands of the repeated rRNA genes could be resolved by gel electrophoresis even in the presence of other sequences, we could determine which strand was the coding strand for the rRNA types without having to clone the rDNA first. This method proved to be successful. We cultured *Flammulina velutipes*, *Agaricus bisporus*, *Coprinus atramentarius*, *Coprinus comatus* and *Coprinus micaceus* in yeast extract + malt extract medium (Rao & Niederpruem, 1969). DNA was extracted using an efficient 'mini-prep' procedure (Zolan & Pukkila, 1986). Lyophilised tissue was ground to a fine powder in a small tube and DNA extracted using CTAB (hexadecyltrimethylammonium bromide) and chloroform/isoamyl alcohol. These DNAs were digested with different restriction endonucleases until one was found which produced one large fragment that hybridised to pCc1, indicating that this recognition site was present once in each repeating rDNA unit. Each genomic DNA sample was then digested with a restriction enzyme which cut once per rDNA repeat, and the digested DNAs were loaded on an agarose gel in the presence of 1M NaOH, which denatured the DNAs. Gel electrophoresis was followed by transfer to nitrocellulose filters. Both 5·8S RNA and 5S RNA were purified by excising the appropriate RNA's from acrylamide gels, and cDNAs to these two types of RNAs were prepared as probes. Duplicate filters containing the denatured DNAs were prepared and reacted with each probe. After autoradiography, the filters were reprobed with pCcl to reveal the location of both separated strands. A typical result is illustrated in Fig. 5.2, which shows the results obtained from *Agaricus bisporus*. Although some degradation of the DNA is evident, the patterns of strand separation were surprisingly clear considering that total genomic DNA, which was not highly purified, was being used. Our results for the five species tested are summarised in Table 5.1. In four of these species (*Flammulina velutipes*, *Agaricus bisporus*, *Coprinus atramentarius*, and *Coprinus micaceus*), the strand which hybridised to the 5·8S probe also hybridised to the 5S probe. This organisation is similar to that in *Coprinus cinereus*. However, in *Coprinus comatus*, the 5S cDNA hybridised to the more slowly migrating strand, while the 5·8S cDNA hybridised to the rapidly migrating one. We carried out more extensive mapping of the repeated rDNA sequence from *Coprinus comatus* and confirmed that the three largest ribosomal RNAs are transcribed in one direction, while the 5S RNA is transcribed in the opposite direction (Cassidy & Pukkila, 1987).

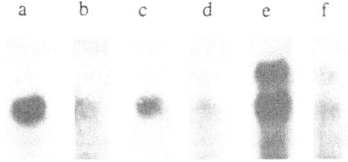

a   b   c   d   e   f

Fig. 5.2. Autoradiographic determination of the direction of transcription of the 5S RNA gene in *Agaricus bisporus*. Lanes a, c, and e contain *Coprinus cinereus* DNA for comparison. Lanes a and b were hybridised with cDNA prepared to 5S RNA and lanes c and d were hybridised with cDNA prepared to 5.8S RNA. The positions of the two complementary strands of the repeated rDNA sequence were revealed in lanes e and f, which were hybridised with the cloned rDNA sequence. Note that all cDNA probes hybridise only to the more rapidly migrating (lower) strand.

Table 5.1. *Arrangement of 5S RNA genes in some basidiomycetes*

Species	Direction of transcription
Coprinus cinereus	same as rRNAs
Coprinus comatus	opposite to rRNAs
Coprinus atramentarius	same as rRNAs
Coprinus micaceus	same as rRNAs
Agaricus bisporus	same as rRNAs
Flammulina velutipes	same as rRNAs

Discussion

Discovering that the 5S gene is in the opposite orientation in members of the same genus was unexpected. It would appear that the 5S RNA gene can become inverted in one or more members of the rDNA array and this inverted pattern can be propagated to one or all of the tandem copies at the rDNA locus. In *Coprinus*, the 5S RNA gene is located in the 'non-transcribed spacer' DNA between the 26S and 18S genes. In many organisms, this spacer DNA contains repetitive elements and alterations in spacer length due to the presence of various numbers of these repeats are common (Gerbi, 1985). Thus, changes in the relative location of the embedded 5S gene might occur as a consequence of changes in spacer length. However, inversions within the spacer have not been reported previously. Their occurrence might reflect the presence of inverted repeating elements flanking the 5S gene since a simple exchange between such elements would lead to inversion of any intervening DNA. Inversions within the spacer may have gone undetected in previous studies due to the lack of a convenient probe. The discovery of inverted 5S genes was also surprising because of the existence of an orientation-dependent 'promoter' element within the spacer DNA of *Saccharomyces cerevisiae* at a considerable distance from the DNA corresponding to the 5' end of the large RNA precursor (Elion & Warner, 1984). The 5S gene actually lies between this element and the 5' end of the precursor. If such an element exists in *Coprinus*, it must lie outside the inverted segment.

The presence of these two different arrangement in related species suggests that an inverted sequence was propagated through the tandem array, replacing the existing genes with copies of itself. Brown, Wensink & Jordan (1972) concluded that the non-transcribed spacer DNA within the tandemly repeated rDNA unit was different in the two *Xenopus* species they studied, although within each species the tandem copies were similar to each other. They proposed that sequence changes were spread from one spacer sequence to the next by a process they called 'horizontal evolution'. Two major mechanisms have been proposed for such homogenisation events: gene conversion (Edelman & Gally, 1970; Nagylaki & Petes, 1982) and unequal crossing over (Smith, 1976; and see Turner, Chapter 2). Gene conversion is non-Mendelian segregation at a heterozygous locus in which sequences on one homologue replace those on the other. This form of information transfer has also been shown to occur between repeated genes on the same chromosome (Klein & Petes, 1981; Jackson & Fink, 1981, 1985), and it is even thought to occur between widely scattered genes (Morzycka-Wroblewska *et al.*, 1985). In the latter case, an RNA-mediated information transfer process was postulated, since only the transcribed portions of the widely scattered 5S genes were found to be similar. However, such a mechanism would not lead to propagation of an inversion through a series of tandem repeats. Furthermore, if conversion proceeded through a heteroduplex intermediate, as is widely supposed (Holliday, 1964), the inversion heterozygote would result in unpaired DNAs which are thought to block symmetric heteroduplex formation (Hamza *et al.*, 1981). It would appear more likely that the inverted variant was fixed by repeated cycles of unequal crossing over (Smith, 1976). If an exchange occurred between misaligned chromatids one resulting recombinant would have more rDNA copies than the other (Turner, Chapter 2, Fig. 2.1). Repeated cycles of unequal exchange could result in the replacement of existing members of the tandem array with copies of the inverted variant. Meiotic recombination appears to be suppressed in this tandem array (Cassidy *et al.*, 1984). The required cycles of exchanges between newly replicated chromatids might occur by allowing these to align out of register by only a few copies. Such an alignment would avoid drastic changes in rDNA copy number which would be deleterious for growth.

Singer (1975) discusses four sections within the genus *Coprinus* and places *C. comatus*, *C. atramentarius* and *C. cinereus* in Section 1, and *C. micaceus* in Section 2. Thus, it appears that quite closely related species have an altered 5S organisation. Examination of this trait and others in additional species should prove useful in establishing correct sub-groupings among closely related species within this genus.

The primary sequence of 5S RNA has been used by a number of authors to attempt to establish phylogenetic relationships among basidiomycetes (Walker & Doolittle, 1982, 1983; Huysmans *et al.*, 1983; Gottschalk & Blanz, 1984, 1985) and the resulting interpretations have been somewhat controversial. The sequence heterogeneity observed among 5S RNAs in *Neurospora crassa* where the genes are scattered throughout the genome (Selker *et al.*, 1981) suggests that genome organisation itself can influence divergence. It seems that the utility of sequence data from a particular gene family for phylogenetic reconstruction will be easier to assess when the molecular processes responsible for mutation, propagation of a variant and selection against deleterious forms are more completely understood.

Acknowledgements We thank Hal Burdsall for providing the strains used (except for *Coprinus cinereus*). This work was supported by the National Science Foundation.

References

Bell, G. I., Degennaro, L. J., Gelfand, D. H., Bishop, R. J., Valenzuela, P. & Rutter, W. J. (1977). Ribosomal RNA genes of *Saccharomyces cerevisiae*. *Journal of Biological Chemistry*, **252**, 8114 − 8125.

Brown, D. D., Wensink, P. C. & Jordan, E. (1972). A comparison of the ribosomal DNA's of *Xenopus laevis* and *Xenopus mulleri*: the evolution of tandem genes. *Journal of Molecular Biology*, **63**, 57 − 73.

Cassidy, J. R., Moore, D., Lu, B. C. & Pukkila, P. J. (1984). Unusual organisation and lack of recombination in the ribosomal RNA genes of *Coprinus cinereus*. *Current Genetics*, **8**, 607 − 613.

Cassidy, J. R. & Pukkila, P. J. (1987). Inversion of 5S ribosomal RNA genes within the genus *Coprinus*. *Current Genetics,* in press.

Edelman, G. M. & Gally, J. A. (1970). Arrangement and evolution of eukaryotic genes. In *Neurosciences: Second Study Program*, ed. Schmitt, F. O., pp. 962 − 972. New York: Rockefeller University Press.

Elion, E. A. & Warner, J. R. (1984). The major promoter element of rRNA transcription in yeast lies 2 kb upstream. *Cell*, **39**, 663 − 673.

Gerbi, S. A. (1985). Evolution of ribosomal RNA. In *Molecular Evolutionary Genetics*, ed. MacIntyre, R. J., pp. 419 − 517. New York: Plenum Publishing.

Gottschalk, M. & Blanz, P. A. (1984). Highly conserved 5S ribosomal RNA sequences in four rust fungi and atypical 5S RNA secondary structure in *Microstroma juglandis*. *Nucleic Acids Research*, **12**, 3951 − 3958.

Gottschalk, M. & Blanz, P. A. (1985). Untersuchungen an 5S ribosomalen ribonukleinsauren als beitrag zur klarung von systematik und phylogenie der basidiomyceten. *Zeitschrift für Mykologie*, **51**, 205 − 243.

Hamza, H., Haedens, A., Mekki-Berrada, A. & Rossignol, J. L. (1981). Hybrid DNA formation during meiotic recombination. *Proceedings of the National Academy of Sciences, USA*, **78**, 7648 − 7651.

Hayward, G. S. (1972). Gel electrophoretic separation of the complementary strands of bacteriophage DNA. *Virology*, **49**, 342 − 344.

Holliday, R. (1964). A mechanism for gene conversion in fungi. *Genetical Research*, **5**, 282 − 303.

Huysmans, E., Dams, E., Vandenberghe., A. & De Wachter, R. (1983). The nucleotide sequences of the 5S rRNAs of four mushrooms and their use in studying the phylogenetic position of basidiomycetes among the eukaryotes. *Nucleic Acids Research*, **11**, 2871 − 2880.

Jackson, J. A. & Fink, G. R. (1981). Gene conversion between duplicated genetic elements in yeast. *Nature*, **292**, 306 − 311.

Jackson, J. A. & Fink, G. R. (1985). Meiotic recombination between duplicated genetic elements in *Saccharomyces cerevisiae*. *Genetics*, **109**, 303 − 332.

Klein, H. L. & Petes, T. D. (1981). Intrachromosomal gene conversion in yeast. *Nature*, **289**, 144 − 148.

Morzycka-Wroblewska, E., Selker, E. U., Stevens, J. N. & Metzenberg, R. L. (1985). Concerted evolution of dispersed *Neurospora crassa* 5S RNA genes: pattern of sequence conservation between allelic and nonallelic genes. *Molecular and Cellular Biology*, **5**, 46 − 51.

Nagylaki, T. & Petes, T. D. (1982). Intrachromosomal gene conversion and the maintenance of sequence homogeneity among repeated genes. *Genetics*, **100**, 315 − 337.

Rao, P. S. & Niederpruem, D. J. (1969). Metabolism during morphogenesis of *Coprinus lagopus* (*sensu* Buller). *Journal of Bacteriology*, **100**, 1222 − 1228.

Selker, E. U., Yanofsky, C., Driftmier, C., Metzenberg, R. L., Alzner-Deweerd, B. & Raj Bhandandary, U. L. (1981). Dispersed 5S RNA genes in *N. crassa*: structure, expression and evolution. *Cell*, **24**, 819 − 828.

Singer, R. (1975). *The Agaricales in Modern Taxonomy*. Vaduz: J. Cramer.

Smith, G. P. (1976). Evolution of repeated DNA sequences by unequal crossover. *Science*, **191**, 528 − 535.

Southern, E. M. (1975). Detection of specific sequences among DNA fragments separated by gel electrophoresis. *Journal of Molecular Biology*, **98**, 503 − 517.

Tabata, S. (1980). Structure of the 5-S ribosomal RNA gene and its adjacent regions in *Torulopsis utilis*. *European Journal of Biochemistry*, **110**, 107 − 114.

Valenzuela, P., Bell, G. I., Venegas, A., Sewell, E. T., Masiarz, F. R., Degennaro, L. J., Weinberg, F. & Rutter, W. J. (1977). Ribosomal RNA genes of *Saccharomyces cerevisiae*. *Journal of Biological Chemistry*, **252**, 8126 − 8135.

Walker, W. F. & Doolittle, W. F. (1982). Redividing the basidiomycetes on the basis of 5S RNA sequences. *Nature*, **299**, 723 − 724.

Walker, W. F. & Doolittle, W. F. (1983). 5S RNA sequences from eight basidiomycete and fungi imperfecti. *Nucleic Acids Research*, **11**, 7625 − 7630.

Wu, M. J., Cassidy, J. R. & Pukkila, P. J. (1983). Polymorphisms in DNA of *Coprinus cinereus*. *Current Genetics*, **7**, 385 − 392.

Zolan, M. E. & Pukkila, P. J. (1986). Inheritance of DNA methylation in *Coprinus cinereus*. *Molecular and Cellular Biology*, **6**, 195 − 200.

6
Molecular variation and evolution

JOHN BARRETT

Department of Genetics, University of Cambridge, Cambridge CB2 3EH, UK

Introduction

Over the last twenty years progress in biochemistry and molecular biology has produced a range of techniques which can be exploited to investigate problems in evolutionary biology. The purpose of this chapter is to introduce some of these and show how they may be used in mycology.

The study of evolution and systematics in fungi has traditionally depended on morphological and physiological characters. To some extent these studies, especially studies of populations, have been limited by the paucity of suitable characters. Although immunotaxonomic and chemotaxonomic methods have been available for some time, they have not been applied extensively (for recent examples see Scott, 1981a; Ianelli *et al.*, 1982; Pérez-Silva & Alfonso, 1983). Mycologists have a tradition of interpreting variation in taxonomic terms and work on the dynamics of intraspecific variation has remained a minority interest within mycology. It is only in the last ten years or so that plant pathologists have begun to interpret changes in fungal populations as problems in evolutionary biology, rather than strictly classificatory problems (e.g. see Wolfe & Schwarzbach, 1978). The ability to investigate proteins and DNA increases the number of characters available to taxonomists and evolutionary biologists and opens up the possibility of population studies and more accurate resolution of phylogenetic relationships.

Analysis of protein variation

Until the development of techniques for the direct examination of DNA, the nearest approach that could be made to examining individual genes was to examine the gene products, which in most cases are proteins. The technique that has found widest application in evolutionary studies is electrophoresis, which allows proteins to be separated by charge and size in a suitable medium, usually starch or acrylamide gels (Hames & Rickwood, 1981). The gel can be stained with histochemical reagents to detect proteins or specific enzyme activities. The pattern of proteins on the gel can be used as a character and comparisons made between different taxa. The analysis of general proteins can offer a quick and cheap diagnostic tool (e.g. Burdon, Seviour & Fripp, 1982; Seviour *et al.*, 1985; Stipes, 1970; Wong & Willetts, 1973). The use of stains for the activity of specific enzymes is a more precise technique because it allows the effects of individual loci to be detected (Koehn, Zera & Hall, 1983). A particular enzyme activity may be controlled by one or more gene loci and, if these loci produce enzymes with slightly different characteristics, the different enzyme forms are referred to as *isozymes*.

Point mutations in a gene can change the amino acid sequence of the protein and so produce changes in charge, or activity, or both. More drastic changes, for example deletions of parts of the gene, can give rise to enzymes with no activity. Where different forms of an enzyme are coded by different alleles at the same locus, the enzyme variants are referred to as *allozymes*. They can be detected by their different rate of migration (mobility), or activity, in the gel. Allozymes behave like normal Mendelian genes and so are amenable to conventional genetic analysis and have the advantage over more conventional genetic markers in diploid (or dikaryotic) species that heterozygotes can almost always be distinguished from homozygotes. This is very useful in studies of population structure and the effects of breeding systems, because each genotype can be recognised without recourse to breeding tests.

Perhaps the most surprising observation made when electrophoretic surveys of populations began in the 1960s was the high level of variation present in populations. In most surveys approximately 25 − 30% of all enzyme systems investigated were polymorphic (Hamrick, 1979; Nevo, 1978) and this variation has provided a useful tool for dissecting the genetic structure of populations. For discussion of the arguments about maintenance of such variation, see Kimura (1983a & b), Lewontin (1974) and Nei (1983).

Not all mutations give rise to changes in amino acid composition because many amino acids are coded by more than one codon and not all amino acid substitutions produce changes in charge or activity. Electrophoresis will only detect about 30% of all amino acid substitutions, so population variation detected in this way will underestimate the real extent of variation (Nei & Chakraborty, 1973).

Some enzyme systems show no variation within taxa or populations (i.e. are monomorphic), but different taxa or populations may be monomorphic for different enzyme variants. In these cases, the allozyme phenotypes displayed by particular isolates or populations can be used as reasonably good criteria for assigning them to particular taxa. A recent example of this approach is a study where three rust species, *Puccinia striiformis*, *P. recondita* and *P. hordei*, were assayed for electrophoretic variation (Newton, Caten & Johnson, 1985). Isolates of yellow rust, *P. striiformis*, were divided into two groups depending on whether they attacked wheat or barley. No variation was detected within either group of isolates, but the two biotypes differed from one another in two of the twelve enzyme systems investigated. This consistent difference led to the proposition that the two forms be recognised as distinct *formae speciales*. Further examples can be found in Baptist & Kurtzman (1976), Burdon & Marshall (1981), Royse & May (1982a), Scala *et al.* (1981), Scott, (1981b), Stipes (1970), Sweetingham, Cruikshank & Wong (1986), Tariq, Gutteridge & Jeffries (1985) and Wong & Willetts (1973, 1975).

Polymorphic loci are not particularly useful for assigning isolates or individuals to particular taxa, but they do have other useful properties.

1. Different populations or sub-populations may differ in their frequencies of the variants. Statistical methods are available which can combine such frequency data from pairs of populations into a single index that can be used as a measure of genetic identity (or its converse, genetic distance) between populations (e.g. Nei, 1975). Apart from studies of fungal pathogens, the study of fungal populations has not been practised to any great extent for the reasons mentioned earlier, but examples can be found which have attempted to apply this approach in a mycological context: Backhouse, Willetts & Adam (1984), Boisselier-Dubayle (1983), Maghrabi & Kish (1985a & b), Royse & May (1982b) and Seviour *et al.* (1985).

2. Where taxa or populations are known to be monomorphic and a polymorphic population or heterozygous isolate is found, this may suggest hybridisation or introgression between related groups or populations. For example, Burdon, Marshall & Luig (1981) found isolates of *Puccinia graminis* which were pathogenic on the wild grass *Agropyron scabrum* and which possessed enzyme variants characteristic of *P. graminis* f. sp. *tritici* and f. sp. *secalis*, each of which were known to be monomorphic. This led them to suggest that the *scabrum* types had arisen by somatic hybridisation.

3. Where differences in the frequencies of variants show a consistent

association with particular features of the environment, this can be taken as suggesting the action of natural selection. Again, this type of investigation has not been undertaken to any great extent by mycologists because of the tendency to classify variation into taxonomic entities. Many examples, however, can be found in other organisms (e.g. Koehn *et al.*, 1983).

4. Isozyme variation can be used to study the breeding system under natural conditions, for example. Where a fungal species is known to be an outbreeder, and the data show a poor fit to Hardy — Weinberg proportions based on random mating, a deficiency of heterozygotes may be indicative of either partial reproductive isolation between two or more subgroups in the same area, a sample composed of a mixture of two or more subpopulations, or unexpectedly high rates of selfing. Conversely, the presence of heterozygotes in populations or species thought to be inbreeding can provide estimates of outbreeding rates (for an example, see Marshall & Allard, 1970).

5. Where extensive population data are available, the breeding structure of the population can be inferred from the distribution of genotype frequencies. For example, the sexual stage of *Puccinia graminis* occurs only on *Berberis* species. In an attempt to eliminate epidemics of stem rust from the Great Plains of North America, barberry was eradicated during the 1920s and 1930s. However, West of the Rocky Mountains, barberry plants were not destroyed. Burdon & Roelfs (1985a & b) collected isolates from both sides of the Rockies and demonstrated greater genetic diversity in the Western population than in the samples from the Great Plains. There was also complete association between virulence and enzyme phenotypes East of the Rockies, but no association in the West. This set of results demonstrates quite clearly the importance of the sexual cycle in the evolutionary dynamics of the fungus.

6. Surveys of populations over time may permit inferences to be made about the evolutionary dynamics of a system. For example, analysis of the virulence spectrum of *Puccinia graminis* in Australia has suggested that the population of the fungus underwent radical evolutionary changes in 1925 and 1954. Two hypotheses have been proposed to account for the observations: (i) the evolution, from pre-existing genotypes, of a novel genotype better adapted to Australian conditions; (ii) the introduction of novel forms from outside Australia. Using collections of stem rust that had been maintained from these periods, Burdon *et al.* (1982) were able to demonstrate that each of the major shifts in virulence patterns was accompanied by major changes in the electrophoretic phenotypes.

Comparison of these patterns with isolates from elsewhere in the world provided strong circumstantial evidence that the change that occurred in 1954 was probably due to the introduction or migration of stem rust of African origin.

Use of the physico-chemical properties of proteins can improve our understanding of some taxonomic relationships and evolutionary processes; but the technique remains a fairly blunt instrument. Although variation can be detected, the actual genetic differences, the primary causes of the different electrophoretic phenotypes, remain unknown. An increase in precision can be obtained by analysis of the amino acid sequence of the protein. The pattern of substitutions of amino acids can then be used to establish the most probable evolutionary tree or, at least, the minimum number of changes required to produce the observed differences (e.g. Fitch & Margoliash, 1967). The biochemical procedures involved in sequencing proteins are considerable more expensive and time-consuming than electrophoresis. It therefore follows that it is not possible to carry out population surveys using polymorphic proteins (like some of the enzyme systems). For amino acid sequences to be of value, the right protein must be chosen because proteins evolve at different rates (Dickerson, 1971). At one end of the spectrum are the polymorphic proteins, e.g. some of the enzyme systems, which cannot be used because it is physically impossible to accumulate enough data. At the other end of the spectrum are proteins (like cytochrome *c*) which are almost invariant, irrespective of the organism from which they were obtained. For the approach to be of use, proteins from somewhere in the middle of this spectrum must be chosen.

By counting the number of amino acid differences between taxa being compared, a reasonable picture of phylogenetic distances can be obtained. However, the number of amino acid changes between two or more taxa may be an underestimate of the phylogenetic distance because many amino acids are coded by more than one codon; e.g. the leucine codons are UUA, UUG, CUU, CUC, CUA and CUG. So, for a more precise estimate of the number of genetic changes between different taxa it is necessary to go to the DNA level, because this is the only way that ambiguities can be resolved.

Analysis of DNA variation

If a solution of DNA is heated, its two complementary strands will dissociate from one another and, if then allowed to cool under appropriate conditions, the strands will reassociate. The kinetics of reassociation allows categorisation of the DNA into unique and repeated, or reiterated, sequences. In contrast to prokaryotes, the genomes of most eukaryotes do not consist solely of unique sequences, but of a mixture of components showing different degrees of repetition. For example, repeated DNA makes up about 16% of the genome of *Agaricus bisporus* (= *A. brunnescens*) (Arthur *et al.*, 1982). These repeated sequences necessarily reassociate faster than unique sequences because there are more copies of them. If the unique sequences are separated from the repeated sequences, the rate of reassociation can become a measure of the identity of a genome. If single stranded DNA from two species is mixed together, then sequences that the species have in common will tend to reassociate, to form heteroduplexes. However, the match between these sequences will not necessarily be exact because the genomes of the two species have evolved independently of one another. When reheated, heteroduplexes will tend to dissociate ('melt') at a lower temperature than do the duplexes formed from DNA from a single species, because the strands of which they are composed are not exactly complementary. The more differences there are between sequences, the lower the temperature at which it will melt. Reduction in melting temperature can, therefore, be used as a measure of affinity or homology between the genomes of different species. Other methods are available for carrying out DNA − DNA hybridisation to measure homology between genomes of different taxa, but they all rely on the assumption that the degree of homology is a measure of phylogenetic distance (see Kurtzman, 1985).

For this approach to be successful, the observed levels of DNA homology must be calibrated against taxa whose phylogenetic relationships are, or can be, reasonably well established; for example, by carrying out simultaneous mating tests and DNA − DNA hybridisations. This approach has been used in several studies with fungi and the data suggest that all species further apart than sibling species show about 10% homology; so, the resolution of relationships in fungi further apart than this would not seem possible using DNA − DNA hybridisation methods (Kurtzman, 1985). Having calibrated the homology between genomes of these species, the relationships between different taxa can be investigated; for example, Bak & Stenderup (1969) demonstrated that three *Candida* species (*C. albicans*, *C. stellatoidea* and *C. clausenii*) show between 65% and 81% homology at the DNA level and on that criterion they can be considered a single species. For further examples, see Kurtzman (1985) and Ellis (1985). One of the more useful applications of this technique in mycology is as a fairly straightforward test of putative perfect and imperfect states of the same fungus; e.g. Kurtzman, Johnson & Smiley (1979) and Manachini (1979) have shown that the imperfect stage of *Hansenula jadinii* is the fungus usually described as *Candida utilis*.

DNA can be cut very specifically into fragments using restriction endonuclease enzymes (Pukkila, Chapter 5). The number of fragments obtained depends on the enzyme used; an enzyme which recognises a target of four base pairs (bp) will cut the molecule more frequently and give more fragments than one which recognises a 6 bp target. The fragments of DNA can be separated electrophoretically on agarose gels and stained with ethidium bromide which fluoresces in UV light (Old & Primrose, 1985).

Mitochondrial DNA can be isolated independently of genomic DNA, as it is a relatively small (about 16 to 100 kb), circular molecule (Grossman & Hudspeth, 1985). When cut with a restriction endonuclease a distinctive pattern of bands is generated on the electrophoretic gel (e.g. Specht, Novotny & Ullrich). Cutting with different enzymes generates different patterns of fragments and comparisons of single and double digests permits the order of cutting sites and, hence, the order of fragments within the molecule to be ascertained, producing a restriction map of the molecule (e.g. Pukkila, Chapter 5, Fig. 5.1). If a point mutation changes a base pair within a restriction enzyme recognition site so that it is no longer recognised as a target, digestion of the mutated molecule will produce a pattern lacking two fragments that are present in the original, but with a single additional band equal in length to the sum of the missing two fragments. Similarly, mutations which produce new target sites, deletions and insertions will produce easily recognised changes in the pattern of fragments. Analysis of these differences between taxa or populations can then be used to build up dendrograms or phylogenetic trees using the numbers, positions and associations of these changes (for examples see Croft, Chapter 21; Kozlowski & Stepień, 1982; Avise, 1986). Where intraspecific variation can be shown to exist, the variants are usually known as restriction fragment length polymorphisms (RFLPs) and these are amenable to analysis in exactly the same way as electrophoretic variants of polypeptides.

In the same way that classical genetic analysis can be done with morphological mutants, even when their genetic basis is not known, a great deal of useful information can be obtained using random DNA fragments of unknown function by effectively using the electrophoretic pattern of DNA fragments as a 'morphological' phenotype. The random fragments can be selected from a cloned 'library' prepared either in a plasmid (e.g. Pukkila, Chapter 5) or viral vector (see Old & Primrose, 1985). Much of the current work in molecular genetics is devoted to understanding how various genes work and how they are controlled, and considerable effort has been put into isolating genes of known function, e.g. ribosomal RNA sequences (e.g. Pukkila, Chapter 5) and various enzymes, particularly in *Saccharomyces cerevisiae, Aspergillus nidulans* and *Neurospora crassa* (Fincham, 1985). Among related organisms, genes of similar function may be related by descent and may, consequently, have homology with each other; a feature that can be detected by probing Southern blots with the radio-labelled sequence of interest.

If a restriction enzyme recognises a site within a repeating unit, it will cut each unit at the same point; but if some of the repeats have mutated in this position, then the fragments will appear as integer multiples of the basic fragment length. Thus, if the probe used is part of a repeated sequence, then the resulting autoradiograph will show a more or less evenly spaced ladder of fragments. Repeated sequences can vary between species and can be used to investigate phylogenetic relationships (e.g. Strachan *et al.*, 1982). Although the work involved in obtaining useful probes is rather protracted, once they have been obtained it should be possible to carry out population surveys for variation in the same way as for allozymes; though the somewhat more lengthy procedures for extracting DNA may limit the amount of data it is possible to obtain.

Having obtained a probe, particularly if it is a sequence in a gene of known function, it can be used to aid isolation of the gene itself. Since all evolutionary change must ultimately be detectable at the DNA level, analysis of DNA sequences should provide the most objective test of phylogenetic relationship. The efficiency of methods for sequencing DNA has improved greatly during the past ten years, so that it is now somewhat faster to sequence DNA than proteins. It is possible to sequence up to about 400 bp per day, but since an average protein is represented by a sequence at least 1000 bp long, it would still take considerable time to accumulate sufficient data for taxonomic purposes, even under the best of conditions. The approach may well, therefore, be impracticable as a method of looking at evolutionary changes on a population scale. The use of DNA sequences comes into the same category as the use of amino acid sequences, though with some increase of precision and efficiency, but the choice of gene to use for phylogenetic analysis is equally critical. The analysis, however, is quite straightforward; the sequences are aligned and differences counted (for examples see Bodmer & Ashburner, 1984; Kreitmann, 1983).

Despite high hopes that the analysis of DNA sequences could provide a more objective method for studying evolutionary relationships a number of problems have been encountered: (i) redundancy of the genetic code results in the different base positions in each codon evolving at different rates; (ii) different parts of the gene product are more constrained than others and this is reflected in the rates at which different parts of the genes evolve; (iii) different genes evolve at different rates; (iv) although use of DNA sequences can overcome ambiguities in amino acid sequences due to synonymous codons, the very simplicity of the genetic code itself puts limits on the level of precision that can be attained. Since there are only four bases — two purines and two pyrimidines — a second mutation at a given position can restore the original base at that position. The further the phylogenetic 'distance' between species, the greater the chance for such back-mutation. The direct consequence is that the number of base position differences between sequences is not linear with respect to phylogenetic distance (Kimura & Ohta, 1972; Neymann, 1970). Therefore, sequences have to be chosen with some care if they are to yield useful information (for discussion of some of the problems see Fitch, 1986).

Among the more surprising discoveries of the past several years have been the observations that eukaryotic genes consist of 'bits' of coding sequence (exons) separated by lengths of intervening DNA, called introns, and that, as well as functional genes, the eukaryotic genome often contains functionless gene copies (pseudogenes). Introns are characteristic features of the DNA level of genetic structure; they are spliced out of the RNA during processing and prior to translation and seemingly have no phenotypic effects. Similarly, pseudogenes appear not to be affected by selection in the same way as their functional relatives. Although the presence of introns raises the evolutionary enigma of why they are there at all, they also provide another tool for looking at evolutionary processes. Ironically, it also appears that pseudogenes may be more useful for studying evolution, in terms of phylogenetic relationships, than functional genes. Freed from selection, these sequences may accumulate mutations at the neutral mutation rate and hence, more clearly reflect true evolutionary time without the distortions introduced by selection (Kimura, 1986). Some indication of this can be gauged from studies of globin pseudogenes. The sequence of this gene seems to show the same rate of substitution at all base positions, in contrast to the functional versions, in which the highest rates of replacement are at the third base position (which has least influence on codon meaning). It is also worth noting that the rate of substitution in the pseudogene appears to be higher even than the rate at the third base position in the functional sequence (Li, Gojobori & Nei, 1981). Molecular biology may offer ways of tackling some of the more elusive and intractable problems in fungal evolution. It must be remembered, however, that these techniques are merely tools for tackling problems, not ends in themselves, and no substitute for a good understanding of mycology and the biology of fungi (see Tariq *et al.*, 1985). On the other hand, the application of these techniques to the study of fungi may require mycologists to reassess their interpretation of variation within the Fungi.

References

Arthur, R., Herr, F., Strauss, N., Anderson, J. & Horgen, P. (1982). Characterisation of the genome of the cultivated mushroom, *Agaricus brunnescens*. *Experimental Mycology*, **7**, 127 − 132.

Avise, J. C. (1986). Mitochondrial DNA and the evolutionary genetics of higher animals. *Philosophical Transactions of the Royal Society, series B*, **312**, 325 − 342.

Backhouse, D., Willetts, H. J. & Adam, P. (1984). Electrophoretic studies of *Botrytis* species. *Transactions of the British Mycological Society*, **82**, 625 − 630.

Bak, A. L. & Stenderup, A. (1969). Deoxyribonucleic acid homology in yeasts. Genetic relatedness within the genus *Candida*. *Journal of General Microbiology*, **59**, 21 − 30.

Baptist, J. N. & Kurtzman, C. P. (1976). Comparative enzyme patterns in *Cryptococcus laurentii* and its taxonomic varieties. *Mycologia*, **68**, 1195 − 1203.

Bodmer, M. & Ashburner, M. (1984). Conservation and change in the DNA sequences coding for alcohol dehydrogenase in sibling species of *Drosophila*. *Nature*, **309**, 425 − 430.

Boisselier-Dubayle, M. -C. (1983). Taxonomic significance of enzyme polymorphism among isolates of *Pleurotus* (Basidiomycetes) from umbellifers. *Transactions of the British Mycological Society*, **81**, 121 − 127.

Burdon, J. J. & Marshall, D. R. (1981). Isozyme variation between species and formae speciales of the genus *Puccinia*. *Canadian Journal of Botany*, **59**, 2628 − 2634.

Burdon, J. J., Marshall, D. R. & Luig, N. R. (1981). Isozyme analysis indicates that a virulent cereal rust pathogen is a somatic hybrid. *Nature*, **293**, 565 − 566.

Burdon, J. J., Marshall, D. R., Luig, N. R. & Gow, D. J. S. (1982). Isozyme studies on the origin and evolution of *Puccinia graminis* f. sp. *tritici* in Australia. *Australian Journal of Biological Sciences*, **35**, 231 − 238.

Burdon, J. J. & Roelfs, A. P. (1985a). Isozyme and virulence variation in asexually reproducing populations of *Puccinia graminis* and *P. recondita* on wheat. *Phytopathology*, **75**, 907 − 913.

Burdon, J. J. & Roelfs, A. P. (1985b). The effect of sexual and asexual reproduction in the isozyme structure of populations of *Puccinia graminis*. *Phytopathology*, **75**, 1068 − 1073.

Burdon, J. J., Seviour, R. J. & Fripp, Y. J. (1982). Electrophoretic patterns of soluble proteins of *Hendersonia* species. *Transactions of the British Mycological Society*, **78**, 551 − 553.

Dickerson, R. C. (1971). The structure of cytochrome *c* and the rates of molecular evolution. *Journal of Molecular Evolution*, **1**, 26 − 45.

Ellis, J. J. (1985). Species and varieties in the *Rhizopus arrhizus-Rhizopus oryzae* group as indicated by their DNA complementarity. *Mycologia*, **77**, 243 − 247.

Fincham, J. R. S. (1985). From auxotrophic mutants to DNA sequences. In *Gene Manipulations in Fungi*, ed. Bennett, J. W. & Lasure, L. L., pp. 3 − 34. Orlando, Florida: Academic Press.

Fitch, W. M. (1986). The estimate of total nucleotide substitutions for pairwise differences is biassed. *Philosophical Transactions of the Royal Society, series B*, **312**, 317 − 324.

Fitch, W. M. & Margoliash, E. (1967). Construction of phylogenetic trees. *Science*, **155**, 279 − 284.

Grossman, L. I. & Hudspeth, M. E. S. (1985). Fungal mitochondrial genomes. In *Gene Manipulations in Fungi*, ed. Bennett, J. W. & Lasure, L. L., pp. 65 − 103. Orlando, Florida: Academic Press.

Hames, B. D. & Rickwood, D. (1981). *Gel Electrophoresis of Proteins*. Oxford: IRL Press.

Hamrick, J. L. (1979). Genetic longevity. In *Topics in Plant Population Biology*, ed. Solbrig, O., Jain, S., Johnson, G. & Raven, P., pp. 84 − 114. New York: Columbia University Press.

Ianelli, D., Capparelli, R., Cristinzio, G., Marzanio, F., Scala, F. & Noviello, C. (1982). Serological differentiation among formae speciales and physiological races of *Fusarium oxysporum*. *Mycologia*, **74**, 313 − 319.

Kimura, M. (1983a). *The Neutral Theory of Molecular Evolution*. Cambridge University Press.

Kimura, M. (1983b). The neutral theory of molecular evolution. In *Evolution of Genes and Proteins*, ed. Nei, M. & Koehn, R. K., pp. 208 − 233. Sunderland, Massachusetts: Sinauer Associates.

Kimura, M. (1986). DNA and the neutral theory. *Philosophical Transactions of the Royal Society, series B*, **312**, 343 − 354.

Kimura, M. & Ohta, T. (1972). On the stochastic model for estimation of mutational distances between homologous proteins. *Journal of Molecular Evolution*, **2**, 87 − 90.

Koehn, R. K., Zera, A. J. & Hall, J. G. (1983). Enzyme polymorphism and natural selection. In *Evolution of Genes and Proteins*, ed. Nei, M. & Koehn, R. K., pp. 115 − 136. Sunderland, Massachusetts: Sinauer Associates.

Kozlowski, M. & Stepień, P. (1982). Restriction enzyme analysis of mitochondrial DNA of the genus *Aspergillus* as an aid in taxonomy. *Journal of General Microbiology*, **128**, 471 − 476.

Kreitmann, M. (1983). Nucleotide polymorphism at the alcohol dehydrogenase locus of *Drosophila melanogaster*. *Nature*, **304**, 412 − 417.

Kurtzman, C. P. (1985). Molecular taxonomy of the fungi. In *Gene Manipulations in Fungi*, ed. Bennett, J. W. & Lasure, L. L., pp. 35 − 63. Orlando, Florida: Academic Press.

Kurtzman, C. P., Johnson, C. J. & Smiley, M. J. (1979). Determination of conspecificity of *Candida utilis* and *Hansenula jadinii* through DNA reassociation. *Mycologia*, **71**, 844 − 847.

Lewontin, R. C. (1974). *The Genetic Basis of Evolutionary Change*. New York: Columbia University Press.

Li, W. -H., Gojobori, T. & Nei, M. (1981). Pseudogenes as a paradigm of neutral evolution. *Nature*, **292**, 237 − 239.

Maghrabi, H. A. & Kish, L. P. (1985a). Isozyme characterization of Ascosphaerales associated with bees. I. *Ascosphaera apis*, *Ascosphaera proliperda* and *Ascosphaera aggregata*. *Mycologia*, **77**, 358 − 365.

Maghrabi, H. A. & Kish, L. P. (1985b). Isozyme characterization of Ascosphaerales associated with bees. II. *Ascosphaera major*, *Ascosphaera atra* and *Ascosphaera asterophora*. *Mycologia*, **77**, 366 − 372.

Manachini, P. L. (1979). DNA sequence similarity, cell wall mannans and physiological characteristics in some strains of *Candida utilis*, *Hansenula jadinii* and *Hansenula petersonii*. *Antonie van Leeuwenhoek*, **45**, 451 − 463.

Marshall, D. R. & Allard, R. W. (1970). Maintenance of isozyme polymorphisms in natural populations of *Avena barbata*. *Genetics*, **66**, 393 − 399.

Nei, M. (1975). *Molecular Population Genetics and Evolution*. Amsterdam, The Netherlands: North-Holland Publishing Company.

Nei, M. (1983). Genetic polymorphism and the role of mutation in evolution. In *Evolution of Genes and Proteins*, ed. Nei, M. & Koehn, R. K., pp. 165 − 190. Sunderland, Massachusetts: Sinauer Associates.

Nei, M. & Chakraborty, R. (1973). Genetic distance and electrophoretic identity of proteins between taxa. *Journal of Molecular Evolution*, **2**, 323 − 328.

Nevo, E. (1978). Genetic variation in natural populations: patterns and theory. *Theoretical Population Biology*, **13**, 121 − 177.

Newton, A. C., Caten, C. E. & Johnson, R. (1985). Variation for isozymes and double stranded RNA among isolates of *Puccinia striiformis* and two other cereal rusts. *Plant Pathology*, **34**, 235 − 247.

Neymann, J. (1970). *Molecular Studies of Evolution: A Source of Novel Statistical Problems*. Berkeley, California: Statistical Laboratory, University of California.

Old, R. W. & Primrose, S. B. (1985). *Principles of Gene Manipulation*. Oxford: Blackwells Scientific Publications.

Pérez-Silva, E. & Alfonso, R. M. A. (1983). Chromatographic and taxonomic evaluation of *Amanita citrina* (Agaricales). *Mycologia*, **75**, 1030 − 1035.

Royse, D. J. & May, B. (1982a). Use of isozyme variation to identify genotypic classes of *Agaricus brunnescens*. *Mycologia*, **74**, 93 − 102.

Royse, D. J. & May, B. (1982b). Genetic relatedness and its application in selective breeding of *Agaricus brunnescens*. *Mycologia*, **74**, 569 − 575.

Scala, F., Cristinzio, G., Marziano, F. & Noviello, C. (1981). Endopolygalacturonase zymograms of *Fusarium* species. *Transactions of the British Mycological Society*, **77**, 587 − 591.

Scott, S. W. (1981a). Serological relationships of three *Sclerotinia* species. *Transactions of the British Mycological Society*, **77**, 674 − 676.

Scott, S. W. (1981b). Separation of *Sclerotinia* isolates collected from herbage legumes. *Transactions of the British Mycological Society*, **76**, 321 − 323.

Seviour, R. J., Pethica, L. M., Soddell, J. A. & Pitt, D. E. (1985). Electrophoretic patterns of sporangiophore proteins of *Rhizopus* isolates as taxonomic characters. *Transactions of the British Mycological Society*, **84**, 701 − 708.

Specht, C. A., Novotny, C. P. & Ullrich, R. C. (1983). Isolation and characterisation of mitochondrial DNA from the basidiomycete *Schizophyllum commune*. *Experimental Mycology*, **7**, 336 − 343.

Stipes, R. J. (1970). Comparative mycelial protein and enzyme patterns in four species of *Ceratocystis*. *Mycologia*, **62**, 987 − 995.

Strachan, T., Coen, E., Webb, D. & Dover, G. (1982). Modes and rates of change of complex DNA families of *Drosophila*. *Journal of Molecular Biology*, **158**, 37 − 54.

Sweetingham, M. W., Cruikshank, R. H. & Wong, D. H. (1986). Pectic zymograms and taxonomy and pathogenicity of the Ceratobasidiaceae. *Transactions of the British Mycological Society*, **86**, 305 − 311.

Tariq, V.-N., Gutteridge, C. S. & Jeffries, P. (1985). Comparative studies of cultural and biochemical characters used for distinguishing species within *Sclerotinia*. *Transactions of the British Mycological Society*, **84**, 381 − 397.

Wolfe, M. S. & Schwarzbach, E. (1978). The recent history of the evolution of barley powdery mildew in Europe. In *The Powdery Mildews*, ed. Spencer, D. M., pp. 129 − 157. London: Academic Press.

Wong, A. L. & Willetts, H. J. (1973). Electrophoretic studies of soluble proteins and enzymes of *Sclerotinia* species. *Transactions of the British Mycological Society*, **61**, 167 − 178.

Wong, A. L. & Willetts, H. J. (1975). Electrophoretic studies of Australasian, North American and European isolates of *Sclerotinia sclerotiorum* and related species. *Journal of General Microbiology*, **90**, 355 − 359.

7

Fungal chromosomes as observed with the light microscope

EVA R. SANSOME

Stonecroft, Post Office Lane, Lighthorne, Warwickshire CV35 0AP, UK

Introduction

Observations on chromosomes have usually been made at metaphase when they are most condensed and arranged in one plane forming the metaphase plate. This also applies to fungal chromosomes, but since metaphase chromosomes are rarely seen during somatic division, observations on fungal chromosomes have mostly been made on meiotic divisions or the mitotic divisions immediately following them. During somatic divisions, fungal chromosomes do not condense to the same extent as in higher plants. The spindle is intranuclear and the nucleolus persists throughout the division. However, the essential features of mitosis, i.e. the longitudinal replication of individual chromosomes and separation of the two daughters to opposite poles, is present. The term 'fungal mitosis' as opposed to 'classical mitosis' has come into use and it is probable that fungal mitosis differs from the classical in having the metabolic and reproductive functions of the chromosomes less well separated, perhaps because of the limited amount of differentiation in the fungi.

In meiosis, which occurs in specialised organs, metaphase chromosomes can be observed, described and counted in many fungi, although this is often difficult because of their small size. Chromosomes are also sometimes observable during metaphase II, and in some cases chromosomes have been observed in the first few divisions after meiosis. The observation of meiosis is, of course, of the utmost importance in establishing the life history of any organism.

The Ascomycetes

I will first consider meiosis in *Neurospora*, a genus which has been studied intensively both cytologically and genetically because it is particularly amenable to genetic study.

The Ascomycetes are haploid, meiosis occurring in the ascus. The young ascus is binucleate. These nuclei fuse, meiosis follows and subsequently, there is an extra mitotic division resulting in eight haploid nuclei, around which the ascospores are delimited. The early stages of meiosis in *Neurospora* differ from those of classical meiosis in that the chromosomes enter prophase in a condensed state. The shortening of the chromosomes starts before nuclear fusion and continues until synapsis. During synapsis the chromosomes extend and at the end of pachytene they are most extended and most clearly distinguishable. Since they are small, it is often possible to trace individual chromosomes even when fully extended.

Fig. 7.1 shows pachytene in *Neurospora* and *Zea mays* at the same magnification; it will be seen that *Neurospora* has extremely small chromosomes. All seven chromosomes can be seen sufficiently well in the nucleus shown in Fig. 7.2(a) to be traced (Fig. 7.2b). Paired chromomeres are visible in the pachytene stage shown in Fig. 7.2(c) in which a nucleolar organiser can also be seen attached to the prominent nucleolus. Successive stages of meiosis are illustrated in Figs. 7.2(d), which shows diplotene, and 7.3 (a), (b) & (c) which show diplotene, diakinesis and metaphase respectively, while stages of the second meiotic division are shown in Fig.

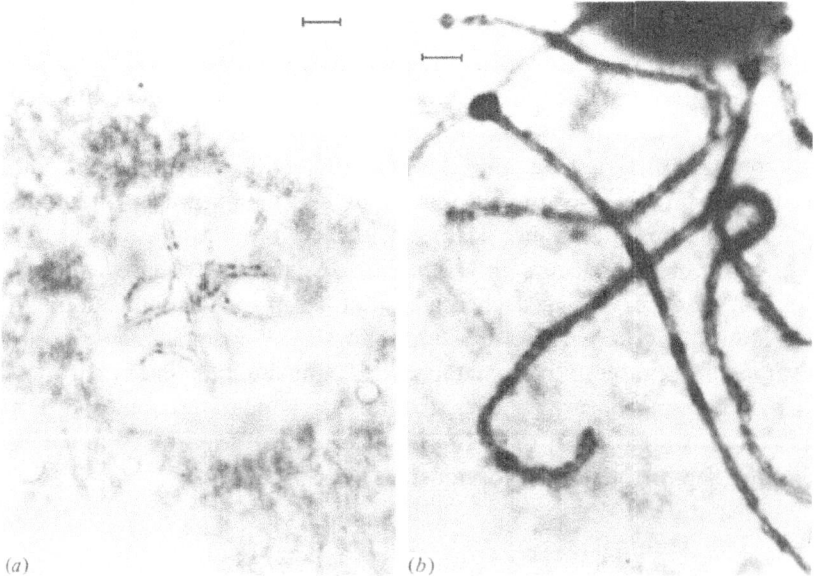

(a)   (b)

Fig. 7.1. Comparison of pachytene in (a) *Neurospora crassa* and (b) *Zea mays* at the same magnification. Scale bars = 1 μm. (Reproduced with permission, from Perkins, 1979.)

7.3 (d) & (e). It will be seen that the chromosomes are very small by metaphase and that they are smaller during the second meiotic division than during the first. Metaphase during the first division of the ascospore

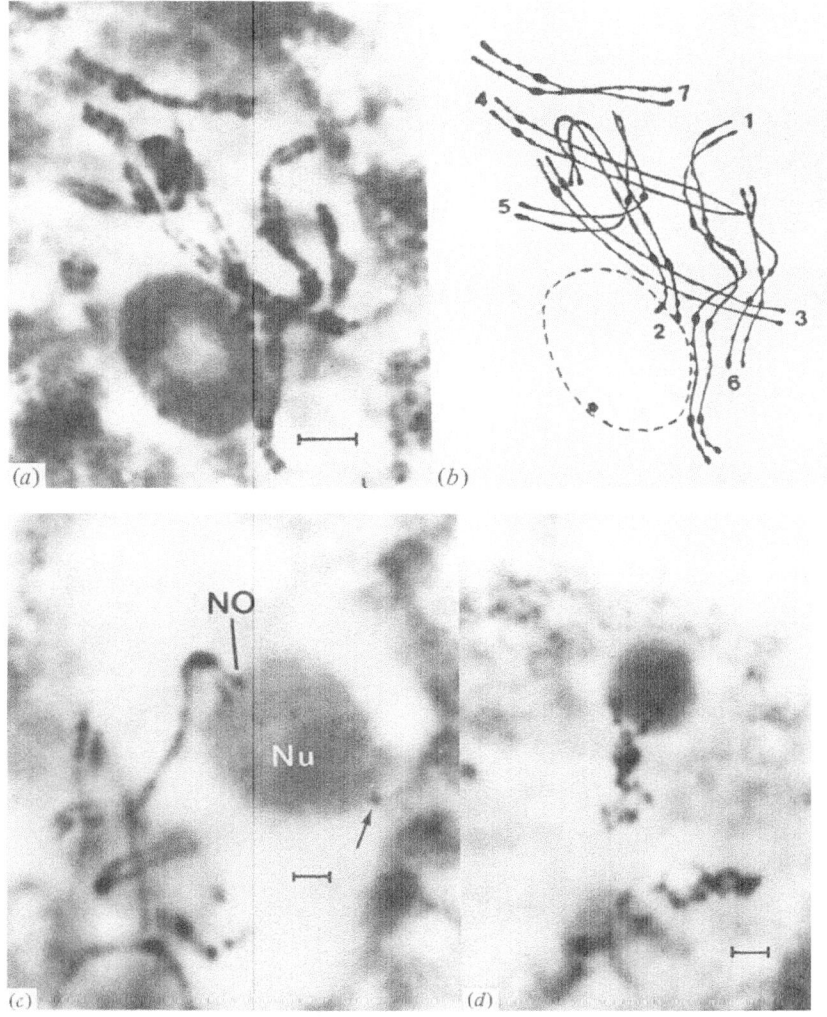

Fig. 7.2. Meiosis in asci of *Neurospora crassa*. (a) Pachytene nucleus; (b) interpretation, showing seven chromosomes; (c) pachytene showing nucleolar organiser (NO) attached to the nucleolus (Nu); (d) beginning of the diffuse diplotene stage, showing the nucleolar organiser attached to the nucleolus. Scale bars = 1 μm. ((a) & (b) reproduced, with permission, from Perkins & Barry, 1977; (c) & (d) reproduced, with permission, from Raju & Barry, 1980.)

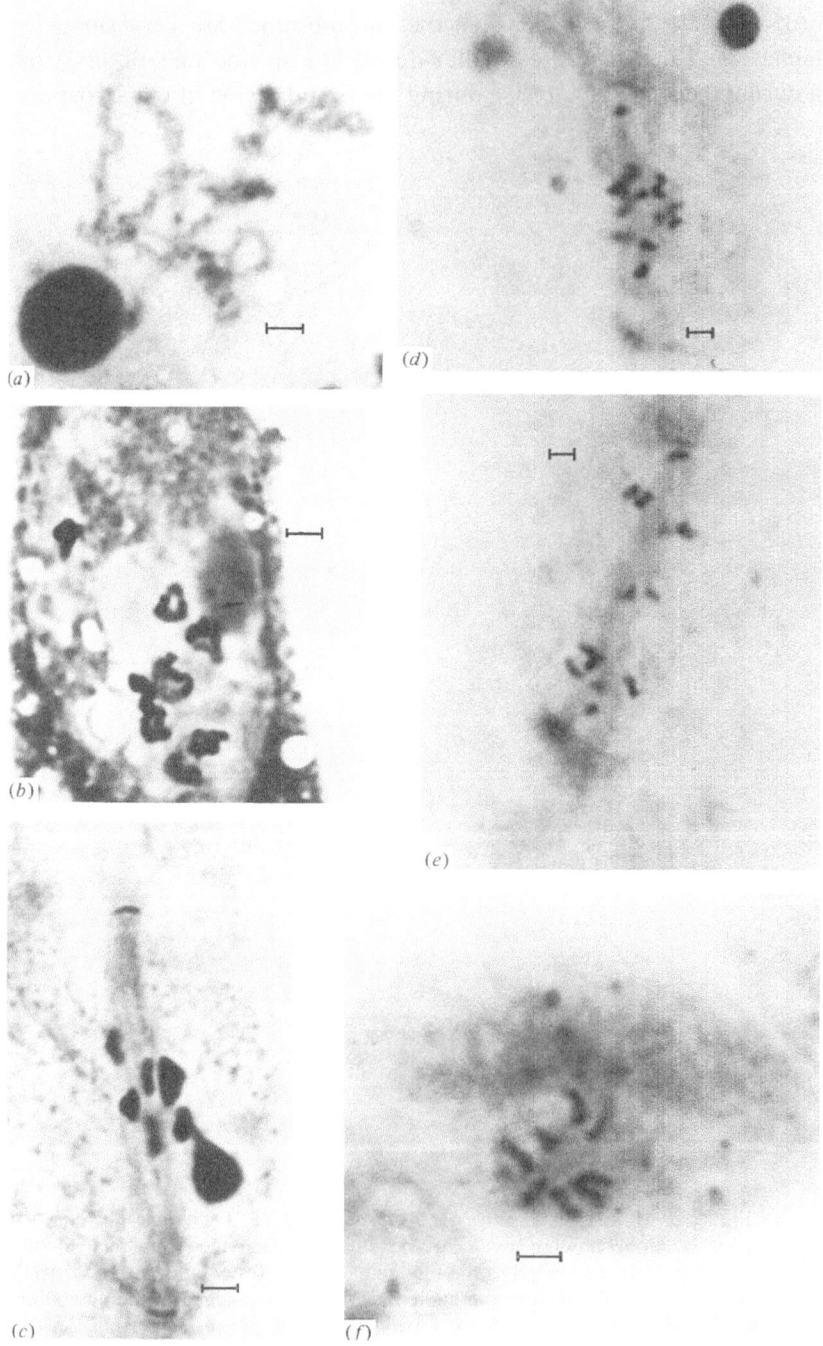

Fig. 7.3. Meiosis and mitosis in asci of *Neurospora*. (a) Diplotene; (b) diakinesis; (c) metaphase; (d) & (e) second meiotic division; (f) mitotic

nucleus, i.e. the second mitosis after meiosis, is depicted in Fig. 7.3(f), which is the clearest fungal mitotic metaphase I have ever seen illustrated and clearly shows all seven chromosomes.

Intensive studies of pachytene and other stages was started by McClintock (1945) and followed up by Fincham (1949), Singleton (1953), Barry (1967), Barry & Perkins (1969), Perkins & Barry (1977), Perkins (1979), Perkins, Raju & Barry (1980) and Raju (1978, 1980), among others. The seven chromosomes were numbered according to their relative lengths at pachytene and to the position of their centromeres, seen most clearly at metaphase II. In the meantime, seven linkage groups were established by genetic experiments. Joint genetical and cytological observations on translocations with genetic markers enabled the two types of map to be correlated (Barry, 1967; Barry & Perkins, 1969).

Electron microscope studies of pachytene have shown the presence of synaptonemal complexes similar to those observed in other organisms. These have been used to determine chromosome numbers, notably in the case of *Saccharomyces cerevisiae* which has 17 pairs of chromosomes in the diploid that are too small to be seen with the light microscope (Zickler, 1981). At the end of pachytene, after a short early diplotene (Figs. 7.2d & 7.3a) the chromosomes become diffuse, entering a stage known as diffuse diplotene. This stage, although not always present, has been observed in numerous plants and animals. The diffuse stage lasts until the ascus is nearly fully grown. Rhoades (1961) assumed diffuse diplotene to be an interruption to the normal course of meiosis caused by unusual metabolic conditions; he considered it not to be an essential feature of meiosis.

The Homobasidiomycetes

In the homobasidiomycetes, *Coprinus*, for example, has been studied by Lu & Raju (1970) and Pukkila & Lu (1985), who found that prophase resembled *Lilium* rather than *Neurospora* in that the chromosomes are fully extended at the beginning and no further extension occurs during zygotene and pachytene. Pukkila & Lu (1985) obtained beautiful preparations using a haematoxylin or a silver staining technique. Pachytene is the most favourable stage to study since the chromosomes are very small (Fig. 7.4a − d). The basic chromosome number is 13, with the

metaphase (metaphase IV) in the ascospore, showing seven chromosomes. Scale bars = 1 μm; a, b, & e = *N. crassa*, c & d = *Neurospora* sp. nov. (Reproduced, with permission, from Raju, 1980.)

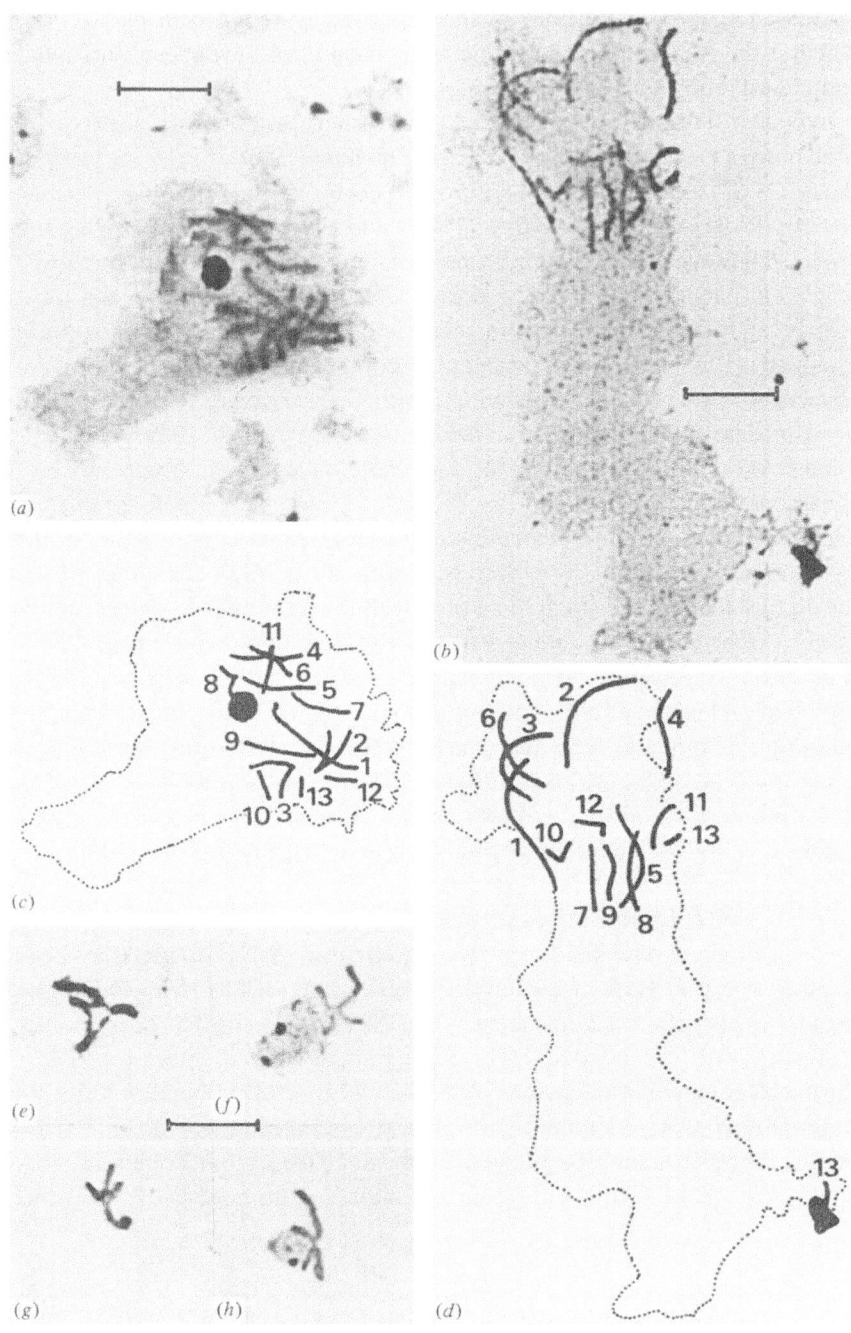

Fig. 7.4. Meiosis in *Coprinus cinereus*. (a) & (b) show pachytene nuclei and

longest chromosome about five times the length of the shortest. Cross-shaped configurations resulting from a reciprocal translocation between two non-homologous chromosomes are shown in Fig 7.4 (e − h); if chiasmata formed in every pairing segment, such configurations would form rings of four chromosomes at metaphase.

The Heterobasidiomycetes

McGinnis (1956) studied the nuclei in the sporidia, i.e. the first divisions after meiosis, in a number of species of rust. He found that the chromosomes had a tendency to be joined at the ends but was able to determine chromosome numbers as follows: *Puccinia graminis*, six, often with the ends attached to form three pairs; *P. coronata*, three (suggesting that *P. graminis* may be polyploid); while the homothallic species *P. malvacearum*, *P. xanthii* and *P. asteris* had a basic chromosome number of four.

Individual chromosomes could not be identified during nuclear division in basidiospores of *Puccinia kraussianna*, parasitic on *Smilax*, because they failed to condense to a metaphase state but it has between 30 and 40. Meiosis commenced in the teleutospore cells and the earliest identified stage was pachytene (Sansome, 1959). The condensed presynaptic stage, described for Ascomycetes, was not observed although stages earlier than pachytene were not intensively examined. The early diplotene stage is shown in Fig. 7.5(a); this stage passed rapidly into a diffuse diplotene (Fig. 7.5b), similar to that observed in Ascomycetes and Basidiomycetes, which lasted through the formation of the promycelium. The nucleus passed into the promycelium in the diffuse state and completed meiosis when the promycelium was fully extended. Association of the diffuse condition with growth of the promycelium suggests that it represents a period of metabolic activity of the nucleus as suggested for *Neurospora* by Rhoades (1961).

The Chytridiomycetes

In the uniflagellatae there is some variation in life history. For example, the gametes are usually flagellate, but may be equal in size or the

(c) & (d) the corresponding interpretations showing the 13 chromosomes. (e) to (h) show four examples of multiple pachytene configurations in reciprocal-translocation heterozygous nuclei. Scale bars = 1 μm. (Reproduced, with permission, from Pukkila & Lu, 1985.)

female may be the larger, while in *Monoblepharis*, the female gamete is non-motile. The timing of meiosis has been established in *Allomyces* where there is an alternation of sporophytic and gametophytic generations. The sporophyte produces thin-walled sporangia which reproduce the sporophyte, and thick-walled resistant sporangia. Meiosis occurs in the latter (Emerson & Wilson, 1949), to yield haploid zoospores which give rise to the gametophyte bearing male and female gametangia. The gametes are all flagellate, but the female (10 μm dia) is almost twice the size of the male (6 μm); the zygote develops into the sporophyte. Emerson & Wilson (1954) found a polyploid series in *Allomyces*. In *A. arbuscula* basic chromosome numbers were 6, 16, 24 and 32, with the majority of natural isolates having the basic number 16. *A. javanicus* var. *macrogyna* had a basic number of 14 and artificial hybrids between the two species had numbers ranging from 22 to 44. Certain strains of *A. arbuscula* have a tendency towards parthenogenesis, accompanied by a doubling of the chromosomes of the female. When crossed with *A. javanicus* two types of hybrid occur; one (with type A meiosis) in which little chromosome pairing occurs, and another (type B meiosis) in which much more pairing occurs. Type B is believed to have resulted from fertilisation of a female gamete with doubled chromosomes, the bivalents being paired *A. arbuscula* chromosomes.

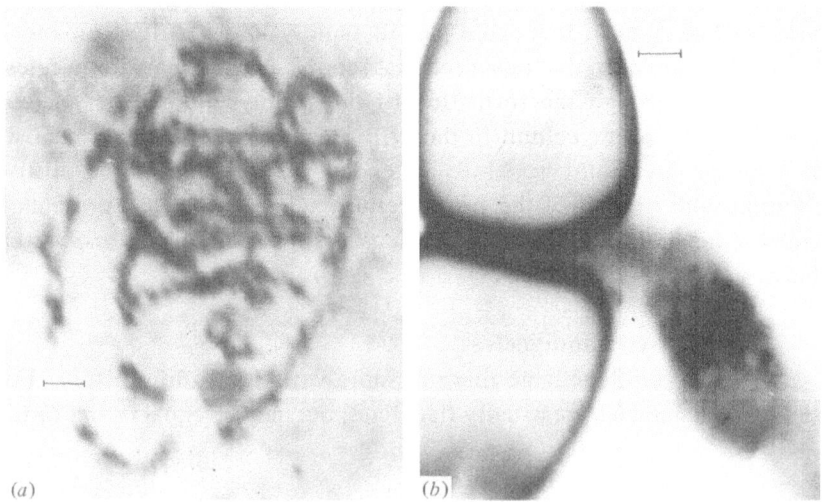

(a)   (b)

Fig. 7.5. Meiosis in *Puccinia kraussiana*. (a) Early diplotene in a squashed-out nucleus showing high chiasma frequency; (b) diffuse diplotene. Scale bars = 1 μm.

The Oomycetes

In the Oomycetes, nuclear divisions and chromosomes were observed in the gametangia of a number of genera of the Peronosporales and Saprolegniales from 1920 onwards, but most workers assumed these divisions were mitotic and thought that meiosis occurred in the oospore. Some workers, notably McDonough (1937), reported the occurrence of two divisions in the gametangia and illustrated the second division nuclei as being smaller than the first, which might be taken as an indication of meiosis, but there was a tendency to regard the fungi as a whole as being haploid. Furthermore, the difficulty of obtaining preparations of divisions in the oospore allowed the belief that meiosis occurred there to persist. The early work has been reviewed by Dick & Tin (1973).

In 1963 I described meiosis in the gametangia of *Pythium debaryanum* (Sansome, 1963) and suggested that the Oomycetes as a whole were diploid in the vegetative phase. Since then, observations on other Oomycetes, both cytological and genetical, have confirmed this view and it is now generally accepted that in Oomycetes meiosis occurs in the gametangia (e.g. Flanagan, 1970; Tin & Dick, 1975; Shaw, 1983).

Stages of meiosis in *Sclerospora graminicola* are shown in Fig. 7.6 (a − d). The chromosomes are larger than those in *Neurospora* and there is no diffuse diplotene stage; this may be because of the many nuclei undergoing meiosis, only one forms the female gamete, the remaining periplasm nuclei contributing to formation of the oospore wall. Fig. 7.6(b) shows metaphase with a ring of four chromosomes probably resulting from the presence of a reciprocal translocation. The fact that two nuclear divisions occur and that the nuclei do not increase in size between these two divisions (Fig. 7.6d) shows that the divisions are meiotic. The presence of multiple associations, which can occur at meiosis but not mitosis, confirms this.

Chromosome complements

It is probable that chromosome configurations at meiosis have been observed and approximate counts made in more Oomycetes than in any other fungal group. In the case of the large genus *Phytophthora*, the approximate chromosome numbers of over 16 taxa have been determined (Table 7.1). Relative chromosome sizes in *Phytophthora* range from large in *P. megakarya* to very small in *P. heveae*. Polyploidy has been observed in the homothallic species *P. megasperma* and in A1 mating type (the presumptive *aa* homozygote) isolates of the heterothallic *P. infestans*. In *P. meadii*, both diploid and tetraploid complements were observed in A1

106   *Eva R. Sansome*

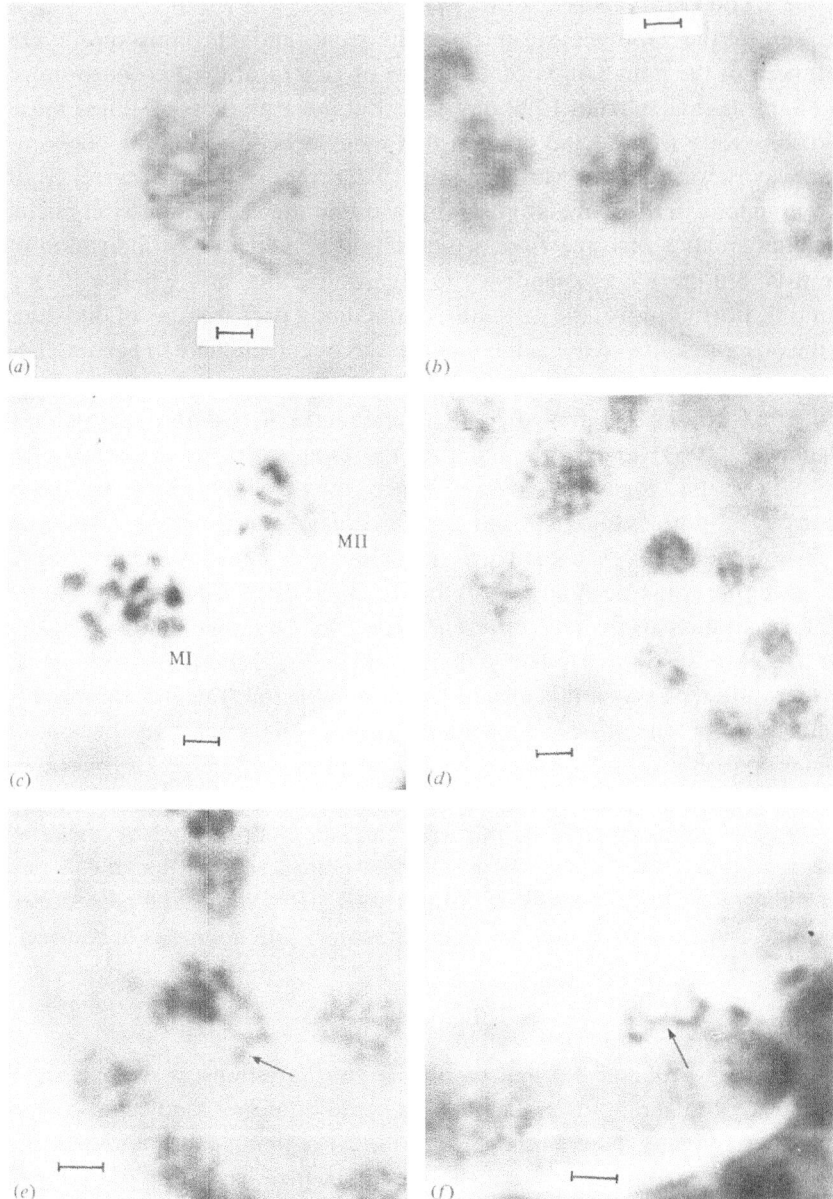

Fig. 7.6. (a) to (d), meiosis in *Sclerospora graminicola*. (a) Pachytene; (b) metaphase I, showing an association of four chromosomes; (c) metaphase I (MI) and II (MII) showing smaller chromosomes in the latter; (d) periplasm nuclei, note tetrad arrangement and small size of nuclei. (e) & (f) show trisomy in *Phytophthora drechsleri*; (e) ring of four chromosomes plus one extra

isolates and in isolates behaving as A2. However, tetraploid forms of the latter are also weakly self fertile (secondarily homothallic) and able to give rise to A1s as well as A2s. It is thought, therefore, that '*P. meadii*' may have arisen as a species hybrid, leading to allopolyploidy (Table 7.1; see also Brasier, Chapter 16).

Chromosome numbers of Oomycetes other than *Phytophthora* (Table 7.2) range from n = 3 in some species of *Achlya*, to n = 25 to 40 in *Albugo candida*. Polyploidy occurs in *Achlya* (n = 3, 6 or 12), the polyploids showing multiple configurations at meiosis, and in *Pythium* (n = 10 or 20). An association of four chromosomes, comparable to that seen in heterothallic *Phytophthora* species (Table 7.1 and see below) has been observed at meiosis in *Sclerospora graminicola* and in the heterothallic *Bremia lactucae* (Table 7.2).

The mating type complex

As more *Phytophthora* species were examined, it became apparent that the heterothallic species all had an association of four chromosomes at metaphase (Table 7.1), indicating that they were heterozygous for a reciprocal translocation (Sansome, 1980). Since the mating type reaction is non-specific, by using A1 mating type (presumptive *aa*) and A2 (*Aa*) isolates of *P. megakarya* (n = 5) as markers it was possible to establish that both the A1s and A2s in each species are heterozygous for the translocation. Since reciprocal translocation heterozygotes would normally segregate the respective homozygotes during sexual reproduction, there must be some mechanism for maintaining heterozygosity. A balanced lethal system, eliminating the homozygous oospores would result in 50% of the oospores being inviable. Alternatively, it is possible that selection takes place at the gametic stage. Thus, if the homologous chromosome segments are labelled AB − BC − CD − DA, oospheres carrying chromosomes AB − CD may attract male nuclei carrying BC − DA and *vice versa*; or, more possibly, one type only may function in the female and the other in the male (see Sansome, 1980).

The assumption that the mating type locus *A* or *a* is located on the chromosomes involved in the translocation was supported when secondarily homothallic isolates obtained from single oospores of *P. drechsleri*

(arrow) in a type I trisomic; (f) complicated chain of bivalents united by a chromosome partly homologous with each of them (arrow) in a type II trisomic. Scale bars = 1 μm.

Table 7.1. *Chromosome complements in* Phytophthora

Species	Chromosome number (n)	Interpretation of chromosome configuration	Authors
Homothallic			
P. cactorum	9 – 10	2n	Sansome, 1965
P. erythroseptica	9 – 10	2n	Sansome, 1965
P. heveae	9 – 10	2n	C. M. Brasier & E. R. Sansome (unpublished)
P. megasperma	11 – 15	2n	Sansome & Brasier, 1974;
	15 – 23	?	Hansen *et al.*, 1986 and
	22 – 28	?	see Hansen, Chapter 22
	26 – 34	4n	
Heterothallic			
P. botryosa	9 – 11	2n; ring of 4 plus 7 – 9 pairs	Sansome, 1980
P. cambivora	10 – 12	2n; ring of 4 plus 8 – 10 pairs	Sansome, 1977, 1980
P. capsici	9 – 10	2n; ring of 4 plus 7 – 8 pairs	Sansome, 1976; Maia Venard & Lavrot, 1976
Phytophthora MF4	9 – 12	2n; ring of 4 plus 7 – 10 pairs	Brasier & Medeiros, 1978
P. cinnamomi	9 – 10	2n; ring of 4 plus 7 – 8 pairs	Brasier & Sansome, 1975; Sansome, 1980
P. cryptogea	9 – 11	2n; ring of 4 plus 7 – 8 pairs	Sansome, 1977, 1980
P. drechsleri	9 – 12	2n; ring of 4 plus 7 – 10 pairs	Brasier & Sansome, 1975 Galindo & Zentmeyer, 1967; Sansome, 1980
P. infestans	9 – 10	2n; either ring of 6 plus 6 – 7 pairs, or ring of 4 plus 7 – 8 pairs	Brasier & Sansome, 1975; Sansome, 1980; Sansome & Brasier, 1973
	18 – 20	4n	Sansome, 1977
P. meadii	8 – 10	2n; ring of 4 plus 6 – 8 pairs, species hybrids?	E. R. Sansome & C. M. Brasier, unpublished; and see Brasier, Chapter 16
	16 – 20	4n; allopolyploid?	
P. megakarya	5	2n; ring of 4 plus 3 pairs	Sansome, 1980; Sansome, Brasier & Griffin, 1975; Sansome, Brasier & Sansome, 1979
P. nicotianae var. *parasitica*	8 – 9	2n; ring of 4 plus 6 – 7 pairs	Sansome, 1977, 1980; 1985
P. palmivora	9 – 12	2n; ring of 4 plus 7 – 10 pairs	Sansome, 1980; Sansome *et al.*, 1975, 1979

The chromosome number, n, is given as a range because of the difficulty of distinguishing between bivalents and early separated univalents at metaphase I/anaphase I.

Table 7.2. *Examples of chromosome complements in Oomycetes other than* Phytophthora

Species	Chromosome number (n)	Interpretation of chromosome configuration	Authors
Saprolegniales			
Achlya sp.	8		Sansome, 1965
A. ambisexualis	3	2n	Barksdale, 1968
Achlya, 8 species	3	2n; no multiples	Tin & Dick, 1975
Achlya, 4 species	6	4n; multiples present	Tin & Dick, 1975
A. klebsiana	12	8n	Flanagan, 1970
Apodachlyella completa	8–12*		Tin & Dick, 1975
Saprolegnia, 2 spp.	8–12		Tin & Dick, 1975
S. ferax	10		Flanagan, 1970
Pythiopsis cymosa	8–12		Tin & Dick, 1975
Peronosporales			
Pythium debaryanum	18	2n	Sansome, 1963
Pythium, 3 species	10	2n	Tin & Dick, 1975
Pythium, 2 species	20	4n	Tin & Dick, 1975
Peronospora parasitica	18–20	4n; ring of 4 plus 16–18 pairs	Sansome & Sansome, 1974; Sherriff & Lucas, 1986
Sclerospora graminicola	14–16	ring of 4 plus 12–14 pairs	Sansome, 1966 and see Fig. 7.6
Bremia lactucae	7–8	2n; ring of 4 plus 5–6 pairs	Michelmore & Sansome, 1982
Albugo candida	25–40	6n or 8n	Sansome & Sansome, 1974

* see note on Table 7.1.

were investigated (Mortimer, Shaw & Sansome, 1977). Single zoospore isolations of the homothallics showed the occurrence of somatic segregation, since A1 and A2 mating types were obtained as well as further secondary homothallics, in different proportions in different lines. Some of the A2s gave A1s and homothallics as well as A2s when similarly subjected to single zoospore analysis. Cytological study showed that homothallics and the A2 examined had an extra chromosome associated with the translocation, which I have therefore named the 'mating type complex' (Sansome, 1980); Fig. 7.6(e) shows a ring of four chromosomes with an extra chromosome attached (a type I trisomic) and Fig. 7.6(f) illustrates the type II trisomic, with two pairs of chromosomes united by an extra chromosome partly homologous with each of them — this type forms complicated chains rather than rings.

The presence of a reciprocal translocation restricts somatic segregation in the chromosomes concerned, since such segregation depends upon a cross-over and a non-crossover chromatid passing into the same nucleus. However, in the case of single cross-overs in the proximal segments (those containing the centromeres) such a combination would give an unbalanced nucleus with one distal segment represented only once and the other three times. It would therefore be expected to be eliminated in competition with nuclei of the original type. Thus, the situation of the mating type locus on the translocation complex enables the heterozygous mating type (*Aa*), widely presumed to be the A2, to survive throughout the vegetative phase without reverting to A1. The behaviour of the trisomic secondarily homothallics and unstable trisomic A2s in which the bar to segregation is removed by the presence of an extra chromosome is in accordance with this hypothesis (Sansome, 1980).

If the *A* allele of the mating type locus is situated on a proximal segment (containing the centromere), crossing over in this segment could lead to the chromosome carrying *A* being duplicated instead of that carrying *a* and thus, to the production of a 'new' A2 mating type *AAa*. If the *A* locus is on the distal segment, crossing over in the larger proximal segment could lead from the trisomic *Aaa* to the disomic *Aa*. Since crossing over in the new type would lead to reversion to the original homothallic type *Aaa*, such A2s are unstable (Sansome, 1985). Crossing over in the other proximal segments of a type I trisomic could lead to the formation of a type II trisomic. Configurations of type II (Fig. 7.6f) were observed in a culture derived from a type I trisomic in *P. drechsleri* (Fig. 7.6e; see also Sansome, 1985).

Secondary homothallism associated with the presence of an extra chromosome belonging to the complex has also been observed in *Bremia lactucae* (Michelmore & Sansome, 1982) and *Peronospora parasitica* (Sherriff & Lucas, 1986). Interestingly, an association of four chromosomes was observed at meiosis in a number of collections of the diploid Myxomycete, *Ceratiomyxa fruticulosa* (Sansome & Dick, 1965). It seems probable that that mating type locus in *C. fruticulosa* is situated on this complex, the translocation heterozygote serving to maintain the mating type factor in the heterozygous condition. These observations serve to emphasise the importance of somatic crossing over in diploid fungi and other organisms.

The problem of the relationship between the homothallic and hetero-thallic *Phytophthora* species is of great interest. It is thought probable that many of the genes connected with sexual reproduction are located on the mating type complex. If so, heterothallism could have arisen concurrently with the reciprocal translocation if the translocation process was assoc-iated with small duplications and deficiencies, or if some of the genes governing sexual reproduction were subject to a position effect, being active when on one chromosome but inactivated when translocated to another.

Acknowledgements I am grateful to the Director, National Vegetable Research Station, Wellesbourne, Warwick for library and photographic facilities; to the late Prof. F. W. Sansome for taking the photographs of *Sclerospora*; to Mr R. B. Sampson for printing the Figures; and to Drs C. M. Brasier and I. R. Crute for comments on the manuscript. The generosity of Drs P. J. Pukkila, N. B. Raju and D. D. Perkins in supplying their original photographs is gratefully acknowledged.

References

Barksdale, A. W. (1968). Meiosis in the antheridium of *Achlya ambisexualis*. *Journal of the Elisha Mitchell Scientific Society*, **84**, 187 – 194.

Barry, E. G. (1967). Chromosome aberrations in *Neurospora*, and the correlation of chromosomes and linkage groups. *Genetics*, **55**, 21 – 32.

Barry, E. G. & Perkins, D. D. (1969). Position of linkage group V markers in chromosome 2 of *Neurospora crassa*. *Journal of Heredity*, **60**, 120 – 125.

Brasier, C. M. & Medeiros, A. G. (1978). Karyotype of *Phytophthora palmivora* Morphological Form 4. *Transactions of the British Mycological Society*, **70**, 295 – 298.

Brasier, C. M. & Sansome, E. (1975). Diploidy and gametangial meiosis in *Phytophthora cinnamomi*, *P. infestans* and *P. drechsleri*. *Transactions of the British Mycological Society*, **65**, 49 – 65.

Dick, M. W. & Tin, W. (1973). The development of cytological theory in the Oomycetes. *Biological Reviews*, **48**, 133 – 158.

Emerson, R. & Wilson, C. M. (1949). The significance of meiosis in *Allomyces*. *Science*, **110**, 86 – 88.

Emerson, R. & Wilson, C. M. (1954). Interspecific hybrids and the cytogenetics and cytotaxonomy of *Euallomyces*. *Mycologia*, **46**, 393 – 434.

Fincham, J. R. S. (1949). Chromosome numbers in species of *Neurospora*. *Annals of Botany*, **13**, 23 – 28.

Flanagan, P. W. (1970). Meiosis and mitosis in Saprolegniaceae. *Canadian Journal of Botany*, **48**, 2069 – 2076.

Galindo, A. J. & Zentmyer, G. (1967). Genetical and cytological studies of *Phytophthora* strains pathogenic to pepper plants. *Phytophthora*, **57**, 1300 – 1304.

Hansen, E., Brasier, C. M., Shaw, D. S. & Hamm, P. (1986). Evidence for emerging biological species in *Phytophthora megasperma*. *Transactions of the British Mycological Society*, in press.

Lu, B. C. & Raju, N. B. (1970). Meiosis in Coprinus. II. Chromosome pairing and the lampbrush diplotene stage of meiotic prophase. *Chromosoma*, **29**, 305 – 316.

Maia, N., Venard, P. & Lavrut, F. (1976). Etude des divisions mitotiques et meiotiques du cycle de *Phytophthora capsici*. *Annales de Phytopathologie*, **8**, 141 – 146.

McClintock, B. (1945). *Neurospora*. I. Preliminary observations of the chromosomes of *Neurospora crassa*. *American Journal of Botany*, **32**, 671 – 678.

McDonough, E. S. (1937). The nuclear history of *Sclerospora graminicola*. *Mycologia*, **29**, 151 – 173.

McGinnis, R. C. (1956). Cytological studies of chromosomes of rust fungi. *Journal of Heredity*, **47**, 257 – 259.

Michelmore, R. W. & Sansome, E. R. (1982). Cytological studies of heterothallism and secondary homothallism in *Bremia lactucae*. *Transactions of the British Mycological Society*, **79**, 291 – 297.

Mortimer, A. M., Shaw, D. S. & Sansome, E. R. (1977). Genetical studies of secondary homothallism in *Phytophthora drechsleri*. *Archives of Microbiology*, **111**, 255 – 259.

Perkins, D. D. (1979). *Neurospora* as an object for cytogenetic research. In *Stadler Symposium*, vol. 2, pp. 145 – 164. Columbia, USA: University of Missouri.

Perkins, D. D. & Barry, E. G. (1977). The cytogenetics of *Neurospora*. *Advances in Genetics*, **19**, 133 – 285.

Perkins, D. D., Raju, N. B. & Barry, E. G. (1980). A chromosome rearrangement in *Neurospora* that produced viable progeny containing two nucleolus organisers. *Chromosoma*, **76**, 255 – 275.

Pukkila, P. J. & Lu, B. C. (1985). Silver staining of meiotic chromosomes in the fungus *Coprinus cinereus*. *Chromosoma*, **91**, 108 – 112.

Raju, N. B. (1978). Meiotic nuclear behaviour and ascospore formation in five homothallic species of *Neurospora*. *Canadian Journal of Botany*, **56**, 754 – 763.

Raju, N. B. (1980). Meiosis and ascospore genesis in *Neurospora*. *European Journal of Cell Biology*, **23**, 208 – 223.

Rhoades, M. M. (1961). Meiosis. In *The Cell*, vol. 3, ed. Brachet, J. & Mirsky, A. E., pp. 1 – 75. New York: Academic Press.

Sansome, E. R. (1959). Pachytene in *Puccinia kraussiana* Cooke, on *Smilax kraussiana*. *Nature*, **184**, 1820 – 1821.

Sansome, E. R. (1963). Meiosis in *Pythium debaryanum* Hesse and its significance in the life history of the Biflagellatae. *Transactions of the British Mycological Society*, **46**, 63 – 72.

Sansome, E. R. (1965). Meiosis in diploid and polyploid sex organs of *Phytophthora* and *Achlya*. *Cytologia*, **30**, 103 – 117.

Sansome, E. R. (1966). Meiosis in the sex organs of the Oomycetes. In *Chromosomes Today*, vol. I, ed. Darlington, C. D. & Lewis, K. R., pp. 77 – 83. Edinburgh: Oliver & Boyd.

Sansome, E. R. (1976). Gametangial meiosis in *Phytophthora capsici*. *Canadian Journal of Botany*, **54**, 1535 – 1545.

Sansome, E. R. (1977). Polyploidy and induced gametangial formation in British isolates of *Phytophthora infestans*. *Journal of General Microbiology*, **99**, 311 – 316.

Sansome, E. R. (1980). Reciprocal translocation heterozygosity in heterothallic species of *Phytophthora* and its significance. *Transactions of the British Mycological Society*, **74**, 175 – 185.

Sansome, E. R. (1985). Cytological studies on *Phytophthora nicotianiae* var. *parasitica* in relation to mating type. *Transactions of the British Mycological Society*, **84**, 87 − 93.

Sansome, E. R. & Brasier, C. M. (1973). Diploidy and chromosomal structural hybridity in *Phytophthora infestans*. *Nature*, **241**, 344 − 345.

Sansome, E. R. & Brasier, C. M. (1974). Polyploidy associated with varietal differentiation in the *megasperma* complex of *Phytophthora*. *Transactions of the British Mycological Society*, **63**, 461 − 467.

Sansome, E. R., Brasier, C. M. & Griffin, M. J. (1975). Chromosome size differences in *Phytophthora palmivora*, a pathogen of cocoa. *Nature*, **255**, 704 − 705.

Sansome, E. R., Brasier, C. M. & Sansome, F. W. (1979). Further cytological studies on the 'L' and 'S' types of *Phytophthora* from cocoa. *Transactions of the British Mycological Society*, **73**, 293 − 302.

Sansome, E. R. & Dixon, P. A. (1965). Cytological studies of the myxomycete *Ceratiomyxa fruticulosa*. *Arkiv für Mikrobiologie*, **52**, 1 − 9.

Sansome, E. R. & Sansome, F. W. (1974). Cytology and life history of *Peronospora parasitica* on *Capsella bursa-pastoris* and of *Albugo candida* on *C. bursa-pastoris* and on *Lunaria annua*. *Transactions of the British Mycological Society*, **62**, 323 − 332.

Shaw, D. S. (1983). The Peronosporales − a fungal geneticists' nightmare. In *Zoosporic Plant Pathogens: A Modern Perspective*, ed. Buczacki, S. T., pp. 85 − 121. London: Academic Press.

Sherriff, C. & Lucas, J. A. (1986). Meiosis in heterothallic and homothallic isolates of *Peronospora parasitica*. *Bulletin of the British Mycological Society*, **20** (supplement 1), 12 (abstract).

Singleton, J. R. (1953). Chromosome morphology and the chromosome cycle in the ascus of *Neurospora crassa*. *American Journal of Botany*, **40**, 124 − 144.

Tin, W. & Dick, M. W. (1975). Cytology of Oomycetes. Evidence for meiosis and multiple chromosome associations in Saprolegniaceae and Pythiaceae, with an introduction to the cytotaxonomy of *Achlya* and *Pythium*. *Arkiv für Mikrobiologie*, **105**, 283 − 293.

Zickler, D. (1981). Ultrastructure of the yeast nucleus. In *The Fungal Nucleus*, British Mycological Society Symposium volume 5, ed. Gull, K. & Oliver, S. G., pp. 63 − 83. Cambridge University Press.

8

Regulation of mycelial organisation and responses

A. D. M. RAYNER AND D. COATES

School of Biological Sciences, University of Bath, Claverton Down, Bath BA2 7AY UK

Introduction

There are two reasons why an understanding of patterns of mycelial development is fundamental to the evolutionary biology of fungi. First, for the majority of fungi, interactions with both biotic and abiotic environmental factors are mediated through mycelial thalli. The evolution of fungal life styles, incorporating adaptation to routine selection pressures (Brasier, 1986a; Chapter 16), must therefore be closely attuned with, if not conditioned by, the properties of mycelia.

Second, as the basic building material of the fungus, the mycelium will contain within its developmental programming certain basic morphogenetic options allowing or restricting the potential for diversification under episodic selection pressures (Brasier, 1986a; Chapter 16). Identification of these options may allow characterisation of the capacity for evolutionary innovation by re-modelling of development, or, as Jacob (1983) has termed it, by molecular tinkering. Moreover, careful observation of, or indeed deliberate tinkering with, developmental processes is likely to yield evidence of evolutionary ancestry.

In this context a mycelium can be likened to an embryo whose development is indeterminate. Far from just being a sort of 'animated cotton wool' (Wood, 1985), the mycelium, particularly in fungi occupying complex niches, must be a subtle and changeable entity capable of response to varied and often contradictory environmental conditions. Correspondingly, it must fulfil diverse functions, often, because of its ability to bridge between them, in different places at the same time. The capacity to switch between functionally distinct states or 'modes' (Gregory, 1984) provides a means of doing this. In this chapter we outline evidence for the existence of such programmable modes, discuss their adaptive significance and relation to self – non-self recognition phenomena, and describe recent studies in which we have 'tinkered' directly with their expression.

Developmental modes and switches

During the development of a mycelium from a point source under initially homogeneous conditions, it is well known that changes in patterns of morphogenesis occur leading to establishment of a marginal growth zone of leader hyphae, behind which are produced morphologically and functionally distinctive mycelial regions. Additionally, when grown on homogeneous media but under different environmental conditions, the same mycelial genotype can adopt correspondingly distinctive patterns of development; even under the same conditions it may produce morphologically distinctive sectors or annulations. Mycelia therefore exhibit considerable developmental plasticity, and this is brought out further by observations of growth patterns in the heterogeneous environments which commonly obtain in nature, e.g. the alternation between diffuse and compacted growth of rhizomorphic and mycelial cord-forming fungi as they ramify between spatially separated food bases in soil (Rayner *et al.*, 1985; Dowson, Rayner & Boddy, 1986).

A major challenge is the determination of how much of this developmental plasticity is attributable to direct interactions between microenvironmental parameters and metabolic functioning, and how much to selection of different modes by pre- or post-transcriptional switch mechanisms. The elucidation of this problem will be difficult for several reasons. First, there has been neglect of many aspects of mycelial development other than in connection with reproduction, trophophase – idiophase transitions and mycelial-yeast dimorphism; second, changes in morphogenesis are often expressed in a continuum, related to the underuse of heterogeneous conditions in experimental work. Finally, overemphasis on the role of exogenous factors has been accompanied by a lack of discrimination between effects of morphogenetic factors as *cues* for rather than *determinants* of developmental pathways.

However, a variety of criteria may help to indicate when mode-switches are involved, including: (i) where response to gradation in an environmental parameter is discontinuous; (ii) where a particular morphogenetic pattern can be correlated with a specific gene or product; (iii) where such patterns are correlated with differences in ploidy or with heterokaryosis; (iv) where a particular pattern is sustained under non-inductive conditions; (v) where a mycelial dimorphism or polymorphism is consistently expressed by the same genotype under specific conditions; (vi) where two

two or more individually expressed morphogenetic patterns can be shown to be components whose co-expression results in another morphogenetic pattern; (vii) where a specific morphogenetic state has a distinct functional role in nature.

Based on such criteria, a working hypothesis may be proposed to the effect that much of the developmental programming of mycelia, and hence of structures elaborated from them, is indeed governed by an hierarchical series of switches. Each switch controls an option between at least two alternative developmental pathways or modes (analogous to the 'creodes' of Waddington, 1940). Phenotypic expression at any one point in space and time will then depend on the particular combination of modes which is selected. This hypothesis is thus in keeping with current theories about epigenetic mechanisms and homoeotic gene function in segmental eukaryotes (e.g. Hall, 1983; Gehring, 1985), but is distinguished by the relative organisational simplicity of the organisms (i.e. the fungi) to which it relates.

Examples of mycelial polymorphisms bearing on this hypothesis will now be described. It should be noted that although they may be affected by gene mutations, they should not be confused with the latter, albeit that their manifestation, for example by emergence of a sector, may often suggest such. Mode switches involve the occurrence of distinct alternate states, interconversion between which has pleiotropic effects and occurs at rates differing radically from those of individual point mutations.

Determinate — indeterminate transitions

Conversion from determinate (cellular) to indeterminate (filamentous) morphogenesis occurs at spore germination and in reverse at sporogenesis. A growing number of fungi are also known to possess the ability to switch between unicellular and mycelial growth forms. Although this has naturally aroused much interest, the accompanying neglect of other switches is epitomised by the frequent use of the term 'dimorphism' to cover only mycelial-yeast dimorphism (Stewart & Rogers, 1983). Here, only the adaptive significance of determinate — indeterminate transitions will be considered briefly, together with evidence for their regulatory control.

With respect to sporogenesis, attainment of competence, whereby a mycelium is only capable of response to a suitable stimulus at a critical phase in its development, has been demonstrated both with respect to conidiation and fruit body initiation, and is clearly a genetically regulated process (Axelrod, Gealt & Pastushok, 1973; Ross, 1982, 1985). With respect to spore germination, of particular interest is the occurrence of multiple modes of germination, the jelly fungi being a prime example. Thus in *Tilletiopsis washingtoniensis* ballistospores germinating on malt agar produce initial yeast-like growth, but those germinating close to an established colony give rise to germ tubes. By contrast, in *Hirneola auricula-judae* and *Auricularia mesenterica* basidiospores on malt agar develop germ tubes, but when stranded on the hymenium they germinate by repetition and, when in close contact with a mycelium on agar, produce conidia (Ingold, 1984). Equally instructive are the variable patterns of germination and morphogenesis, related to ecological requirements, in aquatic *Erynia* and *Entomophthora* species (Descals *et al.*, 1981; Descals & Webster, 1984; Webster, Chapter 13).

The ecological significance of yeast-mycelial dimorphism lies in the advantages of the yeast form in stress-tolerance (cf. Cooke & Whipps, Chapter 9) and in dispersion in mobile media; and of the mycelium in the colonisation of a fixed spatial domain. Thus several vertebrate, arthropod and plant pathogens initiate colonisation as yeasts, for example in the haemocoel or sapstream, reverting to mycelial development as the host becomes debilitated by the effects of pathogenesis or stress. Moreover, the fungi of ambrosia beetles and wood wasps are often carried in unicellular form by their associate prior to introduction into wood. More generally, possession of a unicellular stage may allow establishment in stressful habitats prior to rapid exploitation of an amelioration in conditions by mycelial outgrowth (Cooke & Rayner, 1984).

Whether mycelial-yeast dimorphism should be regarded as 'true' differentiation has been questioned by Turian (1983) because of its ease of reversibility by environmental modification. However, this view is not supported by evidence that cyclic AMP is involved in regulation of dimorphism in Mucorales (Orlowski, 1979, 1980) and the fact that unicellular and mycelial morphogenesis are partitioned between haploid homokaryotic and heterokaryotic states in several Heterobasidiomycetes (see below). Moreover, there is evidence of terminal differentiation in several cases including *Ophiostoma* (*Ceratocystis*) *ulmi* and *Aureobasidium pullulans*. Originally, in *O. ulmi*, it was reported that in a defined liquid medium proline induced yeast formation whereas NH_4^+, arginine or asparagine as nitrogen sources induced mycelial development. Incubation for a suitably prolonged period in one or other medium resulted in terminal differentiation of the corresponding morphogenetic pattern such that reversion on the alternative medium did not occur until stationary phase in yeasts or conidiation in mycelia (Kulkarni & Nickerson, 1981).

However, others have not been able to repeat this finding (A. G. Mitchell, personal communication) and more recently Muthukumar, Kulkarni & Nickerson (1985) have proposed that a Ca^{2+}-calmodulin interaction may underlie differentiation. In *A. pullulans*, mycelial morphogenesis was induced by yeast extract and continued thereafter even in the absence of this substance (Cooper & Gadd, 1984).

Alterations in internode length and branch angle: 'gear changes'

In terms of the channeling of energy resources into more or less polarised growth, mycelia have two basic adjustments open to them. First, internode length (often correlated with hyphal diameter) can be changed, so altering the balance between extension growth and branching. Second, branch-angle can be varied to provide different degrees of alignment of hyphae.

These adjustments are often evident in the maturation of a colony from a germinating spore, where there is a progression from initially divergent, closely spaced branching to increasingly outwardly directed growth. This progression sometimes occurs disjunctly, in a manner which may be likened to 'gear changes' (Fig. 8.1a). This has clear implications for colonisation processes: a high (fast) gear will facilitate exploration and coverage of domain, whilst a low gear will aid in initial establishment, resource exploitation, consolidation of territorial gains and perhaps also tolerance of environmental stress. It is therefore of interest that in a wide range of fungi what may be termed slow-dense/fast-effuse dimorphisms or polymorphisms occur. As detailed below, these are correlated with a variety of factors. Where spontaneous conversion from slow-dense to fast-effuse forms occurs, this has sometimes been referred to as point-growth (Rayner *et al.*, 1985; Fig. 8.1b), and there is evidence of differentiation since reversion rarely occurs directly.

In Basidiomycotina transitions between primary and secondary mycelia may quite commonly be associated with slow-dense/fast-effuse transitions, this perhaps being related to the usual functional role of the homokaryon in establishment, and of the heterokaryon in exploration and extension of domain. In some cases this results in striking dimorphisms, as in certain *Coprinus* species (Casselton, 1978) and in *Coniophora puteana* (Ainsworth, Chapter 19), but in others obvious dimorphisms may be either absent, or occur independently of life cycle stage, as in *Hypholoma fasciculare* (Rayner, unpublished).

In *Volvariella volvacea* and *V. bombycina*, the occurrence of a slow-

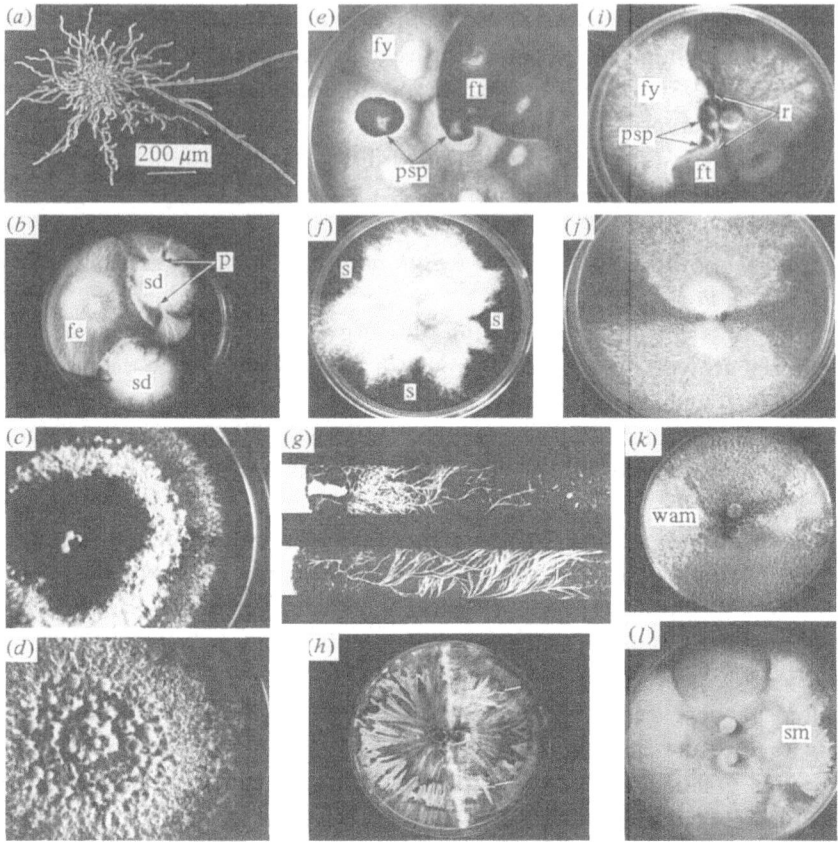

Fig. 8.1. (a) Development of a colony of *Coniophora puteana* from a single basidiospore showing progressive shifts in extension rate of marginal hyphae (after Kemper, 1937). (b) Outgrowth of mycelia from fruit body tissue of *Hypholoma fasciculare*, showing slow-dense (sd)/fast-effuse (fe) dimorphism and 'point-growth' (p) (courtesy of C. G. Dowson). (c) Appressed colony form of *Diplodina acerina*, with conidia produced on hyphae. (d) Sporodochial form of *Diplodina acerina*. (e) Subcultures from a single colony of *Hymenochaete corrugata* which have grown out in 'flat' (ft) and 'fluffy' (fy) non-aerial and aerial forms, the two colony types interacting to produce a pseudosclerotial plate (psp) (from Sharland, Burton & Rayner, 1986). (f) Single ascospore-derived colony of *Hypoxylon multiforme*, showing formation of 'senescent' sectors (s) (courtesy of P. R. Sharland). (g) Outgrowth from wood blocks of mycelial cord systems of *Steccherinum fimbriatum* into 3.2 cm diam glass tubes filled with non-sterile soil, showing fast-effuse (lower tube) and slow-dense reverting to fast-effuse (upper tube) morphogenesis (courtesy of C. G. Dowson). (h) Unilateral outgrowth of mycelial cords (arrowed) of *Hypoxylon serpens* from an interaction zone between two different single ascospore cultures (courtesy of P. R. Sharland). (i) Rejection reaction (r) between two different heterokaryons of *Hymenochaete corrugata* and

dense/fast-effuse dimorphism between single basidiospore cultures has elicited the proposal of a diploid – tetraploid secondary homothallic cycle in which diploid spores heterozygous for mating type yield fast-effuse colonies whereas homozygotes yield slow-dense colonies (Elliott & Challen, 1985). Heteroploidy unconnected with mating type has also been proposed to account for cultural polymorphism in *Phellinus noxius* (Bolland, Griffin & Heather, 1984) and haploidisation of diploid strains is associated with point-growth in *Schizophyllum commune* (Deng, Rubnitz & Leonard, 1985).

Since several plant pathogenic fungi exhibit, or can be induced to exhibit, slow-dense/fast-effuse dimorphism, there is the possibility that the form they are in can affect their pathogenicity. In *Rhizoctonia cerealis* and *R. solani* a switch to slow-dense growth is induced by the antibiotic validamycin A: hyphal density increases at the expense of extension but without overall change in biomass increment. On solid medium this switch occurs more rapidly with increased antibiotic concentration, but final extension rate is independent of the latter – an all-or-none effect. In the field, validamycin controls *R. solani* by preventing spread from lower to upper portions of the plant (Trinci, 1985).

In *Ophiostoma ulmi* strains of the EAN aggressive race produce either fibrous-striate and relatively rapidly extending 'wild-type', or more slowly extending uniform '*up-mut*' colonies (Brasier, 1986b). The latter often revert to the former by point-growth, and one form may change to the other following subculture. This slow-dense/fast-effuse dimorphism is thought to influence, via growth rate, the pathogenicity of EAN race isolates. A similar dimorphism occurs in the non-aggressive subspecies, but is absent from the more uniformly highly pathogenic NAN race (Brasier, 1986b). In *Diaporthe phaseolorum* f. sp. *caulivora*, which causes stem canker of soya bean, slow-extending ascospore cultures are highly pathogenic while fast-extending cultures have lesser and variable pathogenicity (M. M. Kulik, personal communication).

Slow-dense colony forms may generally exhibit more rapid commitment

associated changes in morphogenesis (cf. Fig. 8.1e) (from Sharland, Burton & Rayner, 1986). (j) 'Bow-tie' reaction between sib-related mating-type incompatible homokaryons of *Stereum hirsutum* (k) 'Bow-tie' reaction between single ascospore cultures of *Hypoxylon nummularium*, showing formation of white aerial mycelium (wam) (courtesy of P. R. Sharland). (l) 'Compatible bow-tie' reaction between two homokaryons of *Stereum insignitum*, showing formation of a secondary mycelium (sm) (courtesy of A. M. Ainsworth). All cultures shown grown on 2% malt agar in 9 cm Petri dishes.

to conidiogenesis than corresponding fast-effuse types. In Basidio-mycotina this not only often applies to homokaryon – heterokaryon transitions, but also applies to dimorphisms within individual homo-karyons or heterokaryons, as in *Hypholoma fasciculare*. In the ascomycete *Hypoxylon serpens* a major dimorphism occurs between slowly extending conidiogenous mycelia and rapidly extending non-conidiogenous mycelia (Sharland & Rayner, unpublished), but this is also associated with an aerial-appressed dimorphism (see below).

A good example of the influence of environmental stress is provided by the promotion of point-growth in *Serpula lacrimans* by low water potential (Clarke, Jennings & Coggins, 1980; Jennings, 1982). In *Coprinus disseminatus*, formation of clamped marginal hyphae only occurs when extension rate is above a threshold imposed by environmental constraints such as temperature and chemical inhibitors (Butler, 1984).

Aerial versus appressed or submerged growth

Production of aerial mycelium acts both as a drain on resources from hyphae in contact with the substratum, and as a means of freeing growing hyphae from constraints imposed by the physical boundaries of the substratum, staling products and poor aeration. Aerial and non-aerial growth may in different fungi be closely coupled, partially uncoupled (resulting in endogenously or exogenously regulated rhythmicity), or largely uncoupled, resulting in distinctive sequential growth phases, dimorphisms or polymorphisms.

Amongst Deuteromycotina, species of *Fusarium* are notorious for their phenotypic plasticity, and in, for example, *F. oxysporum*, five common morphological variants occur of which *sporodochial* (*sp*), *cottony* (*co*) and *ropy* (*ro*) have abundant aerial mycelium, whereas *slimy pionnotal* (*slp*) and *shorn* (*sh*) do not (see Burnett, 1984). Although often fairly stable, these forms can interconvert, e.g. from *sp* to *co* or *slp*, especially when aged inocula are used, and a genetic regulatory mechanism is clearly implicated. Somewhat parallel behaviour has been discovered in the Coelomycete *Diplodina acerina* (Rayner, unpublished), where two forms with reciprocal morphogenetic patterns were isolated from *Acer pseudo-platanus* bark (Fig. 8.1c & d). Subculturing of single conidia from the sporodochial and appressed forms eventually produced intermediate, concentrically zoned types. In the ascomycete, *Endothia parasitica*, a widespread variant, *flat*, apparently regulated by a single nuclear gene, contrasts with 'normal' types in producing dense, appressed, highly pigmented mycelium with conidia produced on hyphae rather than within pycnidia (Anagnostakis, 1984).

Amongst Basidiomycotina, striking dimorphisms occur in the wood-decaying Hymenochaetaceae. In *Hymenochaete corrugata* the two colony forms have similar extension rates, but the appressed form is yellow – brown pigmented whereas the aerial form is white. Interconversion between the two colony forms is common and where the two forms are juxtaposed, a brown pseudosclerotial plate develops (Fig. 8.1e). Both forms develop on the natural substratum, the appressed type being associated with more decayed regions (Sharland, Burton & Rayner, 1986) a feature which is of particular interest because only this type possesses the tyrosinase and laccase activities which are associated with ligninolytic activity and secondary metabolism (P. R. Sharland, personal communication; cf. Ander & Eriksson, 1978; Kirk & Fenn, 1982; Lewis, Chapter 11). A very similar dimorphism occurs in *Phellinus tremulae*, except that here the appressed, phenoloxidase-producing, pigmented form has a slower extension rate and can grow at higher temperatures than the peroxidase-producing aerial form (Niemela, 1977; Hiorth, 1965). A similar dimorphism also appears to occur in the ectotrophic root pathogen of rubber, *Rigidoporus lignosus* (Boisson, 1968). In this, as in several other root pathogens, colonisation of the internal wood is preceded by development of external mycelium which ramifies the bark. This may be correlated with the partitioning of laccase and peroxidase activity which has been shown within and in advance of the colonised wood (Geiger, Nandris & Goujon, 1976).

Compacted versus diffuse morphogenesis

At critical times during mycelial development in higher fungi the initially divergent growth of hyphae is superseded by convergent growth, perhaps mediated by positive autotropisms (Moore, 1984; Ainsworth & Rayner, 1986), resulting in hyphal fusion and aggregation. In this way plectenchymatous structures are generated whose exact form is dictated by how localised or generalised is the compaction process and by the mycelial form on which it operates. Localised proliferation and compaction of hyphal branches initiated by the slow-dense and aerial mycelial switches may lead to development either of sclerotia or fruit body primordia (cf. Moore, 1981). More generalised expression of the same mechanism probably results in the crust-like plates which when formed within substrata delimit what have been termed pseudosclerotia (Campbell, 1933; cf.

Fig. 8.1e & i). Compaction of outwardly extending, collaterally aligned hyphae results in the formation of linear vegetative organs such as mycelial cords and rhizomorphs (Rayner *et al.*, 1985). In *Steccherinum fimbriatum* such linear organs have been found to exhibit a slow-dense/fast-effuse switch of their own (Fig. 8.1f).

Evidence for genetic and endogenous regulation of mycelial compaction is reviewed elsewhere (e.g. Elliott, 1985; Rayner *et al.*, 1985). Here only the mound (*mnd*) allele in *Schizophyllum commune* will be mentioned. This allele causes monokaryons and heteroallelic dikaryons to produce hyperplastic coherent aerial growths. In the dikaryons this appears to be due to localised somatic recombination at the *mnd*+ locus following internuclear transfer of an *mnd*-containing chromosomal segment which either replaces or is added to the *mnd*+ nucleus (Gaber & Leonard, 1981; Deng *et al.*, 1985).

'Juvenility' and 'senescence'

In some fungi, early growth following spore germination is consistently of a particular, usually aerially well-developed, type commonly referred to as 'juvenile'. This phase eventually becomes curtailed or superseded by a switch to what have variously been termed 'senescent' or 'differentiated' states, often associated with pigmentation and suppression of radial and/or aerial growth. This switch thus has common features, or may be co-expressed with some of the aerial/non-aerial transitions mentioned previously.

In several Deuteromycotina and Ascomycotina there is evidence that the transition to senescent states (use of 'differentiated' will be abandoned in the present discussion) is effected by extrachromosomal elements. These are transmissible to juvenile forms via hyphal fusions, and are also subject to nuclear control since mutants at one or more loci can be obtained which maintain strains permanently in one phase or another. Examples include *Hypomyces ipomoeae* (Boissonnet-Menes, 1969), *Pestalozzia annulata* (Chevaugeon, 1974), *Nectria haematococca* (Daboussi-Bareyre, 1980) and *Podospora anserina* (Esser *et al.*, 1984). In *N. haematococca*, two types of differentiated state develop randomly from small brown areas in the submarginal zone of juvenile mycelia, each associated with a specific cytoplasmic determinant. 'Ring' differentiation occurs circumferentially, associated with transmission at twenty times the rate of hyphal extension, whereas 'sector' differentiation involves transmission at twice the rate of extension. Two nuclear loci, A and S, appear to control differentiation, the A locus existing in at least three allelic forms. Similar patterns occur in certain Xylariaceae (Fig. 8.1g), but have not yet been analysed. In *P. anserina*, senescence has been attributed to release of a mitochondrial plasmid, and can be permanently prevented by the combined action of the nuclear genes *incoloris* and *vivax*.

A general functional role of these transitions may concern the ability locally or generally to re-direct growth preparatory to adoption of a new morphogenetic mode; the activation by light of phenol oxidase activity prior to fruit body initiation in *Coprinus congregatus* being just one possible illustration (Ross, 1985). Moreover, melanisation itself is of major significance in the morphogenesis of many fungal structures, including fruit bodies, sclerotia, pseudosclerotia and rhizomorphs, as well as being an effective means of damage limitation.

Relationships between development and recognition responses

Interactions between different fungal thalli are fundamental determinants of natural population and community structure and dynamics (Rayner & Todd, 1979; Brasier, 1984; Rayner, 1986). These interactions are mediated by recognition responses which allow (i) acceptance or rejection of non-self respectively in mating and somatic incompatibility reactions, (ii) combat (defence or attack) between somatic adversaries, and (iii) establishment of parasitic associations. As will now be detailed, these responses appear in turn to involve modulation of morphogenetic switches, hence establishing a vital link between recognition, development and the evolution of fungal life styles.

Responses within species

Currently, three major classes of somatic reaction appear to be distinguishable between thalli of different genotypes: rejection, inhibition or appression extending beyond the immediate contact zone, and secondary mycelium formation. Any one or combination of all three of these reactions may occur in any particular interaction between strains (Rayner, 1986).

Rejection reactions These result in the formation of typically narrow demarcation zones between colonies, both in culture and in natural substrata. Commonly the demarcation zones contain sparse mycelium, are pigmented, and at the hyphal level vacuolation and coagulation of cytoplasm may be evident in compartments made contiguous by anasto-mosis. Although details of biochemical events are available in only a few cases, a tenable interpretation of rejection reactions is that they involve activation of senescence pathways associated with autophagic vacuolation and protease and phenol oxidase activity (cf. Lane, 1981; Aylmore & Todd, 1986; Rayner, 1986). They may result in a deadlock interaction, with neither mycelium able to invade the other's domain, or be breached by access of nuclei (see below) or lead to invasion by mycelium growing in a mode differing from that of the original strains, e.g. in the 'penetration effect' in *Ophiostoma ulmi* (Brasier, 1984), in invasive mycelial fronts produced by *Nectria coccinea* (I. C. Parrett, personal communication) and invasive mycelial cords of *Hypoxylon serpens* (Fig. 8.1h). Mode switches adjacent to the interaction zone may also result in initiation of sporophore, sclerotium or pseudosclerotial plate production (Fig. 8.1i), as often occurs following damage to the mycelium. In the penetration effect in *O. ulmi*, lysis is thought to be involved in the production of lines of synnemata by invading isolates (Brasier, 1984).

In Ascomycotina, persistent rejection reactions often occur directly between homokaryons, even when these are mating type compatible, but in Basidiomycotina rejection reactions are usually confined to interactions involving heterokaryons or homokaryons incapable of mating. This reciprocity has led to the proposition that mating temporarily or permanently overrides rejection of non-self following somatic interchange in these fungi (Rayner *et al.*, 1984; Coates, Rayner & Boddy, 1985). It may be significant here that on grounds of their lack of pigment compared with secondary mycelia, homokaryons of at least some fungi, including *Armillaria* species (Korhonen & Hintikka, 1974; Korhonen, Chapter 20) and *Coniophora puteana* (Ainsworth, Chapter 19), may have different competence with respect to activation of their phenol oxidase systems. This also recalls the possible role of such systems in basidiocarp initiation (Ross, 1985).

Inhibition or appression reactions These result either in the inhibition of marginal extension of one or both thalli, or in production of appressed mycelium beyond the interaction interface (Fig. 8.1i & j). Proliferation of hyphal branches, often forming characteristic knots and with abnormal morphology, typically occurs in the appressed/inhibited regions. The latter are often either sector-shaped, or extend around the circumference of participant thalli, respectively indicating slow or rapid migration of nuclei or cytoplasmic factors, and reminiscent of A and S differentiation in *Nectria haematococca*. An example of a sector-shaped reaction is the bow-tie reaction in the basidiomycete *Stereum hirsutum* (Coates &

Rayner, 1985a), where a symmetric or asymmetric, unilaterally or bilaterally migrating outwardly flared zone of appressed mycelium develops, within which nuclear migration occurs and from which slow-dense point subcultures can be made. Migration is evidently via anastomoses and predominantly proximal, i.e. towards tips of recipient hyphae, a route which has also been proposed for cytoplasmic transmission of the dsRNA associated disease or d-factor in *Ophiostoma ulmi* (Brasier, 1983, 1986c). Cessation of migration is usually associated with development of a pigmented rejection zone, either within or at one margin of the bow-tie region, followed by regeneration of aerial mycelium which is often associated with partial replacement of one genotype by the other.

Essentially similar reactions to the *S. hirsutum* bow-tie occur in certain xylariaceous Ascomycotina (Sharland & Rayner, 1986). These involve development of non-conidiogenous interaction regions which are often brightly coloured and delimited by brown zones from below, whilst supporting white aerial mycelial growth containing straight, acutely-branched hyphae from above (Fig. 8.1k). It may be significant that these reactions are expressed between single ascospore cultures from out-crossing populations, but occur constitutively in non-outcrossing strains of *Hypoxylon multiforme* (Fig. 8.1f).

Secondary mycelium formation This involves the emergence of a mycelium within which nuclei contributed from both original participants are stably associated. It is fundamental in the outcrossing cycles of Basidiomycotina, where it follows diverse patterns, and often involves one or more major changes of mode.

In certain Ustilaginales and Tremellales, hormonal or pheromonal substances have been implicated both in conjugation between yeast-like cells, which represent the primary thallus, and emergence of a secondary mycelial phase. Thus, in *Tremella mesenterica*, tremerogen a-13 and tremerogen A-10 are constitutively produced and induce conjugation tube formation by the respective mating types (Yoshida *et al.*, 1981). In *Rhodosporidium toruloides* a constitutive hormone, rhodotorucine A, induces production of the reciprocal rhodotorucine a in cells of mating type 'a' (Abe, Kusada & Fukui, 1975). In *Ustilago violacea*, α-tocopherol (vitamin E) stimulates both conjugation and mycelium formation between complementary mating types (Castle & Day, 1984).

Amongst Homobasidiomycetes, secondary mycelium emergence appears to follow two distinctive pathways (Rayner, 1986). First, associated with rapid nuclear migration, emergence occurs around the perimeter, and often also through the body of mated homokaryons. This pattern has been attributed to 'acceptor migration'. Second, much more localised emergence preceded by appression or inhibition reactions and often curtailed by a rejection response, has been deemed to follow 'access migration' (Rayner *et al.*, 1984). Here the secondary phase often occupies a bow-tie shaped region between closely paired homokaryons, this effect being enhanced by a slow-dense/fast-effuse switch, inhibition of marginal extension of the homokaryons, or both (Butler, 1972; Rayner, 1986).

Secondary mycelium is not a prerequisite for sexual outcrossing in Ascomycotina. However, that the same processes may be present as in Basidiomycotina is indicated by the temporarily heterokaryotic nature of the white aerial mycelium in bow-tie reactions of the Xylariaceae (Fig. 8.1k), whose formation appears to be under a form of multiallelic control (Sharland & Rayner, 1986). Here the absence of stabilisation of the heterokaryon seems to be the only major difference from a typical basidiomycete mating reaction.

Responses between species

Many interspecific interactions seem also to involve changes in developmental regulation (see Rayner, 1986). These include long range inhibition and short range hyphal interference reactions, together with enhanced production of aerial mycelium and compaction of pseudosclerotial plates and linear organs. Some forms of parasitism may depend on tropic reactions similar to those which precede hyphal fusions, and on non-activation of rejection responses. An interesting case is provided by hyphae of *Mycotypha microspora* which direct the growth of the parasite *Piptocephalis fimbriata* toward them. However, germinating spores of *M. microspora* give rise, in the presence of the mycoparasite, to a persistent yeast phase in which the attractant factor is sequestered (Evans & Cooke, 1982), this thus being an effective defensive developmental response.

A 'tinkering' element and its effects in *Stereum hirsutum*

Our discovery of an apparently cytoplasmically mobile regulatory factor in the Aphyllophoraceous basidiomycete, *Stereum hirsutum* (Coates & Rayner, 1985b), has reinforced the concepts of mode switches and interactions between recognition and development. Furthermore, intervention of the factor seems to have exposed, in some strains, differentiation pathways which are not normally expressed, and which may have been inactive in the fungus for a long time. However, these putative atavisms still require rigorous substantiation. The molecular nature of the factor has not yet been established, but there is preliminary evidence that it may be a DNA moiety (J. R. Pryke, personal communication).

The properties and effects of the factor so far discovered are as follows. (1) It is transmissible, with or without accompanying nuclear migration, to both homokaryotic and heterokaryotic British strains by direct pairing. Recipient strains become 'transformed' and capable of 'transforming' other strains. (2) 'Transformed' strains become orange − pink and sporulate prematurely when exposed to daylight, producing basidia in an initially verticillate arrangement directly on the vegetative hyphae. (3) Basidiospore progeny from 'transformed' homokaryotic strains segregate into two mating types, indicative of a mating type switch. The mating type complementary to the original gave rejection responses against homo-karyons with which the original was mating-compatible. (4) Exposure of 'transformed' strains to near-UV light results in them acquiring a yellow-pigmented 'senescent' morphology, similar to that of colonies derived by subculture from somatic rejection zones. 'Untransformed' strains are unaffected. (5) Single-basidiospore progeny lines from 'transformed' strains exhibit a range of abnormalities. These include: (i) premature sporulation as in (2) above; (ii) slow-dense/fast-effuse dimorphism (Fig. 8.2a), in some lines 1:1 segregation occurs and succeeding generations alternate between the two colony types; (iii) slow-dense types can spon-taneously produce fast-effuse sectors (Fig. 8.2b) − following reversion some strains have developed mounds and rhizomorphic organs (Fig. 8.2c; Coates & Rayner, 1985c); (iv) diverse hymenial structures, some of which contain structural elements like acanthocystidia and chondrocystidia not normal in *S. hirsutum*, or even the genus *Stereum*; (v) crosses between unrelated lines with complementary mating alleles have in some cases resulted in somatic rejection instead of mating; (vi) several have sectored into two distinctive phases, a slow-dense white type composed of attenuate hyphae with verticillate conidiogenous cells or densely branched wide hyphae with clamp connections and a fast-effuse buff form composed mostly of wide diameter hyphae (Fig. 8.2d − f). The slow-dense forms appear to be specifically parasitic, either via lysis (Fig. 8.2g) or hyphal coiling, on the effuse forms, other strains of *S. hirsutum* being resistant to such effects associated with expression of a rejection response. One slow-dense line became highly 'self-parasitic', containing closely entwined hyphae (Fig. 8.2h) as well as some hyphae with abnormal and large numbers of hook cells (Fig. 8.2i). 'Untransformed' homokaryons with similar morphology to the self-parasitic line have occasionally been isolated, and preliminary evidence indicates a two-gene system controlling

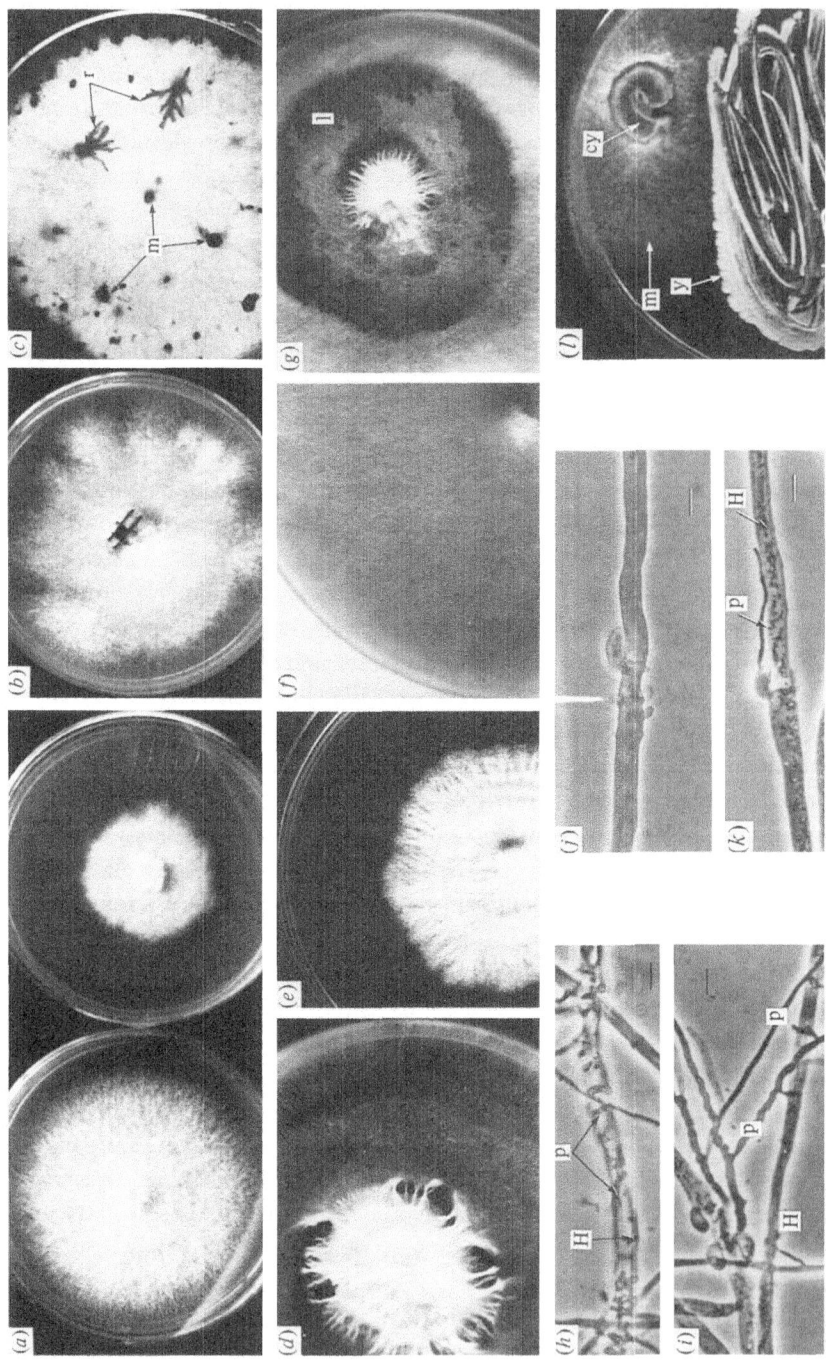

Fig. 8.2. Properties of the *Stereum hirsutum* 'transformed' strain and its progeny. (a) Colonies of the same age exhibiting slow-dense/fast-effuse dimorphism. (b) Production of fast-effuse sectors by a slow-dense colony. (c) Production of 'mounds' (m) and rhizomorphic organs (r), as seen in silhouette. (d) Sectoring of an effuse colony type from a slowly extending, dense white colony form. (e) Subculture of the white form from Fig. 8.2(d). (f) Subculture of the effuse form from Fig. 8.2(d). (g) Interaction between the white and effuse forms showing extensive lysis of the latter. (h) – (k) Photomicrographs of a highly 'self-parasitic' strain (H, 'host' hyphae; p, 'parasite' hyphae). (l) Yeasts (y), mycelium (m) and a yeast colony which has become crusty and yellow pigmented (cy) in response to the presence of mycelium. (a) and (b) from Coates & Rayner, 1985b; (c) from Coates & Rayner, 1985c.

this behaviour. The two colony forms look like components whose co-expression would give rise to the normal mycelial phenotype of *S. hirsutum*, a possibility with major implications but needing very thorough substantiation. (6) 'Transformed' strains and their basidiospore progeny can be induced to produce a characteristic multiply-budding yeast phase. The yeasts are stable in subculture, but can be reverted by contact with a 'transformed' mycelium, whence they turn yellow and produce mycelial sectors. However, recent attempts to repeat this observation have been unsuccessful (A. M. Ainsworth, C. G. Dowson, personal communication).

Conclusions and outlook

What some might regard as the treacherous variability of mycelial thalli is in fact the hallmark of their remarkable versatility, both with respect to short term adaptation and long term innovation. The mechanisms underlying their developmental plasticity are mostly still obscure but there are clear indications for a central role of wall morphogenesis, perhaps associated with modulation of ion fluxes (Harold, Kropf & Caldwell, 1985), of activation of phenol oxidase and associated systems, and for mobile controlling elements. Clearly there is both a great urgency and enormous scope for further investigation of mycelial morphogenesis and responses.

Dedication We would like to dedicate this chapter to the memory of our late friend and colleague, Dr Norman Todd.

References

Abe, K., Kusada, I. & Fukui, S. (1985). Morphological changes in the early stages of the mating process of *Rhodosporidium toruloides*. *Journal of Bacteriology*, **122**, 710 − 718.
Ainsworth, A. M. & Rayner, A. D. M. (1986). Responses of living hyphae associated with self and non-self fusions in the basidiomycete *Phanerochaete velutina*. *Journal of General Microbiology*, **132**, 191 − 201.
Anagnostakis, S. L. (1984). The mycelial biology of *Endothia parasitica*. I. Nuclear and cytoplasmic genes that determine morphology and virulence. In *The Ecology and Physiology of the Fungal Mycelium*, British Mycological Society Symposium volume 8, ed. Jennings, D. H. & Rayner, A. D. M., pp. 353 − 366. Cambridge University Press.
Ander, P. & Eriksson, K. -E. (1978). Lignin degradation and utilization by micro-organisms. *Progress in Industrial Microbiology*, **14**, 1 − 58.
Axelrod, D. E., Gealt, M. & Pastushok, M. (1973). Gene control of developmental competence in *Aspergillus nidulans*. *Developmental Biology*, **34**, 9 − 15.

Aylmore, R. C. & Todd, N. K. (1986). Cytology of non-self hyphal fusions and somatic incompatibility in *Phanerochaete velutina*. *Journal of General Microbiology*, **132**, 581 – 591.

Boisson, C. (1968). Mise en evidence de deux phases myceliennes successives au cours du développement du *Leptoporus lignosus* (Kl.) Heim. *Comptes Rendus de l'Académie des Sciences, Paris, Série D*, **266**, 112 – 115.

Boissonnet-Menet, A. (1969). Intervention du génome dans un phénomène extra-chromosomique chez l'Hypomyces ipomoeae Wr. *Comptes Rendus de l'Académie des Sciences, Paris, Série D*, **268**, 1593 – 1596.

Bolland, L., Griffin, D. M. & Heather, W. A. (1984). Polyploidy in *Phellinus noxius*. In *Proceedings of the 6th International Conference on Root and Butt Rots of Forest Trees*, ed. Kile, G. A., pp. 12 – 27. Melbourne, Australia: CSIRO.

Brasier, C. M. (1983). A cytoplasmically transmitted disease of *Ceratocystis ulmi*. *Nature*, **305**, 220.

Brasier, C. M. (1984). Inter-mycelial recognition systems in *Ceratocystis ulmi*: their physiological properties and ecological importance. In *The Ecology and Physiology of the Fungal Mycelium*, British Mycological Society Symposium volume 8, ed. Jennings, D. H. & Rayner, A. D. M., pp. 451 – 497. Cambridge University Press.

Brasier, C. M. (1986a). The population biology of Dutch elm disease: its principal features and some implications for other host-pathogen systems. *Advances in Plant Pathology*, **5**, 55 – 118.

Brasier, C. M. (1986b). Comparison of pathogenicity and cultural characteristics in the EAN and NAN aggressive sub-groups of *Ophiostoma ulmi*. *Transactions of the British Mycological Society*, **87**, 1 – 13.

Brasier, C. M. (1986c). The d-factor in *Ceratocystis ulmi* – its biological characteristics and implications for Dutch elm disease. In *Fungal Virology*, ed. Buck, K. W., pp. 177 – 208. Boca Raton, Florida: CRC Press.

Burnett, J. H. (1984). Aspects of *Fusarium* genetics. In *The Applied Mycology of Fusarium*, British Mycological Society Symposium volume 7, ed. Moss, M. O. & Smith, J. E., pp. 39 – 69. Cambridge University Press.

Butler, G. M. (1972). Nuclear and non-nuclear factors influencing clamp connection formation in *Coprinus disseminatus*. *Annals of Botany*, **36**, 263 – 279.

Butler, G. M. (1984). Colony ontogeny in the basidiomycetes. In *The Ecology and Physiology of the Fungal Mycelium*, British Mycological Society Symposium volume 8, ed. Jennings, D. H. & Rayner, A. D. M., pp. 53 – 71. Cambridge University Press.

Campbell, A. H. (1933). Zone lines in plant tissues. I. The black lines formed by *Xylaria polymorpha* (Pers.) Grev. in hardwoods. *Annals of Applied Biology*, **20**, 123 – 145.

Casselton, L. A. (1978). Dikaryon formation in higher basidiomycetes. In *The Filamentous Fungi, vol. 3, Developmental Mycology*, ed. Smith, J. E. & Berry, D. R., pp. 275 – 297. London: Edward Arnold.

Castle, A. J. & Day, A. W. (1984). Isolation and identification of α-tocopherol as an inducer of the parasitic phase of *Ustilago violacea*. *Phytopathology*, **74**, 1194 – 1200.

Chevaugeon, J. (1974). Stability of the differentiated state in *Pestalozzia annulata*. *Transactions of the British Mycological Society*, **63**, 371 – 379.

Clark, R. W., Jennings, D. H. & Coggins, C. R. (1980). Growth of *Serpula lacrimans* in relation to water potential of the substrate. *Transactions of the British Mycological Society*, **75**, 271 – 280.

Coates, D. & Rayner, A. D. M. (1985a). Genetic control and variation in expression of

the 'bow-tie' reaction between homokaryons of *Stereum hirsutum.*
Transactions of the British Mycological Society, **84**, 191 − 205.

Coates, D. & Rayner, A. D. M. (1985b). Evidence for a cytoplasmically transmissible factor affecting recognition and somato-sexual differentiation in the basidiomycete, *Stereum hirsutum. Journal of General Microbiology,* **131**, 207 − 219.

Coates, D. & Rayner, A. D. M. (1985c). Induction of rhizomorphic organs in a non-rhizomorphic fungus, *Stereum hirsutum,* via a mobile regulatory factor. *Transactions of the British Mycological Society,* **84**, 527 − 530.

Coates, D., Rayner, A. D. M. & Boddy, L. (1985). Interactions between mating and somatic incompatibility in the basidiomycete, *Stereum hirsutum. New Phytologist,* **99**, 473 − 483.

Cooke, R. C. & Rayner, A. D. M. (1984). *Ecology of Saprotrophic Fungi.* London: Longman.

Cooper, L. A. & Gadd, G. M. (1984). The induction of mycelial development in *Aureobasidium pullulans* (IMI 45533) by yeast extract. *Antonie van Leeuwenhoek,* **50**, 249 − 260.

Daboussi-Bareyre, M. J. (1980). Heterokaryosis in *Nectria haematococca. Journal of General Microbiology,* **116**, 425 − 433.

Deng, R. C., Rubnitz, J. E. & Leonard, T. J. (1985). Somatic recombination of the *mnd* chromosomal region in diploids and dikaryons of *Schizophyllum commune. Experimental Mycology,* **9**, 122 − 132.

Descals, E. & Webster, J. (1984). Branched aquatic conidia in *Erynia* and *Entomophthora sensu lato. Transactions of the British Mycological Society,* **83**, 669 − 682.

Descals, E., Webster, J., Ladle, M. & Bass, J. A. B. (1981). Variations in asexual reproduction in species of *Entomophthora* on aquatic insects. *Transactions of the British Mycological Society,* **77**, 85 − 102.

Dowson, C. G., Rayner, A. D. M. & Boddy, L. (1986). Outgrowth patterns of mycelial cord-forming basidiomycetes from and between woody resource units in soil. *Journal of General Microbiology,* **132**, 203 − 211.

Elliott, T. J. (1985). Developmental genetics − from spore to sporophore. In *Developmental Biology of Higher Fungi,* British Mycological Society Symposium volume 10, ed. Moore, D., Casselton, L. A., Wood, D. A. & Frankland, J. C., pp. 451 − 465. Cambridge University Press.

Elliott, T. J. & Challen, M. P. (1985). The breeding system of the silver-silk straw mushroom, *Volvariella bombycina. Mushroom Newsletter for the Tropics,* **6**, 3 − 8.

Esser, K., Kuck, U., Stahl, U. & Tudzynski, P. (1984). Senescence in *Podospora anserina* and its implications for genetic engineering. In *The Ecology and Physiology of the Fungal Mycelium,* British Mycological Society Symposium volume 8, ed. Jennings, D. H. & Rayner, A. D. M., pp. 343 − 352. Cambridge University Press.

Evans, G. H. & Cooke, R. C. (1982). Studies on mucoralean mycoparasites. III. Diffusible factors from *Mortierella vinacea* Dixon-Stewart that direct germ growth in *Piptocephalis fimbriata* Richardson and Leadbetter. *New Phytologist,* **91**, 245 − 253.

Gaber, R. F. & Leonard, T. J. (1981). Unilateral internuclear gene transfer and cell differentiation in *Schizophyllum. Nature,* **291**, 342 − 344.

Gehring, W. J. (1985). The homeo box: a key to the understanding of development? *Cell,* **40**, 3 − 5.

Geiger, J. P., Nandris, D. & Goujon, M. (1976). Activitié des laccases et des peroxidases au sein de vacines d'Hevea attaquees par le pourridie blanc (Leptoporus lignosus (K. L.) Heim). *Physiologie Végétale*, **14**, 271 — 282.

Gregory, P. H. (1984). The First Benefactors' Lecture. The fungal mycelium: an historical perspective. *Transactions of the British Mycological Society*, **82**, 1 — 11.

Hall, B. K. (1983). Epigenetic control in development and evolution. In *Development and Evolution*, ed. Goodwin, B. C., Holder, N. & Wylie, C. C., pp. 353 — 379. Cambridge University Press.

Harold, F. M., Kropf, D. L. & Caldwell, J. H. (1985). Why do fungi drive electric currents through themselves? *Experimental Mycology*, **9**, 183 — 186.

Hiorth, J. (1965). The phenoloxidase and peroxidase activities of two culture types of *Phellinus tremulae* (Bond.) Bond. & Boriss. *Meddelelser Norske Skoforsöksvesen*, **20**, 249 — 272.

Ingold, C. T. (1984). Patterns of ballistospore germination in *Tilletiopsis*, *Auricularia* and *Tremella*. *Transactions of the British Mycological Society*, **83**, 583 — 591.

Jacob, F. (1983). Molecular tinkering in evolution. *Evolution from Molecules to Men*, 131 — 144.

Jennings, D. H. (1982). The movement of *Serpula lacrimans* from substrate to substrate over nutritionally inert surfaces. *Decomposer Basidiomycetes: Their Biology and Ecology*, British Mycological Society Symposium volume 4, ed. Frankland, J. C., Hedger, J. N. & Swift, M. J., pp. 91 — 108. Cambridge University Press.

Kemper, W. (1937). Zur Morphologie und Cytologie der Gattung *Coniophora*, insbesondere des sogenannten Kellerschwammes. *Zentralblatt fuer Bakteriologie Parasitenkunde Infektionskrankheiten und Hygiene Abteilung II*, **97**, 100 — 124.

Kirk, T. K. & Fenn, P. (1982). Formation and action of the ligninolytic system in basidiomycetes. In *Decomposer Basidiomycetes: Their Biology and Ecology*, British Mycological Society Symposium volume 4, ed. Frankland, J. C., Hedger, J. N. & Swift, M. J., pp. 67 — 90. Cambridge University Press.

Korhonen, K. & Hintikka, V. (1974). Cytological evidence for somatic diploidization in dikaryotic cells of *Armillariella mellea*. *Archiv für Mikrobiologie*, **95**, 187 — 192.

Kulkarni, R. K. & Nickerson, K. W. (1981). Nutritional control of dimorphism in *Ceratocystis ulmi*. *Experimental Mycology*, **5**, 148 — 154.

Lane, E. B. (1981). Somatic incompatibility in fungi and myxomycetes. In *The Fungal Nucleus*, British Mycological Society Symposium volume 5, ed. Gull, K. & Oliver, S. G., pp. 239 — 258. Cambridge University Press.

Moore, D. (1981). Developmental genetics of *Coprinus cinereus*: genetic evidence that sclerotia and carpophores share a common pathway of initiation. *Current Genetics*, **3**, 145 — 150.

Moore, D. (1984). Positional control of development in fungi. In *Positional Controls in Plant Development*, ed. Barlow, P. W. & Carr, D. J., pp. 107 — 135. Cambridge University Press.

Muthukumar, G., Kulkarni, R. K. & Nickerson, K. W. (1985). Calmodulin levels in the yeast and mycelial phases of *Ceratocystis ulmi*. *Journal of Bacteriology*, **162**, 47 — 49.

Niemela, T. (1977). The effects of temperature on two culture types of *Phellinus tremulae* (Fungi, Hymenochaetaceae). *Annales Botanici Fennici*, **14**, 21 — 24.

Orlowski, M. (1979). Changing pattern of cyclic AMP-binding proteins during germination of *Mucor racemosus* sporangiospores. *Biochemical Journal*, **182**,

547 — 554.

Orlowski, M. (1980). Cyclic adenosine 3′, 5′ monophosphate and germination of sporangiospores from the fungus *Mucor*. *Archives of Microbiology*, **126**, 133 — 140.

Rayner, A. D. M. (1986). Mycelial interactions — genetic aspects. In *Natural Antimicrobial Systems*, ed. Rhodes-Roberts, M. E., Gould, G. W., Charnley, A. K., Cooper, R. M. & Board, R. G., pp. 277 — 296. Bath, UK: Bath University Press.

Rayner, A. D. M., Coates, D., Ainsworth, A. M., Adams, T. J. H., Williams, E. N. D. & Todd, N. K. (1984). The biological consequences of the individualistic mycelium. In *The Ecology and Physiology of the Fungal Mycelium*, British Mycological Society Symposium volume 8, ed. Jennings, D. H. & Rayner, A. D. M., pp. 509 — 540. Cambridge University Press.

Rayner, A. D. M., Powell, K. A., Thompson, W. & Jennings, D. H. (1985). Morphogenesis of vegetative organs. In *Developmental Biology of Higher Fungi*, British Mycological Society Symposium volume 10, ed. Moore, D., Casselton, L. A., Wood, D. A. & Frankland, J. C., pp. 249 — 279. Cambridge University Press.

Rayner, A. D. M. & Todd, N. K. (1979). Population and community structure and dynamics of fungi in decaying wood. *Advances in Botanical Research*, **7**, 333 — 420.

Ross, I. K. (1982). Location of carpophore initiation in *Coprinus congregatus*. *Journal of General Microbiology*, **128**, 2755 — 2762.

Ross, I. K. (1985). Determination of the initial steps in differentiation in *Coprinus congregatus*. In *Developmental Biology of Higher Fungi*, British Mycological Society Symposium volume 10, ed. Moore, D., Casselton, L. A., Wood, D. A. & Frankland, J. C., pp. 353 — 373. Cambridge University Press.

Sharland, P. R., Burton, J. L. & Rayner, A. D. M. (1986). Mycelial dimorphism, interactions and pseudosclerotial plate formation in *Hymenochaete corrugata*. *Transactions of the British Mycological Society*, **86**, 158 — 163.

Sharland, P. R. & Rayner, A. D. M. (1986). Mycelial interactions in *Daldinia concentrica*. *Transactions of the British Mycological Society*, **86**, 643 — 649.

Stewart, P. R. & Rogers, P. J. (1983). Fungal dimorphism. In *Fungal Differentiation*, ed. Smith, J. E., pp. 267 — 313. New York: Marcel Dekker.

Trinci, A. P. J. (1985). Effect of validamycin A and L-sorbose on the growth and morphology of *Rhizoctonia cerealis* and *Rhizoctonia solani*. *Experimental Mycology*, **9**, 20 — 27.

Turian, G. (1983). Concepts of fungal differentiation. In *Fungal Differentiation*, ed. Smith, J. E., pp. 1 — 18. New York: Marcel Dekker.

Waddington, C. H. (1940). *Organisers and Genes*. Cambridge University Press.

Wood, R. K. S. (1985). (Review) The Ecology and Physiology of the Fungal Mycelium (1984) edited by D. H. Jennings and A. D. M. Rayner, 514 pp. Cambridge University Press. *Journal of Experimental Botany*, **36**, 858.

Yoshida, M., Sakagami, Y., Isogai, A. & Suzuki, A. (1981). Isolation of tremerogen a-13, a peptidal sex hormone of *Tremella mesenterica*. *Agricultural and Biological Chemistry*, **45**, 1043 — 1044.

9

Saprotrophy, stress and symbiosis

R. C. COOKE AND J. M. WHIPPS*

*Botany Department, University of Sheffield, Sheffield S10 2TN, and *Plant Pathology and Microbiology Department, GCRI, Littlehampton, West Sussex BN17 6LP, UK*

Introduction

It would seem to be widely considered, on the basis of relative specialism, that saprotrophy is a primitive condition from which the more exalted states of necrotrophy and biotrophy have arisen. Yet, a vast range of fungi of disparate origins and with a great variety of life-styles are either obligately saprotrophic or are able to adopt saprotrophy as a regular alternative to necrotrophy or biotrophy. This indicates that the possession or acquisition of saprotrophy has, at various times, conferred selective advantages and that evolutionary relationships between saprotrophy and other nutritional modes may be more subtle and complex than generally supposed.

Here, saprotrophy and its relation to other modes will be re-evaluated within the conceptual framework of life strategies and against the background of symbiosis. This allows behavioural traits to be examined free from phylogenetic speculations and also focuses some attention on those commonly occurring circumstances in which fungi exhibit nutritional shifts during association with their hosts. In an earlier publication on the evolution of nutritional modes, the authors of this chapter argued strongly that speculations in this field should not ignore phylogeny (Cooke & Whipps, 1980). However, the recent emergence of strategy theory as a strongly unifying functional theme in the eco-physiology of both autotrophic and heterotrophic organisms relegates phylogenetic detail to a secondary issue.

Concepts of nutritional modes and life strategies

Three nutritional modes are recognised according to the manner in which fungi are able to utilise external resources. These are sapro-

Table 9.1. *Eco-nutritional groups of fungi based on nutritional modes and possible ecological behaviour (after Lewis, 1973, 1974)*

Obligate saprotrophs	no capacity for necrotrophy or biotrophy
Facultative necrotrophs	either normally saprotrophic with an ability to become necrotrophic or with an equal ability for both saprotrophy and necrotrophy
Obligate necrotrophs	normally necrotrophic, having severely limited saprotrophic ability restricted to their survival in dead host tissue
Facultative biotrophs	normally biotrophic but with some, albeit limited, saprotrophic ability
Obligate biotrophs	no saprotrophic ability

trophy, in which non-living organic materials, other than those killed by the fungus itself, are utilised; necrotrophy, in which tissues are first killed then utilised saprotrophically; and biotrophy, in which only living cells are exploited, their death curtailing biotrophic activity. This division has been extended to five eco-nutritional categories of behaviour on the basis of whether, in nature, a fungus is obligately dependent on a particular mode for completion of its life cycle, or has the capacity to adopt another mode (Table 9.1). Given that obligate saprotrophs can be both facultative and obligate symbionts, then this scheme becomes a behavioural classification of symbiotic fungi. However, it is by intent hierarchical and equates specific nutritional modes with particular states of behavioural advancement: saprotrophs are primitive, biotrophs are advanced, and necrotrophs occupy an intermediate position (Lewis, 1974; Cooke, 1977). Additional bias is evident in the failure to consider using an alternative term 'facultative saprotrophy' for fungi with an equal ability for saprotrophy and necrotrophy. Possibly this reflects the long-standing pre-occupation of plant pathologists with categorising pathogens on the basis of their degree of nutritional specialisation. Behaviour within the host has been given pre-eminence, the remainder of the life-cycle, no matter how ecologically important, being regarded as having only a subsidiary interest. Such an approach not only lacks balance but also fails to appreciate that nutritional modes should be used to classify behaviour in an ecologically complete context and not fungi as entities within a narrow one.

A less direct route to the question of inter-relationships between nutritional modes, but perhaps a more objective one, can be made through a consideration of strategies. With respect to fungi, a strategy may be loosely defined as a grouping of similar or analogous physiological characteristics occurring between species or within communities which cause them to exhibit ecological similarities. The importance of strategy theory in the search for a unified classification of organisms into functional types is widely accepted in plant and animal ecology but has only recently been applied to fungi (Pugh, 1980; Andrews & Rouse, 1982; Cooke & Rayner, 1984).

The fundamental premise of strategy theory is the occurrence, as a result of natural selection, of a spectrum of strategies for survival at the opposite poles of which are two types of organisms (Harper & Ogden, 1970; Pianka, 1970; Gadgil & Solbrig, 1972). The first, r-selected, have a short life expectancy and rapidly commit much or all of their available resources to reproduction. The second, K-selected, have a long life expectancy and either devote only a small proportion of available resources to reproduction at any one time or only commit themselves to reproduction at the end of their life span. Within the r-K continuum three distinct forms of selection have operated to give three primary strategies (Grime, 1977, 1979). These also form a continuum of overlapping domains. Selection forces and the distribution of species within strategy domains can be represented in a simple model (Fig. 9.1). Thus, C-selection has resulted in

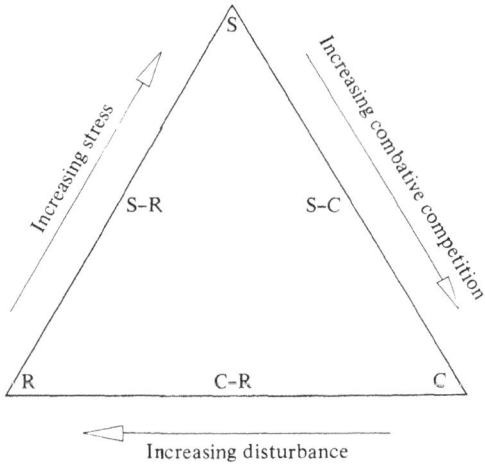

Fig. 9.1. Model of primary and secondary strategies in relation to selective forces. Modified from Grime (1977) and rotated to locate S-selection at the apex. Primary strategies: C, combative; S, stress-tolerant; R, ruderal. Secondary strategies: $C-R$, combative ruderal; $S-R$, stress-tolerant ruderal; $S-C$, stress-tolerant combative.

a combative or competitive strategy which maximises the ability to occupy and exploit resources in conditions of low stress and low disturbance. A ruderal strategy, emerging via *R*-selection, is characterised by a short life span associated with high reproductive capacity and determines success in highly disturbed, but nutrient-rich conditions. Finally, *S*-selection has culminated in a stress-tolerant strategy in which there has been adaptation to conditions of continuous environmental stress. Where primary strategies merge secondary strategies may arise that combine their features (Fig. 9.1).

Another theoretical model which, whilst being compatible with that just described, stems from a rather different approach, is the habitat templet. This has been used as an aid with which to study ecological strategies in animals and consists of a rectangular figure defined by quantifiable attributes of habitat (Southwood, 1977). The proportions of the templet will depend on the scaling of the axes and on the relationships between the chosen variables. A simple version of it with some relevance to fungi consists of a square (Fig. 9.2) with axes representing the favourableness or predictability of habitat conditions, and a diagonal vector indicating the complexity of community properties and processes, and hence biotic

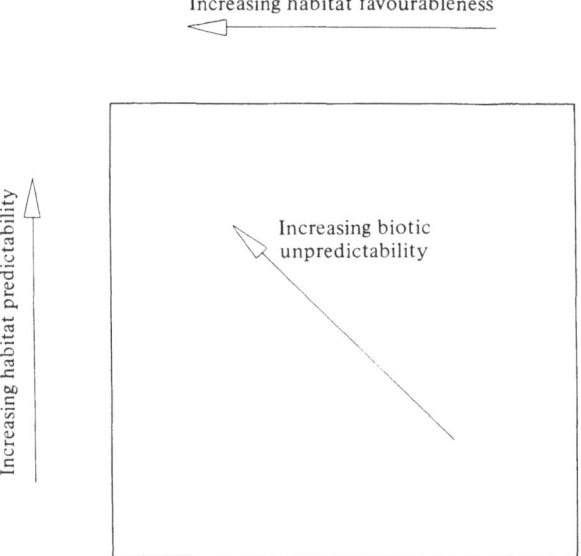

Fig. 9.2. The habitat templet showing habitat conditions and biotic properties operating (Greenslade, 1983). Reproduced from *American Naturalist* by permission of The University of Chicago Press.

unpredictability (Greenslade, 1983). Certain areas of the templet are dominated by particular selection processes (Fig. 9.3). Habitats with low predictability, that is prone to disturbance, will favour *r*-selected species, whilst *K*-selected species will predominate in habitats where predictability and favourableness and biotic unpredictability are all high. Predictably unfavourable habitats, that is those characterised by continuous stress, encourage adversity selection, *A*-selection, which results in the conservation of adaptation to consistently adverse environments (Whittaker, 1975; Southwood, 1977). Although somewhat differently defined, there would seem to be agreement that *S*-selection and *A*-selection are virtually synonymous and it is now pertinent to examine stress and *S*- or *A*-selection with respect to its relevance to saprotrophic symbiosis.

Stress and symbiosis

As it relates to fungi, stress has been defined as any form of continuously imposed environmental extreme which tends to restrict biomass production by most organisms in question (Cooke & Rayner, 1984). Stress factors, therefore, confer an environmental stability and predictability which, whilst allowing certain populations to become closely adapted to them, by their very severity also reduce the fitness of less well-adapted populations that may include potential combative competitors. Thus, severe environments have 'protective' properties (Pugh, 1980). Not only will there be low species diversity in stressful habitats but also, in many instances, a low density of *S*-selected individuals further reduces competition between inhabitants. The latter will correspondingly have a highly specialised life style, will frequently − but by no means always − exhibit slow or sparse vegetative development and will

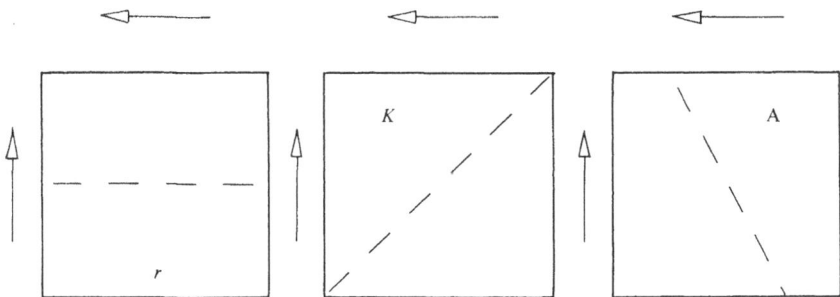

Fig. 9.3. Location of dominant selection processes on the habitat templet (Greenslade, 1983). For axes and explanation, see text. Reproduced from *American Naturalist* by permission of The University of Chicago Press.

occupy very narrow and precisely-defined realised niches which can often be predicted from the nature of the stress factors operating.

Depending on habitat, and within a single habitat at different times, stress may be imposed by either a single factor or several factors and may or may not have a nutrient basis. Regarding the latter, organic nutrients may be absent, or insufficient, or may be present as refractory substrates. Where available organic nutrients are present in a good supply their utilisation may be retarded or prevented by adverse temperatures, extremes of pH, low water potential and other physical or chemical factors. Despite such rigours, a number of highly successful symbionts of plants and animals are obligately saprotrophic *S*-strategists. Symbiosis is here regarded as being any reasonably intimate association with another organism, irrespective of whether the relationship is mutually or unilaterally beneficial. Some symbionts are obligately dependent either directly on their hosts or, as in some groups associated with animals, on the activities of their partners. Saprotrophy confines them either to non-living components of host tissues or to dead organic materials made available to them by associates with which they share a habitat (Table 9.2).

It seems evident that the vast majority of symbiotic obligate saprotrophs that dwell attached to or within their hosts, with the possible exception of trichomycetous fungi of arthropod guts, are *S*-strategists for which their hosts act as extreme environments. Such a state is in accord with a notion that occupants of any habitat that demands a highly specialised way of life, contains few other inhabitants and is stable, will probably be *A*-selected (Greenslade, 1983). It follows that symbiotic fungi at large, irrespective of their nutritional mode, may be *A*-selected and their hosts viewed as providing extreme and relatively stable environments for them. This suggestion merits examination because it further implies that symbiotic saprotrophs, biotrophs — both mutualistic and antagonistic, and necrotrophs are all specialists that face problems when living in association with, and deriving nutrients from, host cells and tissues. The manner in which each overcomes these problems may then give some fresh indications of the evolutionary relationships between them.

Plant cells and tissues as extreme environments

Fungal symbioses with autotrophs vastly outweigh those with heterotrophs, both numerically and probably also ecologically, and have certainly been studied in far greater depth and detail. As a result, views on most important aspects of the evolution of fungal symbioses are based firmly on the nature of associations with plants or cyanobacteria. Para-

Table 9.2. *Some ecological groupings of symbiotic saprotrophic fungi in relation to their habitats and major stress factors*

Host or associated organism	Ecological group	Stress factors
Endothermic vertebrates	dermatophytes	refractory substrates fungistatic secretions
	epidermal yeasts	desiccation fungistatic secretions
	mucosal yeasts	temperature above 35°C low pO_2 digestive enzymes
	rumen and caecum flagellates	temperature of 39−40.5°C anaerobic conditions digestive enzymes
Invertebrates	exoskeleton inhabitants	refractory substrates desiccation fungistatic secretions
	mycetangial fungi food fungi of ants and termites	fungistatic secretions elevated temperatures high pCO_2 fungistatic and fungitoxic secretions
Herbaceous and woody plants	yeasts of sugar-rich exudates and fluxes early phylloplane colonisers	low water potential osmotic stress low nutrient levels high thermal radiation and insolation desiccation
	latent invaders of bark and wood, heart-rot fungi	low nutrient availability fungistatic or fungitoxic compounds

doxically, the possibility that the host cells might be looked upon as an extreme environment, which has arisen from studies on animal parasitology, has tended to be overlooked by mycologists. The contention is that biotic stresses imposed by a cell on potential inhabitants resemble abiotic stress factors in that they limit diversity and restrict inhabitants to those possessing specialised fitness traits (Moulder, 1974). It might be added that the environment so provided is also relatively stable and therefore predictable. If the habitat afforded by a living cell is thus projected onto the habitat templet (Figs. 9.2, 9.3) it is clear that it has similar properties to other habitats which favour A-selection. Expanding on this theme, it has then been suggested that biotic, diversity-limiting factors operate successively and differentially at various times during attempted colonisation of an animal cell by a potential microbial inhabitant. As a result, a host cell will have at least three different roles *vis à vis* colonisation (Moulder, 1979). As a fortress it will bar entry to most would-be inhabitants; as a killer it will destroy many that do manage to invade; as a competitor of any successful invader it will deny them free access to its metabolites. Much the same concepts can be applied to plant cells and tissues. Although the scale and intensity of effects may differ between them, plant cell and tissue characteristics, including some dynamic defence or repair reactions, that discourage or prevent colonisation can be considered, at a gross level, as imposing stress. Symbiosis, whether antagonistic, neutral, or mutualistic, then comes about through successful avoidance or confrontation of stress.

Stress barriers and nutritional shifts

To summarise, it must be emphasised that a symbiont's realised niche and physical location are determined by interplay between the stresses it experiences and its nutritional mode. The gradation of the conditions obtaining from the external to the internal environment of a cell or tissue can be visualised as constituting successive strata of single stress factors, or group factors, within which symbionts come to be accommodated at various levels; and hence with different degrees of physical and physiological intimacy with their hosts. Taking the broad view, obligate saprotrophs will occupy an extracellular position and will commonly experience conditions of low nutrient availability. Diffusates emanating from the host will be metabolised and use may be made of some of its refractory structural materials. However, whilst such activities may increase nutrient flux from the host, access to the source is prevented by an inability to breach its defences; the fortress state is largely, but perhaps not totally, maintained. Necrotrophs have the capability for invasion but the rate of their subsequent progress through captured tissues may be modified by the imposition of stress via the presence of either pre-formed or symbiont-activated chemical defences; the host's killing ability may be potentiated. Biotrophs may obviate these and other problems consequent upon their intimate association with host cells but will still experience stress arising from the ability of the host to compete for available internal resources; even in ostensibly well-balanced biotrophy a competitive state must exist.

All this tends to re-inforce the proposal, implicit in the classification shown in Table 9.1, that biotrophy has been derived from saprotrophy through an intermediate state of necrotrophy. Saprotrophs first acquired competence to exploit host tissues, so reaching a state of facultative necro-trophy. Some hosts then developed mechanisms to repress wall-degrading enzyme synthesis by necrotrophs which then came to occupy relatively undamaged host tissues. Within these biotrophy was then achieved by the evolution of mechanisms for hormonal control of host metabolism and the loss of catastrophically pathogenic capability (Lewis, 1973, 1974).

Interfaces between nutritional modes

Many symbiotic fungi are not fixed in a single nutritional mode during their symbiotic phase but show some degree of flexibility through-out their life-cycle. Regular nutritional shifts can occur, most notably amongst hemibiotrophs (Luttrell, 1974). These, predominantly asco-mycetous, leaf inhabitants have been described as being initially biotro-phic, often for a considerable period, this phase then giving way to one of necrotrophy which may be followed by one in which the fungus is entirely saprotrophic on dead tissues. This nutritional sequence has been used as a basis to argue for the evolution of saprotrophy from biotrophy (Cooke & Whipps, 1980). Whether nutritional evolution has occurred in this or the opposite direction, and both may have taken place at different times and in varying situations, necrotrophy is still seen as occupying an intermediate position. Contingent on this, saprotrophy must be considered as being considerably distant from necrotrophy and while, in evolutionary terms, interfaces can be identified between necrotrophy and saprotrophy, and between necrotrophy and biotrophy, no such gradation seems possible between saprotrophy and biotrophy. However, applying the rule of parsimony to the evolution of nutritional modes, the acquisition by necro-trophs of major physiological characteristics which are then subsequently lost would seem to be a highly wasteful process. In the light of this it is possible that, in at least some instances, a state of biotrophy may have arisen directly from saprotrophy and that evidence for such a transition may be found.

The currently accepted definition of biotrophy is of necessity broad since biotrophs have a wide range of life-styles and morphology. They derive their organic nutrients only from living host cells and should the host or its occupied cells die then the biotroph also dies or is forced into a period of inactivity in the form of resting mycelium or propagules until a new association can be initiated (Lewis, 1973). An additional important characteristic is that most have the capability for intracellular penetration with minimal damage to the host cell. Over-emphasis on this feature, which is by no means universal amongst biotrophs, has to some extent obscured other aspects of their behaviour that have a direct bearing on possible relationships between biotrophy and saprotrophy.

Saprotrophy is an essential preliminary to the establishment of many biotrophic symbioses and there is abundant evidence that diffusible host nutrients are utilised, and their release even stimulated, during the pre-biotrophic phase (Cooke & Rayner, 1984). Furthermore, some outstandingly successful biotrophs are permanently extracellular, so that acquisition of host nutrients, coupled with non-fatal modifications of host metabolism, can be achieved without the necessity for the invasion of host cells. The supposedly sharp disjunction between saprotophy and biotrophy thus becomes blurred; even more so if the biotrophic phase of hemibiotrophy is considered. During this period, while infection is cryptic or latent, there may be significant erosion of host wall material which releases soluble nutrients to the hemibiotroph. It might be argued that such apparently controlled utilisation of host components is biotrophy but, since they are non-living, might equally be considered as being saprotrophy.

Difficulty in assigning a distinct nutritional mode to physiological states of this kind is also encountered with yeasts that can maintain occupancy of animal cells and the symptomless endophytes of grasses and other plants (see Cooke & Rayner, 1984; Carroll, 1986; Clay, 1986). Although few physiological investigations have so far been made, their cryptic habit and host-dependence strongly suggests that many of them occupy a borderline area between saprotophy and biotrophy. The closeness with which specialised, saprotrophic symbiosis resembles extracellular biotrophy can be expressed diagrammatically (Fig. 9.4a). Both involve a high degree of adaptation to the environment so provided, exploitation of the host's macromolecular synthetic systems and infliction of minimal damage.

Whilst in both cases the fortress state of the host cell is largely maintained it is compromised to various degrees depending on the ability of the fungus to compete with it for endogenous metabolites and, in some instances, also exogenous nutrients. The step from extracellular saprotrophy to intracellular biotrophy then would seem to be a relatively small one in that few additional stress barriers require to be crossed (Fig. 9.4b). Certainly, it seems unnecessary to postulate an intermediate necrotrophic phase; in fact, such an interpolation would be a retrograde step.

Conclusion

Undoubtedly, the recognition of eco-nutritional behavioural categories based on the three nutritional modes has given great impetus to studies on many aspects of symbiosis (Lewis, 1973, 1974). Unfortunately, the rigidly hierarchical structure of the original classification has also to some extent fettered further speculation of the ways in which such eco-nutritional states may have been achieved. However, it is not the purpose here, nor would it be helpful, to immediately re-define the limits either of nutritional modes or the ecological groupings arising from them. But it does seem timely to make three important points.

First, it is still not widely enough appreciated that individual fungi are not unchanging entities. There is rapidly increasing evidence for huge developmental and physiological shifts during the life-span of many symbiotic fungi which go far beyond those which can be accommodated in present eco-nutritional schemes. These movements frequently directly relate to changes in the quality and intensity of stresses experienced as the host − fungus relationship develops and are not easily explained on the basis of an hierarchical sequence of saprotrophy − necrotrophy − biotrophy. Second, the notion of increasing specialisation embodied within this ranking should be questioned. Each mode has its specialists; and the ability of, and indeed the necessity for, many fungi to shift nutritional ground with changing circumstances argues strongly for an equation of the modes. Finally, and contingent upon this, interfaces between all three modes − rather than between only necrotrophy and the other two − become theoretically possible. Critical examination of these

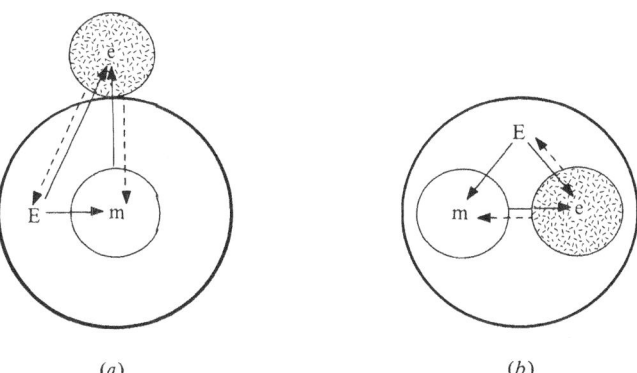

(a)   (b)

Fig. 9.4. Positional and functional relationships between a symbiotic fungus (stippled) and environmental components of a host cell (open circles). Based on a diagram in Moulder (1979). *(a)*, symbiotic saprotrophy and extracellular biotrophy; *(b)*, intracellular biotrophy; E, host cell's external and internal environment plus its metabolic pools. m = macromolecular synthetic systems of the host cell, e = all energy and other requirements of the fungus.

interfaces, in both theory and practice, may lead to a beneficial enlargement of current views on the causes and consequences of nutritional evolution amongst the fungi.

References

Andrews, J. H. & Rouse, D. I. (1982). Plant pathogens and the theory of r- and K-selection. *American Naturalist*, **120**, 283 – 296.

Carroll, G. C. (1986). The biology of endophytism in plants with particular reference to woody perennials. In *Microbiology of the Phyllosphere*, ed. Fokkema, N. & van den Heuvel, J., in press. Cambridge University Press.

Clay, K. (1986). Grass endophytes. In *Microbiology of the Phyllosphere*, ed. Fokkema, N. & van den Heuvel, J., in press. Cambridge University Press.

Cooke, R. C. (1977). *The Biology of Symbiotic Fungi*. Chichester & New York: John Wiley.

Cooke, R. C. & Rayner, A. D. M. (1984). *Ecology of Saprotrophic Fungi*. London & New York: Longman.

Cooke, R. C. & Whipps, J. M. (1980). The evolution of modes of nutrition in fungi parasitic on terrestrial plants. *Biological Reviews*, **55**, 341 – 362.

Gadgil, M. & Solbrig, O. T. (1972). The concept of r- and K-selection: evidence from wild flowers and some theoretical considerations. *American Naturalist*, **106**, 14 – 31.

Greenslade, P. J. M. (1983). Adversity selection and the habitat templet. *American Naturalist*, **122**, 352 – 365.

Grime, J. P. (1977). Evidence for the existence of three primary strategies in plants and its relevance to ecological and evolutionary theory. *American Naturalist*, **111**, 1169 – 1194.

Grime, J. P. (1979). *Plant Strategies and Vegetation Processes*. Chichester & New York: John Wiley.

Harper, J. L. & Ogden, J. (1970). The reproductive strategy of higher plants. I. The concept of strategy with special reference to *Senecio vulgaris* L. *Journal of Ecology*, **58**, 681 – 689.

Lewis, D. H. (1973). Concepts in fungal nutrition and the origin of biotrophy. *Biological Reviews*, **48**, 261 – 278.

Lewis, D. H. (1974). Micro-organisms and plants: the evolution of parasitism and mutualism. *Symposia of the Society for General Microbiology*, **24**, 367 – 392.

Luttrell, E. S. (1974). Parasitism of fungi on vascular plants. *Mycologia*, **66**, 1 – 15.

Moulder, J. W. (1974). Intracellular parasitism: life in an extreme environment. *Journal of Infectious Diseases*, **130**, 300 – 306.

Moulder, J. W. (1979). The cell as an extreme environment. *Proceedings of the Royal Society of London, series B*, **204**, 199 – 210.

Pianka, E. R. (1970). On r- and K-selection. *American Naturalist*, **104**, 592 – 597.

Pugh, G. J. F. (1980). Strategies in fungal ecology. *Transactions of the British Mycological Society*, **75**, 1 – 14.

Southwood, T. R. E. (1977). Habitat, the templet for ecological strategies? *Journal of Animal Ecology*, **46**, 337 – 365.

Whittaker, R. H. (1975). The design and stability of plant communities. In *Unifying Concepts in Ecology*, ed. van Dobben, W. H. & Lowe-McConnell, R. H., pp. 169 – 181. The Hague: Junk.

10
Evolution of parasitism in the fungi

MICHÈLE C. HEATH

Botany Department, University of Toronto, Canada M5S 1A1

Introduction and basic assumptions

To discuss the evolution of parasitism is to enter a world in which there are few undisputed facts but many theories. Numerous people have trod this speculative path before me (Cooke & Whipps, 1980; Lewis, 1973, 1974; Massee, 1905; McNew, 1960; Savile, 1955), and I hope that solid data will soon emerge to help direct the footsteps of those that follow. Conceivably, recombinant DNA technology may reveal information about the nature of parasitism that was previously unobtainable (Heath, 1986), and, although it remains to be seen how many questions related to the evolution of parasitism can be answered by a molecular genetical approach, currently the future looks relatively bright.

For the moment, however, the evolution of parasitism remains shrouded in mystery. This chapter will present my thoughts on certain areas of this topic as applied to fungal parasites of plants. These thoughts are based on two, probably controversial, assumptions, which, since they are important to the rest of my thesis, will now be discussed in some detail.

Assumption 1: parasitism requires a degree of specialisation greater than that needed for saprotrophy

Parasitism, to my mind, represents a specialised ecological niche. As discussed elsewhere (Heath, 1986), most modern fungal plant parasites must have the ability to (a) respond to the plant surface in such a way that they attempt to penetrate it, (b) succeed in penetrating the tissue, and (c) successfully colonise the tissue and sporulate. In addition, the parasite must ensure that it eventually finds a new host. What specific attributes are needed depends on the combination of fungus and plant, but, for many parasites, these will involve the ability to respond to specific regions of the plant surface by forming appressoria, to produce cutinase and other enzymes to allow penetration of the plant epidermis, to secrete suitable enzymes and other materials (e.g. permeability-inducing factors) to facilitate fungal spread through the tissue or to release nutrients necessary for growth, and the ability to overcome, often actively (Heath, 1982), the multitude of basic defence mechanisms possessed by every plant. As discussed in detail later, all plant parasites need at least the latter attribute, including obligately biotrophic fungi (sensu Lewis, 1973) that normally grow only in association with living tissue.

Arguments that the production of extracellular enzymes necessary for saprotrophy is a characteristic more difficult to evolve than biotrophic parasitism (Cooke & Whipps, 1980; Savile, 1955) seem to imply that the latter can easily be achieved by merely avoiding causing host cell damage or death. As discussed later, keeping the host alive probably requires special adaptations, as does the ability to avoid the plant's other defence mechanisms. Moreover, all modern fungi that have long-lasting biotrophic relationships with their hosts show signs (e.g. in their ultra-structure, biochemistry, and/or host specificity) of significant special-isation. In the case of rust fungi, for example, their responses to the topography of the plant surface (Heath, 1977), the ultrastructural specialisation of the haustorium (Harder & Chong, 1984; Heath & Allen, 1985; Littlefield & Heath, 1979), and the means by which plant defences are overcome (Heath, 1981a) seem to indicate particularly complex interactions with the plant. Furthermore, biotrophic parasites seem as capable as saprotrophs and other parasites of secreting wall-degrading enzymes, although they often do so in a more controlled and localised manner (e.g. only when they penetrate plant walls). Indeed, the wide-spread production of such enzymes by certain prokaryotes (Glenn, 1976), and the obvious benefit of external digestion to walled heterotrophs, suggests that secretion of enzymes and other substances may pre-date the origin of the eukaryotic cell.

> *Assumption 2: Defence mechanisms in autotrophs pre-date the move of plants to land and have always been a deterrent against parasitism*

Organisms heterotrophic for carbon must obtain this element in organic form from other organisms. There is no inherent reason why these carbon sources should be dead, rather than living, so that, with the advent of heterotrophy, all organisms must have been 'fair game' as a potential source of carbon, *provided that the predator or parasite could gain access to it*. In sedentary autotrophs that could not rely on their motility to escape, constitutive and inducible defence mechanisms against predation or parasitism by heterotrophic micro-organisms must have evolved early during the evolution of life, possibly in conjunction with repair mechanisms to protect against mechanical damage. These mechanisms may not have been as elaborate as those of today, since a successful defence mechanism need only reflect the *current* capabilities of both host and potential parasite. For example, the plant cell wall will be an effective physical barrier as long as the potential parasite lacks the means to cross it. This example also illustrates the fact that a defence mechanism may have other roles to play in the plant and may not have evolved specifically as an infection deterrent.

Significantly, antibiotic production is widespread in eukaryotic marine algae (Horney & Hide, 1974) whose ancestors are assumed to have given rise to the terrestrial vascular and non-vascular plants. Moreover, the fact that the algae are still with us suggests that they have reasonably successful defence mechanisms against parasites. In addition, I know of no evidence to indicate that the more ancient groups of land plants (i.e. bryophytes, lycopods, cycads, ferns) have less effective defence mechanisms against fungi than more recently evolved groups. Consequently, I see no reason to doubt that the first land plants, and their ancestors, were well equipped with an array of mechanisms to deter the invasion of co-existing bacteria and algae, as well as predation by aquatic and terrestrial invertebrates. It seems likely that at least some of these mechanisms would also be effective against invasion by the first fungi, although more specific deterrents may subsequently have evolved.

If potential fungal pathogens have always had to contend with defence mechanisms in their host plants, then one of the prime adaptations necessary for parasitism must be the ability to specifically tolerate, negate, avoid or otherwise overcome those defences that would normally prevent successful infection. Biotrophic parasitism would seem to require even greater specialisation than other forms since there are abundant experimental data to suggest that in extant plants potential defence mechanisms, including cell death, are elicited by factors likely to be common to virtually all fungi (Darvill & Albersheim, 1984; Heath, 1980); consequently, the lack of cell death and other obvious defence responses in interactions with biotrophs indicates their specific adaptation to avoid triggering, or to suppress, these responses.

Modern vascular plants seem to possess many potential defence mechanisms (Heath, 1980, 1985) whose biochemical nature and mode of elicitation may not be identical in all plant species (Heath, 1981b). In consequence, each parasite must adapt specifically to the features of its host or hosts, leading to host species specificity (Heath, 1981c, 1985). Nutritional specialisation by the fungus seems to me to be of lesser importance in determining this specificity. The vast majority of fungal plant pathogens can be grown on relatively simple media with no more difficulty than most saprotrophs (Yarwood, 1956). Killed plant tissue is commonly an excellent medium for strict saprotrophs (Yarwood, 1956), and such fungi have been shown to grow extensively in living plant tissue that has had its active defence mechanisms incapacitated by heat (Klarman, 1965) or by a toxin (Yoder & Scheffer, 1969). Moreover, in spite of a general belief to the contrary, I know of no good evidence to prove that obligately biotrophic fungi are nutritionally more specialised than other parasites or most saprotrophs. The culturable strains of rust fungi do not seem to require particularly unusual nutrients when grown axenically (Coffey & Allen, 1973), and the general inability of most rust fungi to grow in culture may rather reflect the fact that their development is closely regulated by signals from the plant. Even the lack of growth in dead plant tissue may be more related to a high sensitivity of rust fungi to secondary metabolites and degradative enzymes, usually safely stored in the vacuole of living cells, than to any requirement for a special nutrient. The over-riding importance of active defence mechanisms in determining host specificity is exemplified by experiments using nonhost plants and the bean and cowpea rust fungi. When prehaustorial defence responses are delayed by a pre-inoculation heat treatment of the nonhost leaf, these fungi can establish an appreciable mycelium in many of the nonhost plants tested (Heath, 1979 and unpublished). Thus, even for these highly specialised, obligately biotrophic parasites, the ability to grow in a plant is more a reflection of whether it can overcome prehaustorial defence barriers (Heath, 1981a) than whether the fungus can establish a sufficient nutritional relationship with the plant.

The origin of plant parasitism

There is a widely publicised view that the first terrestrial fungi, and perhaps the first aquatic fungi, were biotrophic parasites (Cain, 1972; Cooke & Whipps, 1980; Raper, 1968; Savile, 1955) and/or were derived from biotrophic, parasitic red algae (Demoulin, 1974; Kohlmeyer, 1975). I find this unconvincing for two primary reasons. The first is based on the assumption, general to all areas of biology except, it seems, mycology, that all forms of parasitism, and particularly biotrophic parasitism, represent a highly specialised ecological niche. Moreover, the specialisation required to use living plants as a food source is well recognised by entomologists, with the result that the first insects are generally assumed to be saprophagous, rather than herbivorous (Edwards & Wratten, 1980). Thus, the derivation of the fungi from parasitic ancestors contradicts Savile's (1955) first principle of evolution that new major groups do not arise from specialised climax groups but rather from unspecialised groups of greater genetic plasticity.

My second argument against any of the main terrestrial fungal groups (Zygomycotina, Ascomycotina and Basidiomycotina) being derived from parasitic ancestors is that I can see no need for such an hypothesis. Certainly, the association of fungi with other living organisms is very old (Pirozynski, 1976). However, the distribution of parasites among the major groups of fungi strongly indicates that parasitism has arisen many times during evolution (Yarwood, 1956) and that, in itself, it cannot be considered a primitive characteristic. Moreover, the truly terrestrial fungi may not have evolved in the same environment as the primarily aquatic fungi. In the warm intertidal pools or similarly wet areas in which the first vascular plants presumably evolved (see Pirozynski & Malloch, 1975), there could have been sufficient build-up of debris from vascular and nonvascular plants to support heterotrophic organisms without the need for any close association with living plants to supply nutrients or to protect against desiccation. As Denison & Carroll (1966) point out, walled heterotrophs with an invasive mycelium, such as the fungi, would have an advantage over bacteria in their ability to gain quick access to nutrients buried within the dead plant body. In such an environment, a diverse array of saprotrophic mycelial organisms could have arisen (possibly from *free-living* autotrophic progenitors) to provide the ancestors of the modern terrestrial fungi. Such a hypothesis does not exclude the simultaneous evolution of fungus-plant associations. Some of the first saprotrophic fungi may have had the ability to penetrate cell walls of moribund, as well as dead tissues. Such fungi would be in a favourable position to invade living tissue if mutation produced attributes that overcame existing defence mechanisms.

These first fungus-plant associations, as well as the growth of free-living fungi, probably were initially restricted to the relatively moist and protected environment of the soil. In such an environment mutualistic mycorrhizal associations between plant and fungus would be of benefit to the plant, increasing its access to inorganic nutrients and breakdown products of the surrounding debris. It is not surprising, therefore, that associations of fungi with subterranean vascular plant structures, resembling the vesicular − arbuscular endomycorrhizas of today, evolved early in the history of vascular plants and were probably influential in their evolution (Pirozynski & Malloch, 1975). However, the ubiquity of endomycorrhizal associations between fungi and terrestrial plants does not mean that they evolve easily; indeed the restriction of such associations to a rather odd group of fungi with uncertain taxonomic affiliations suggests a degree of specialisation that other fungi have not been able to achieve. Therefore, endomycorrhizal fungi seem no more likely than biotrophic parasites to be the ancestors of the main terrestrial fungal groups. Nevertheless, it should be borne in mind that unlike many parasites, particularly of aerial plant parts, mycorrhizal fungi retain a saprotrophic relationship with the rhizosphere. Therefore, it is possible that less specialised forms could, during evolution, relinquish the plant association for a truly saprotrophic life. It should also be remembered that roots and fungi have evolved together under selection pressures that favour tolerance of fungal invasion and the development of mutualistic associations. Moreover, the evolution of root defences may have been influenced by the fact that such associations can protect the root against subsequent invasion by parasites (Pirozynski, 1983). In consequence, roots may differ from aerial plant parts in the ways in which parasitic associations with fungi have evolved, and in the fungal adaptations required for successful parasitism.

As pointed out by Lewis (1973, and see Chapter 11), fungi in association with aerial plant parts are not in a position to supply the plant with nutrients abstracted from the environment; thus, selection is likely to favour anti-fungal defence mechanisms that are particularly vigorous and effective. The first fungi may also have found aerial plant parts more difficult to parasitise than roots because of the problems of reaching the plant surface and surviving long enough to enter the tissue. It is perhaps significant that foliar pathogens were apparently not abundant until the Cretaceous period (Pirozynski, 1976). One explanation for this observation could be that, at least for the Ascomycotina (Pirozynski, 1976), air-borne spores were not initially abundant and most species lacked a reliable means, other than systemic infection, by which the mycelium could easily reach the foliage. Alternatively, the leaves of the vascular plants evolving at the beginning of the Cretaceous period may have had less effective defence mechanisms than previously existing relatives. It is noticeable that extant cycads and *Ginkgo*, for example, have fewer known fungal parasites than conifers or flowering plants (Seymour, 1929). However, it is difficult to assess the significance of such uneven distribution of parasites between major plant taxa. It is often suggested that the antiquity of the host reflects that of the parasite (see Savile, 1955), in which case the lack of parasites on cycads and *Ginkgo* may merely reflect the paucity of potential parasites available when these groups evolved. Such an explanation assumes that new parasitic relationships are primarily established when the host is undergoing rapid evolution (Savile, 1955, 1971). Although there is no inherent reason why such plants should have less effective defence mechanisms than members of climax groups, it is conceivable that during a time of great genetic diversity, slight variations in these defence mechanisms may favour the invasion by certain fungi that fortuitously have the ability to overcome variant forms of resistance (Savile, 1971). Obviously, however, whether a parasitic association develops depends not only on the potential parasite being able to evolve the necessary adaptations, but also on the existence of suitable selection pressures, such as the stable co-existence of plant and fungus (Savile, 1971) and the relative abundance of the potential host.

The evolution of different forms of parasitism

The possible routes by which a new plant − parasite combination may arise have been discussed in detail elsewhere (Heath, 1985). However, I should like to reiterate that even if fungal parasitism primarily involves the negation of the host's defences, there are many mechanisms by which this could be achieved depending on the way in which the genetic capabilities of the fungus have been constrained by its evolutionary history, and on the types of defences exhibited by the plant (Heath, 1981c). In turn, these mechanisms will affect the range of susceptible species and determine the type of nutritional relationship the fungus can develop with its host. For example, if massive secretion of non-specific cell wall-degrading enzymes is used to kill cells before they can react in a resistant manner, the fungus may be able successfully to infect a number of plant species and will, perforce, be a necrotroph. It seems unlikely that such a fungus would eventually evolve any sort of biotrophic relationship with its host, because any reduction in wall degradation is likely to permit the expression of, or even induce (see Darvill & Albersheim, 1984), active defence mechanisms. However, if active resistance is prevented by means of a host-selective toxin, the host range of the fungus is likely to be narrow, and it is conceivable that a biotrophic relationship could evolve by lowering the toxin concentration, so that the cell is narcotised rather than killed (Heath, 1985). On the whole, I feel that grouping fungi into categories defined only by their source of nutrients (dead vs. living cells, e.g. Cooke & Whipps, 1980; Lewis, 1973, 1974) is of dubious value in discussions of the evolution of parasitism as it ignores the multitude of plant − fungus interactions that can take place within each category. When considered in detail these categories are not, in fact, distinct, as mentioned by others (e.g. Rayner, Watling & Frankland, 1985). Even the boundary between saprotrophs and parasites is not clear, as there may be relatively little difference in terms of nutrient requirements or mode of nutrient acquisition between fungi that invade already dead, but intact, plant tissues and relatives that invade these same tissues when they are senescing and incapable of showing active resistance. Defining a biotroph is particularly difficult. Biotrophs are usually regarded as those organisms that derive their nutrition from living cells (Lewis, 1973). However, in practice they are recognised by the fact that invaded tissue does not die. One could argue, particularly for non-obligate biotrophs, that it is the lack of cell death that really defines 'biotrophy', and the fungus has no choice but to obtain its nutrients from living tissue.

The absence of tissue necrosis is particularly significant because necrosis is the most frequent response to attempted fungal invasion that is associated with resistance in host and nonhost plants (Heath, 1980). Significantly, recent work suggests that the amount of necrosis elicited by a fungal plant pathogen in a nonhost plant is related to that elicited in its susceptible host (Fernandez & Heath, 1986). It is possible, therefore, that biotrophs are unique, not because of their mode of nutrition but because they lack necrosis elicitors, directly suppress plant responses that lead to necrosis, or have their necrosis-eliciting activities suppressed by the plant (for an example of the latter see Cooke & Whipps, 1980). The fact that there are several ways in which the lack of necrosis may be achieved emphasises that 'biotrophy' may cover a number of phenomena, including many ways in which a biotroph obtains its nutrients. Intercellular biotrophs may merely utilise materials present in the apoplast (Hancock & Huisman, 1981), or may induce the release of nutrients from adjacent cells. Biotrophs that penetrate plant cells may have an even greater control over plant metabolism, but the degree of control may vary with the length of time that the invaded cell stays alive. Possibly a less elaborate fungal − plant interaction is required if the invaded cell survives for only a short time. *Colletotrichum lindemuthianum*, for example, is a hemibiotroph that has a 'biotrophic' phase lasting for less than 24 h (O'Connell, Bailey & Richmond, 1985). For this short period of time, the fungus may do little to maintain the vitality of the cell which may stay alive merely because death is not elicited. Subsequent death of the cell could result from the stress put upon it by the nutritional demands of the fungus. This form of 'biotrophy' may require relatively little adaptation on the part of the fungus in comparison to that shown by obligately biotrophic parasites.

Necrotrophy too could represent a range of interactions between plant and parasite. Necrotrophic fungi, by definition, gain their nutrients from cells they have killed (Lewis, 1973), but there are many ways in which such death may be elicited. For fungi that kill cells in advance of invasion, their characterisation as necrotrophs is unarguable. However, for many so-called necrotrophs, proof of the timing of cell death is lacking. Careful cytological studies to show the relationship between fungal growth and cell death commonly reveal that many fungi, even those that secrete host selective toxins (Yoder & Scheffer, 1969), kill cells *after* penetration. Consequently, the distinction between these fungi and the hemibiotrophic *C. lindemuthianum* becomes one of the relative timing of cell death.

If biotrophy and necrotrophy, as currently defined, represent a multitude of molecular interactions between plant and parasite, it is useless to try and relate them, in terms of evolution, in any all-encompassing hypothesis. The evolutionary history of any given extant parasite is going to depend on what specific adaptations it has made to achieve success in invading its host. In some cases, this may mean that a biotrophic parasite has passed, during evolution, through what would be defined currently as a necrotrophic stage; but in other cases no such stage may have existed (Lewis, 1974). However, in each specific plant − parasite combination, the molecular adaptations required by the fungus to achieve its parasitic status are going to be unique and only when we know what is involved in a specific type of host − parasite interaction can we knowledgeably discuss its evolution.

Summary

Plant parasitism represents a highly specialised ecological niche that requires special adaptations on the part of the fungus that occupies it. What these adaptations represent in molecular terms must depend on the type of interactions that take place between fungus and host, but the negation of plant defences seems to be of prime importance. There appears to be no reason to assume that there has been any time during which terrestrial plants or their ancestors totally lacked such defences. Therefore, fungal plant parasites, and particularly biotrophs that elicit minimal defence responses, have always been relatively specialised organisms and for this reason are unlikely to be the progenitors of the major groups of modern free-living terrestrial fungi. Although modern obligately biotrophic fungi appear to have a particularly complex relationship with their hosts, this may not be true of all biotrophs and especially hemibiotrophs, since the recognition of biotrophy hinges on the lack of plant cell death, not the mode of fungal nutrition. Necrotrophy and biotrophy, as currently defined, cover a multitude of capabilities on the part of the fungus and probably are a consequence of the way in which host defences are overcome. In consequence, it is unlikely that there is any universal pattern in the evolution of nutritional relationships between plants and fungal parasites. The mode of evolution of any specific plant – parasite combination will depend on the type of molecular interactions that take place. Only when these are elucidated will there be any relevant data on which to base hypotheses on the evolution of parasitism in the fungi.

Acknowledgements I am indebted to Dr D. W. Malloch for numerous discussions during the preparation of this manuscript, and to Dr Malloch, Dr K. A. Pirozynski and Dr D. B. O. Savile for critical reviews of the original version.

References

Cain, R. F. (1972). Evolution of the fungi. *Mycologia*, **64**, 1 – 14.

Coffey, M. D. & Allen, P. J. (1973). Nutrition of *Melampsora lini* and *Puccinia helianthi*. *Transactions of the British Mycological Society*, **60**, 245 – 260.

Cooke, R. C. & Whipps, J. M. (1980). The evolution of modes of nutrition in fungi parasitic on terrestrial plants. *Biological Reviews*, **55**, 341 – 362.

Darvill, A. G. & Albersheim, P. (1984). Phytoalexins and their elicitors – a defense against microbial infection in plants. *Annual Review of Phytopathology*, **35**, 243 – 275.

Demoulin, V. (1974). The origin of ascomycetes and basidiomycetes. The case for a red algal ancestry. *Botanical Review*, **40**, 315 – 345.

Denison, W. C. & Carroll, G. C. (1966). The primitive ascomycete: a new look at an old problem. *Mycologia*, **58**, 249 – 269.

Edwards, P. J. & Wratten, S. D. (1980). *Ecology of Insect – Plant Interactions*. London: Edward Arnold Ltd.

Fernandez, M. R. & Heath, M. C. (1986). Cytological responses induced by five phytopathogenic fungi in a nonhost plant *Phaseolus vulgaris*. *Canadian Journal of Botany*, **64**, 648 – 657.

Glenn, A. R. (1976). Production of extracellular proteins by bacteria. *Annual Review of Microbiology*, **30**, 41 – 62.

Hancock, J. G. & Huisman, O. C. (1981). Nutrient movement in host – pathogen systems. *Annual Review of Phytopathology*, **19**, 309 – 331.

Harder, D. E. & Chong, J. (1984). Structure and physiology of haustoria. In *The Cereal Rusts. Vol. 1, Origins, Specificity, Structure and Physiology*, ed. Bushnell, W. R. & Roelfs, A. P., pp. 431 – 476. Orlando: Academic Press.

Heath, M. C. (1977). A comparative study of non-host interactions with rust fungi. *Physiological Plant Pathology*, **10**, 73 – 88.

Heath, M. C. (1979). Effects of heat shock, actinomycin D, cycloheximide and blastocidin S on nonhost interactions with rust fungi. *Physiological Plant Pathology*, **15**, 211 − 218.

Heath, M. C. (1980). Reactions of nonsuscepts to fungal pathogens. *Annual Review of Phytopathology*, **18**, 211 − 236.

Heath, M. C. (1981a). Resistance of plants to rust infection. *Phytopathology*, **71**, 971 − 974.

Heath, M. C. (1981b). Nonhost resistance. In *Plant Disease Control: Resistance and Susceptibility*, ed. Staples, R. C. & Toenniessen, G. H., pp. 201 − 217. New York: John Wiley & Sons Inc.

Heath, M. C. (1981c). A generalized concept of host − parasite specificity. *Phytopathology*, **71**, 1121 − 1123.

Heath, M. C. (1982). The absence of active defense mechanisms in compatible host − pathogen interactions. In *Active Defence Mechanisms in Plants*, ed. Wood, R. K. S., pp. 143 − 156. London: Plenum Press.

Heath, M. C. (1985). Implications of nonhost resistance for understanding host − parasite interactions. In *Genetic Basis of Biochemical Mechanisms of Plant Disease*, American Phytopathological Society Symposium Book no. 4, ed. Groth, J. V. & Bushnell, W. R., pp. 25 − 42. St Paul, Minnesota: APS Press.

Heath, M. C. (1986). Fundamental questions related to plant − fungal interactions: can recombinant DNA technology provide the answers? In *Biology and Molecular Biology of Plant − Pathogen Interactions*, ed. Bailey, J. A., pp. 15 − 27. Berlin: Springer-Verlag.

Heath, M. C. & Allen, F. H. E. (1985). Morphology, element composition, and responses to acids and oxidizing agents of the haustorial neckband of *Uromyces phaseoli* var. *vignae*. *Canadian Journal of Botany*, **63**, 463 − 473.

Hornsey, I. S. & Hide, D. (1974). The production of antimicrobial compounds by British marine algae. *British Phycological Journal*, **9**, 353 − 361.

Klarman, W. L. (1965). Heat-induced susceptibility of soybeans to nonpathogenic fungi. *Phytopathology*, **55**, 505 (abstract).

Kohlmeyer, J. (1975). New clues to the possible origin of ascomycetes. *Bioscience*, 25, 86 − 93.

Lewis, D. H. (1973). Concepts in fungal nutrition and the origin of biotrophy. *Biological Reviews*, **48**, 261 − 278.

Lewis, D. H. (1974). Micro-organisms and plants: the evolution of parasitism and mutualism. *Symposia of the Society for General Microbiology*, **24**, 367 − 392.

Littlefield, L. J. & Heath, M. C. (1979). *Ultrastructure of Rust Fungi*. New York: Academic Press.

Massee, G. (1905). On the origin of parasitism in fungi. *Philosophical Transactions of the Royal Society of London, series B*, **197**, 7 − 24.

McNew, G. L. (1960). The nature, origin, and evolution of parasitism. In *Plant Pathology, An Advanced Treatise. Vol. II, The Pathogen*, ed. Horsefall, J. G. & Dimond, A. E., pp. 19 − 69. New York: Academic Press.

O'Connell, R. J., Bailey, J. A. & Richmond, D. V. (1985). Cytology and physiology of infection of *Phaseolus vulgaris* by *Colletotrichum lindemuthianum*. *Physiological Plant Pathology*, **27**, 75 − 98.

Pirozynski, K. A. (1976). Fossil fungi. *Annual Review of Plant Pathology*, **14**, 237 − 246.

Pirozynski, K. A. (1983). Pacific mycogeography: an appraisal. *Australian Journal of Botany, Supplementary Series*, **10**, 137 − 159.

Pirozynski, K. A. & Malloch, D. W. (1975). The origin of land plants: a matter of mycotrophism. *BioSystems*, **6**, 153 − 164.

Raper, J. R. (1968). On the evolution of fungi. In *The Fungi, An Advanced Treatise. Vol III, The Fungal Population*, ed. Ainsworth, G. C. & Sussman, A. S., pp. 677 − 693. New York: Academic Press.

Rayner, A. D. M., Watling, R. & Frankland, J. C. (1985). Resource relations − an overview. In *Developmental Biology of Higher Fungi*, British Mycological Society Symposium volume 10, ed. Moore, D., Casselton, L. A., Wood, D. A. & Frankland, J. C., pp. 1 − 40. Cambridge University Press.

Savile, D. B. O. (1955). A phylogeny of the basidiomycetes. *Canadian Journal of Botany*, **33**, 60 − 104.

Savile, D. B. O. (1971). Coevolution of the rust fungi and their hosts. *The Quarterly Review of Biology*, **46**, 211 − 218.

Seymour, A. B. (1929). *Host Index of the Fungi of North America*. Cambridge, Mass.: Harvard University Press.

Yarwood, C. E. (1956). Obligate parasitism. *Annual Review of Plant Physiology*, **7**, 115 − 142.

Yoder, O. C. & Scheffer, R. P. (1969). Role of toxin in early interactions of *Helminthosporium victoriae* with susceptible and resistant oat tissue. *Phytopathology*, **59**, 1954 − 1959.

11

Evolutionary aspects of mutualistic associations between fungi and photosynthetic organisms

DAVID H. LEWIS

Department of Botany, The University, Sheffield S10 2TN, UK

Introduction

At the outset, it is necessary to delimit precisely the concept of mutualism, the taxonomic limits encompassed by the organisms discussed in this chapter and the nature of the symbioses considered. 'Symbiosis' is used in its literal meaning of 'living together', with no overtones of time or nature of contact (Lewis, 1985).

Notwithstanding the problems of quantitatively assessing the fitness of fungal individuals, mutualism is defined as the outcome of those symbioses in which the fitness of *both* symbionts is increased as a result of the association (Law & Lewis, 1983; Lewis, 1985). Following Margulis & Schwartz (1982), the fungi are limited to the Zygo-, Asco- and Basidiomycota (= -mycotina) and their imperfect forms. The autotrophs are either prokaryotic cyanobacteria or eukaryotic protoctistan algae (principally Chlorophyta) and land plants (Bryophyta and Tracheophyta). Table 11.1 lists the three kinds of mutualistic association between fungi and photosynthetic organisms to be considered. The reasons for regarding these as mutualistic are set out below. It must be stressed that the nature of the outcome of a symbiosis (see Lewis, 1985 for the continua involved) is important because selective forces act differently on the symbionts depending on this outcome (Law & Lewis, 1983; Law, 1985).

Mycorrhizas

The mutualistic mycorrhizas

There are good reasons for substituting the now rightly defunct classification of mycorrhizas into ecto- and endo-trophic kinds with another dichotomy. This is between those where net movement of carbon (especially as carbohydrate) has been demonstrated or can be reliably inferred to be *usually* from plant to fungus, and those where the opposite obtains (Fig. 11.1). Only the former kinds are mutualistic (Table 11.1). The qualification 'usually' is needed because of the demonstration of the marked movement of carbon between plants of the same or different species when connected by ectomycorrhizal or V-A mycorrhizal fungi (Brownlee *et al.*, 1983; Francis & Read, 1984; Finlay & Read, 1986).

To the eight kinds of mycorrhiza noted in Fig. 11.1 may be added those found in *Pisonia grandis* (Ashford & Allaway, 1982, 1985) and *Carex* species (Haselwandter & Read, 1982). The direction of net flux of carbon in these associations has not been determined (Table 11.2) but it could be from fungus to plant (Lewis, 1986). Although the description, Nyctaginaceous, has been applied to mycorrhizas of *Pisonia* (Lewis, 1986), it is not yet known whether their characteristic transfer cells consistently occur in the ectomycorrhizas of other genera of the Nyctaginaceae, e.g. those of *Neea* species (see Alexander & Hogberg, 1986). The occurrence of transfer cells may be a generic rather than a familial character.

Mycorrhizal types below the dotted line in Fig. 11.1 and in the right-hand column of Table 11.2 are classified as agonistic (+ /-) (Lewis, 1985) because the fungi supply both carbon *and* mineral nutrients to the plant and, as yet, there is no clear evidence that their own fitness is increased by symbiosis. As such, these mycorrhizas are beyond the scope of this chapter, except with regard to their evolutionary relationships to mutualistic kinds. The five kinds of mycorrhiza placed above the line in Fig. 11.1 and in the mutualistic column of Table 11.2 are delimited as types I, II and III. Their general biological features have most recently been considered by Harley & Smith (1983) and Gianinazzi-Pearson &

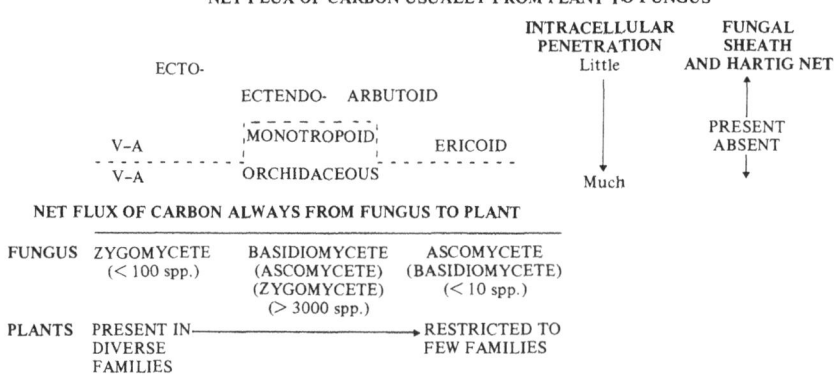

Fig. 11.1. Characteristics of the principal kinds of mycorrhiza.

Table 11.1. *Mutualistic symbioses between fungi and photosynthetic organisms*

Photosynthetic organisms	Symbioses
A. Whole organism	
(i) Cyanobacteria	Lichens
(ii) Algae	Lichens
B. Photosynthetic tissue of land plants	Endophytic symbioses
C. Non-photosynthetic tissue of land plants	Some mycorrhizas

Table 11.2. *Summary of types of mycorrhiza in relation to net flux of carbon between the symbionts (cf. Fig. 11.1)*

	Net carbon flux	
Mycorrhizal type	Plant to fungus (mutualistic)	Fungus to plant (agonistic)
I V−A	*	*
Ecto	*	
Ectendo	*	
II Arbutoid	*	
Monotropoid		*
Nyctaginaceous	?	
III Ericoid	*	
IV Orchidaceous		*
IV? Cyperaceous	?	

Gianinazzi (1986) whilst their ecological features are summarised in Fig. 11.2. These features will now be examined in an attempt to delimit the selective forces controlling evolution of the three main types of mycorrhiza. An alternative approach via strategy theory (Grime, 1974; Cooke & Whipps, Chapter 9) deserves attention but has not yet been attempted.

Origin and evolution of Type I (vesicular − arbuscular) mycorrhizas

Vesicular − arbuscular (V − A) mycorrhizas, the most widespread extant fungal symbioses, involve only six zygomycetous genera and perhaps 200 species (Trappe, 1982; Trappe & Schenck, 1982). They may represent an isolated relict taxon (Law & Lewis, 1983). All appear to be obligately dependent on plants as sources of carbon. Although both Bryophyta and Tracheophyta possess V − A mycorrhiza, only the latter will be considered because the early history of the former is so sparse and the same argument would, in any case, apply.

The adaptations necessary to transform a characean green alga into a land plant have been discussed by Raven (1977, 1984, 1985, 1986). The observation that early vascular land plants had Type I mycorrhizas (see Harley, 1969; Nicolson, 1975; Pirozynski, 1976) led Pirozynski & Malloch (1975) to conclude that their very origin was intimately dependent on the mycorrhizal habit (see also Malloch, Pirozynski & Raven, 1980; Raven, 1977; Raven, Smith & Smith, 1978; Pirozynski, 1981). The first macrofossils of land plants (*Cooksonia*), some of which are clearly vascular, are Silurian (Edwards & Fanning, 1985) but earlier Ordovician microfossils have also been attributed to a land flora (Gray, 1985). These early land plants would have been subjected to precisely the same selective forces that extant plants currently respond to by association with fungal partners; viz., acquisition of minerals and water, maintenance of compatible acid − base balance and exclusion of toxic elements (Read, 1984; Raven *et al.*, 1978). Furthermore, it has been argued that lignification was primarily a defence reaction, only later exploited in vascular tissue and

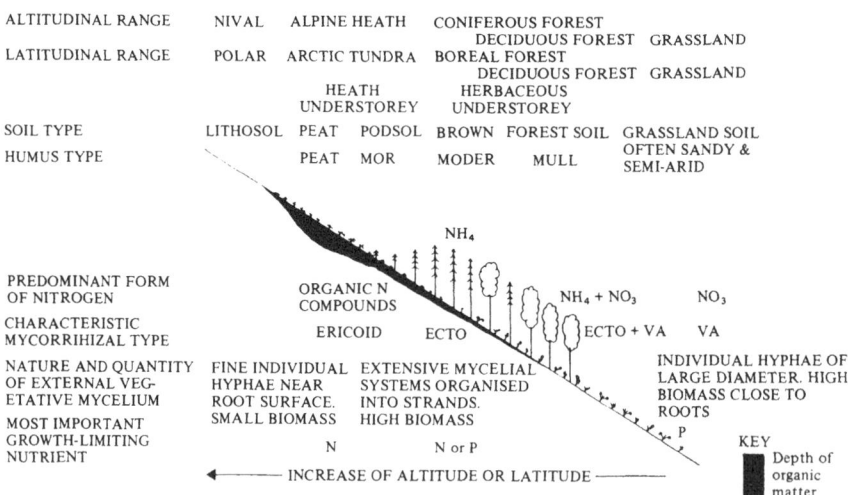

Fig. 11.2. Diagrammatic presentation of the postulated relationship between latitude or altitude, climate, soil and mycorrhizal type, and development of the vegetative mycelium associated with mycorrhizas. (Slightly modified, with permission, from Read, 1984.)

still later for support (Raven, 1977, 1984). Fungal infection of the pre-cursor charophyte may therefore have stimulated lignification and been responsible for acquisition of resources from soil.

The suggestion that the first land plants were obligately mycorrhizal is reinforced by the fact that even early Devonian plants, e.g. *Rhynia* and *Asteroxylon*, had no roots. Unless elements were present in Ordovician, Silurian and Devonian soils in much greater concentrations than in extant soils, it is likely that phosphorus would often have been growth limiting (see Fig. 11.2). Not only is phosphorus, as phosphate, present in soil solutions at concentrations some two to three orders of magnitude lower than other macro-nutrients (Epstein, 1972) but it is extremely immobile (Nye & Tinker, 1977). Land plants need to grow towards phosphate rather than wait for it to diffuse to them! Clearly, the early vascular plants achieved this via mycorrhizas, a situation which, 400 million years later, persists in over two hundred thousand species.

During the 120 million years of the Devonian and Carboniferous, vascular plants evolved in several directions by increase of size, to about 40 m, and by becoming woody (Raven, 1986). Before considering the origin of the ectomycorrhizal habit, the following features need to be emphasised. Firstly, if the gametophytes of some early trimerophytes, pteridophytes and lycopsids were similar to those of *Psilotum* and some extant ferns and clubmosses (see Harley, 1969), they were heterotrophic and agonistically dependent on their type I mycorrhizal associates. Secondly, the Carboniferous is named after its massive deposits of coal, much derived from wood.

Origin and evolution of Type II mycorrhizas

Extant ectomycorrhizas are characteristic of mull and moder soils (Fig. 11.2). However, although such soils had developed by the early Carboniferous (Wright, 1985), the ectomycorrhizal habit probably had not, only really coming into its own during the Cretaceous (Pirozynski & Malloch, 1975; Malloch *et al.*, 1980). Nonetheless, it was during the Carboniferous that the stage was set for these symbioses. The massive deposits of carbon in this and later Eras were due to low rates of decay (Schopf, 1952). This suggests either lack of appropriate degradative organisms, or deficiency of oxygen, the co-substrate for rapid decay, or both.

White rot basidiomycetes degrade lignin most rapidly (Reddy, 1984; see also Whalley & Edwards, Chapter 28, for xylariaceous ascomycetes). As Dennis (1970) demonstrated clamp connections on hyphae in wood of a

Carboniferous fern, these basidiomycetes were thought to have originated in the Carboniferous itself (Pirozynski, 1976). However, Stubblefield, Taylor & Beck (1985) have now provided convincing evidence for the existence of wood-inhabiting fungi, likely to be basidiomycetes, in the late Devonian, so appropriate organisms for rapid decay of wood existed at the start of the Carboniferous. Attention must therefore be focussed on the availability of oxygen.

While much deposition of carbon undoubtedly took place under acidic, waterlogged conditions, it is not clear what the precise atmospheric concentration of oxygen was during the period when the basidiomycetes emerged. To understand the relevance of this, it is necessary to digress to the mechanism by which lignin is degraded. This is because the process is a highly oxidative one and, in the best-studied organism, *Phanerochaete chrysosporium*, which may not be representative of the full range of ligninolytic fungi, occurs optimally at $40 - 60\%$ O_2. Rates at 100% O_2 are two to three-fold higher than in air (Kirk & Fenn, 1982). The precise role of oxygen is not clear. The 'ligninase' initiating breakdown is an extracellular, haem-containing peroxidase which, after reaction with hydrogen peroxide, oxidises lignin, thereby generating cationic free radicals. These, in turn, cause further degeneration of the polymer by a chain reaction (J.M. Palmer, personal communication; see Paterson & Lundquist, 1985). As 'ligninase' may also catalyse *polymerisation* of fragments, stabilisation of these by oxidations catalysed by phenol oxidases utilising molecular oxygen (e.g. laccase and polyphenol oxidases) may be an important adjunct to the initial degradative steps. The oxidases, which also catalyse further degradation, have a low affinity for oxygen so that the network of reactions, as well as the synthesis of the enzymes themselves, is highly aerobic. Also, since lignin occurs in wood as lignocellulose, enzymes which break down cellulose, including a cellobiose:quinone oxidoreductase, are intimately concerned with the overall degradation. Lastly, the ability to degrade lignin is a feature of secondary metabolism, a metabolic phase induced by low availability of carbon, sulphur or, especially, nitrogen (Reddy, 1984), a situation especially relevant in naturally N-poor wood.

The history of changes in atmospheric oxygen concentration in the Phanerozoic is much disputed (Holland, 1984; Raven, 1985). Although much evidence is geochemical, three biologically related phenomena do set limits; the concurrent appearance of flying insects and fossil charcoal in the Devonian and the bulkiness of plant tissues in the late Carboniferous (Raven, 1985, 1986). Taken together, they suggest that atmospheric oxygen concentration was not less than $0.3 - 0.6$ of the extant value when basidiomycetes first appeared. For further details and alternatives, see Broecker (1970), Walker (1977), Hart (1978), Cope & Chaloner (1980, 1981), Clark & Russell (1981), Holland (1984) and Holland, Laza & McCaffrey (1986). One alternative view (Walker, 1977), envisages a concentration during the Carboniferous much higher than at present. If confirmed, this might account for the low affinities of phenol oxidases for molecular oxygen in that these enzymes first evolved in an oxygen-rich environment.

From the above, it is clear that basidiomycetes capable of destroying wood either as saprotrophs or necrotrophs were present by the start of the Carboniferous. As such, they may have posed a threat to woody plants, perhaps thereby acting as an important selective force in the origins of sinapyl-rich (angiospermous) lignin and, indeed, of the re-emergence of the herbaceous (low-lignin) habit among angiosperms. [See Rayner (1986) for a discussion of factors affecting the *internal* concentration of oxygen in woody tissues. He has also pointed out (personal communication) that, with regard to ligninolysis and mycorrhizal infection, concentration of oxygen around roots in the soil environment is more important than that around shoots in the air.] Despite this threat, however, basidiomycetes were not 'tamed' into the ectomycorrhizal state until much later. Fossils of Pinaceae, the principal extant ectomycorrhizal gymnospermous taxon, do not appear until the late Triassic, and *Pinus* itself not until the Cretaceous, contemporary with ectomycorrhizal angiospermous taxa (see Malloch *et al.*, 1980).

As ligninolysis is a feature of secondary metabolism, it follows that maintenance of potentially ligninolytic fungi in a primary metabolic state will suppress this potential. This could have been achieved by absorption of nitrogen and phosphorus from organic soils, concomitant with carbon from a woody host in a biotrophic manner. Ectomycorrhizal utilisation of organic nitrogen is discussed by Abuzinadah, Finlay & Read (1986) and Abuzinadah & Read (1986a & b). Geological and physiological consider-ations therefore suggest that ectomycorrhizal biotrophy is derived from the saprotrophic/necrotrophic direction (cf. Lewis, 1973). It is also likely that ectomycorrhizal plants are geologically more recent than their fungi from the fact that fungal genera are common to both Northern and Southern hemisphere plants (Trappe, 1962). This is consistent with establishment of the fungi before the split of the Pangean land mass (Pirozynski, 1981; cf. Miller & Watling, Chapter 29) and with the propensity of ectomycorrhizal symbioses to be characteristic of the higher latitudes (Fig. 11.2). However, the greater potential for dispersal exhibited by fungal spores should not be overlooked. The taxonomic and phylogenetic relationships of wood-destroying and ectomycorrhizal basidiomycetes are discussed by Watling (1982), Miller & Watling (Chapter 29), and Kuhner (1984).

The above discussion is not, however, meant to imply that the dichotomy, wood-destroying vs. ectomycorrhizal, is absolute. There is much evidence for a continuum between the two extremes as has been reinforced by recent evidence of utilisation by ectomycorrhizal fungi of lignin, lignocellulose and cellulose from Trojanowski, Haider & Hüttermann (1984) who cite earlier studies to support this view.

From the compilations of Malloch *et al.* (1980) and Harley & Smith (1983), supplemented by more recent studies of tropical and Australian species (Warcup & McGee, 1983; Alexander, 1986; Alexander & Hogberg, 1986), the ectomycorrhizal habit clearly occurs widely in taxonomically unrelated orders of angiosperms (including herbaceous species) many of which only date to late Cretaceous or even early (Paleogene) Tertiary. The fungi are, therefore, probably still extending their range of angiospermous hosts.

Like V − A mycorrhizal fungi, ectomycorrhizal species exploit their hosts for carbon whereas the plants gain more varied and extensive physiological attributes, notably an ability to obtain nitrogen from sources not otherwise available (Abuzinadah *et al.*, 1986; Abuzinadah & Read, 1986a & b). The fungi also provide water, particularly where seasonal aridity occurs (Alexander, 1986). Indeed, K. A. Pirozynski and D. C. Smith (personal communication) link the development of the ectomycorrhizal habit to the onset of pronounced climatic fluctuations. This is in line with the relationship to plate tectonic events already mentioned.

Origin and evolution of Type III (Ericoid) mycorrhizas
Unlike mycorrhizas considered in the previous two sections, the type III, ericoid, kind is taxonomically restricted to a close-knit group of families in the Ericales (Malloch *et al.*, 1980; Read, 1983). Their overall geographical and ecological range is indicated in Fig. 11.2. Ericoid mycorrhizal fungi must not only be able to utilise complex sources of nitrogen and phosphorus and tolerate high concentrations of elements, such as manganese and aluminium, more soluble at low pH, but also must be well adapted to the polyphenolic environment of mor humus. Moreover, since ericaceous plants are often rich in phenolic glycosides (Franz, 1982), any tolerance of soil phenolics by fungi with access to complex

nitrogen and phosphorus compounds would preadapt them to becoming mutualistic endophytes. These attributes are found in the ascomycetous *Hymenoscyphus* (= *Pezizella*) *ericae*, the imperfect *Oidiodendron* species and related strains and, possibly, the basidiomycetous *Clavaria argillacea* (Englander, 1982; Read, 1983; Dalpé, 1983; Miller & Watling, Chapter 29). This mutualistic symbiosis is, however, geologically recent, fossil records of the Ericaceae dating back only to the late Eocene whereas the Epacridaceae, from its biogeographical distribution, is possibly Cretaceous (Raven & Axelrod, 1974; K. A. Pirozynski, personal communication). This recent origin probably means that there has been insufficient time for any agonistic versions to develop.

Evolutionary relationships between mutualistic and agonistic mycorrhizas

Implicit in the last sentence is that for type I and type II mycorrhizas there *is* an evolutionary relationship between mutualism and agonism. Furthermore, if the initial vascular plants were mycorrhizal, the non-mycorrhizal condition is necessarily derived. The most important interactive selective forces bringing about the mutualistic condition consequent upon the sharing of resources are improvement of acquisition of carbon by the fungi, and of other resources by the plants, plus ease of infection of plants by the fungi.

Conservation of carbon by the plant appears to be important in the evolution both of agonistic mycorrhizas and the non-mycorrhizal condition. In agonism, carbon is not only conserved but is derived partially or wholly from fungi in a hemibiotrophic manner, whereby biotrophy is superseded by necrotrophy. Hence, the fungi lose their ability to derive resources from plants whilst the latter retain their susceptibility to infection. Three kinds of fungi are involved (Fig. 11.1), (a) typical V − A endophytes in heterotrophic archegoniate gametophytes and Gentianaceae, (b) typical ectomycorrhizal endophytes in *Monotropa* and its allies, and in some orchids and (c) facultatively necrotrophic fungi (e.g. *Rhizoctonia* and *Armillaria*) (Harley, 1969; Harley & Smith, 1983). The last group appears to be the result of direct 'taming' of pathogens, there being no mutualistic equivalent. To exist in the non-mycorrhizal condition, plants must be able to acquire resources, especially phosphate, independently of fungi. This can be achieved by restriction to nutrient-rich, disturbed habitats and by extrusion of protons into the rhizosphere, so solubilising otherwise unavailable sources of phosphate (Grinsted, *et al.*, 1982; Hedley, Nye & White, 1982, 1983; Hedley, White & Nye, 1982).

At the same time, fungi are excluded by increased resistance to infection.

Conservation of carbon can be achieved by synthesis of carbohydrates which are either not readily available as carbon sources to invading organisms or which, more effectively, are actually toxic. Orders considered to be evolutionarily advanced among both di- and monocotyledons contain a range of soluble sugars. In these, the basic molecule of sucrose is extended by addition of units of fructose, galactose or glucose to yield the various sucrosyl − fructans and sucrosyl − galactans and the single sucrosyl − glucan, gentianose (Lewis, 1984). Fructans are not toxic to fungi as shown by the occurrence of V − A mycorrhizas in many species of Asterales and Poales. However, Lewis (1986) has speculated that sucrosyl − galactans and gentianose may give rise to reducing disaccharides (melibiose and gentiobiose) which, by acting as agents capable of sequestering phosphate, could alter permeability of membranes as do other analogues of glucose (Herold & Lewis, 1977). Indeed, one function of the extracellular, wall-bound invertase of vascular plants (ap Rees, 1984) could be the hydrolysis of non-toxic, non-reducing trisaccharides such as raffinose and gentianose to protective, toxic reducing disaccharides. Thus, depending on toxicity and concentration of such sugars, potentially pathogenic or mutualistic fungi could either be prevented from infecting completely or be overcome after infection. Fungi in the latter situation would be induced to release nutrients including carbon compounds to plants over a substantial period, the agonistic mycorrhizal situation. This testable hypothesis should be viewed in the light of the effects of exogenous mannose on release of electrolytes from leaf discs of spinach beet over a period of days (Herold, 1978).

Here, it is relevant that the taxonomically isolated, largely non-mycorrhizal order, the Caryophyllales, includes families such as the Caryophyllaceae and Chenopodiaceae which store sucrosyl − galactans. Also, within the order is the Nyctaginaceae with its peculiar mycorrhizas. Whether these are related to the occurrence of peculiar sugars is not yet known (Lewis, 1986). Mycorrhizas of the Gentianaceae, some genera of which contain gentianose and gentiobiose, are also unusual (Jacquelinet-Jeamougin & Gianinazzi-Pearson, 1983; McGee, 1985).

Another way of sequestering sugars is as phenolic and other glycosides. Although over 50 kinds of disaccharide exist in various glycosides (Avigad, 1982), a role for this diversity has yet to be suggested. Could they be anti-fungal in the manner suggested above for more familiar disaccharides? With their extensive range of mycorrhizal types, the Ericales would be an ideal order in which to investigate these possible interrelationships.

Endophytic symbioses between fungi and the aerial parts of plants

The concept that symbioses between endophytic fungi and shoots, previously often referred to as symptomless parasitism, are both abundant and mutualistic is rapidly gaining ground following work by Carroll, Clay and Müller (e.g. Carroll, 1984; Carroll & Carroll, 1978; Clay, 1984; Luginbühl & Müller, 1980a & b; Petrini & Dreyfus, 1981; Petrini, Stone & Carroll, 1982; Riesen & Sieber, 1985).

This upsurge in interest warrants a more concise terminology for these associations which I propose to call 'mycophylla' (singular: 'mycophyllon'; adjective: 'mycophyllous'). Just as mycorrhizas also encompass infection of rhizoids and rhizomes, mycophylla would include infections of both aerial stems and leaves. In line with the all-embracing definition of mycorrhizas of Harley (1959), mycophylla are associations of fungi and photosynthetic organs constant in structure and development and consistently present and functional under natural conditions. Consequently, infections of heterotrophic archegoniate gametophytes remain as mycorrhizas whereas those of photosynthetic gametophytes become mycophylla.

Bradshaw (1959) showed that infections of *Agrostis capillaris* (= *A. tenuis*) by *Epichloe typhina* could be ' . . . of advantage to plants growing under conditions where vegetative growth is important . . . '. Infection of other grasses by species from other clavicipitaceous genera (*Atkinsonella*, *Balansia*) and the anamorph of *Epichloe*, *Acremonium*, also enhance vegetative vigour (see Clay, 1986). This phenomenon was investigated experimentally by Clay (1984) who showed that infected ramets had higher rates of survival and growth, but lower reproductive capacities, than uninfected ones. This increased fitness of ramets at the expense of production of genets has been demonstrated to be due to production of alkaloids in infected tissues which deter both invertebrate and vertebrate predators. Work in New Zealand, reviewed by Clay (1986), has been especially revealing. Movement of photosynthetically fixed carbon from grass to fungus has been demonstrated by Thrower & Lewis (1973) for the *Agrostis stolonifera/E. typhina* combination. The fungi may well be obligate biotrophs in the sense of Lewis (1973). It would be of interest to determine whether the mycophylla of leafy liverworts (Pocock & Duckett, 1985) are less palatable to predators than axenically raised gametophytes.

It seems probable that mycophylla arose early in the history of the land flora in response to predatory pressure and may even pre-date vascular plants. As noted above, Gray (1985) believes that a bryophyte-like flora existed in the Ordovician and, certainly, endophytic hyphae were present in the aerial parts of plants from the Devonian onwards (Pirozynski,

1976). Whether the outcome of these associations was mutualistic or not cannot, of course, be assessed from the fossil record.

Lichens

Almost 50% of ascomycetous species (ca. 13 500) are associated with cyanobacteria, Chlorophyta (or both) or, in a few cases, eukaryotic algae from other phyla as lichens (Hawksworth & Hill, 1984). There are also a few basidiolichens (less than 2% of all lichens). In terms of the outcome of these symbioses, most lichens are clearly mutualistic for the associations are ecologically obligate, i.e. the fitness of both partners approaches zero when they are not together.

Pirozynski (1976) asserted that ascomycetes probably originated in the Mesozoic but more recent evidence (Sherwood-Pike & Gray, 1985) indicates that this group of fungi is much more ancient (Silurian). A land flora which included septate fungi was therefore probably contemporaneous with the origin of vascular plants. Indeed, it is possible that these vascular plants emerged from a soil surface encrusted and stabilised by cyanobacteria, algae and lichens (cf. Wright, 1985; and Schulten, 1985). Biogeographical arguments also suggest that lichens are pre-Mesozoic in origin (e.g. Sipman, 1983; Tehler, 1983; Hawksworth & Hill, 1984).

Ahmadjian (1970) has discussed the origin and evolution of lichens, Barrett (1983) co-evolutionary aspects and Smith, Muscatine & Lewis (1969) pointed out some similarities of the physiology of the Capnodiaceae and green-algal lichens. It seems probable that, as for mycorrhizas, there are several different kinds of lichen with polyphyletic origins within the ascomycetes (e.g. see Honegger, 1986 for a range of haustorial types in trebouxioid lichens). The taxonomic positions and morphological versatility of the fungi involved are well documented but a closer integration with the ecology of lichens as a group and the relationships of their fungi to free-living and agonistic relatives as well as to those showing other mutualistic modes is necessary to understand their evolution.

Epilogue

The above discussion has necessarily been selective in its approach and, often, speculative in its outcome. A major omission has been any quantitative or theoretical consideration of the costs and benefits involved (see Keeler, 1985). Assessment of fitness of fungi under natural conditions is especially difficult owing to their indeterminate growth form (cf. Rayner & Coates, Chapter 8; Caten, Chapter 15). Linked with this is the nature and frequency of sexual reproduction and other methods of generating genetic change. Law & Lewis (1983) and Law (1985) have stressed the preponderance of asexual reproduction among intracellular mutualistic symbionts including the fungi of V − A and ericoid mycorrhizas. Does the trend towards asexuality of the fungi in mycophyllous associations (e.g. the *Epichloe/Acremonium* relationship) indicate a relatively recent origin of this particular kind of association? How does the frequently observed suppression of sex in the host plant affect the evolution of these symbioses? The suggestion that soluble carbohydrates are involved in the origin of both agonistic mycorrhizas and the non-mycorrhizal condition is testable in so far as the study of the effects of sundry reducing disaccharides on the permeability of fungal membranes is amenable to experiment although, as always, interpretation of results in an evolutionary context should be made with care.

It is evident that there are many kinds of mycorrhizas. This should alert all to the probability of diverse kinds of mycophylla and lichens. In this way, broad generalisations should be avoided and, as is implicit in the above discussion, solutions to evolutionary enigmas will only emerge from inter-disciplinary approaches to particular problems. Evidence from taxonomy, anatomy, morphology, physiology and biochemistry must be integrated with that from ecology, biogeography, paleoclimatology, paleopedology and plate tectonics.

Acknowledgements I am especially grateful to K. A. Pirozynski, J. A. Raven, D. J. Read, D. C. Smith, R. Watling, A. D. M. Rayner and J. M. Palmer for much enlightening discussion. I also wish to thank them and the following for helpful comments, references, reprints and access to unpublished work: I. J. Alexander, R. C. Cooke, J. P. Grime, G. A. F. Hendry, A. J. E. Lyon, E. Müller, J. M. Trappe and M. Whitfield. I am indebted to Mrs Patricia Miles for her advice on derivation of terms from Greek. None of these must bear any responsibility for the views expressed.

References

Abuzinadah, R. A., Finlay, R. D. & Read, D. J. (1986). The role of proteins in the nitrogen nutrition of ectomycorrhizal plants. II. Utilization of protein by mycorrhizal plants of *Pinus contorta. New Phytologist*, **103**, 495 − 506.

Abuzinadah, R. A. & Read, D. J. (1986a). The role of proteins in the nitrogen nutrition of ectomycorrhizal plants. I. Utilization of peptides and proteins by ectomycorrhizal fungi. *New Phytologist*, **103**, 481 − 493.

174   *David H. Lewis*

Abuzinadah, R. A. & Read, D. J. (1986b). The role of proteins in the nitrogen nutrition of ectomycorrhizal plants. III. Protein utilization by *Betula, Picea* and *Pinus* mycorrhizal association with *Hebeloma crustuliniforme. New Phytologist*, **103**. 507 – 514.

Ahmadjian, V. (1970). The lichen symbiosis: its origin and evolution. *Evolutionary Biology*, **4**, 163 – 184.

Alexander, I. J. (1987). Systematics and ecology of ectomycorrhizal legumes. In *Advances in Legume Biology*, ed. Stirton, C. H. & Zarruchi, J. L., in press. Missouri, USA: Missouri Botanical Garden.

Alexander, I. J. & Högberg, P. (1986). Ectomycorrhizas of tropical angiospermous trees. *New Phytologist*, **102**, 541 – 549.

ap Rees, T. (1984). Sucrose metabolism. In *Storage Carbohydrates in Vascular Plants: Distribution, Physiology and Metabolism*, ed. Lewis, D. H., pp. 53 – 73. Cambridge University Press.

Ashford, A. E. & Allaway, W. G. (1982). A sheathing mycorrhiza on *Pisonia grandis* R. Br. (Nyctaginaceae) with development of transfer cells rather than Hartig net. *New Phytologist*, **90**, 511 – 519.

Ashford, A. E. & Allaway, W. G. (1985). Transfer cells and Hartig net in the root epidermis of the sheathing mycorrhiza of *Pisonia grandis* R. Br. from Seychelles. *New Phytologist*, **100**, 595 – 612.

Avigad, G. (1982). Sucrose and other disaccharides. In *Plant Carbohydrates, I. Intracellular Carbohydrates*, ed. Loewus, F. A. & Tanner, W., pp. 217 – 347. Berlin: Springer-Verlag.

Barrett, J. A. (1983). Plant – fungus symbioses. In *Coevolution*, ed. Futuyama, D. J. & Slatkin, R. M., pp. 137 – 160. Sunderland, Massachusetts: Sinauer Associates.

Bradshaw, A. D. (1959). Population differentiation in *Agrostis tenuis* Sibth. II. The incidence and significance of infection by *Epichloe typhina. New Phytologist*, **58**, 310 – 315.

Broeker, W. S. (1970). A boundary condition on the evolution of atmospheric O_2. *Journal of Geophysical Research*, **75**, 3553 – 3557.

Brownlee, C., Duddridge, J. A., Malibari, A. & Read, D. J. (1983). The structure and function of mycelial systems of ectomycorrhizal roots with special reference to their role in forming inter-plant connections and providing pathways for assimilate and water transport. *Plant and Soil*, **71**, 433 – 443.

Carroll, G. C. (1987). The biology of endophytism in plants with particular reference to woody perennials. In *Microbiology of the Phyllosphere*, ed. Fokkema, N. & van den Heuval, J., Cambridge University Press.

Carroll, G. C. & Carroll, F. E. (1978). Studies on the incidence of coniferous needle endophyte in the Pacific North-West. *Canadian Journal of Botany*, **56**, 3034 – 3043.

Clark, F. R. S. & Russell, D. A. (1981). Fossil charcoal and the paleoatmosphere. *Nature*, **290**, 428.

Clay, K. (1987). Grass endophytes. In *Microbiology of the Phyllosphere*, ed. Fokkema, N. & van den Heuval, J., Cambridge University Press.

Clay, K. (1984). The effect of the fungus *Atkinsonella hypoxylon* (Clavicipitaceae) on the reproductive system and demography of the grass *Danthonia spicata. New Phytologist*, **98**, 165 – 175.

Cope, M. J. & Chaloner, W. G. (1980). Fossil charcoal as evidence of past atmospheric composition. *Nature*, **283**, 647 – 649.

Cope, M. J. & Chaloner, W. G. (1981). Reply to Chark and Russell. *Nature*, **290**, 428.

Dalpé, Y. (1986). Axenic synthesis of ericoid mycorrhiza in *Vacinum angustifolium* Ait.

by *Oidiodendron* species. *New Phytologist*, **103**, 391 − 396.

Dennis, R. L. (1970). A middle Pennsylvanian basidiomycete mycelium with clamp connections. *Mycologia*, **62**, 578 − 584.

Edwards, D. & Fanning, U. (1985). Evolution and environment in the late Silurian − early Devonian: the rise of the pteridophytes. *Philosophical Transactions of the Royal Society, series B*, **309**, 147 − 165.

Englander, L. (1982). Taxonomy of the fungi forming endomycorrhizae. B. Endomycorrhizae by septate fungi. In *Methods and Principles of Mycorrhizal Research*, ed. Schenck, N. C., pp. 11 − 13. St. Paul, Minnesota: The American Phytopathological Society.

Epstein, E. (1972). *Mineral Nutrition of Plants: Principles and Perspectives*. New York: John Wiley.

Finlay, R. D. & Read, D. J. (1986). The structure and function of the vegetative mycelium of ectomycorrhizal plants. I. Translocation of ^{14}C-labelled carbon between plants interconnected by a common mycelium. *New Phytologist*, **103**, 143 − 156.

Francis, R. & Read, D. J. (1984). Direct transfer of carbon between plants connected by vesicular − arbuscular mycorrhizal mycelium. *Nature*, **307**, 53 − 56.

Franz, G. (1982). Glycosylation of heterosides (glycosides). In *Plant Carbohydrates, I. Intracellular Carbohydrates*, ed. Loewus, F. A. & Tanner, W., pp. 386 − 393. Berlin: Springer-Verlag.

Gininazzi-Pearson, V. & Gianinazzi, S. E. (1986). *Proceedings of the First European Congress on Mycorrhizae*. Dijon: INRA.

Gray, J. (1985). The microfossil record of early land plants: advances in understanding of early terrestrialisation, 1970 − 1984. *Philosophical Transactions of the Royal Society, series B*, **309**, 167 − 195.

Grime, J. P. (1974). Vegetation classification by reference to strategies. *Nature*, **250**, 26 − 31.

Grinsted, M. J., Hedley, M. J., White, R. E. & Nye, P. H. (1982). Plant-induced changes in the rhizosphere of rape (*Brassica napus* var. Emerald) seedlings. I. pH change and the increase in P concentration in the soil solution. *New Phytologist*, **91**, 19 − 29.

Harley, J. L. (1959). *The Biology of Mycorrhiza*. 1st edition. London: Leonard Hill.

Harley, J. L. (1969). *The Biology of Mycorrhiza*. 2nd edition. London: Leonard Hill.

Harley, J. L. & Smith, S. E. (1983). *Mycorrhizal Symbiosis*. London: Academic Press.

Hart, M. H. (1978). The evolution of the atmosphere of the Earth. *Icarus*, **33**, 23 − 39.

Haselwandter, K. & Read, D. J. (1982). The significance of a root-fungus association in two *Carex* species of high-alpine communities. *Oecologia*, **53**, 352 − 354.

Hawksworth, D. L. & Hill, D. J. (1984). *The Lichen-Forming Fungi*. Glasgow & London: Blackie.

Hedley, M. J., Nye, P. H. & White, R. E. (1983). Plant-induced changes in the rhizosphere of rape (*Brassica napus* var. Emerald) seedlings. IV. The effect of rhizosphere phosphorus status on the pH, phosphatase activity and depletion of soil phosphorus fractions in the rhizosphere and on the cation − anion balance in the plants. *New Phytologist*, **95**, 69 − 82.

Hedley, M. J., Nye, P. H. & White, R. E. (1982). Plant-induced changes in the rhizosphere of rape (*Brassica napus* var. Emerald) seedlings. II. Origin of the pH change. *New Phytologist*, **91**, 31 − 44.

Hedley, M. J., White, R. E. & Nye, P. H. (1982). Plant-induced changes in the rhizosphere of rape (*Brassica napus* var. Emerald) seedlings. III. Changes in L value, soil phosphate fractions and phosphatase activity. *New Phytologist*, **91**, 45 − 56.

Herold, A. (1978). Induction of wilting by mannose in spinach beet leaves. *New Phytologist*, **81**, 299 − 305.

Herold, A. & Lewis, D. H. (1977). Mannose and green plants: occurrence physiology and metabolism, and use as a tool to study the role of orthophosphate. *New Phytologist*, **79**, 1 − 40.

Holland, H. D. (1984). *The Chemical Evolution of the Atmosphere and Oceans*. Princeton: Princeton University Press.

Holland, H. D., Lazar, P. & McCaffrey, M. (1986). Evolution of the atmosphere and oceans. *Nature*, **320**, 27 − 33.

Honegger, R. (1986). Ultrastructural studies in lichens. I. Haustorial types and their frequencies in a range of lichens with trebouxioid photobionts. *New Phytologist*, **103**, 785 − 795.

Jacquelinet-Jeanmougin, S. & Gianinazzi-Pearson, V. (1983). Endomycorrhizas of the Gentianaceae. I. The fungi associated with *Gentiana lutea* L. *New Phytologist*, **95**, 663 − 666.

Kühner, R. (1984). Some main lines of classification of the gill fungi. *Mycologia*, **76**, 1059 − 1074.

Keeler, K. H. (1985). Cost: benefit models of mutualism. In *The Biology of Mutualism: Ecology and Evolution*, ed. Boucher, D. H., pp. 100 − 127. Beckenham: Croom Helm.

Kirk, T. K. & Fenn, P. (1982). Formation and action of the ligninolytic system in basidiomycetes. In *Decomposer Basidiomycetes: Their Biology and Ecology*, British Mycological Society Symposium volume 7, ed. Frankland, J. C., Hedger, J. N. & Swift, M. J., pp. 67 − 90. Cambridge University Press.

Law, R. (1985). Evolution in a mutualistic environment. In *The Biology of Mutualism: Ecology and Evolution*, ed. Boucher, D. H., pp. 145 − 170. Beckenham: Croom Helm.

Law, R. & Lewis, D. H. (1983). Biotic environments and the maintenance of sex − some evidence from mutualistic symbioses. *Biological Journal of the Linnean Society*, **20**, 249 − 271.

Lewis, D. H. (1973). Concepts in fungal nutrition and the origin of biotrophy. *Biological Reviews*, **48**, 261 − 278.

Lewis, D. H. (1985). Symbiosis and mutualism: crisp concepts and soggy semantics. In *The Biology of Mutualism: Ecology and Evolution*, ed. Boucher, D. H., pp. 29 − 39. Beckenham: Croom Helm.

Lewis, D. H. (1986). Inter-relationships between carbon nutrition and morphogenesis in mycorrhizas. In *Proceedings of the First European Symposium on Mycorrhizae*, ed. Gianinazzi-Pearson, V. & Gianinazzi, S., pp. 85 − 100. Dijon: INRA.

Lewis, D. H. (1984). Occurrence and distribution of storage carbohydrates in vascular plants. In *Storage Carbohydrates in Vascular Plants: Distribution, Physiology and Metabolism*, ed. Lewis, D. H., pp. 1 − 52. Cambridge University Press.

Luginbühl, M. & Müller, E. (1980b). Untersuchungen über endophytische Pilze. I. Infektionswege von Endophyten bei *Hedera helix* L. *Berichte der Schweizerischen Botanischen Gesellschaft*, **90**, 244 − 250.

Luginbühl, M. & Müller, E. (1980a). Endophytische Pilze in den oberirdischen Organen von 4 gemeinsam an gleichen Standorten wachsenden Pflanzen (*Buxus, Hedera, Ilex, Ruscus*). *Sydowia, Annales Mycologici Ser. II*, **33**, 185 − 209.

Malloch, D. W., Pirozynski, K. A. & Raven, P. H. (1980). Ecological and evolutionary significance of mycorrhizal symbioses in vascular plants (a review). *Proceedings of the National Academy of Sciences, USA*, **77**, 2113 − 2118.

Margulis, L. & Schwartz, K. V. (1982). *Five Kingdoms: An Illustrated Guide to the Phyla of Life on Earth.* San Francisco: Freeman.

McGee, P. A. (1985). Lack of spread of endomycorrhizas of *Centaurium* (Gentianaceae). *New Phytologist*, **101**, 451 − 458.

Nicholson, T. (1975). Evolution of vesicular − arbuscular mycorrhizas. In *Endomycorrhizas*, ed. Sanders, F. E., Mosse, B. & Tinker, P. B., pp. 25 − 34. London: Academic Press.

Nye, P. H. & Tinker, P. B. (1977). *Solute Movement in the Soil − Root System.* Oxford: Blackwell Scientific Publications.

Peterson, A. & Lundquist, K. (1985). Radical breakdown of lignin. *Nature*, **316**, 575 − 576.

Petrini, O. & Dreyfuss, M. (1981). Endophytische Pilze in Epiphytischen Araceae, Bromeliaceae und Orchidaceae. *Sydowia, Annales Mycologici Ser. II*, **34**, 135 − 148.

Petrini, O., Stone, J. K. & Carroll, F. E. (1982). Endophytic fungi in evergreen shrubs in Western Oregon: a preliminary study. *Canadian Journal of Botany*, **660**, 789 − 796.

Pirozynski, K. A. (1976). Fossil fungi. *Annual Review of Phytopathology*, **14**, 237 − 246.

Pirozynski, K. A. (1981). Interactions between fungi and plants through the ages. *Canadian Journal of Botany*, **59**, 1824 − 1827.

Pirozynski, K. A. & Malloch, D. W. (1975). The origin of land plants: a matter of mycotrophism. *BioSystems*, **6**, 153 − 164.

Pocock, K. & Duckett, J. G. (1985). The alternative mycorrhizas: fungi and hepatics. *Bulletin of the British Bryological Society*, **45**, 10 − 11.

Raven, J. A. (1977). The evolution of vascular plants in relation to supracellular transport processes. *Advances in Botanical Research*, **5**, 153 − 219.

Raven, J. A. (1985). Comparative physiology of plant and arthropod land adaptation. *Philosophical Transactions of the Royal Society, series B*, **309**, 273 − 288.

Raven, J. A. (1986). Evolution of plant life forms. In *On the Economy of Plant Form and Function*, ed. Givnish, T., pp. 421 − 492. Cambridge University Press.

Raven, J. A. (1984). Physiological correlates of the morphology of early vascular plants. *Botanical Journal of the Linnean Society*, **88**, 105 − 126.

Raven, J. A., Smith, S. E. & Smith, F. A. (1978). Ammonium assimilation and the role of mycorrhizas in climax communities in Scotland. *Transactions of the Botanical Society of Edinburgh*, **43**, 27 − 35.

Raven, P. H. & Axelrod, D. I. (1974). Angiosperm biogeography and past continental movements. *Annals of the Missouri Botanical Garden*, **61**, 539 − 673.

Rayner, A. D. M. (1986). Water and the origins of decay in trees. In *Water, Fungi and Plants*, British Mycological Society Symposium volume 11, ed. Ayres, P. G. & Boddy, L., pp. 321 − 341. Cambridge University Press.

Read, D. J. (1984). The structure and function of the vegetative mycelium of mycorrhizal roots. In *The Ecology and Physiology of the Fungal Mycelium*, British Mycological Society Symposium volume 8, ed. Jennings, D. H. & Rayner, A. D. M., pp. 215 − 240. Cambridge University Press.

Read, D. J. (1983). The biology of mycorrhiza in the Ericales. *Canadian Journal of Botany*, **61**, 985 − 1004.

Reddy, C. A. (1984). Physiology and biochemistry of lignin degradation. In *Current Perspectives in Microbial Ecology*, ed. Klug, J. & Reddy, C. A., pp. 558 − 571. Washington, DC: American Society for Microbiology.

Riesen, T. & Sieber, T. (1985). *Endophytic Fungi in Winter Wheat* (Triticum aestivum

L.). Zurich: ADAG Administration and Druck AG.

Schopf, J. M. (1952). Was decay important in origin of coal? *Journal of Sedimentary Petrology*, **22**, 61 − 69.

Schulten, J. A. (1985). Soil aggregates by cryptogams of a sand prairie soil. *American Journal of Botany*, **72**, 1657 − 1661.

Sherwood-Pike, M. A. & Gray, J. (1985). Silurian fungal remains: probable records of the class Ascomycetes. *Lethaia*, **18**, 1 − 20.

Sipman, H. J. M. (1983). A monograph of the lichen family Megalosporaceae. *Bibliotheca Lichenologica*, **18**, 1 − 241.

Smith, D., Muscatine, L. & Lewis, D. (1969). Carbohydrate movement from autotrophs to heterotrophs in parasitic and mutualistic symbiosis. *Biological Reviews*, **44**, 17 − 90.

Stubblefield, S. P., Taylor, T. N. & Beck, C. B. (1985). Studies on Paleozoic fungi. IV. Wood-decaying fungi in *Callixylon newberryi* from the Upper Devonian. *American Journal of Botany*, **72**, 1765 − 1774.

Tehler, A. (1983). The genera *Dirina* and *Roccellina* (Roccellaceae). *Opera Botanica*, **70**, 1 − 86.

Thrower, L. B. & Lewis, D. H. (1973). Uptake of sugars by *Epichloe typhina* (Pers. ex Fr.) Tul. in culture and from its host, *Agrostis stolonifera* L. *New Phytologist*, **72**, 501 − 508.

Trappe, J. M. (1962). Fungus associates of ectomycorrhizae. *Botanical Reviews*, **28**, 538 − 606.

Trappe, J. M. (1982). Synoptic keys to the genera and species of zygomycetous fungi. *Phytopathology*, **72**, 1102 − 1108.

Trappe, J. M. & Schenck, N. C. (1982). Taxonomy of the fungi forming endomycorrhizae. A. Vesicular − arbuscular mycorrhizal fungi (Endogonales). In *Methods and Principles of Mycorrhizal Research*, ed. Schenck, N. C., pp. 1 − 9. St. Paul, Minnesota: The American Phytopathological Society.

Trojanowski, J., Haider, K. & Hüttermann, A. (1984). Decomposition of ^{14}C-labelled lignin, holocellulose and lignocellulose by mycorrhizal fungi. *Archives of Microbiology*, **139**, 202 − 206.

Walker, J. C. G. (1977). *Evolution of the Atmosphere*. London: Collier Macmillan.

Warcup, J. H. & McGee, P. A. (1983). The mycorrhizal associations of some Australian Asteraceae. *New Phytologist*, **95**, 667 − 672.

Watling, R. (1982). Taxonomic status and ecological identity in the basidiomycetes. In *Decomposer Basidiomycetes*, British Mycological Society Symposium volume 7, ed. Frankland, J. C., Hedger, J. N. & Swift, M. J., pp. 1 − 32. Cambridge University Press.

Wright, V. P. (1985). The precursor environment for vascular plant colonization. *Philosophical Transactions of the Royal Society, series B*, **309**, 143 − 145.

12

The evolution and adaptation of sexual reproductive structures in the Ascomycotina

DAVID L. HAWKSWORTH

CAB International Mycological Institute, Ferry Lane, Kew, Surrey TW9 3AF, UK

Introduction

Sutton (1986) recently re-iterated the view of Gregory (1952) that the possession of any organ, substance, or character in a fungus is evidence that it has some function. We can be confident that this supposition is correct when similar characters arise in what appear to be phylogenetically remote fungi, even if plausible theories as to their functions elude us. With this in mind it is vital in the circumscription of orders, families, and even genera, as well as phylogenetic reconstructions, to be aware of features for which there may be strong selection pressures and which consequently may evolve repeatedly. Over-emphasis on such convergently arising characters has resulted in taxa which later prove to be polyphyletic, as in the Plectomycetes (Malloch, 1981), and, in view of their importance in the classification of Ascomycotina, it is surprising that few authors have questioned the significance of these characters in this group; Eriksson (1981) and Sherwood (1981) are stimulating exceptions. By contrast, in Basidiomycotina, where the functional convergence of fruitbody shape and hymenial configuration used to delimit Friesian genera and families has long been recognised, more natural schemes were developed much earlier (Donk, 1971; Heim, 1971).

This chapter, therefore, draws attention to a selection of recurring characters in ascomycete sexual structures which must thereby have an adaptive value. Some, for example, the production of non-ostiolate ascomata or retention of hyaline non-septate ascospores, are a result of 'paedomorphosis', the retention of youthful ancestral features (Eriksson, 1981); others, such as modification of ascospore number, are a result of convergent evolution.

Table 12.1. *Numbers of the 43 orders of Ascomycotina including different types of ascomata*

Types of ascoma	Number of orders	Percentage of orders
Cupuliform (apothecioid)	18	38
Effuse	5	11
Globose (non-ostiolate; cleistotchecioid)	12	26
Lirelliform	7	15
Pyriform (ostiolate; perithecioid)	20	43
Setose	13	28
Stipitate	10	21
Stromatic	14	30

Ascomata

The ascus bearing structure of the Ascomycotina, the ascoma, shows a remarkable range of morphological diversity. This diversity has provided a basis for delimitation of orders, families and genera, but since it is at least partially underlain by convergent evolution this has contributed to considerable instability in classification (Hawksworth, 1985). The convergence arises because particular types of ascomata evolve in response to distinctive strategies for spore liberation and dispersal which occur independently in different orders (Table 12.1), and give rise to recognisable series of intergrading forms (Fig. 12.1).

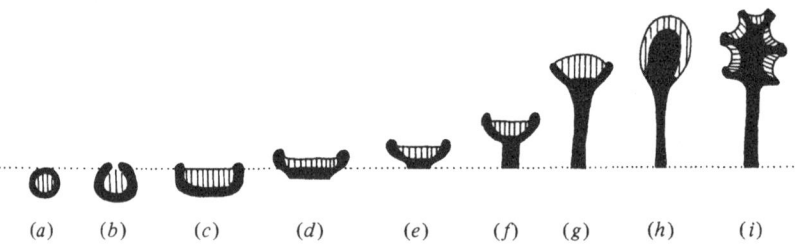

(a)   (b)   (c)   (d)   (e)   (f)   (g)   (h)   (i)

Fig. 12.1. Schematic representation of selected ascomatal types. (a) non-ostiolate (cleistothecioid), (b) ostiolate (perithecioid), (c) immersed cupulate, (d) erumpent cupulate, (e) sessile cupulate, (f) short-stipitate, (g) & (h) large-stipitate types, (i) stromatic-stipitate. (c) to (i) are all apothecioid. The dotted line indicates the surface of the substratum.

Table 12.2. *Examples of pairs of genera with ostiolate and non-ostiolate ascomata in different orders of Ascomycotina*

Order	Ostiolate	Non-ostiolate
Dothideales	*Caryospora*	*Zopfia*
	Sporomiella	*Preussia*
Hypocreales	*Pyxidiophora*	*Mycorhynchidium*
Microascales	*Lophotrichus*	*Kernia*
	Petriella	*Pseudallescheria*
Sordariales	*Chaetomium*	*Chaetomidium*
	Coniochaeta	*Coniochaetidium*
	Gelasinospora	*Anixiella*
	Melanospora	*Microthecium*
	Strattonia	*Zopfiella*

Ostiolate and non-ostiolate ascomata

The occurrence of ostiolate (perithecioid) vs. non-ostiolate (cleistothecioid) ascomata was formerly taken as a fundamental difference meriting separation at the rank of order or above, but the evidence that non-ostiolate taxa are often secondarily derived is now overwhelming. However, whilst the view of von Arx (1973) that this difference is, therefore, often not even a suitable generic, or in some cases a species, criterion is now generally accepted, non-ostiolate ascomata alone are seen in the Ascosphaerales, Elaphomycetales, Eurotiales and Onygenales and exclusively hypogeous families of Pezizales (e.g. Terfeziaceae). On the other hand, where 'genera' are separated only in this feature the character must be recently derived and the list of pairs of such 'genera' is considerable (Table 12.2); many are being united.

The secondary development of closed ascomata is associated almost exclusively with hypogeous, root- or dung-inhabiting, or food spoilage fungi and is related to a loss of active spore discharge and often also the production of thick-walled ascomata able to withstand desiccation (e.g. *Elaphomyces, Eupenicillium*).

Non-ostiolate ascomata sometimes have walls composed of angular plates which break apart at maturity. These 'cephalothecoid' plates are seen in a wide variety of ascomycetes (Table 12.3), but do not always originate in the same way (P. F. Cannon, unpublished). An intermediate stage is seen in certain *Cercophora* species and *Weddellomyces* where only the upper parts of the flask-shaped ascomata are cephalothecoid.

Table 12.3. *Genera of Ascomycotina including species with ascomata composed of 'cephalothecoid' plates*

Dothideales	*Eremomyces*	*Richonia*
	Kirschsteiniothelia	*Testudina*
	Lepidosphaeria	*Weddellomyces*
	Neotestudina	*Zopfia*
	Phaeotrichum	*Zopfiofoveola*
	Rechingeriella	
Eurotiales	*Aporothielavia*	*Fragosphaeria*
	Cephalotheca	*Leuconeurospora*
	Cryptendoxyla	*Samarospora*
Microascales	*Batistia*	
Onygenales	*Albertiniella*	
Sordariales	*Cercophora*	*Monosporascus*
	Chaetomidium	*Rhytidiospora*
	Diplogelasinospora	*Zopfiella*
Uncertain	*Argynna*	

Cupuliform and Pyriform ascomata

The assumed primary distinction between disc- or cupuliform and flask-shaped or pyriform ascomata was evident in the separation of Discomycetes and Pyrenomycetes. However, a gradual widening of the ostiole and increase in the number of asci and hamathecial elements can lead from perithecioid to apothecioid forms (Fig. 12.1), and in a variety of orders both types can be found, e.g. the Ostropales and Pertusariales, whilst in others the opening can be extremely irregular, e.g. the Nitschkiaceae (Sordariales). Moreover, the tendency to form carbonaceous apothecia which can close on drying and re-open on wetting is especially frequent in lignicolous ascomycetes in xeric situations (Sherwood, 1981) and marginal setae can play a key role in this mechanism in several groups (e.g. *Skyttea*).

The production of cup-like structures has been related to rain-splash dispersal as have the cupuliform podetia ('scyphi') of certain *Cladonia* species (Brodie, 1951). However, in the latter case it is soredia that are splashed out and this mechanism is unlikely to be relevant to actively discharged spores, for which the main advantage may lie in the production of double eddy currents within which they are caught up (Brodie & Gregory, 1953).

Lirelliform ascomata

Elongate ascomata opening by a narrow or broad slit could be derived from rounded apothecioid ascomata by unequilateral growth. They may have the same function as some closable discoid apothecia as most taxa producing such lirelliform ascomata occur in xeric situations. Examples from a variety of orders are *Aulographum*, *Graphina*, *Lophodermium*, *Opegrapha* and *Wadeana*. *Opegrapha* seems to be a lirelliform counterpart of *Lecanactis* as there are some intermediate species (e.g. *O. lyncea*). In *Colpoma quercinum* the ascomata can vary from elongate discs to slit-like lirellae. Lirelliform ascomata also tend to lie parallel to the circumference of the bark on which they occur: this may enable them to grow concurrently with the bark.

Stipitate ascomata

The stipitate ascoma is a means of raising the point at which spores are discharged, so increasing their chances of entry into more turbulent air and reducing the probability of their obstruction by fungal tissue or other objects. The extent of development of the stipe can vary considerably within particular genera (e.g. *Dasyscyphus pygmaeus* vs. *D. bicolor*) or even within single species (e.g. *Calicium corynellum*). The largest stipes occur in the Helvellaceae, Humariaceae and Morchellaceae, attaining 1 m in *Geopyxis cacabus*. In the Lecanorales, the apothecial stipes of the Cladoniaceae commonly contain algal cells and the resultant podetia (i.e. lichenised stipes; Ahti, 1982) may represent the total thallus at maturity.

Since in all orders containing stipitate species, sessile representatives are also found, it is evident that stipes have developed independently in response to the evolutionary advantage at which it places terricolous Ascomycotina.

Stromatic ascomata

Ascomata are often grouped together in a communal stroma. Where the stromata are vertical, the resultant effect is functionally comparable to that achieved by the elongation of apothecial stipes (e.g. *Barya*, *Claviceps*, *Cordyceps*, *Podosordaria*, *Trichoglossum*, *Xylaria*). Stromata uniting ascomata may also be globose, as in *Cyttaria*, *Daldinia* and *Stromatothecia*, but lateral unification is most common, as in *Camarops*, *Dothidea*, *Hypocreopsis*, *Hypoxylon* and *Ustulina*. Circles of ascomata united by their elongate ostioles are seen in numerous members of the Diatrypaceae and Valsaceae. Corner (1964) regarded forms with extensive stromata (e.g. *Thamnomyces*) as retaining ancestral traits compared to related taxa where the sterile tissue is much reduced, but while this may be so in the Xylariales there is little supporting evidence for this in most other non-lichenised orders. With respect to lichenised orders, if stromata are defined as vegetative tissues uniting discrete ascomata, the lichenised thalli of numerous foliose and fruticose genera can be interpreted as lichenised stromata just as podetia are equivalent to lichenised stipes.

Setose ascomata

The production of setae on the outer walls of ascomata occurs in a wide variety of fungi including the Dothideales, Erysiphales, Helotiales, Hypocreales, Onygenales, Ostropales, Pezizales, Sordariales, Trichosphaeriales, and Xylariales. The function of setae is obscure but in some instances they may deter grazing by invertebrates (Wicklow, 1979) or aid dispersal by attaching whole ascomata to animals brushing against them.

In the Ostropales and *Trichothyrina*, setae around the ostiole have a role in closing the opening during dry conditions, whereas ostiolar setae in *Ceratocystis*, *Chaetomium* and *Melanospora* support a mass of spores, so that they are more likely to be detached and adhere to passing invertebrates.

Hamathecial tissues

The nature of the interascal tissues, the hamathecium, relates to the ontogeny of the ascomata and so is of particular systematic importance. However, superficially rather similar hamathecial elements can originate in different ways, as in the case of pseudoparaphyses and paraphysoids (Eriksson, 1981). Periphyses may function in closing globose or flask-shaped ascomata to reduce water loss in dry periods, but the role of the elements in the cavity of such ascomata is obscure. In apothecioid types, however, the tips of the paraphyses form a layer over the asci which can close in dry conditions by a variety of mechanisms. These include the production of swollen (e.g. *Orbilia*) or apically branched and expanded (e.g. *Buellia*) tips, or the deposition of amorphous or crystalline deposits between the tips or even above them to form an 'episamma' layer (e.g. *Sarea*). Supporting the theory that such elaborations are an adaptation to more xeric conditions is the fact that simple paraphyses not thickened or branched at the tips are almost always found in genera of constantly humid habitats.

Asci

Structure

Whilst the number of types of asci is much greater than recognised in earlier decades, trends within some of these main types can be related to modifications in spore liberation or dispersal and too much emphasis on ascus types without regard to these modifications can again lead to closely related taxa being placed in orders remote from one another (von Arx, 1979). Foremost of these trends is the loss of apical apparatus associated with a change from active to passive discharge of the ascospores where the asci either deliquesce at maturity to form a mass of ascospores oozing out of the ostiole for insect dispersal (e.g. *Chaetomium, Melanospora, Monosporascus, Neocosmospora*), or form a dry mazaedial mass suited to wind dispersal (e.g. *Chaenotheca, Onygena, Roesleriella*). Ascus structure and function is also simplified in genera with closed non-ostiolate ascomata (e.g. *Genea, Pseudeurotium, Talaromyces*) or where there is an adaptation to aquatic or marine environments (e.g. Halosphaeriales).

Extension of asci to a distant ostiole or through an epithecium can be facilitated by asci with elongated stalks (e.g. *Bertia, Diatrype*), asci forming in chains and becoming detached (e.g. *Diaporthe*), extension of one or more wall layers in either a fissitunicate (Dothideales) or rostrate (Lecanorales) manner, or by oozing of the contents (Verrucariales).

Number of spores

Two trends in spore number can be distinguished, the production of larger numbers than eight or reductions from that to even the monosporous condition. A large number of spores increases the probability of a suitable substratum being found and can be achieved by the fragmentation of the original eight ascospores (e.g. *Creopus, Preussia, Raciborskiella, Schizoxylon*), or by being cut simultaneously from the protoplasm (e.g. *Acarospora, Caloplaca, Muellerella, Sarea, Thelebolus*), the failure of all but a few spores to mature (e.g. *Nectria parmeliae*) or by the formation of conidia within the ascus (e.g. *Claussenomyces atrovirens, Nectria aquifolii, Rhamphoria pyriformis*). *Thelebolus nanus* produces a single ascus at a time, each with 256 ascospores which it projects as one propagule to a vertical distance of up to 7 cm (Ingold, 1953).

In contrast, the formation of fewer larger ascospores leads to propagules with substantially more nutritional resources which are consequently able to survive for longer while awaiting contact with the appropriate substratum or impose greater inoculum potential thereafter. Progressive reductions in spore number can be seen in different species of *Megalo-spora*, *Nectria*, *Paranectria*, *Pertusaria*, and *Solorina*. Single-spored asci are found in, for example, '*Lecania*' *sulphureorufa*, *Monosporascus cannonballus*, *Megalospora gompholoma*, *Mycoblastus sanguinarius*, *Pertusaria velata* (to 310 x 90 μm), and *Xerotrema megalospora*. Larger propagules can also be projected for increased distances (Ingold, 1965). Increased ascospore size may arise through polyploidy in *Neocosmospora* (Cannon & Hawksworth, 1984).

Ascospores

The colour, size, shape and texture of spores should be viewed as probable functional adaptations modified in evolution by requirements of liberation, of flotation in fluids, and ultimately of deposition and survival (Ingold, 1971; Gregory, 1973). Such functional adaptations in aquatic fungi are discussed by Webster (Chapter 13); here they will only be considered in the context of air-borne dispersal.

Shape

Whilst bipolar symmetrical ascospores tend to be correlated with passive dispersal, bipolar asymmetric spores are related to active dis-charge, the upper part of the spore always being the widest, so that at the moment of discharge all the hydrostatic pressure of the ascus is con-centrated to discharge and not to stretching (Ingold, 1965). Further, asymmetric spores tend to spin on their axis or move sideways rather than fall vertically (Gregory, 1973) and so may drift into crevices, underhangs, etc. Ellipsoidal spores may also be preferred to globose ones of the same volume as their rate of settling will tend to be lower (A. Fonda in Ingold, 1965).

Filiform spores tend to coil on release and this may be of considerable advantage in attachment to appropriate substrata and appear in represent-atives of a wide range of orders, for example, *Acrospermum*, *Bacidia*, *Belonia*, *Claviceps*, *Epichloe*, *Geoglossum*, *Lophodermium*, *Ophiobolus*, and *Trichonectria*.

Septation and colour

In many families, ranges from simple or uniseptate to multiseptate or muriform ascospores are found. These derived features have, together with colour, in some cases led to generic concepts which overlook fundamental characters of the ascomata themselves (Table 12.4). Further-more, the development of septation may follow different lines (Eriksson,

Table 12.4. *Examples of families in which genera have been delimited on the basis of ascospore colour and septation*

Family	Colourless		Brown	
	Trans-septate	muriform	Trans-septate	muriform
Graphidaceae	*Graphis*	*Graphina*	*Phaeographis*	*Phaeographina*
Lophiostomaceae	*Lophiotrema*	*Lophidiopsis*	*Lophiostoma*	*Platystomum*
Thelotremataceae	*Ocellularia*	*Thelotrema*	*Phaeotrema*	*Leptotrema*

1981) and different wall layers may be involved (euseptate vs. distoseptate). The development of a larger number of cells each of which can germinate independently at different times may clearly bring advantages in colonisation, while deposition of pigments in the spore walls may serve to reduce ultraviolet sensitivity and desiccation and so enhance longevity.

Appendages

Although optimally developed in the Lasiosphaeriaceae, gelatinous appendages of various types appear in a wide range of orders and families, for example, *Anthostomella, Caudospora, Discodiaporthe, Guignardia, Haplocystis, Pleospora scirpicola,* and *Rosellinia*. In most cases their function is to facilitate attachment to substrata, whereas in *Ascobolus immersus* whole groups of ascospores are bound by mucilage and can be discharged as single projectiles to 60 cm (Buller, 1909).

Ornamentation

The role of ascospore ornamentation is often obscure but may relate to increasing longevity by the production of additional wall layers, the patterns of spore germination, palatability to browsing invertebrates, or their attachment to appropriate substrata for germination. In general, basic ornamentation types vary little within genera, but parallel themes between genera are often evident: e.g. presence of a reticulum (as in *Ascobolus, Sphaerodes, Testudina*), pores (e.g. *Gelasinospora, Persiciospora*), lumpy excrescences (e.g. *Copromyces, Rechingeriella*), germ slits (e.g. *Coniochaeta, Delitischia, Hypoxylon, Lopadostoma, Scopinella*), parallel striae (e.g. species of *Astrosphaeriella, Caryospora, Didymosphaeria, Lophiostoma, Kirschsteiniothelia, Nectria, Neurospora*), and spines (e.g. species of *Humaria, Lamprospora, Peziza* and *Tuber*).

Reproductive strategies

The preceding sections have concentrated on the sexual reproductive bodies and propagules themselves, but phylogenetic traits can also be seen in reproductive strategies. Separate spermatia and ascogonia on a single thallus are almost certainly primitive and are especially common in Dothideales, Lecanorales and Peltigerales. The subsequent establishment of dikaryotic hyphal systems giving rise to asci which develop in a crozier-like manner is widespread in lichenised and non-lichenised groups and enables hymenia to be formed. Paired ascogonia and spermatia (e.g. *Pyronema*), dimorphic gametangia from the same or adjacent cells (e.g. *Endomyces*, Gymnoascaceae), and isomorphic gametangia (e.g. *Schizosaccharomyces*) tend to be associated with species with short life cycles in ephemeral habitats (i.e. ones with *R*-selected strategies; see Cooke & Whipps, Chapter 9) and are highly derived. In *Ascosphaera*, *Monascus* and the Thelebolaceae individual ascogonial cells may develop into asci (Kimbrough, 1984).

Further, progressive reduction may lead to the parasexual cycle itself with genetic recombination occurring without the production of specialised fruit bodies.

Alternatively, emphasis may be switched from sexual reproductive systems to asexual mechanisms for propagation and dispersal. These can be based on elaborations from the spermatial phase leading to lichenised or non-lichenised Coelomycetes and Hyphomycetes with no extant teleomorph, or vegetative hyphae forming specialised propagules in association with algal cells in the isidia, soredia and similar propagules of many Lecanorales (Bowler & Rundel, 1975). Conidia assume particular importance in dispersal where ascospore discharge is passive (Malloch, 1979) as in *Penicillium* where 'sclerotia' may represent relict non-functional *Eupenicillium*-like ascomata. In the absence of parasexual systems, such strategies are presumably evolutionary dead-ends but facilitate the rapid exploitation of particular hosts or ecological niches.

References

Ahti, T. (1982). The morphological interpretation of cladoniiform thalli in lichens. *Lichenologist*, **14**, 105 − 113.

von Arx, J. A. (1973). Ostiolate and non-ostiolate pyrenomycetes. *Proceedings Koninklijke Nederlandre Akademie van Wetenschappen C*, **76**, 289 − 296.

von Arx, J. A. (1979). Ascomycetes as Fungi imperfecti. In *The Whole Fungus*, vol. 1, ed. Kendrick, W. B., pp. 201 − 213. Ottawa: National Museums of Canada.

Bowler, P. A. & Rundel, P. W. (1975). Reproductive strategies in lichens. *Botanical Journal of the Linnean Society*, **70**, 325 – 340.

Brodie, H. J. (1951). The splash-cup dispersal mechanism in plants. *Canadian Journal of Botany*, **29**, 224 – 234.

Brodie, H. J. & Gregory, P. H. (1953). The action of wind in the dispersal of spores from cup-shaped plant structures. *Canadian Journal of Botany*, **31**, 402 – 410.

Buller, A. H. R. (1909). *Researches on Fungi*. Vol. 1. London: Longmans.

Cannon, P. F. & Hawksworth, D. L. (1984). A revision of the genus *Neocosmospora*. *Transactions of the British Mycological Society*, **82**, 673 – 688.

Corner, E. J. H. (1964). *The Life of Plants*. London: Weidenfeld & Nicolson.

Donk, M. A. (1971). Progress in the study of the classification of the higher Basidiomycetes. In *Evolution in the Higher Basidiomycetes*, ed. Petersen, R., pp. 3 – 25. Knoxville, Tenn.: University of Tennessee Press.

Eriksson, O. (1981). The families of bitunicate ascomycetes. *Opera Botanica*, **60**, 1 – 209.

Gregory, P. H. (1952). Fungus spores. *Transactions of the British Mycological Society*, **35**, 1 – 18.

Gregory, P. H. (1973). *Microbiology of the Atmosphere*, 2nd ed. Aylesbury: L. Hill.

Hawksworth, D. L. (1985). Problems and prospects in the systematics of the Ascomycotina. *Proceedings of the Indian Academy of Sciences, Plant Sciences*, **94**, 319 – 339.

Heim, R. (1971). The interrelationships between the Agaricales and Gasteromycetes. In *Evolution in the Higher Basidiomycetes*, ed. Petersen, R., pp. 505 – 534. Knoxville, Tenn.: University of Tennessee Press.

Ingold, C. T. (1953). *Dispersal in Fungi*. Oxford: Clarendon Press.

Ingold, C. T. (1965). *Spore Liberation*. Oxford: Clarendon Press.

Ingold, C. T. (1971). *Fungal Spores: Their Liberation and Dispersal*. Oxford: Clarendon Press.

Kimbrough, J. W. (1984). Life cycles and natural history of ascomycetes. In *Fungus – Insect Relationships*, ed. Wheeler, Q. & Blackwell, M., pp. 184 – 210. New York: Columbia University Press.

Malloch, D. (1979). Plectomycetes and their anamorphs. In *The Whole Fungus*, vol 1, ed. Kendrick, W. B., pp. 153 – 165. Ottawa: National Museums of Canada.

Malloch, D. (1981). The plectomycete centrum. In *Ascomycete Systematics*, ed. Reynolds, D. R., pp. 73 – 91. New York: Springer.

Sherwood, M. A. (1981). Convergent evolution in discomycetes from bark and wood. *Botanical Journal of the Linnean Society*, **82**, 15 – 34.

Sutton, B. C. (1986). Improvisations on conidial themes. *Transactions of the British Mycological Society*, **86**, 1 – 38.

Wicklow, D. T. (1979). Hair ornamentation and predator defence in *Chaetomium*. *Transactions of the British Mycological Society*, **72**, 107 – 110.

13

Convergent evolution and the functional significance of spore shape in aquatic and semi-aquatic fungi

JOHN WEBSTER

Department of Biological Sciences, University of Exeter, Exeter EX4 4PS, UK

Introduction

Spores of fungi are remarkably diverse in form, size, colour, surface coating and ornamentation. Despite this, there are a number of recurrent themes in their morphology (Sutton, 1986; Hawksworth, Chapter 12), and spores of similar shape have evolved independently in unrelated taxa. Some good examples are seen amongst Ingoldian Hyphomycetes which colonise twigs and leaves in freshwater and in marine fungi, most of which grow on wood or on seaweeds.

Tetraradiate spores

Many freshwater fungi have branched, tetraradiate or multi-radiate conidia. These spores, which are large, often spanning over 100 μm, accumulate in foam. They are abundant in Autumn (Iqbal & Webster, 1973b, 1977; Aimer & Segedin, 1985) so that a leaf which falls into a stream is quickly colonised. Ingold (1966, 1975a & b, 1979, 1984) has suggested that such propagules result from convergent evolution, for two reasons. Firstly, the details of conidial ontogeny vary, e.g. *Lemonniera*, *Heliscus* and *Clavatospora* produce phialoconidia, whilst *Clavariopsis*, *Articulospora* and *Tricladium* produce blastoconidia. Secondly, teleomorphs of fungi with tetraradiate anamorphs have been shown to belong to Zygomycete, Ascomycete and Basidiomycete taxa (Table 13.1).

Some of the Ingoldian Hyphomycetes are aquatic indwellers (Park, 1972), rather than having a transient existence in water. Many form conidia under water and fail to form them in air. In other cases, Ingoldian Hyphomycetes colonise terrestrial litter (Bandoni, 1972, 1974, 1981; Sanders & Webster, 1978; Park, 1974). It is best to regard some as

Table 13.1. *Relationships of some aquatic fungi with branched propagules*

Group	Anamorph	Teleomorph	Reference
Zygomycotina		*Erynia conica*	Descals & Webster (1984)
		E. plecopteri	Descals & Webster (1984)
		E. rhizospora	Descals & Webster (1984)
		Entomophaga cf. *thaxteri*	Descals & Webster (1984)
Ascomycotina			
Pezizales	*Actinospora megalospora*	*Miladina lechithina*	Descals & Webster (1978)
Hypocreales	*Heliscus lugdunensis*	*Nectria lugdunensis*	Webster (1959b)
Helotiales	*Varicosporium* sp.	*Hymenoscyphus varicosporoides*	Tubaki (1966)
	Tricladium splendens	*H. splendens*	Abdullah et al. (1981)
	Articulospora tetracladia	*H. tetracladius*	Abdullah et al. (1981)
	Geniculospora grandis	*H. africanus*	Descals et al. (1984)
Loculascomycetes	*Clavariopsis aquatica*	*Massarina* sp.	Webster & Descals (1979)
Halosphaeriaceae	*C. bulbosa*	*Corollospora pulchella*	Shearer & Crane (1971)
Basidiomycotina	*Taeniospora gracilis*	*Leptosporomyces galzinii*	Nawawi et al. (1977)
	Ingoldiella hamata	*Sistotrema hamatum*	Nawawi & Webster (1982)
	I. fibulata		Nawawi (1985)
	Dendrosporomyces prolifer		Nawawi (1985)
	D. splendens		Nawawi (1985)
	Tricladiomyces malaysianum		Nawawi (1985)
	Vanrja aquatica		Moore (1980)
		Nia vibrissa	Doguet (1969)
		Digitatospora marina	Doguet (1969)

amphibious, with the capacity to colonise immersed and emergent substrata. This is shown by the fact that many of the teleomorphs so far discovered in nature develop on woody substrata incubated in moist chambers, rather than on submerged materials. The asci are capable of discharging spores in air (Webster & Descals, 1979; Willoughby & Archer, 1973).

A further group of species with branched spores grow in the phylloplane of deciduous tree leaves and collect in the run-off from dew-laden foliage (Ando & Tubaki, 1983, 1984a & b, 1985; Tubaki, Tokumasu & Ando, 1985). These include some species referable to established anamorph genera, such as *Tripospermum*, and some newly described genera, e.g. *Titaeella capnophila*, which has clamped conidia. Where trees bearing these fungi overhang streams, it is possible for their spores to be trapped in foam since air bubbles are especially effective in removing tetraradiate spores from suspension (Iqbal & Webster, 1973a). It is therefore, unsafe to assume that spores in foam have necessarily come from an underwater source.

Tetraradiate conidia are unusual in marine fungi, but do occur in *Orbimyces spectabilis*, *Asteromyces cruciatus*, *Clavariopsis bulbosa* and *Varicosporina ramulosa* (Kohlmeyer & Kohlmeyer, 1979; Kohlmeyer, 1981). However, many marine Ascomycetes produce ascospores with extensions of the spore wall (Jones & Moss, 1978; Rees & Jones, 1984).

Three suggestions have been made on the significance of the tetraradiate shape. Firstly it has been suggested that this shape could result in slower sedimentation, and improved dispersal. However, Webster (1959a) found no evidence for this and the observed rates of sedimentation were negligible compared with the rates of water flow in streams. By contrast, the aquatic yeast, *Vanrija aquatica*, produces propagules which are common in mountain tarns (Jones & Sloof, 1965) and whose morphology is much affected by nutrient availability (Webster & Davey, 1975). Under starvation conditions, the propagules develop long tapering arms, and sediment much more slowly than propagules from well-nourished cultures. Possibly, under conditions of starvation, the propagules are planktonic. Moreover, working with marine fungi, Rees (1980) showed that spores from which appendages had been disrupted by sonication settled more slowly than untreated spores, sometimes by a factor or two or more.

Secondly, it has been suggested that branched spores facilitate anchorage to underwater objects. Using mixed spore suspensions of tetraradiate and sigmoid spores, Webster (1959a) showed that the trapping

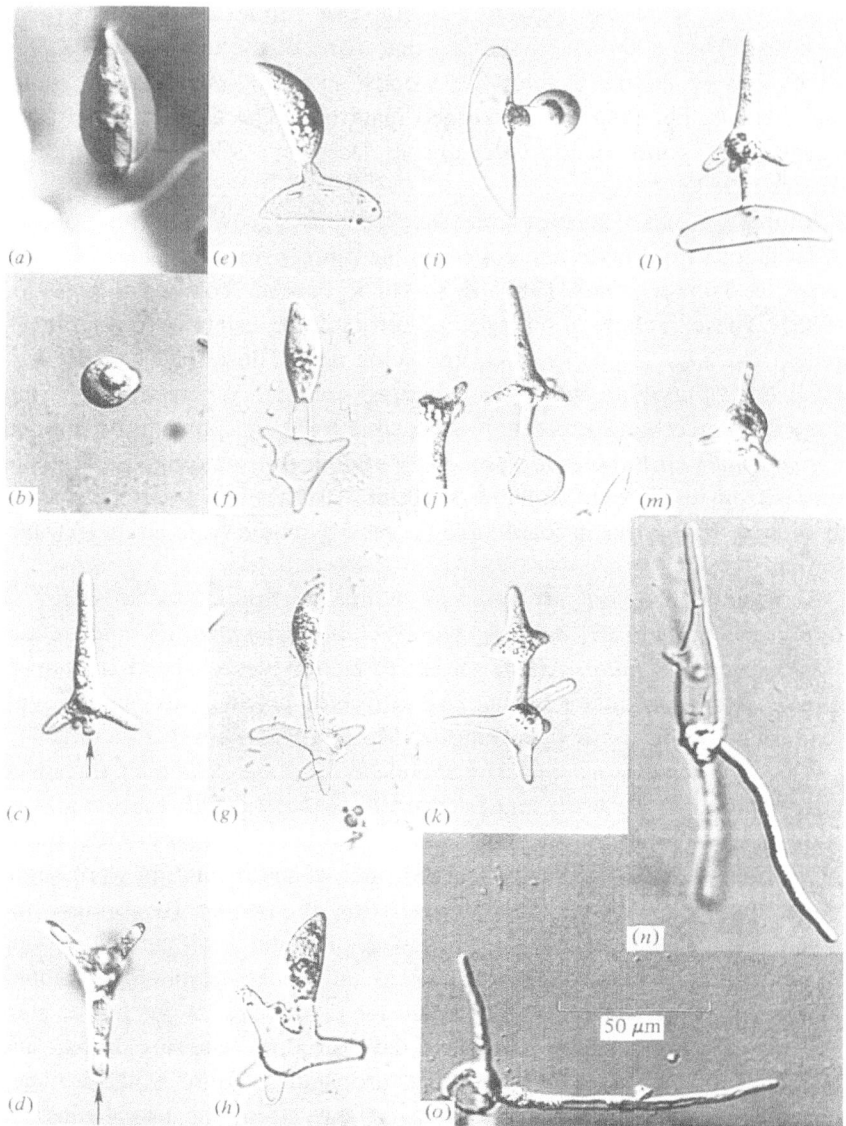

Fig. 13.1. Conidial types of *Erynia conica*. (a) Primary cornute aerial conidium (type 1) showing the bitunicate wall; the outer wall has separated from the body of the conidium. (b) Secondary globose aerial conidium (type 2). (c) Secondary stellate aquatic conidium (type 3); the arrow indicates the scar of attachment, between the backwardly directed arms. (d) Primary coronate aquatic conidium (type 4); the arrow points to the attachment scar, at the end of one of the arms. (e) − (l) Conidial interconversions: (e) type 1 to type 1, (f) & (g) type 3 to type 1, (h) type 4 to type 1, (i) type 1 to type

efficiency of tetraradiate spores on collodion-coated glass rods was many times higher than for spores of other shapes. Possible reasons for this are, firstly, that tetraradiate spores make a very stable tripedal contact with a plane surface, and secondly, there is a contact stimulus to the formation of attachment structures: the three arms in contact with the surface develop adhesive pads from which germ-tubes emerge, whilst the fourth arm, not in contact with the surface, does not. The pads aid adhesion and diminish the effect of current flow in removing spores. Webster & Davey (1984) have filmed the movements of spores of aquatic fungi flowing through capillary tubes. In the boundary layer, tetraradiate spores showed a tumbling, or cartwheeling, motion in which the arms were brought successively into contact with the capillary wall. If repeated contact with surfaces results in the development of adhesive at the tips of the arms, this may result in arrest of the spore. Once the spore has been arrested, the development of adhesoria and germ-tubes increases the firmness of the attachment. For ascospores and basidiospores and tetraradiate conidia of marine fungi, there is also excellent visual evidence of entanglement of spore appendages within cells of wood, and experimental evidence that attached spores are resistant to detachment (Rees & Jones, 1984).

The third suggestion is that tetraradiate spores are adapted for dispersal in surface films (Bandoni, 1974, 1975; Bandoni & Koske, 1974), this being of special relevance in phylloplanes. Tubaki *et al.* (1985) showed that addition of water droplets to cultures of *Tripospermum acerinum* both markedly enhanced conidium production and was also very effective in liberating conidia, which are probably weakly attached to their conidiophores. To quote: ' . . . The conidial shape is such that the surface area to volume ratio is high, and surface tension forces of water droplets applied against this surface are sufficient to remove conidia from their conidiophores. Similar observations have been made with conidia of *Ingoldiella hamata* Shaw. After release, the conidia can float free and can be dispersed by water movement . . . '.

Several Entomophthorales occurring on insects with aquatic larval stages have tetraradiate conidia (Table 13.1). *Erynia conica* causes epizootics on females of *Simulium* spp. at egg-laying sites on moss-covered boulders on the River Teign in Devon (Hywel-Jones, 1984). As shown in Fig. 13.1, there are four distinct conidial types and germination

2, (j) type 2 to type 3, (k) type 3 to type 3, (l) type 1 to type 3. (m) Multipolar germination of a type 2 conidium. (n) Multipolar germination of a type 1 conidium. (o) Germinating type 2 conidium.

by repetition or by interconversion of one type to another is frequent. Descals *et al.* (1981) speculated that cornute and globose aerial conidia might infect terrestrial adult insects, while the aquatic tetraradiate (i.e. coronate and stellate) types might have an underwater role in infecting larvae, either following ingestion or impaction on the cuticle. These ideas have since been investigated by Hywel-Jones (1984) and Hywel-Jones & Webster (1986a & b). Although conidia of all four kinds, including some with germ tubes, occur in larval guts, no evidence was found to suggest that infection was through the gut wall, nor was it possible to demonstrate impaction on the cuticle. When adults were exposed to numerous conidia of all types, the only kind which resulted in successful infection was the secondary globose. What then are the functions of primary cornute conidia? It appears that they are dispersal and the production of further cornute, globose or stellate conidia. The boat-like shape of cornute conidia is probably no coincidence, but an adjustment to dispersal at the surface of water. When discharged onto water, they soon show an inflated, outer envelope (buoyancy aid?) (Fig. 13.1) by the separation of two wall-layers. Within about 6 h at 15°C in the dark, many of the primary cornute conidia develop secondary globose conidia from what appears to be a very precisely determined site near the middle of the concave side (Fig. 13.1a − i). The degree of synchrony in the development of the secondary globose conidia is remarkable, so that, when viewed from above, the 'germinating' cornute conidia appear as a flotilla of surface-to-air missile launchers. The primary cornute conidia appear to be the vehicle for conveying the infective secondary globose conidium. The role of the tetraradiate conidia is not fully understood. Hywel-Jones & Webster (1987) have suggested that they may be held in surface films in moss on water-splashed boulders, where they may develop into other spore stages. There is little evidence of germ-tube formation. Attempts to germinate them on agar media have failed.

Sigmoid spores

The second common spore form amongst freshwater and marine fungi is the worm-shaped or sigmoid spore. These spores are often large, and can develop as conidia, as ascospores or more rarely, as basidiospores. Webster & Davey (1984) have tabulated the affinities of some aquatic fungi with sigmoid spores. An interesting addition to this list is *Mycaureola dilseae*, a parasite of the red seaweed *Dilsea carnosa* which has recently been rediscovered by Porter & Farnham (1986). Although originally interpreted as a Pyrenomycete by Maire & Chemin (1922),

Porter has shown that the 'perithecia' are basidiomata lined by basidia, and that the supposed asci are filiform or sigmoid basidiospores (Fig. 13.2). The mycelium also contains dolipore septa.

Webster & Davey (1984) concluded that sigmoid spores are an adaptation to aid attachment. When in suspension during flow, sigmoid spores do not show any characteristic attitude, but as they approach a boundary layer, they tend to roll along or tip end-over-end as a result of the slower velocity of the end of the spore nearer the surface. As the velocity of the spore decreases in the boundary layer, one end of the spore becomes arrested, and the body of the spore swings parallel to the current. Once this orientation has been achieved, any increase in current velocity tends to push the spore more firmly into contact with the surface. As with aquatic tetraradiate conidia, contact with a surface stimulates the development of appressoria at the tip of the spore, and germination is also stimulated by contact. In some 'sigmoid' marine ascospores, e.g. *Kohlmeyerella tubulata* and *Lulworthia medusa*, a distinct mucilaginous pad can be seen at each end of the spore (Rees & Jones, 1984). Because of its shape, a sigmoid spore makes a two-point contact with a plane surface, usually at one end and at an intermediate point along its length. The sigmoid spore may thus represent an alternative solution to the problem of attachment to

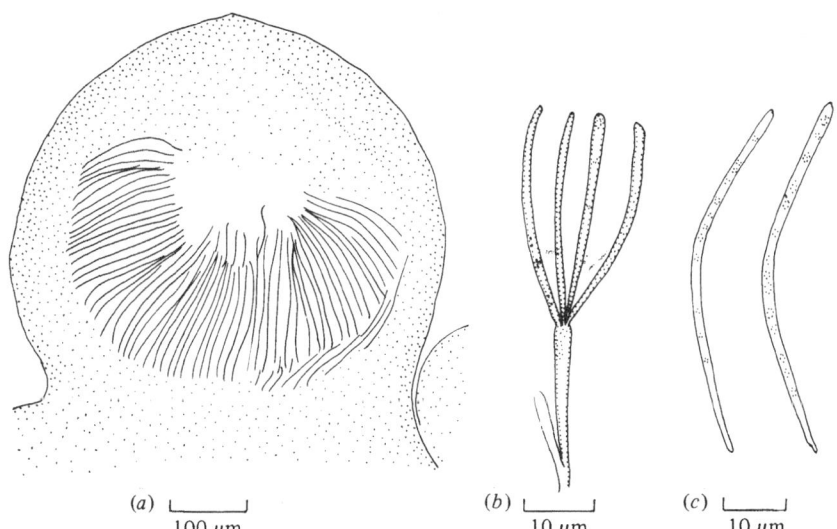

(a) └──────┘
    100 μm

(b) └──────┘
    10 μm

(c) └──────┘
    10 μm

Fig. 13.2. *Mycaureola dilseae*; (a) basidiocarp in section, (b) basidium with four basidiospores, (c) two filiform basidiospores. (Traced, by permission, from photographs kindly supplied by Dr D. Porter.)

an underwater surface. As with tetraradiate spores, there is also a tendency for sigmoid spores to accumulate in surface films, possibly related to their relatively large surfaces.

Emphasis has been laid on the large size range of aquatic tetraradiate and sigmoid spores. Possibly large size is an adaptation to impaction. Cox (1983) has pointed out that the pollen grains of aquatic Angiosperms with surface pollination are often long (up to 1 − 6 mm) and thread-like, and that the possibility of target encounter increases with increasing length. This, coupled with some stickiness at the spore tip, may be an additional reason for elongate shape in aquatic fungal spores.

There are numerous examples of sigmoid or scolecosporic spores in terrestrial fungi, e.g. the ascospores of *Cochliobolus*, the conidia of *Pseudocercosporella*, *Septoria*, *Phaeoseptoria*, etc. Kollmorgen, Owczarzak & Trione (1980) have shown that *Tilletia tritici* may develop secondary filiform (sigmoid) conidia. The sigmoid conidia of *Pseudocercosporella herpotrichoides* are effectively dispersed in splash droplets (Fitt & Bainbridge, 1983; Fitt & Nijman, 1983; Fatemi & Fitt, 1983), whilst the ascospores of *Cochliobolus cymbopogonis*, which are extruded in tendrils, can also be spread artificially by water drops (El Shafie & Webster, 1980). The relationship between droplet dispersal and spore shape is not understood, and it is possible that scolecosporic and sigmoid form may be related to attachment at the substratum.

References

Abdullah, S. K., Descals, E. & Webster, J. (1981). Teleomorphs of three aquatic hyphomycetes. *Transactions of the British Mycological Society*, **77**, 475 − 483.

Aimer, R. D. & Segedin, B. P. (1985). Fluctuation in spore numbers of aquatic hyphomycetes in a New Zealand stream. *Botanical Journal of the Linnean Society*, **91**, 61 − 66.

Ando, K. & Tubaki, K. (1983). Ordus, a new genus of Hyphomycetes. *Transactions of the Mycological Society of Japan*, **24**, 271 − 276.

Ando, K. & Tubaki, K. (1984a). Some undescribed hyphomycetes in rain drops from the intact leaf-surface. *Transactions of the Mycological Society of Japan*, **25**, 1 − 37.

Ando, K. & Tubaki, K. (1984b). Some undescribed hyphomycetes in rainwater draining from intact trees. *Transactions of the Mycological Society of Japan*, **25**, 39 − 47.

Ando, K. & Tubaki, K. (1985). Three new Hyphomycetes from Japan: *Anthopsis microspora*, *Scutiporus brunneus*, and *Titaeella capnophila*. *Transactions of the Mycological Society of Japan*, **26**, 151 − 160.

Bandoni, R. J. (1972). Terrestrial occurrence of some aquatic hyphomycetes. *Canadian Journal of Botany*, **50**, 2283 − 2288.

Bandoni, R. J. (1974). Mycological observations on the aqueous films covering decaying leaves and other litter. *Transactions of the Mycological Society of Japan*, **15**, 309 − 315.

Bandoni, R. J. (1975). Significance of the tetraradiate form in dispersal of terrestrial fungi. *Reports of the Tottori Mycological Institute, Japan*, **12**, 105 − 113.

Bandoni, R. J. (1981). Aquatic hyphomycetes in terrestrial litter. In *The Fungal Community: Its Role in the Ecosystem*, ed. Wicklow, D. T. & Carroll, G. C., pp. 693 − 703. New York: Marcel Dekker.

Bandoni, R. J. & Koske, R. E. (1974). Monolayers and microbial dispersal. *Science*, **183**, 1079 − 1081.

Cox, P. A. (1983). Search theory, random motion, and the convergent evolution of pollen and spore morphology in aquatic plants. *The American Naturalist*, **121**, 9 − 31.

Descals, E., Fisher, P. J. & Webster, J. (1984). The *Hymenoscyphus* teleomorph of *Geniculospora grandis*. *Transactions of the British Mycological Society*, **83**, 541 − 546.

Descals, E. & Webster, J. (1978). *Miladina lechithina* (Pezizales), the ascigerous state of *Actinospora megalospora*. *Transactions of the British Mycological Society*, **70**, 466 − 472.

Descals, E. & Webster, J. (1984). Branched aquatic conidia in *Erynia* and *Entomophthora sensu lato*. *Transactions of the British Mycological Society*, **83**, 669 − 682.

Descals, E., Webster, J., Ladle, M. & Bass, J. A. B. (1981). Variations in asexual reproduction in species of *Entomophthora* on aquatic insects. *Transactions of the British Mycological Society*, **77**, 85 − 102.

Doguet, G. (1962). *Digitatispora marina* n.g., n.sp., Basidiomycete marin. *Compte rendu Hebdomadaire des séances de l'Académie des Sciences, Série D*, **254**, 4336 − 4338.

Doguet, G. (1969). *Nia vibrissa* Moore et Meyers, remarquable Basidiomycete marin. *Compte rendu Hebdomadaire des séances de l'Académie des Sciences, Série D*, **265**, 1780 − 1783.

El Shafie, A. K. & Webster, J. (1980). Ascospore liberation in *Cochliobolus cymbopogonis*. *Transactions of the British Mycological Society*, **75**, 141 − 146.

Fatemi, F. & Fitt, B. D. L. (1983). Dispersal of *Pseudocercosporella herpotrichoides* and *Pyrenopeziza brassicae* in splash droplets. *Plant Pathology*, **32**, 401 − 404.

Fitt, B. D. L. & Bainbridge, A. (1983). Dispersal of *Pseudocercosporella herpotrichoides* spores from infected wheat straw. *Phytopathologische Zeitschrift*, **106**, 214 − 225.

Fitt, B. D. L. & Nijman, D. J. (1983). Quantitative studies on dispersal of *Pseudocercosporella herpotrichoides* spores from infected wheat straw by simulated rain. *Netherlands Journal of Plant Pathology*, **89**, 198 − 202.

Hywel-Jones, N. L. (1984). Infections of *Simulium* spp. by *Erynia conica*. Ph. D. thesis, University of Exeter.

Hywel-Jones, N. L. & Webster, J. (1986a). A scanning electron microscope study of the external development of *Erynia conica* on *Simulium*. *Transactions of the British Mycological Society*, in press,

Hywel-Jones, N. L. & Webster, J. (1986b). Mode of infection of *Simulium* by *Erynia conica*. *Transactions of the British Mycological Society*, **87**, 381 − 387.

Hywel-Jones, N. L. & Webster, J. (1987). The function of tetraradiate conidia of Entomophthoraceae in the riverine environment with reference to *Erynia conica*. *Transactions of the British Mycological Society*, in press,

200   *John Webster*

Ingold, C. T. (1966). The tetraradiate aquatic fungal spore. *Mycologia*, **58**, 43 − 56.
Ingold, C. T. (1975a). *An Illustrated Guide to Aquatic and Water-borne Hyphomycetes (Fungi Imperfecti) with Notes on their Biology.* Scientific Publication No. 30. Freshwater Biological Association.
Ingold, C. T. (1975b). Hooker Lecture 1974. Convergent evolution in aquatic fungi: the tetraradiate spore. *Biological Journal of the Linnean Society*, **7**, 1 − 25.
Ingold, C. T. (1979). Advances in the study of so-called aquatic Hyphomycetes. *American Journal of Botany*, **66**, 218 − 226.
Ingold, C. T. (1984). Aquatic Hyphomycetes with tetra-radiate conidia. In *Taxonomy of Fungi: Proceedings of the International Symposium on Taxonomy of Fungi, University of Madras, 1973*, Vol. 2, 353 − 364.
Iqbal, S. H. & Webster, J. (1973a). The trapping of aquatic hyphomycete spores by air bubbles. *Transactions of the British Mycological Society*, **60**, 37 − 48.
Iqbal, S. H. & Webster, J. (1973b). Aquatic Hyphomycete spora of the River Exe and its tributaries. *Transactions of the British Mycological Society*, **61**, 331 − 346.
Iqbal, S. H. & Webster, J. (1977). Aquatic Hyphomycete spora of some Dartmoor streams. *Transactions of the British Mycological Society*, **69**, 233 − 241.
Jones, E. B. G. & Moss, S. T. (1978). Ascospore appendages of marine ascomycetes: an evaluation of appendages as taxonomic criteria. *Marine Biology*, **49**, 11 − 26.
Jones, E. B. G. & Sloof, W. (1965). *Candida aquatica* sp.nov. isolated from water scums. *Antonie van Leeuwenhoek*, **32**, 223 − 228.
Kohlmeyer, J. (1981). Distribution and ecology of conidial fungi in marine habitats. In *Biology of Conidial Fungi*, Vol. 1, ed. Cole, G. T. & Kendrick, B., pp. 357 − 372. London: Academic Press.
Kohlmeyer, J. & Kohlmeyer, E. (1979). *Marine Mycology: The Higher Fungi.* New York: Academic Press.
Kollmorgen, J. F., Owczarzak, A. & Trione, E. J. (1980). Morphology and timing of secondary sporidial mating in a wheat-bunt fungus *Tilletia caries*. *Transactions of the British Mycological Society*, **75**, 461 − 471.
Maire, R. & Chemin, E. (1922). Un nouveau Pyrenomycete marin. *Compte rendu Hebdomadaire des séances de l'Académie des Sciences*, **175**, 319 − 321.
Moore, R. T. (1980). Taxonomic proposals for the classification of marine yeasts and other yeast-like fungi including the smuts. *Botanica Marina*, **23**, 361 − 373.
Nawawi, A. & Webster, J. (1982). *Sistotrema hamatum* sp.nov., the teleomorph of *Ingoldiella hamata*. *Transactions of the British Mycological Society*, **78**, 287 − 291.
Nawawi, A. & Webster, J. (1985). Basidiomycetes with branched water-borne conidia. *Botanical Journal of the Linnean Society*, **91**, 51 − 60.
Nawawi, A. & Webster, J. (1977). *Leptosporomyces galzinii*, the basidial state of a clamped conidium from freshwater. *Transactions of the British Mycological Society*, **68**, 31 − 36.
Park, D. (1972). On the ecology of heterotrophic organisms in fresh water. *Transactions of the British Mycological Society*, **58**, 291 − 299.
Park, D. (1974). Aquatic hyphomycetes in non-aquatic habitats. *Transactions of the British Mycological Society*, **63**, 183 − 187.
Porter, D. & Farnham, W. F. (1986). *Mycaureola dilseae*, a marine basidiomycete parasite of the red alga, *Dilsea carnosa*. *Transactions of the British Mycological Society*, **87**, 575 − 582.
Rees, G. (1980). Factors affecting the sedimentation rate of marine fungal spores. *Botanica Marina*, **23**, 375 − 385.

Rees, G. & Jones, E. B. G. (1984). Observations on the attachment of spores of marine fungi. *Botanica Marina*, **27**, 145 − 160.

Sanders, P. F. & Webster, J. (1978). Survival of aquatic hyphomycetes in terrestrial situations. *Transactions of the British Mycological Society*, **71**, 601 − 605.

Shearer, C. A. & Crane, J. L. (1971). Fungi of the Chesapeake Bay and its tributaries. I. Patuxent River. *Mycologia*, **63**, 237 − 260.

Sutton, B. C. (1986). Improvizations on conidial themes. *Transactions of the British Mycological Society*, **86**, 1 − 38.

Tubaki, K. (1966). An undescribed species of *Hymenoscyphus*, a perfect stage of *Varicosporium*. *Transactions of the British Mycological Society*, **49**, 345 − 349.

Tubaki, K., Tokumasu, S. & Ando, K. (1985). Morning dew and *Tripospermum* (Hyphomycetes). *Botanical Journal of the Linnean Society*, **91**, 45 − 50.

Webster, J. (1959a). Experiments with spores of aquatic Hyphomycetes. I. Sedimentation and impaction on smooth surfaces. *Annals of Botany* (NS), **23**, 595 − 611.

Webster, J. (1959b). *Nectria lugdunensis* sp.nov., the perfect stage of *Heliscus lugdunensis*. *Transactions of the British Mycological Society*, **42**, 322 − 327.

Webster, J. & Davey, R. A. (1975). Sedimentation rates and trapping efficiency of cells of *Candida aquatica*. *Transactions of the British Mycological Society*, **64**, 437 − 440.

Webster, J. & Davey, R. A. (1984). Sigmoid conidial shape in aquatic fungi. *Transactions of the British Mycological Society*, **83**, 43 − 52.

Webster, J. & Descals, E. (1979). The perfect states of waterborne hyphomycetes from fresh water. In *The Whole Fungus*, ed. Kendrick, W. B., pp. 419 − 451. Ottawa: National Museums of Canada.

Willoughby, L. G. & Archer, J. F. (1973). The fungal flora of a freshwater stream and its colonization pattern on wood. *Freshwater Biology*, **3**, 219 − 239.

14

Genetic exchange and gene flow: their promotion and prevention

MICHAEL J. CARLILE

Department of Pure and Applied Biology, Imperial College at Silwood Park, Ascot, Berkshire SL5 7PY, UK

Introduction

'Unless individual advantage can be shown, natural selection affords no explanation of structures or instincts which appear to be beneficial to the species' (Sir Ronald Fisher, 1941).

Geneticists regard species as being divided into *populations* within which individuals readily interbreed, exchanging genes. Movement of genes between populations is less common, and is termed *gene flow*. Populations may be easy to recognise, as with individuals occupying an island, but are often less clearly defined. The concept is, however, a valuable one, permitting the development of population genetics.

The rate and extent that mutant genes can spread through and between populations is of importance in determining how fast a species can evolve, and whether it will split into two or more species. The topic is not only of interest with respect to evolutionary theory, but of practical significance to the plant pathologist concerned with the spread of fungal genes influencing pathogenicity. Among the factors affecting the spread of genes are the distance between individuals and populations, the number of spores produced and the efficiency with which they are dispersed. Also of crucial importance is the ease with which genes are transferred between individuals. Sexual fusion between genetically different individuals followed by meiosis is widespread in fungi, and results in *genetic exchange*, the recombination of nuclear genes to give new genotypes. Fusion between vegetative cells is common during colony development in higher fungi (Ascomycetes, Basidiomycetes and Deuteromycetes) and in the production of large plasmodia in Myxomycetes. When the cells that fuse differ genetically, heterokaryons will be formed and parasexual recombination between nuclear genes may occur. Both sexual and vegetative fusions may permit the spread of mitochondrial DNA, plasmids and viruses. The mechanisms that may promote or prevent the transfer of genes between individuals and hence within and between populations are considered below.

Vegetative incompatibility

Cell fusion readily occurs between Myxomycete plasmodia and between the hyphae of many fungi if the cells involved are genetically identical. If plasmodia or hyphae are not genetically identical, fusion may be prevented, terminated or limited in its effects by *vegetative incompatibility*, also known as *somatic incompatibility* or *heterokaryon incompatibility*. Since vegetative incompatibility is based on genetic differences, the term *heterogenic incompatibility* has been employed. This term is also applicable to some reactions affecting sexual fusion. Reviews devoted wholly or in part to vegetative incompatibility include those of Carlile & Gooday (1978), Collins (1981), Esser & Blaich (1973), Lane (1981), Rayner *et al.* (1984) and Schrauwen (1984).

Morphological and physiological manifestations of vegetative incompatibility

Vegetative incompatibility can be divided into *fusion incompatibility* and *post-fusion incompatibility*. In the former, cell fusion between the two incompatible strains does not occur. It is well known in Myxomycete plasmodia and occurs in *Rhizoctonia solani* (Anderson, 1984) but may be rare in fungal species as usually demarcated. If fusion incompatibility is lacking, and cell fusion between two strains commences, post-fusion incompatibility reactions are possible. In Myxomycete plasmodia very fast reactions can terminate fusion so rapidly that the effect is difficult to distinguish from fusion incompatibility. Slower reactions may permit extensive protoplasmic mixing and after some hours cause the death of large areas of plasmodium with the elimination of one strain, the nuclei (Lane & Carlile, 1979) and DNA (Schrauwen, 1981) of which are destroyed. With some conditions and strains, there may be no striking reactions but after about a day one genotype has disappeared. In fungi post-fusion reactions can cause 'barrage' effects and demarcation zones between colonies (Anagnostakis, 1984; Brasier, 1984) and death of hyphal compartments adjacent to anastomoses (Garnjobst & Wilson, 1956). Alternatively one genotype may be eliminated without any striking symptoms (Pittenger & Brawner, 1961).

The genetic basis of vegetative incompatibility

Although fusion incompatibility seems rare within fungal species, post-fusion incompatibility is common and has received detailed study in several species. In each species numerous vegetative compatibility (vc) groups are found. Any two strains from within a vc group will show cell fusion without adverse reaction; strains from different groups always give incompatibility reactions. Anagnostakis (1984) found 73 vc groups in 258 isolates of *Endothia parasitica*, the chestnut blight fungus. Jamil, Buck & Carlile (1984) found 18 vc groups in 31 isolates of *Gaumannomyces graminis* var. *tritici*, the wheat take-all fungus, from a single field. For *Ophiostoma* (*Ceratocystis*) *ulmi*, the Dutch elm disease fungus, see Brasier (Chapter 16, Fig. 16.1). In Myxomycetes, both fusion and post-fusion vc groups are numerous.

Genetic studies on post-fusion incompatibility in fungi and on both fusion and post-fusion incompatibility in Myxomycetes have yielded similar results – there are many vegetative incompatibility (vc) loci with two alleles at each locus. An incompatibility reaction occurs when strains differing at one or more vc loci are brought together. Incompatibility due to the interaction of the alleles of a vc gene is termed *allelic incompatibility*. If a species has n loci at which allelic incompatibility occurs, then 2^n vc groups are possible. Thus, the 10 *het* (heterokaryon incompatibility) loci known in *Neurospora crassa* (Perkins *et al.*, 1982) allow 2^{10} or 1024 vc groups. In *Podospora anserina*, Esser & Blaich (1973) found not only allelic incompatibility but also *non-allelic incompatibility* due to interaction between alleles at different vc loci. Non-allelic incompatibility has not been found in any other species.

In Myxomycetes both fusion and post-fusion incompatibility are allelic (Carlile & Gooday, 1978; Collins, 1981 & Chapter 18; Schrauwen, 1984). Myxomycete plasmodia are diploid, allowing vc alleles to be recognised as dominant or recessive. When a reaction occurs between two plasmodia differing at a single post-fusion vc locus, the plasmodium carrying the dominant allele destroys the nuclei of that carrying the recessive allele (Lane & Carlile, 1979). With differences at more than one locus, both strains may have dominant vc alleles and bilateral damage then occurs.

The consequences and possible role of vegetative incompatibility

When a cell undergoes vegetative fusion with a genetically different cell its genome may be endangered. Nuclei containing genes enabling them to replace other nuclei in a heterokaryon could be introduced (Pittenger & Brawner, 1961). So might mitochondria capable of normal replication but defective in their respiratory performance, or plasmids or viruses with deleterious effects. The role of vegetative incompatibility may be protection against such alien genetic material (Caten, 1972; Hartl, Dempster & Brown, 1975).

Nuclei and mitochondria cannot be transmitted between cells without cell fusion, and there is no evidence that plasmids, DNA fragments or viruses can pass through intact fungal cell walls. Hence, fusion incompatibility, where it occurs in fungi, will be a complete protection against alien nucleic acid. Post-fusion vc reactions usually prevent the transmission of nuclei but the extent to which virus movement is restricted varies. A virus in *Endothia parasitica* that reduces the growth rate and pathogenicity of the fungus spreads readily between strains of the same vc group by hyphal anastomosis (Anagnostakis, 1984). The extent to which spread occurs between strains of different vc groups depends on the number of vc loci at which the groups differ, and is negligible with differences at five loci. Similar results have been obtained with *Ophiostoma ulmi* (Brasier, 1984). Vegetative incompatibility is defensive in the above instances but sometimes the old adage that attack is the best form of defence seems to apply. Thus, in *Neurospora crassa* nuclei carrying the *het-I* gene may eliminate those with *het-i* from a heterokaryon (Pittenger & Brawner, 1961). The devastating vc reactions and nuclear destruction that occur in Myxomycetes (Lane & Carlile, 1979) seem aggressive, as well as defensive; in *Didymium iridis*, cultures initially containing plasmodia of several postfusion vc genotypes tend ultimately to contain only plasmodia of the strain dominant at all the relevant loci (Clark, 1980).

Although vegetative incompatibility is widespread in fungi, it is possible that it is absent in populations of some species capable of vegetative fusion. Absence might occur where a fungus has colonised a new area or a new host, as the initial colonists might be of a single haploid genotype and the resulting population of a single vc group. Ultimately, however, genetic diversity would arise and with it the need for individuals to defend themselves against alien nucleic acid and to compete with other individuals. At this point selection for mutations disrupting the complex process of cell fusion and thus creating vc loci would be expected (see also Brasier, Chapter 16).

Although the *role* of vegetative incompatibility may be the protection of individuals, its occurrence may have *consequences* for the species. In fungi it is likely that fusion incompatibility will affect the walls of sexual as well as vegetative cells and thus prevent all genetic exchange. It will, therefore, delimit a genospecies. In Myxomycetes, the cell envelopes of plasmodia and amoebae differ and the vc loci that affect plasmodial fusion are without effect on the sexual fusion of amoebae. Fusion incompatibility is, hence, common within Myxomycete species. Post-fusion vc reactions usually prevent the transmission of nuclei between cells. They will, thus, prevent heterokaryon formation and parasexual recombination. Hence, in fungi with numerous vc loci it is unlikely that heterokaryosis and the parasexual cycle are important in nature. Thus, Perkins, Turner & Barry (1976) found, in an extensive conidial sampling of *Neurospora* colonies, evidence that almost all colonies had originated from single ascospores. The parasexual cycle is of immense value for genetic analysis (Fincham, Day & Radford, 1979); it may, as a result of vegetative incompatibility, be of much less importance in nature, although differences between species are likely.

Sex and genetic exchange

Through most of the life of a fungus, fusion between genetically different strains is prevented or severely limited by heterogenic vc systems. When, however, the sexual process occurs, these systems are largely or entirely suppressed, so outcrossing is possible. Indeed, in many fungi the sexual process is initiated only as a result of interaction between cells that differ genetically with respect to mating type and probably in other ways, thus promoting genetic exchange.

Systems promoting outcrossing

Fungi in which the co-operation of genetically different strains is required for initiating or completing the sexual process are referred to as *self-sterile* or, since different strains need to be involved, as *heterothallic*. The controlling mechanisms have been classed as *homogenic incompatibility*, since mating fails if strains are identical with respect to mating type. Mating systems and their genetics have been reviewed for fungi by Burnett (1975, 1976), Esser (1971) and Fincham *et al.* (1979), and for Myxomycetes by Carlile & Gooday (1978), Collins (1981), Dee (1982) and Schrauwen (1984), so discussion here can be brief.

Breeding systems with two mating types Many fungi have two mating types based on alternative alleles at a single locus, well studied examples being *Neurospora crassa* and *Saccharomyces cerevisiae*. This prevents mating between the genetically identical progeny of a single haploid cell. However, the probability of two unrelated haploid cells being compatible (the *outbreeding potential*) is 50% and so is that for random pairs of cells from the meiosis of a diploid cell (the *inbreeding potential*). Thus, although selfing is prevented, the likelihood of mating between close relatives is not reduced. In Oomycetes such as *Phytophthora infestans*, sexual interaction is between diploid mycelia, posing the problem of how a mating system based on two alternative alleles can operate in a diploid mycelium (Sansome, Chapter 7; Shaw, 1983).

Unifactorial incompatibility systems with many mating types Mating in some Homobasidiomycetes, such as *Coprinus comatus*, is based on multiple alleles at a single locus. Two of the spores borne on a basidium will be of one mating type, and two of a second, so such systems are often termed *bipolar*. The inbreeding potential is 50%, as with a two-allele system. The outbreeding potential, however, can approach 100%, since a population can contain a large number of mating type alleles. Unifactorial incompatibility with many alleles occurs also in Myxomycetes such as *Didymium iridis* (Collins, 1981).

Bifactorial incompatibility systems Incompatibility based on two unlinked factors, A and B, occurs in *Schizophyllum commune* and many other Basidiomycetes. Haploid individuals mate only if both factors differ, so basidia will be heterozygous for both factors, e.g. A_1A_2/B_1B_2 and can produce spores of four mating types (e.g. A_1B_1, A_1B_2, A_2B_1, A_2B_2), so bifactorial incompatibility is sometimes termed *tetrapolar incompatibility*. Since *A* and *B* factors are both numerous, the outbreeding potential, as with bipolar incompatibility, can approach 100%. However, with any spore genotype from a fruit body only one of the four possible genotypes will differ from it at both loci and be compatible. This gives an inbreeding potential of 25%, half of that in a bipolar system and hence a greater outbreeding bias. The *A* and *B* factors each consist of two sub-units, and recombination between sub-units will create new factors compatible with the parental ones, increasing the inbreeding potential. In *Schizophyllum*, the frequency of recombination between sub-units varies between strains and ranges from less than 1% to nearly 20%. The sub-unit structure hence permits flexibility in outbreeding bias within a species. In the Myxomycete *Physarum polycephalum* mating occurs most efficiently if the two haploid strains differ at both a *mat A* and a *mat B* locus, at each of which multiple alleles occur (Dee, 1982; Schrauwen, 1984). Although it has been questioned as to whether *mat B* should be regarded as a mating type locus (Collins, 1981), in essence a system with effects analogous to that of the tetrapolar system has evolved independently in a Myxomycete. In *Ustilago maydis* and some other Heterobasidiomycetes there are many *B* factors but only two *A* factors (Wong & Wells, 1985). This resembles the tetrapolar system in having an inbreeding potential of 25 %, but the two mating type system in an outbreeding potential of only 50 %. The system is sometimes called *modified bifactorial incompatibility* or modified tetrapolar incompatibility *although there is no evidence that it evolved from the usual Basidiomycete tetrapolar system.*

Sexual differentiation and outcrossing In some Oomycetes such as *Achlya ambisexualis* self-fertility is prevented by the inability of a strain to produce antheridia (male) and oogonia (female) simultaneously. Strains range from strong female to strong male, the position in the spectrum being determined by the amount of the hormone antheridiol a strain produces, a strong female producing most. Antheridiol induces antheridia in a more male strain which then releases the hormone oogoniol inciting sexual differentiation in the more female strain (Griffin, 1981). An understanding of the extent to which outcrossing is promoted will need the elucidation of the genetics of hormone production and sexuality in *Achlya*.

Self-fertility
Self-fertility or *homothallism* occurs in all major groups of fungi. It does not prevent outcrossing but reduces the likelihood of it occurring. Homothallism is referred to as *primary* if there is no evidence for a heterothallic ancestor. If it is clear that earlier heterothallism has been circumvented, the terms *secondary homothallism* or *pseudocompatibility* are used. Secondary homothallism can result from the inclusion of two nuclei of different mating type in a single spore, and the subsequent growth of a mycelium heterokaryotic for mating type. In the heterothallic species *Neurospora crassa* formation of such a mycelium is prevented, since the mating type alleles *A* and *a* which briefly cooperate in the sexual process act at other times as vc alleles. In a few strains, however, a gene *tol* (tolerant) suppresses the vegetative incompatibility associated with the mating type locus without affecting its role in mating, allowing mating type heterokaryons (Perkins *et al.*, 1982). Such a development must have occurred in the evolution of the closely related but secondarily homothallic *N. tetrasperma* in which ascospores normally contain two nuclei of different mating type. In *Agaricus bisporus* there is no vegetative incompatibility between different mating types, and a random inclusion of two nuclei in each basidiospore often gives spores heterokaryotic for

mating type (secondary homothallism). Some fungi, such as *Saccharomyces cerevisiae*, have both heterothallic and homothallic strains. Haploid cells carry the two mating type alleles *a* and α at storage loci plus a copy of one at the mating type locus where it is expressed to determine the mating type. In heterothallic strains, a change in the allele at the mating type locus is very rare. In homothallic strains, the allele is frequently replaced by an *a* or α allele from a storage locus. This occurs in the progeny of a single cell in a precise pattern (Nasmyth, 1983). Hence, if the immediate progeny of a germinated spore do not succeed in mating with urelated cells, subsequent generations of cells are able to mate with each other as a result of mating type switching. The fission yeast *Schizosaccharomyces pombe* also shows mating type switching (Johnson, Calleja & Zuker, 1984). It may well be widespread in filamentous fungi, for example *Ophiostoma ulmi* (Brasier & Gibbs, 1975), but less easy to detect than in unicells.

Apomixis

Apomixis is the production of the morphological features that accompany the sexual process but without the usual cell fusion, nuclear fusion and meiosis. It is rare in fungi (Burnett, 1975). In Myxomycetes, however, plasmodial production in amoeba clones is usually by apomixis, rather than homothallic mating. In *Physarum polycephalum* haploid amoeba strains carrying an *mt-h* allele remain capable of mating with strains having other *mt* alleles but also are able to form plasmodia without fusion and change of ploidy (Dee, 1982). In *Didymium iridis* many amoeba strains are heterothallic, many apomictic and a few homothallic (Collins, 1981, Chapter 18). Here, the apomictic amoebae are diploid and may carry two different mating type alleles in a nucleus. Hence, apomixis in Myxomycetes may be facultative as in *P. polycephalum* or potentially capable of reversion to heterothallism as in *D. iridis*.

Restrictions on outbreeding

The widespread occurrence of systems that encourage outcrossing indicates that mating between genetically different individuals often has beneficial effects. This, however, will only be so if the individuals are not too distantly related. Efficient metabolism and normal development are dependent not only on the action of individual genes but on harmonious interaction throughout the genome. If there is little gene flow between two populations they may diverge to the point where, in any hybrid, harmonious interaction between genes from the parental strains is improbable. Mating between distantly related strains may, therefore, result in progeny that have reduced vigour or are sterile. Such effects are better regarded as a result of *genetic disharmony* rather than incompatibility. When the risks in outbreeding, such as disharmony or infection by harmful genetic elements, outweigh the advantages of outbreeding, *sexual heterogenic incompatibility* may be expected. Such incompatibility could result from vegetative incompatibility systems that are usually suppressed in the sexual phase, or from systems that have evolved independently. Sexual heterogenic incompatibility, which when completely effective between two populations will result in speciation, is discussed by Anderson (1984), Brasier (1984, Chapter 16), Esser & Blaich (1973), and Rayner *et al.* (1984).

Units of selection and the role of sex

Favourable mutations are rare and the accumulation of a sequence of favourable mutations in an asexual line will take a long time. Nuclear fusion between genetically different individuals followed by meiosis will generate genetically diverse progeny and can enable favourable mutations occurring in different lines to be brought together in a single individual. Hence, a traditional view is that the primary role of sex is to produce the variability that will enable a species to respond to environmental change, including change in the nature of competition from other organisms. Such a view seems untenable (Maynard Smith, 1978). The value of features that give immediate advantage to an individual is far greater than that of features that benefit the species, remote posterity or even quite close kin. Group selection may well not exist, and the strength of kin selection decreases rapidly as relatedness diminishes. Selection acts most strongly on individuals, and perhaps on genes within individuals (Dawkins, 1982). A further reason for doubting that the generation of diversity is the primary role of sex is the widespread occurrence of self-fertility. It is sometimes argued that self-fertility is due to selection for the sporulation associated with sex, self-fertility enabling sporulation to occur even when a complementary mating type is absent. This is unconvincing. Sporulation could be obtained in such circumstances by apomixis, which, however, seems rare in fungi. Nuclear fusion and meiosis must have an important role, even when variability does not result. The reasons for the origin and persistence of sex must be sought in its advantages for individuals, genes within individuals or the immediate progeny of individuals. The same applies to outbreeding where it occurs.

The nature of the selective forces that maintain nuclear fusion and meiosis, the basic features of the sexual process, remain unclear (Holliday, 1984). One possibility is the repair of double-stranded damage to DNA (Bernstein *et al.*, 1985). The occurrence of outcrossing presents a still more complex problem, as it must be maintained against a variety of forces that favour self-fertility (Maynard Smith, 1978). Given that nuclear fusion and meiosis is of value, there must be at least a short dikaryophase and diplophase. Bernstein *et al.* (1985) suggest that the role of outcrossing is the masking by unmutated genes of deleterious mutations affecting these phases. Another possibility is the achievement in the diploid phase of heterozygous advantage, an effect well known in animals and plants, in which the heterozygote performs better than either corresponding homozygote. Consistent with these views is the fact that among the fungi the most efficient means for securing outbreeding occur in the predominantly dikaryotic Basidiomycetes and the Myxomycetes with their striking diplophase. Whatever the *role* of outcrossing, a *consequence* is the promotion of variability and gene flow with profound effects on evolution.

Conclusions

Genetic exchange and gene flow in fungi are prevented by vegetative incompatibility and promoted by a variety of mating systems. These mating systems are perhaps more flexible than has been appreciated, and the extent to which outbreeding occurs may be capable of rapid evolutionary adjustment. It is possible that a heterothallic system may readily be masked, and when conditions favour outbreeding once more expressed (Brasier, Chapter 16). The limits to natural gene flow are set by genetic disharmony and heterogenic sexual incompatibility. Mycological thought has often been group selectionist, and progress in understanding fungal sexuality, and perhaps other aspects of mycology, requires a more careful consideration of the colony, the individual and the gene as units of selection.

Acknowledgement I wish to thank Dr Bernard Lamb for valuable discussion.

References

Anagnostakis, S. L. (1984). The mycelial biology of *Endothia parasitica*. II. Vegetative incompatibility. In *The Ecology and Physiology of the Fungal Mycelium*, British Mycological Society Symposium Volume 8, ed. Jennings, D. H. & Rayner, A. D. M., pp. 499 – 507. Cambridge University Press.
Anderson, N. A. (1984). Variation and heterokaryosis in *Rhizoctonia solani*. In *The*

Ecology and Physiology of the Fungal Mycelium, British Mycological Society Symposium Volume 8, ed. Jennings, D. H. & Rayner, A. D. M., pp. 367 – 382. Cambridge University Press.

Bernstein, H., Byerly, H. C., Hopf, F. A. & Michod, R. E. (1985). Genetic damage, mutation, and the evolution of sex. *Science*, **229**, 1277 – 1281.

Brasier, C. M. (1984). Inter-mycelial recognition systems in *Ceratocystis ulmi*: their physiological properties and ecological importance. In *The Ecology and Physiology of the Fungal Mycelium*, British Mycological Society Symposium Volume 8, ed. Jennings, D. H. & Rayner, A. D. M., pp. 451 – 497. Cambridge University Press.

Brasier, C. M. & Gibbs, J. N. (1975). Highly fertile form of the aggressive strain of *Ceratocystis ulmi*. *Nature*, **257**, 128 – 131.

Burnett, J. H. (1975). *Mycogenetics*. London: Wiley.

Burnett, J. H. (1976). *Fundamentals of Mycology*. London: Arnold.

Carlile, M. J. & Gooday, G. W. (1978). Cell fusion in myxomycetes and fungi. In *Membrane Fusion*, Cell Surface Reviews, volume 5, ed. Poste, G. & Nicolson, G. L., pp. 219 – 265. Amsterdam: North-Holland Publishing Company.

Caten, C. E. (1972). Vegetative incompatibility and cytoplasmic infection in fungi. *Journal of General Microbiology*, **72**, 221 – 229.

Clark, J. (1980). Competition between plasmodial incompatibility phenotypes of the myxomycete *Didymium iridis*. II. Multiple clone crosses. *Mycologia*, **72**, 512 – 522.

Collins, O. R. (1981). Myxomycete genetics, 1960 – 1981. *The Journal of the Elisha Mitchell Society*, **97**, 101 – 125.

Dawkins, R. (1982). *The Extended Phenotype; the Gene as the Unit of Selection*. Oxford: Freeman.

Dee, J. (1982). Genetics of *Physarum polycephalum*. In *Cell Biology of Physarum and Didymium*, vol. I, ed. Aldrich, H. C. & Daniel, J. W., pp. 211 – 251. New York: Academic Press.

Esser, K. (1971). Breeding systems in fungi and their significance for genetic recombination. *Molecular and General Genetics*, **110**, 86 – 100.

Esser, K. & Blaich, R. (1973). Heterogenic incompatibility in plants and animals. *Advances in Genetics*, **17**, 107 – 152.

Fincham, J. R. S., Day, P. R. & Radford, A. (1979). *Fungal Genetics, 4th ed*. Oxford: Blackwells.

Fisher, R. A. (1941). Average excess and average effect of a gene substitution. *Annals of Eugenics*, **11**, 53 – 63.

Garnjobst, L. & Wilson, J. F. (1956). Heterokaryosis and protoplasmic incompatibility in *Neurospora crassa*. *Proceedings of the National Academy of Sciences, USA*, **42**, 613 – 618.

Griffin, D. H. (1981). *Fungal Physiology*. New York: Wiley.

Hartl, D. L., Dempster, E. R. & Brown, S. W. (1975). Adaptive significance of vegetative incompatibility in *Neurospora crassa*. *Genetics*, **81**, 553 – 569.

Holliday, R. (1984). The biological significance of meiosis. In *Controlling events in Meiosis*, Society for Experimental Biology Symposium volume 38, ed. Evans, C. W. & Dickinson, H. C., pp. 381 – 394. Cambridge: Company of Biologists.

Jamil, N., Buck, K. B. & Carlile, M. J. (1984). Sequence relationships between double-stranded RNA from isolates of *Gaumannomyces graminis* in different vegetative compatibility groups. *Journal of General Virology*, **65**, 1741 – 1747.

Johnson, B. F., Calleja, G. B. & Zuker, M. (1984). Mating-type gene switching in a homothallic fission yeast. *Journal of Theoretical Biology*, **110**, 299 − 312.

Lane, E. B. (1981). Somatic incompatibility in fungi and myxomycetes. In *The Fungal Nucleus*, British Mycological Society symposium volume 5, ed. Gull, K. & Oliver, S. G., pp. 239 − 354.

Lane, E. B. & Carlile, M. J. (1979). Post-fusion somatic incompatibility in plasmodia of *Physarum polycephalum*. *Journal of Cell Science*, **35**, 339 − 354.

Maynard Smith, J. (1978). *The Evolution of Sex*. Cambridge University Press.

Nasmyth, K. (1983). Molecular analysis of a cell lineage. *Nature*, **302**, 670 − 676.

Perkins, D. D., Radford, A., Newmeyer, D. & Bjorkman, M. (1982). Chromosomal loci of *Neurospora crassa*. *Microbiological Reviews*, **46**, 426 − 570.

Perkins, D. D., Turner, B. C. & Barry, E. G. (1976). Strains of *Neurospora* collected from nature. *Evolution*, **30**, 281 − 313.

Pittenger, T. H. & Brawner, T. G. (1961). Genetic control of nuclear selection in *Neurospora* heterokaryons. *Genetics*, **46**, 1645 − 1663.

Rayner, A. D. M., Coates, D., Ainsworth, A. M., Adams, T. J. H., Williams, E. N. D. & Todd, N. K. (1984). The biological consequences of the individualistic mycelium. In *The Ecology and Physiology of the Fungal Mycelium*, British Mycological Society Symposium Volume 8, ed. Jennings, D. H. & Rayner, A. D. M., pp. 509 − 540. Cambridge University Press.

Schrauwen, J. A. M. (1981). Post-fusion incompatibility in *Physarum polycephalum*. The involvement of DNA. *Archives of Microbiology*, **129**, 257 − 260.

Schrauwen, J. A. M. (1984). Cellular interaction in plasmodial slime moulds. In *Cellular Interactions*, Encyclopedia of Plant Physiology, New Series volume 17, ed. Linskens, H. F. & Heslop-Harrison, J., pp. 291 − 308. Berlin: Springer-Verlag.

Shaw, D. S. (1983). The cytogenetics and genetics of *Phytophthora*. In *Phytophthora: Its Biology, Taxonomy, Ecology and Pathology*, ed. Erwin, D. C., Bartnicki-Garcia, S. & Tsao, P. H., pp. 81 − 94. St Paul, Minnesota: American Phytopathological Society.

Smith, M. J. (1978). *The Evolution of Sex*. Cambridge University Press.

Wong, G. J. & Wells, K. (1985). Modified bifactorial incompatibility in *Tremella mesenterica*. *Transactions of the British Mycological Society*, **84**, 95 − 109.

15

The genetic integration of fungal life styles

C. E. CATEN

Department of Genetics, University of Birmingham, Birmingham B15 2TT, UK

Introduction

Fungi possess a great diversity of life styles with major variation in structure, life cycle, mode of nutrition, dispersal and ecology. Many have a filamentous organisation and it is tempting to forget unicellular and plasmodial forms and phases and consider the mycelial habit as a unifying theme. This, however, would be to overlook the large difference in physiology and genetics between the unified mycelial organisation of species with active cytoplasmic streaming and frequent hyphal anastomosis (e.g. *Neurospora crassa*) and the essentially cellular organisation of species which lack these features (e.g. *Cephalosporium acremonium*). With this diversity of life styles it is not surprising that fungi possess a range of genetic processes whose occurrence and biological role vary both between species and between environments within the same species.

Evolution depends upon the genetic variability contained in a population or species and upon how that variation is distributed between individuals. These in turn are influenced by many factors, including ploidy, life cycle, breeding system, chromosome number, chiasma frequency and reproductive potential. Darlington (1939) referred to the integrated complex of these properties and processes as the genetic system. This chapter reviews some aspects of fungal variation and fungal genetic systems and considers how ideas have changed over the last few decades.

Variation in fungi

Fungi have long been considered as highly variable organisms (Brierly, 1931; Buxton, 1960). Variation is apparent both in collections from the field (e.g. Brasier, 1970; Perkins, Turner & Barry, 1976; Roelfs & Groth, 1980) and from the behaviour of single isolates in the laboratory (e.g. Brown, 1926; Caten & Jinks, 1968). In recent years there has been increasing use of quantitative methods for documentation. These provide convenient summaries of large bodies of data and so facilitate comparisons between species, between populations within a species, between reproductive systems and between characters (e.g. Caten & Jinks, 1968; Brasier, 1970).

Present knowledge of fungal variability stems principally from study of morphological and pathological traits. In many cases these are controlled by polygenic systems and the observed phenotypic variation can not be directly related to differences in genotype. Even where variation is determined by major genes, as for example with virulence and vegetative compatibility, several genes are frequently simultaneously involved and determination of the precise genotype requires elaborate genetic analysis (e.g. Croft, 1985). Because of these problems we may expect to see increasing use of molecular characters, such as isozymes and restriction fragment length polymorphisms, in the study of fungal variation. These have the advantage of being discrete variables which are naturally polymorphic and either at, or very close to, the level of the genes. As such they are more amenable to analysis by the methods of population genetics than the complex traits previously considered (Barrett, Chapter 6).

Although frequently impressive, the amount of phenotypic variation amongst a collection of field isolates reflects only part of the genotypic variation actually present. Hybridisation of field-collected isolates in the laboratory generally releases variation exceeding that seen in nature. While this potential variation is apparent from the first generation of such crosses, its full extent is often revealed only by artificial selection (e.g. Connolly & Simchen, 1973; Merrick, 1975). Comparison of the phenotypes isolated from the field with those present in progenies produced in the laboratory gives an indication of the direction and strength of natural selection. For example, dikaryons of *Schizophyllum commune* with growth rates equivalent to the extreme high and low lines that can be produced by selection in the laboratory (Connolly & Simchen, 1973) are not found in natural populations (Brasier, 1970, and see Fig. 16.2), suggesting that intermediate growth rates are favoured in nature, i.e. that stabilising selection operates.

The origins of variation

Genetic variation arises from the presence of two or more different alleles of a gene in a population. The number of genes in fungal nuclei (about 10^4) is such that even if only a proportion are naturally polymorphic the potential number of genotypes is very large. The different alleles arise by mutation which is, therefore, the ultimate source of all genetic variation. However, mutation is a rare event and the contribution of new mutants to the variation in any one generation will generally be insignificant in comparison with that resulting from the reshuffling of pre-existing mutations into new combinations. Laboratory studies have revealed a variety of processes leading to reassociation and recombination of nuclear genes in fungi, including heterokaryosis, parasexuality and sexual reproduction. Although the mechanisms underlying these are well understood in a few species (Burnett, 1975; Fincham, Day & Radford, 1979), their significance in fungal evolution is not clear. As well as nuclear genes, fungi possess additional genomes located in the cytoplasm (Fincham *et al.*, 1979; Scazzocchio, Chapter 4). These cytoplasmic systems are transmitted both asexually and sexually in a different manner to the nuclear system but cytoplasmic genes can vary in copy number, mutate and recombine; hence they are an additional source of variation.

The literature of the 1950s and 1960s contains much speculation about, and some extravagant claims for, the natural role of the various genetic processes in fungi (e.g. Pontecorvo, 1946, 1956; Buxton, 1960; Snyder, 1961; Raper, 1966). A reappraisal in the light of recent advances and current thinking in fungal and population genetics is appropriate.

Mutation

Although mutants for any one gene arise very rarely (the spontaneous mutation frequency per gene in fungi is estimated by Fincham *et al.* (1979) to be about 10^{-6}), several factors make the contribution from current mutation of greater potential significance for fungal evolution than for evolution of higher eukaryotes. First, the large size of fungal populations means that mutant alleles arise continuously, despite the rarity of mutation (Person, Groth & Mylyk, 1976). Second, the absence of a distinct germ line in fungi ensures that mutations arising in virtually any tissue can be transmitted to subsequent generations either sexually or asexually. Third, the dominant haploid phase of many fungal life cycles means that new mutations are expressed immediately they are segregated into a spore or hypha. Fourth, although when they first arise new mutant alleles are so rare that they generally pass undetected, the high asexual reproductive rates of most fungi permit a rapid increase in number and frequency under selection. Clonal evolution *via* mutation in the absence of recombination is clearly shown by changes in the virulence structure of populations of rust fungi (Burdon, Luig & Marshall, 1983; Burdon & Roelfs, 1985a; Newton, Caten & Johnson, 1985).

Mutation is also an important factor in the variability and degeneration of fungal cultures in the laboratory. The average culture is the product of an enormous number of nuclear divisions and contains a vast population of nuclei, some of which will carry recent mutations. Mycological culture media are not natural substrates and a proportion of mutants may be expected to grow better in culture than the wild type and hence be selected. It is not surprising, therefore, that so many fungal cultures sector. This explanation for much commonly observed sectoring is supported by the single nuclear gene determination of the common 'felty' sector of *Ophiostoma* (*Ceratocystis*) *ulmi* (Brasier & Gibbs, 1975) and the recent demonstration, through a heterokaryon test (Jinks, 1966), that two cultural variants of *Septoria nodorum* were of nuclear origin (A. E. Osbourn, personal communication).

Heterokaryosis

Hyphal anastomosis and nuclear migration and hence, the potential for natural formation of heterokaryons between unrelated genotypes, appear to be properties unique to filamentous fungi. There has been much speculation about the evolutionary benefits of heterokaryosis as a means of exploiting hybrid vigour, storing genetic variation and adjusting somatically to a changing environment (Pontecorvo, 1946; Buxton, 1960; Snyder, 1961). However, it was pointed out twenty years ago (Caten & Jinks, 1966) that heterokaryons are rarely isolated from nature and few examples have been reported since. The significance of heterokaryosis as a natural genetic process will depend not only upon its occurrence, but also upon the diversity of the associated genomes. Attempts to synthesise heterokaryons between unrelated isolates have revealed the presence of vegetative incompatibility systems which restrict or block heterokaryon formation between genetically dissimilar strains and which are now seen as common, if not universal, mechanisms in fungi (Caten & Jinks, 1966; Esser & Blaich, 1973; Rayner *et al.*, 1984; Brasier, 1984; Carlile, Chapter 14). From these observations of vegetative incompatibility, the concept of fungal individualism has been developed and the emphasis now is on exposing genotypes to selection, rather than on sheltering variation within a communal effort (Todd & Rayner, 1980; Rayner *et al.*, 1984).

Parasexuality

The discovery of the parasexual cycle in the 1950s aroused

considerable interest because it offered an explanation for the success of the many imperfect species which posed a serious dilemma to a community convinced of the evolutionary benefits of sex (Pontecorvo, 1956; Raper, 1966). Subsequent work, however, has failed to substantiate the idea that the parasexual cycle can replace the sexual cycle where the latter has been lost. Although diploid strains of habitually haploid fungi have been isolated from nature, they occur at a frequency of less than 1% in most species (but see Croft, Chapter 21). Furthermore, haploidisation of these natural somatic diploids indicates that they are either completely homozygous or heterozygous at only a few loci, suggesting an origin through mitotic or meiotic nondisjunction or from fusion of like nuclei (see Caten, 1981 for review). Such lack of heterozygosity of natural diploids would be predicted from the limitations on the diversity of genetic associations imposed by vegetative incompatibility. These considerations, together with the fact that even under ideal conditions completion of the parasexual cycle requires a sequence of individually rare events (Pontecorvo, 1956; Caten, 1981), suggest that parasexuality makes an insignificant contribution to natural variation in any one generation.

The many claims for the formation of new physiologic races of rusts at relatively high frequency in artificial mixtures of dikaryons (for review see Day, 1974) suggest that these fungi may be an exception where parasexuality is important. However, given the technical difficulties of working with rusts these claims should be interpreted with caution unless supported by evidence from independent markers. For example, scoring of isozyme and dsRNA markers indicated that all of five variants of *Puccinia striiformis* selected for possession of a novel combination of spore colour and pathogenicity arose by mutation or contamination, rather than by somatic hybridisation (Newton, Johnson & Caten, 1986). That parasexuality is not a significant phenomenon among rusts in nature is indicated by the association between virulence and isozyme phenotypes observed in asexually reproducing populations of *Puccinia graminis* f. sp. *tritici* (Burdon & Roelfs, 1985a), which contrasts with the independence of these two characters in sexually reproducing populations (Burdon & Roelfs, 1985b). This difference would not exist if, in practice, parasexuality provided an effective alternative to sex.

Sexual recombination
The diversity of life cycles and breeding systems in fungi has led to speculation, not just about the importance of sex in their variation and

microevolution, but also about the benefits of regulating the amount of outbreeding (Whitehouse, 1949; Burnett, 1975; Raper, 1966; Carlile, Chapter 14). Speculation on the latter reached its highest sophistication with consideration of the effects on the outbreeding bias of the control of recombination between the linked α and β subunits within each of the A and B incompatibility factors in Homobasidiomycetes (Stamberg, 1969; Koltin, Stamberg & Lemke, 1972). However, these arguments should be viewed against a background of the current questioning, by leading evolutionary biologists, of whether the release of genetic variation is the primary biological role of sexual reproduction (Maynard Smith, 1978; Bernstein *et al.*, 1985). With such uncertainty over the function of sex, is it justifiable to continue accounting for the diversity of fungal incompatibility systems solely as an evolutionary series favoured by their promotion of outbreeding?

The problem with the release of genetic variation as the principal biological function of sex arises from the short term disadvantage associated with this explanation. The reward for producing variable offspring lies in the future as an insurance against a changing environment. This behaviour favours the species or group rather than the individual who pays a cost for this insurance through the breakup of successful gene combinations, through the risks involved in finding a mate and through the energy drain involved in sexual differentiation and meiosis. Calculations suggest that these short term disadvantages outweigh the long term advantages and that natural selection should favour those individuals who either never possessed sex or who abandon it (Maynard Smith, 1978). Yet sexual reproduction is widely distributed among eukaryotes and this apparent paradox has led evolutionary biologists to look for a role for sex of benefit to the individual (Bernstein *et al.*, 1985). After decades of wondering how imperfect fungi survive, mycologists should now be asking why so many species retain sexuality alongside such efficient means of asexual reproduction.

Whatever the *raison d'être* for sex, there can be no doubt that it has pronounced effects on the composition of fungal populations. For example, sexually reproducing populations of rust fungi are genetically more diverse and show less linkage disequilibrium than asexually reproducing populations (Burdon & Roelfs, 1985a & b). It is difficult to believe that these effects are not of profound importance in the population biology and evolution of these fungi. Which reproductive strategy is better suited to the common agricultural practice of widespread monoculture of a few select host genotypes is debatable, however. While sexual recombination may ensure the rapid appearance of a particular virulence combination, asexual reproduction allows the subsequent full exploitation of that genotype.

Despite the general acceptance among mycologists of the evolutionary importance of sexual reproduction, there is, apart from the higher fungi, little information on just how frequently it occurs in nature and on its importance relative to asexual reproduction. Certain sexually competent species appear to survive very well without it, e.g. *Puccinia graminis* f. sp. *tritici* in Australia (Burdon *et al.*, 1983). However, in many micro-fungi sexual reproduction could be easily overlooked, since it usually requires special environmental conditions and is relatively inconspicuous. Hence, it may be more common and of greater significance than is generally realised. For example, while the *Monilia* stage of *Neurospora* species is frequently observed, the natural formation of perithecia has seldom been reported. Nevertheless, sampling of *Monilia*-state colonies from burnt substrata gave both mating types in approximately equal numbers and with every culture possessing a unique genotype, indicating the each natural colony originated from a single ascospore (Perkins *et al.*, 1976). Another example is that of the wheat pathogen, *Septoria nodorum*. This has a recognised sexual stage, *Leptosphaeria nodorum*, but most plant pathologists have never seen it and do not consider it of importance in the epidemiology of the disease, at least in Britain. Spread is believed to be by asexual pycnidiospores originating each season on debris in the soil (King, Cook & Melville, 1983). In 1985 isolates were collected from small areas (about 12 m²) in two wheat fields in Staffordshire and compared in the laboratory. Virtually every isolate was unique, as was most clearly shown by their vegetative incompatibility types (Morris, Newton & Caten, unpublished). This diversity of genotypes suggests a population established from sexual products, not the spread of splash-dispersed clones (cf. Brasier, Chapter 16).

Another component of the sexual system is the amount of recombination occurring at each meiotic division, which is determined both by chromosome number and frequency of crossing over. Because of the small size of fungal chromosomes (see Sansome, Chapter 7) there is only limited information on these parameters. However, accumulated data from linkage mapping in genetically well-characterised fungi suggest that, although chromosome numbers lie within the normal range for eukaryotes, the frequency of meiotic crossing over is exceptionally high. For the species shown in Table 15.1, the average number of centiMorgans (cM) per kilobase (kb) of DNA for the three fungi is 90 times that for the three

Table 15.1. Frequency of meiotic recombination in three fungi and three higher eukaryotes

Species	n	Meiotic length (cM)	DNA content (kb)	cM per kb
Saccharomyces cerevisiae	17	4500	1.5×10^4	0.3000
Aspergillus nidulans	8	1800	4.1×10^4	0.0439
Neurospora crassa	7	620	4.3×10^4	0.0144
Drosophila melanogaster	4	284	1.8×10^5	0.0016
Lycopersicon esculentum	12	1100	6.8×10^5	0.0016
Zea mays	10	1120	1.5×10^6	0.0007

higher eukaryotes. Clearly, when sexual reproduction does occur only genes in close physical proximity will remain linked and, assuming outcrossing, the meiotic products are likely to be highly variable.

Cytoplasmic inheritance

The existence of cytoplasmic genomes in fungi was revealed by studies of variant phenotypes which were not transmitted in the manner expected for a nuclear controlled difference (Jinks, 1966). Cytoplasmic genomes can contribute to natural variation through copy number differences, mutation, heteroplasmosis and recombination, but their importance relative to the nucleus is difficult to assess from classical genetic studies. In the last 15 years, however, there have been great advances in knowledge of the molecular organisation of cytoplasmic genomes which allow us to approach this question from the other end and ask just how much, and what, genetic information is located in fungal cytoplasms. Three classes of cytoplasmic genomes have been identified: (i) double stranded DNA mitochondrial genomes (Tzagoloff, 1982; Scazzocchio, Chapter 4); (ii) double stranded RNA mycoviral genomes (Buck, 1980); and (iii) double stranded DNA plasmid genomes (Gunge, 1983; Nargang, 1985). While there is considerable variation from system to system, it is clear that the information content of these genomes is limited (Table 15.2). In comparison, the nuclear genome of *Saccharomyces cerevisiae* contains about $1 \cdot 5 \times 10^4$ kb of unique sequence DNA giving a coding potential of around 10,000 genes of which 568 have been identified and mapped (Mortimer & Schild, 1985). The role of the cytoplasmic genomes in the variation and evolution of fungi is, therefore, likely to be small in comparison with that of the nucleus. Nevertheless, recent molecular studies of mitochondrial genomes have demonstrated natural intraspecific cytoplasmic genetic diversity (Taylor, Smolich & May, 1986) and instances where cytoplasmic genomes have a pronounced effect on fungal populations can be found (Anagnostakis, 1984).

As with nuclear associations, the formation of heteroplasmons is subject to control. Cytoplasmic mixing during sexual reproduction is prevented by uniparental transmission. Although vegetative incompatibility influences somatic cytoplasmic exchange, the effect is usually incomplete and some transmission occurs between incompatible strains (Caten, 1972; Anagnostakis, 1983; Brasier, 1984). Any heteroplasmons formed are liable to segregate rapidly into their homoplasmic components (Turner & Rowlands, 1977) and so the opportunity for recombination between cytoplasmic genomes is limited. However, the available data indicate that,

Table 15.2. *Summary of the information content of fungal cytoplasmic genomes*

Genome	Size (kb)	Coding potential[a]	Open reading frames	Identified genes
Mitochondrial	18–110[b]	18–110	40 approx.	33[c]
dsRNA	3–15	3–15	?	2–4[d]
plasmids	1–100[e]	1–6	0–5	0–3[f]

[a]Based on 1 kb of unique sequence DNA = 1 coding unit.
[b]Size varies greatly between species.
[c]Data apply to *Saccharomyces cerevisiae*. Includes two rRNA genes, 24 tRNA genes and 7 genes coding for polypeptides (Tzagoloff, 1982).
[d]Genes for capsid polypeptide, RNA-dependent RNA polymerase, toxin production and resistance (Buck, 1980).
[e]Size varies greatly both within and between species. Many consist of amplified sequences from the mitochondrial genome (Nargang, 1985).
[f]Three polypeptides coded by the *S. cerevisiae* 2μ plasmid (Gunge, 1983).

given the opportunity, the frequency of mitochondrial recombination is high (Croft & Dales, 1984).

Other processes

Notwithstanding the range of genetic processes in the fungi discussed above, we should remain open to the possibility that other processes exist and contribute to the variation observed both in nature and the laboratory. Certainly, not all observed phenomena can be readily explained by the accepted processes; e.g. the spontaneous somatic variation of *Phytophthora* species (Caten & Jinks, 1968; Shaw, 1983). Transposable elements occur in yeast and may be important in genome rearrangement (Oliver, Chapter 3). Their occurrence and significance in other fungi remain to be established, however.

Conclusions

While fungi are undoubtedly variable, we may have over-emphasised their variability and accepted too readily the image of the ' . . . mutable and treacherous tribe . . . '. Most can be grown and cloned under controlled conditions, which focusses attention on the differences between isolates, rather than on their similarities and so encourages taxonomic splitting. Yet we should hardly be surprised that two individuals of the same species are seldom identical when we accept it readily enough in our own species. Indeed, one of the most significant recent developments in fungal population biology is the realisation of just how frequent and widespread individual clones may be and how little variation some populations contain. Thus, it now appears that, despite being isolated hundreds or even thousands of miles apart, isolates which belong to the same vegetative compatibility group or physiologic race are near isogenic and probably clonally related (Croft & Jinks, 1977; Burdon & Roelfs, 1985a; Newton *et al.*, 1985).

Central to much thinking about fungal evolution has been the belief that all variation is beneficial. Processes which promote variability have been considered to be evolutionarily advantageous, without critical consideration of their possible disadvantages or of alternative biological roles. The ready acceptance of parasexuality as an alternative to sex and the emphasis on the outcrossing bias of sexual systems are examples of this. Such thinking is group selectionist, placing the advantage to the group above that to the individual and while a qualitatively seductive concept, it is difficult to substantiate quantitatively (Maynard Smith, 1978). Greater attention should be focussed on the selective forces affecting an individual and its immediate descendants. The very existence of an individual is testimony to the fitness of its genotype. Why discard a tested set of genes in an orgy of variation on the chance that some of the products may do well in the future? It makes more sense to exploit proven fit co-adapted genotypes by spreading them as rapidly and as widely as possible. This is something many fungi, with their efficient asexual reproduction and dispersal mechanisms, are very good at; as is most clearly seen in the success of individual physiologic races of biotrophic pathogens closely adapted to particular host genotypes. Such genetic specialisation would not be possible if outcrossing and recombination were a regular feature of the life history (Caten, 1987).

It can be argued that variation is not always beneficial and that we need to reconsider the biological roles of processes like hyphal anastomosis, sexuality and incompatibility. Above all, we should start to think about fungal evolution in terms of modern population genetics with its greater emphasis on chance processes (Gale, 1987). Fitness, selection, population size, population growth rate, dispersal, migration, population extinction, gene flow, etc., are concepts central to population genetics and are easy to appreciate in theory. But a full understanding of population dynamics and microevolution requires not just a qualitative feel for the effects of these processes, but a quantitative explanation based on reliable field estimates of the relevant parameters. The acquisition of these estimates should be one of the major aims of fungal ecology and fungal genetics in the coming decades.

References

Anagnostakis, S. L. (1983). Conversion to curative morphology in *Endothia parasitica* and its restriction by vegetative compatibility. *Mycologia*, **75**, 777 – 780.

Anagnostakis, S. L. (1984). The mycelial biology of *Endothia parasitica*. I. Nuclear and cytoplasmic genes that determine morphology and virulence. In *The Ecology and Physiology of the Fungal Mycelium*, British Mycological Society Symposium volume 8, ed. Jennings, D. H. & Rayner, A. D. M., pp. 353 – 366. Cambridge University Press.

Bernstein, H., Byerly, H. C., Hopf, F. A. & Michod, R. E. (1985). Genetic damage, mutation and the evolution of sex. *Science*, **229**, 1277 – 1281.

Brasier, C. M. (1970). Variation in a natural population of *Schizophyllum commune*. *American Naturalist*, **104**, 191 – 204.

Brasier, C. M. (1984). Inter-mycelial recognition systems in *Ceratocystis ulmi*: their physiological properties and ecological importance. In *The Ecology and Physiology of the Fungal Mycelium*, British Mycological Society Symposium volume 8, ed. Jennings, D. H. & Rayner, A. D. M., pp. 451 – 497. Cambridge University Press.

Brasier, C. M. & Gibbs, J. N. (1975). Highly fertile form of the aggressive strain of *Ceratocystis ulmi. Nature*, **257**, 128 − 131.

Brierly, W. B. (1931). Biological races in fungi and their significance in evolution. *Annals of Applied Biology*, **40**, 420 − 434.

Brown, W. (1926). Studies on the genus *Fusarium*. IV. On the occurrence of saltations. *Annals of Botany*, **40**, 223 − 244.

Buck, K. W. (1980). Viruses and killer factors of fungi. In *The Eukaryotic Microbial Cell*, Society for General Microbiology Symposium volume 30, ed. Gooday, G. W., Lloyd, D. & Trinci, A. P. J., pp. 329 − 375. Cambridge University Press.

Burdon, J. J., Luig, N. H. & Marshall, D. R. (1983). Isozyme uniformity and virulence variation in *Puccinia graminis* f. sp. *tritici* and *P. recondita* f. sp. *tritici* in Australia. *Australian Journal of Biological Science*, **36**, 403 − 410.

Burdon, J. J. & Roelfs, A. P. (1985a). Isozyme and virulence variation in asexually reproducing populations of *Puccinia graminis* and *P. recondita* on wheat. *Phytopathology*, **75**, 907 − 913.

Burdon, J. J. & Roelfs, A. P. (1985b). The effect of sexual and asexual reproduction on the isozyme structure of populations of *Puccinia graminis. Phytopathology*, **75**, 1068 − 1073.

Burnett, J. H. (1975). *Mycogenetics*. London: Wiley.

Buxton, E. W. (1960). Heterokaryosis, saltation and adaptation. In *Plant Pathology*, vol. II, ed. Horsfall, J. G. & Dimond, A. E., pp. 359 − 405. New York: Academic Press.

Caten, C. E. (1972). Vegetative incompatibility and cytoplasmic infection in fungi. *Journal of General Microbiology*, **72**, 221 − 229.

Caten, C. E. (1981). Parasexual processes in fungi. In *The Fungal Nucleus*, British Mycological Society Symposium volume 5, ed. Gull, K. & Oliver, S. G., pp. 191 − 214. Cambridge University Press.

Caten, C. E. (1987). The concept of race in plant pathology. In *Populations of Plant Pathogens: their Dynamics and Genetics*, ed. Wolfe, M. S. & Caten, C. E., pp. 21 − 37. Oxford: Blackwell Scientific Publications.

Caten, C. E. & Jinks, J. L. (1966). Heterokaryosis: its significance in wild homothallic Ascomycetes and Fungi Imperfecti. *Transactions of the British Mycological Society*, **49**, 81 − 93.

Caten, C. E. & Jinks, J. L. (1968). Spontaneous variability of single isolates of *Phytophthora infestans*. I. Cultural variation. *Canadian Journal of Botany*, **46**, 329 − 348.

Connolly, V. & Simchen, G. (1973). Two-environment selection with inbreeding in *Schizophyllum commune. Genetical Research*, **22**, 25 − 36.

Croft, J. H. (1985). Protoplast fusion and incompatibility in *Aspergillus*. In *Fungal Protoplasts: Applications in Biochemistry and Genetics*, ed. Peberdy, J. F. & Ferenczy, L., pp. 225 − 240. New York: Marcel Dekker.

Croft, J. H. & Dales, R. B. G. (1984). Mycelial interactions and mitochondrial inheritance in *Aspergillus*. In *The Ecology and Physiology of the Fungal Mycelium*, British Mycological Society Symposium volume 8, ed. Jennings, D. H. & Rayner, A. D. M., pp. 433 − 450. Cambridge University Press.

Croft, J. H. & Jinks, J. L. (1977). Aspects of the population genetics of *Aspergillus nidulans*. In *Genetics and Physiology of Aspergillus*, British Mycological Society Symposium volume 1, ed. Smith, J. E. & Pateman, J. A., pp. 339 − 360. London: Academic Press.

Darlington, C. D. (1939). *Evolution of Genetic Systems*. Edinburgh: Oliver & Boyd.

Day, P. R. (1974). *Genetics of Host − Parasite Interaction*. San Francisco: Freeman.

Esser, K. & Blaich, R. (1973). Heterogenic incompatibility in plants and animals. *Advances in Genetics*, **17**, 107 − 152.

Fincham, J. R. S., Day, P. R. & Radford, A. (1979). *Fungal Genetics*. 4th edition. Oxford: Blackwell Scientific Publications.

Gale, J. S. (1987). Factors delaying the spread of a virulent mutant of a fungal pathogen: some suggestions from population genetics. In *Populations of Plant Pathogens: their Dynamics and Genetics*, ed. Wolfe, M. S. & Caten, C. E., pp. 55 − 62. Oxford: Blackwell Scientific Publications.

Gunge, N. (1983). Yeast DNA plasmids. *Annual Review of Microbiology*, **37**, 253 − 276.

Jinks, J. L. (1966). Mechanisms of inheritance. 4. Extranuclear inheritance. In *The Fungi, An Advanced Treatise*, vol. II, ed. Ainsworth, G. C. & Sussman, A. S., pp. 619 − 660. New York: Academic Press.

King, J. E., Cook, R. J. & Melville, S. C. (1983). A review of *Septoria* diseases of wheat and barley. *Annals of Applied Biology*, **103**, 345 − 373.

Koltin, Y., Stamberg, J. & Lemke, P. A. (1972). Genetic structure and evolution of the incompatibility factors in higher fungi. *Bacteriological Reviews*, **36**, 156 − 171.

Maynard Smith, J. (1978). *The Evolution of Sex*. Cambridge University Press.

Merrick, M. J. (1975). The inheritance of penicillin titre in crosses between lines of *Aspergillus nidulans* selected for increased productivity. *Journal of General Microbiology*, **91**, 287 − 294.

Mortimer, R. K. & Schild, D. (1985). Genetic map of *Saccharomyces cerevisiae*, ed. 9. *Microbiological Reviews*, **49**, 181 − 212.

Nargang, F. E. (1985). Fungal mitochondrial plasmids. *Experimental Mycology*, **9**, 285 − 293.

Newton, A. C., Caten, C. E. & Johnson, R. (1985). Variation for isozymes and double-stranded RNA among isolates of *Puccinia striiformis* and two other cereal rusts. *Plant Pathology*, **34**, 235 − 247.

Newton, A. C., Johnson, R. & Caten, C. E. (1986). Attempted somatic hybridization of *Puccinia striiformis* f. sp. *tritici* and *P. striiformis* f. sp. *hordei*. *Plant Pathology*, **35**, 108 − 113.

Perkins, D. D., Turner, B. C. & Barry, E. G. (1976). Strains of *Neurospora* collected from nature. *Evolution*, **30**, 281 − 313.

Person, C., Groth, J. V. & Mylyk, O. M. (1976). Genetic change at the populational level in host − parasite systems. *Annual Review of Phytopathology*, **14**, 177 − 188.

Pontecorvo, G. (1946). Genetic systems based on heterokaryosis. *Cold Spring Harbor Symposia on Quantitative Biology*, **9**, 193 − 201.

Pontecorvo, G. (1956). The parasexual cycle in fungi. *Annual Review of Microbiology*, **10**, 393 − 400.

Raper, J. R. (1966). Life cycles, basic patterns of sexuality and sexual mechanisms. In *The Fungi, An Advanced Treatise*, vol. II, ed. Ainsworth, G. C. & Sussman, A. S., pp. 473 − 511. New York: Academic Press.

Rayner, A. D. M., Coates, D., Ainsworth, A. M., Adams, T. J. H., Williams, E. N. D. & Todd, N. K. (1984). The biological consequences of the individualistic mycelium. In *The Ecology and Physiology of the Fungal Mycelium*, British Mycological Society Symposium volume 8, ed. Jennings, D. H. & Rayner, A. D. M., pp. 509 − 540. Cambridge University Press.

Roelfs, A. P. & Groth, J. V. (1980). A comparison of virulence phenotypes in wheat stem rust populations reproducing sexually and asexually. *Phytopathology*, **70**, 855 − 862.

Shaw, D. S. (1983). The cytogenetics and genetics of *Phytophthora*. In *Phytophthora, its Biology, Taxonomy, Ecology and Pathology*, ed. Erwin, D. C., Bartnicki-Garcia, S. & Tsao, P. H., pp. 81 − 94. St. Paul, Minnesotta: American Phytopathological Society.

Snyder, W. C. (1961). Heterokaryosis as a natural phenomenon. *Recent Advances in Botany*, **1**, 371 − 374.

Stamberg, J. (1969). Genetic control of recombination in *Schizophyllum commune*: the occurrence and significance of natural variation. *Heredity*, **24**, 361 − 368.

Taylor, J. W., Smolich, B. D. & May, G. (1986). Evolution and mitochondrial DNA in *Neurospora crassa*. *Evolution*, **40,** 716 − 739.

Todd, N. K. & Rayner, A. D. M. (1980). Fungal individualism. *Science Progress*, **66**, 331 − 334.

Turner, G. & Rowlands, R. T. (1977). Mitochondrial genetics of *Aspergillus nidulans*. In *Genetics and Physiology of Aspergillus*, British Mycological Society Symposium volume 1, ed. Smith, J. E. & Pateman, J. A., pp. 319 − 337. London: Academic Press.

Tzagoloff, A. (1982). *Mitochondria*. New York: Plenum Press.

Whitehouse, H. L. K. (1949). Heterothallism and sex in the fungi. *Biological Reviews*, **24**, 411 − 447.

16

The dynamics of fungal speciation

C. M. BRASIER

Forest Research Station, Alice Holt Lodge, Farnham, Surrey GU10 4LH, UK

Introduction

Opening his Presidential Address to the British Mycological Society, Burnett (1983) stated 'Mycology and mycologists, on the whole, have contributed very little to the mainstream of ideas concerning the modes of origin of species'. Mycologists should view this statement with unease, if only because an understanding of the natural taxonomic relationships of fungi must depend on knowledge of their speciation processes. Moreover, the failure of mycology to contribute significantly to wider speciation theory contrasts sharply with the fact that the fungi provide superb material with which to study evolution in action on account of their short generation times, flexibility of response to environmental change, relatively simple organisation, often haploid-based genetic systems and ease of laboratory manipulation. Indeed, their very suitability seems to underline the tardiness of our response to the challenge.

This chapter will attempt to explore the main ingredients of fungal speciation in the light of classic speciation theory, consider the role of fungal population structure in the speciation process, and present a speculative model for rapid speciation in the fungi with special emphasis on the influence of man.

The ingredients of speciation

The operational definition of speciation used in this chapter will be simply 'the emergence of reproductively isolated sub-populations from or within an original interbreeding population'. While the emergence of such sub-populations will often be associated with the accumulation of genetic differences, they may not always be expressed in terms of morphological divergence. This is partly because of the relatively simple

Table 16.1. Subgroups within fungal morphospecies and their status

Original morphospecies	Subgroups	Postulated[a] speciation process	Degree of[b] reproductive isolation	Authors
Basidiomycetes				
Armillaria mellea	Europe, 5 biological species; N. America, 10+ biological species;	E E	●●	Korhonen 1978a; Chapter 20 Anderson & Ullrich, 1979
Auricularia auricula	*A. ostoyae*: groups A and B European vs. North American; N. American deciduous vs. coniferous	E G EH	● ○	Morrison *et al.*, 1985 Duncan & Macdonald, 1967
Collybia dryophila	*C. dryophila, C. subsulphurea, C. earleae* and *C. brunneola*	E	●	Vilgalys & Miller, 1983
Coniphora puteana	Three inter-sterile groups	SU	●	Ainsworth, Chapter 19
Coprinus callinus	Inter-sterile groups	SU	●	Lange, 1952
Coprinus subimpatiens	Inter-sterile groups	SU	●	Lange, 1952
Coprinus lagopus	Six inter-sterile groups	SU	●	Kemp, 1983
Coprinus section *Lanatuli*	Eighteen inter-sterile groups	SU	●	Kemp, 1983, 1985
Coprinus macrorhizus	Two inter-sterile groups	SU	●	Kimura, 1952
Fomes (Fomitopsis) pinicola	North American Groups A and B and European group C	G,E	●○	Mounce & Macrae, 1938
Fomes (Phellinus) ignarius	Three inter-sterile groups	EH	●	Verrall, 1937
Gloeocystidium tenue	Bipolar groups 1 and 2 and homothallic group 3	SU	●	Boidin, 1951
Heterobasidion annosum	S (spruce) and P (pine) groups; and homothallic group	EH	○●●	Korhonen, 1978b; Chapter 20
Hirschioporus abietinus	*H. abietinus, H. fuscoviolaceous* and *H. laricinus*	G,E	○	Chase & Ullrich, 1985 Macrae, 1967

Species	Description	Code	Symbol	Reference
Laccaria laccata	*L. bicolor, L. proxima, L. laccata, L. amethystina*	E	●	Fries & Mueller, 1984
Mycocalia denudata	Groups I and II	SU	●	Burnett & Boulter, 1963; Fries, 1985
Paxillus involutus	Intersterility groups I–III	EH	●	McKeen, 1952
Peniophora mutata	Groups A and B	EH	●	Gibbs & Greig, (1986)
Peridermium pini	Two developmental forms in UK		○	
Puccinia graminis	*f.sp. tritici, f.sp. secalis* *f.sp. avenae, f.sp. recondita* etc.	EH	—●	Eriksson & Henning, 1896; Johnson, 1949; Burdon & Marshall, 1981
Puccinia recondita	*f.sp. tritici, f.sp. bromina, f.sp. recondita* etc.	EH	—	Burdon & Marshall, 1981; Newton, Caten & Johnson, 1985
Puccinia striiformis	*f.sp. tritici, f.sp. hordei*	EH	—	Newton, Caten & Johnson, 1985
Rhizoctonia solani	11 anastomosis groups within *Thanetophorus cucumeris* and *Ceratobasidium cornigerum*	SU	●	Parmeter, Sherwood & Platt, 1969; Burpee *et al.*, 1980
Sistotrema brinkmanii	Homothallic group; bipoplar group – five biological species – tetrapolar group – two biological species	SU	●	Biggs, 1937; Lemke, 1969; Ullrich, 1973
Stereum hirsutum	North American, UK and Russian groups	G,E	●	Ainsworth, Chapter 19
Tilletia sp.	*T. foetida, T. caries*	G, EH	○ ●	Fischer & Holton, 1957
Typhula ishikariensis	USA: *T. ish*; Japan: *T. ida*; *T. ida* biotypes A, B & C	E	○ ● ○	Christen & Bruehl, 1979; Matsumoto, Sato & Arak, 1982

Table 16.1. Subgroups within fungal morphospecies and their status (contd)

Original morphospecies	Subgroups	Postulated[a] speciation process	Degree of[b] reproductive isolation	Authors
Ascomycetes and Fungi imperfecti				
Aspergillus nidulans	19+ Heterokaryon-compatibility groups	E	○	Jinks *et al.*, 1966; Croft, Chapter 21
Calonectria rigidiscula	Homothallic–saprophytic, heterothallic–parasitic and non-fertile–parasitic groups	EH	●	Ford, Bourret & Snyder, 1977
Erisyphe graminis	Seven formae speciales	EH	○	Marchal 1902; Hiura, 1962, 1978
Gibberella fujikuroi (*Fusarium moniliforme*)	vars. *fujikuroi, subglutinans, moniliforme* and *intermedium*	E, EH	●	Hsieh, Smith & Synder, 1972; Kuhlman, 1982
Lophodermium pinastri	*L. pinastri, L. conigenum, L. pina-excelsae, L. seditiosum*	EH	—	Minter & Millar, 1980
Nectria haematococca (*Fusarium solani*)	Mating populations I–VII (parasitic); Homothallic forms (saprophytic)	EH	●	Matuo & Snyder, 1973
Neurospora sitophila group	*N. sitophila, N. crassa N. intermedia, N. tetrasperma*		○	Perkins, Turner & Barry, 1976
Ophiostoma (*Ceratocystis*) *ulmi*	NAN, EAN and non-aggressive groups	EH	○	Brasier, 1977, 1978a, 1979, 1986a
Phomopsis oblonga	Forms 1 and 2	E	●	Webber & Gibbs, 1982; Brayford, 1984

Organism	Subgroups/Types	Interpretation[a]	Isolation[b]	Reference
Pyricularia oryzae	Types I–III	EH	—	Matsuyama & Kozaka, 1971
Sclerotinia sclerotiorum	*S. sclerotiorum, S. trifoliorum, S. minor*		—	Willets & Wong, 1980
Sphaeropsis sapinea	A and B types	EH	●	Wang *et al.*, 1986
Verticillium dahliae	Populations P1–P16	SU	●	Puhalla, 1979; Puhalla & Hummel, 1983
Chytridiomycetes, Oomycetes, Zygomycetes and Myxomycetes				
Coelomyces dodgei	*C. dodgei* and *C. punctatus*	E	●	Federici, 1982
Phytophthora megasperma	6–9 subgroups	EH	—	Hansen *et al.*, 1987; Hansen Chapter 22
Phytophthora palmivora	Cacao and rubber forms	EH	—	Brasier & Griffin, 1969
Mucor hiemalis	*M. hiemalis, M. corticola, M. lausanniensis, M. luteus, M. sylvaticus*	E	● ○	Schipper, 1973
Didymium iridis	Heterothallic and non-heterothallic groups	E	●	Betterley & Collins, 1983; Collins, Chapter 18

[a] Either original author's or present author's interpretation, as follows: G, geographical speciation; E, sympatric speciation; EH, as E, showing some form of host plant specialisation; SU, probably sympatric speciation, no apparent cause.

[b] Reproductive isolation between subgroups: ●, total; ○, partial; —, no information.

fruiting structures and considerable developmental plasticity of fungi (Brasier, 1983a) and partly because, in the fungi, reproductive isolation may sometimes be virtually instantaneous (see below). In the latter circumstances, taxonomically useful morphological differences between the resulting closely related or 'sibling species' (Brasier & Rayner, Chapter 25) may be lacking or may develop only after, sometimes long after, the initial speciation event (Kemp, 1977). Thus, there are at least two reasons why one could expect to find morphologically almost identical species in the fungi.

It is not surprising, therefore, that the literature contains many examples of fungal morphospecies within which partially or totally reproductively isolated sub-groups have been identified (Table 16.1). These are the ideal raw material for the study of speciation. Their discovery dates back a century to Eriksson & Henning's (1896) report of *formae speciales* in *Puccinia graminis*. Indeed, for obvious reasons, plant pathogens are prominent in such a list as Table 16.1. So too are the Hymenomycetes, a reflection in part of the relative ease of detecting interfertility or intersterility between their monosporous mycelia, and partly of the alertness of several outstanding early North American and European workers to the sub-group phenomenon. The discovery of further sub-groups as population methods are increasingly applied to the fungi is undoubtedly raising the temperature of debate regarding fungal species concepts and the level at which species or species units should be recognised and defined (Brasier & Rayner, Chapter 25).

As emphasised by such prominent post-Darwinian evolutionists as Dobzhansky, Mayr and Stebbins, the four main elements contributing to speciation are the original interbreeding population, the genetic system, natural selection, and reproductive isolation. How these elements apply to fungi will now be considered, although the diverse genetic or breeding systems of the fungi will not be examined specifically, since these are reviewed by Carlile and Caten (Chapters 14 and 15) and because their significance will emerge under other headings in this chapter.

The original interbreeding population

A population can be defined in terms of numerical size, geographic size and genetic structure. Its geographic size may vary from local to global and is likely to be conditioned by the mode and dispersal range of propagules, the size and distribution of the substratum and of the ecological niche occupied. Even freely interbreeding global populations such as that of *Schizophyllum commune* (Raper, Krongelb & Baxter, 1958) may be partially fragmented into smaller locally adapted units within which most gene flow is local, but with sufficient gene recruitment from long range dispersal to preserve the overall genetic integrity of the species (see Simchen, 1967).

An appreciation of the genetic structure of a population is particularly important for elucidating speciation phenomena. Most local samples of regularly outcrossing populations of Ascomycetes, Basidiomycetes and Myxomycetes occurring under relatively undisturbed conditions contain large numbers of individual genotypes, often represented as vegetatively incompatible individuals (Fig. 16.1). The vegetative incompatibility or vc system (see Carlile, Chapter 14) maintains the territorial integrity of these individuals and may also promote mutual antagonism and even territorial invasion between them, as in *Ophiostoma* (*Ceratocystis*) *ulmi* (Fig. 16.1). Recognition of this pattern of population structure in the fungi appears fleetingly in the work of Verral (1937) with *Phellinus ignarius*, resurfaces

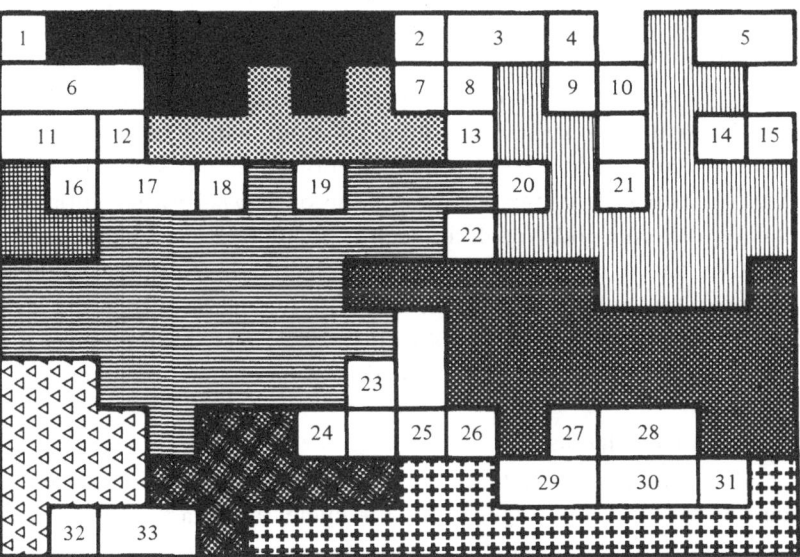

Fig. 16.1. The spatial distribution of vegetative compatibility (vc) groups or genotypes of *Ophiostoma ulmi* in an 11 × 16 cm piece of inner bark from a diseased elm. Each pattern or number represents a different vc genotype (a blank = no *O. ulmi*). The discrete territories occupied by the different genotypes are probably defined by the vegetative incompatibility interactions along their boundaries. These boundaries may not be fixed, since in fully vegetatively incompatible interactions hyphae of one genotype may invade the territory of another via the 'penetration effect' (Brasier, 1984). (From Webber, Brasier & Mitchell, 1986; see also Brasier 1984.)

with Adams & Roth (1967) with *Fomotopsis cajanderi*, is encapsulated in reports in the 1970s on heterokaryon incompatibility in various fungi such as *Neurospora crassa* (e.g. Mylyck, 1976) and is brought to its full significance with the emphasis by Todd & Rayner (1979) on the role of vegetative incompatibility and fungal individualism in Basidiomycetes such as *Coriolus versicolor*. Equally important, other studies show that individuals in such populations are often highly heterogeneous for protein and enzyme polymorphisms as in *N. intermedia* (Spieth, 1975) and *Didymium* (see Collins, Chapter 18); for mating-type factors as in *Coriolus versicolor* (Burnett & Partington, 1957; Williams & Todd, 1986); and for continuously variable characters such as growth rate and fruiting ability, as in *S. commune* (Brasier, 1970; Fig. 16.2). Many Oomycetes and Zygomycetes probably have similar heterogeneous population structures although the involvement of vegetative incompatibility is less certain.

For the present purpose, such a highly heterogeneous local population structure within which regular sexual outcrossing and vegetatively incompatibility play a major role will be termed *basal population structure*. Two points about this are pertinent to speciation. Firstly, it is this basal population structure within or from which speciation will commonly occur. Moderate to strong distortions of it may signal incipient speciation. Secondly, laying aside the issue of asexual propagation, the genetically different individuals in such a population are fundamentally competitive, and so, as in animals, natural selection in the fungi should be seen as operating at least at the level of the individual or even at the level of the gene (cf. 'selfish genes', Dawkins, 1976; and see Rayner *et al.*, 1984). This raises the question of selection as an element of speciation.

Natural selection

Natural selection acts on the interbreeding population via many aspects and during many stages of the life cycle; e.g. via the substrate or host, vectors, climate, competitors and predators; during spore germination, colonisation, consolidation, fruiting, dispersal and resting stages. The sum of these components of selection under conditions of relative ecological stability i.e. where none of the individual components varies so strongly that longer term ecological balance is disturbed, can be called *routine selection* (Brasier, 1986a). Routine selection is, hence, the totality of selection that favours maintenance of a stable population structure. By contrast, *episodic selection* (Brasier, 1986a) results from any form of sudden and extreme ecological disturbance likely to lead to a breakdown of basal population structure. Thus defined, routine and episodic selection encompass the ecological context of selection within which the population as a whole is organised and condition the fate of individuals, rather than of individual phenotypic traits. The latter is covered by the well established terms stabilising, directional and disruptive selection (Mather, 1953) which can be basic dynamic constituents of both routine and episodic selection; with substratum, host, vector, climate, other organism, etc. being the material components. Indeed, episodic selection would often involve a sudden change in the selection intensity imposed by one or more major material components — such as the host plant — to the point where its influence became strongly directional with respect to certain traits in the fungal population. It could also involve the sudden release of previous routine selection constraints.

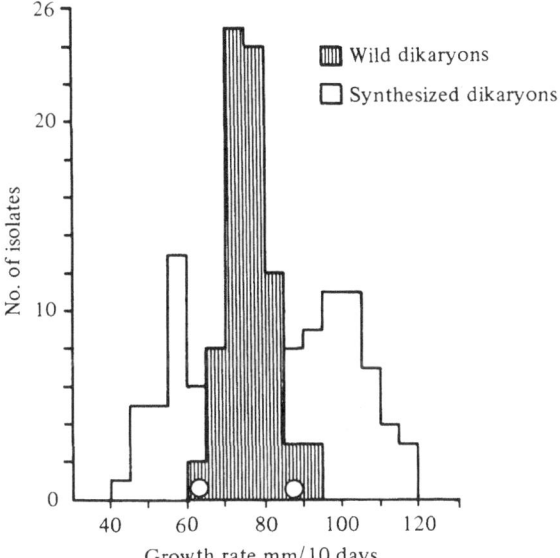

Fig. 16.2. The growth rate distributions of a sample of 77 wild dikaryons of *Schizophyllum commune* from a single location, compared with those of two sets of 72 synthesised F_1 dikaryons. Each F_1 set was derived by *inbreeding* from a single wild dikaryon, represented by open circles. The growth rate distribution of the combined F_1s greatly exceeded that of the whole sample, indicating the impact of selection. Since the wild dikaryons were derived from fruit bodies, they represent survivors following the action of routine selection on a single generation, i.e. from spore dispersal and colonisation to senescence and fruiting. The apparent elimination of extreme (fast and slow growing) phenotypes indicates that growth rate is subject to stabilising selection. (Compiled from data of Brasier (1970) and S. Williams & C. M. Brasier, unpublished; see also Williams *et al.* (1976).)

Hence, at a lower hierarchical level of terms, routine and episodic selection relate to stabilising, directional and disruptive selection, which particularly concern effects on phenotypic traits in populations. At higher hierarchical levels they interact with the concepts of r- and K-selection. The latter are essentially concerned with fungal behaviour within communities and are better viewed within the general context of life history strategies (cf. Andrews & Rouse, 1982; Cooke & Rayner, 1984; Cooke, Chapter 9). Organisms exhibiting either r or K strategies will generally be subject to routine selection and only in unusual circumstances to episodic selection. Although both routine and episodic, and r- and K-selection are defined in terms of environmental disturbance they differ with respect to the element of time. Thus, r-selection can be considered a component of routine selection in that it concerns adaptation to recurrent cycles of disturbance within an ecosystem. Episodic selection, however, is marked by a particular temporal event which leads to a change in the overall spectrum of selection previously experienced by a population.

The influence of routine and also stabilising selection on a population of *Schizophyllum commune* is shown in Fig. 16.2. A local sample of 77 dikaryons was found to be highly heterogeneous for growth rates, fruiting characters, mating types and subsequently (C. M. Brasier, unpublished), vegetative compatibility types; it had basal population structure. When two dikaryons near the extremes of the distribution for growth rates at 30° were individually fruited and inbred, the two resulting sets of 72 progeny dikaryons showed much greater variation than the entire population sample. Since the original sample was taken from fruit bodies, i.e. following colonisation and vegetative development, the discrepancy between the variability of the sample and that of the progeny indicates the impact of selection on a single field generation.

Gradual niche divergence and disruptive selection
Much evidence (Table 16.1), points to the divergence of original interbreeding populations at the same geographic location, i.e. sympatric speciation (*sensu* Mayr, 1942). Several modern evolutionists, including Mather (1955) and Maynard Smith (1966) argue that sympatric speciation is initiated when the original population is confronted with two (or more) ecological niches. If the niches are sufficiently large and distinct a gradual shift will occur, as a result of disruptive selection (Mather, 1953), from a single population with an optimal phenotype for one niche to two or more populations each with an optimal phenotype for a different niche. Maynard Smith (1966) has proposed that the crucial step in sympatric speciation is

the establishment of a stable polymorphism in a heterogeneous environment, which can be achieved so long as population size is separately regulated in the two niches and the respective selective advantages are large.

If one accepts this as a mechanism for gradual sympatric divergence, what, in the fungi, are the most prominent material influences involved? The evidence in Table 16.1, which is biased towards economically important organisms, points not surprisingly to three major factors: climate, living substrates and non-living substrates. Locally strong climatic gradations or discontinuities certainly seem likely to promote divergence. For example, Matsumoto, Sato & Arak (1982) have described a case involving three biotypes of the snow mould, *Typhula ishikariensis*, in Japan. Biotypes A and B are totally reproductively isolated; A occurs mainly in inland areas with deeper, longer-lasting, snow cover, and B mainly in coastal regions. Biotype C is similar to B but is irregular in distribution. It seems difficult to escape the conclusion that biotypes A and B are climatically adapted, perhaps sympatrically divergent, forms in which speciation appears to have gone to completion. By contrast, the warm − and cool − adapted environmental races of *Helminthosporium maydis* described by Hill & Nelson (1976) appear to represent a case of incipient speciation under climatic influence.

Opportunistic nutritional strategies may often lead to divergence among saprotrophs. Probable examples are to be found in *Mucor* and *Neurospora*. *M. mucedo* commonly inhabits rodent dung whilst the morphologically very similar *M. pyriformis* is common on rotting fruits (Schipper, 1975). Although evidence of reproductive isolation has been experimentally unobtainable, sympatric speciation seems likely. In *N. intermedia*, Perkins, Turner & Barry (1976) have reported two subpopulations, 'yellow giant' and 'standard', the former occurring on food and food wastes in east and southeast Asia, the latter on burned substrata in the tropics generally. Individuals from the two populations will hybridise but, significantly, the hybrids probably do not survive in nature.

In view of the selection constraints imposed by a host's defence mechanisms (cf. Heath, Chapter 10) it is not surprising to find strong evidence of niche influences among the plant pathogens. Population divergence in response to relatively narrow host niches is suggested by the occurrence of partially or totally reproductively isolated *formae speciales* of powdery mildews and rusts on different cereal and grass genera, e.g. in *Erisyphe graminis* (Hiura, 1962, 1978) and *Puccinia graminis* (Johnson, 1949). Broader niche specialisation may be represented in such instances

Table 16.2. *Sub-groups or sibling species exhibiting combinations of narrow and broad host specialisation*

Morphospecies, sub-group and host	Other features
Nectria haematococca (Fusarium solani): mating populations	
MP I on cucurbits	Each MP is heterothallic,
II on potato	and reproductively
III on mulberry	isolated.
IV on Japanese pepper	Homothallic forms also
V on cucurbits	exist, but are saprotrophic
VI on pea, mulberry, ginseng, chick pea	(Matuo & Snyder, 1973).
Giberella fujikuroi (Fusarium moniliforme): mating groups	
Group A on corn, pine, sorghum, cotton	Each group is heterothallic, and reproductively isolated.
B on corn, pine, sugar cane	The groups also show certain
C on rice	morphological differences,
D on corn, pine, sugar cane, copra, *Cymbidium, Hibiscus*	and have been designated varieties (Kuhlman, 1982).
Phellinus ignarius: North American groups	
Aspen group on *Populus*	Each group is heterothallic and reproductively isolated. The
Birch group on *Betula*	groups also show
Miscellaneous group on other tree genera including *Betula*	morphological and physiological differences (Verrall, 1937).

as the reproductively isolated pine (P) and spruce (S) forms of *Heterobasidion annosum* (Korhonen, Chapter 20), and yet broader still in the totally reproductively isolated 'deciduous' and 'coniferous' forms of *Auricularia auricula* occurring within North America (Duncan & MacDonald, 1967).

Reproductively isolated sub-groups exhibiting both broad and narrow host ranges are often found within a single morphospecies, as in *Nectria haematococca, Gibberella fujikuroi* and *Phellinus ignarius sensu lato* (Table 16.2) and in *Phytophthora megasperma* (Hansen, Chapter 22). It is tempting to speculate that in such instances the broad host range group, e.g. MPVI in *N. haematococca* (Table 16.2), represents the ancestral form, a less specialised freely interbreeding population adapted to a wider variety of host types, but lacking the specialised genetic systems such as sets of polygenes (Mather, 1953) conferring a higher level of adaptation to a particular host. Such an omnivorous strategy may have been the stock-in-trade of many currently more specialised plant pathogens before man's disturbance and cultivation of their hosts. Experimental evidence for this phenomenon comes from *Erisyphe graminis* and *Puccinia graminis*, where hybrids between certain *formae speciales* have wider graminaceous host ranges but tend to be less pathogenic on an individual host than the parent types (e.g. Hiura, 1978; Johnson, 1949); and from evidence that isolates of *E. graminis* from wild grass communities in Israel have wider host ranges than isolates from elsewhere (Wahl *et al.*, 1978).

Ecological influences such as climate and substratum are clearly not mutually exclusive. Indeed, the divide between what is a living and a non-living substrate is not only conceptually vague and contentious; in terms of the presence or absence of host resistance, it has probably been a common catalyst of sympatric speciation events. The three physiologically distinct species of *Lophodermium* on pine identified by Minter & Millar (1980; see Table 16.1) may be evidence of this. One species infects green needles, i.e. is parasitic; one occurs mainly on killed needles and the other on needle litter. Since they also have similar geographical distributions, sympatric speciation seems likely. Research on the comparative biology of the organisms involved in this and similar cases (such as the parasitic and saprotrophic forms of *Calonectria rigidiscula* and *Nectria haematococca*, Table 16.1) might throw further valuable light on the process of evolution of parasitism from saprotrophy or *vice versa*.

A particularly intriguing and challenging problem is found in the apparently large number of fungal morphospecies which contain reproductively isolated sub-groups for which there is no *evident* niche relationship. Many examples occur in the Basidiomycetes. Some of these are plant pathogens e.g. the eleven intersterile anastomosis groups within *Thanetophorus cucumeris* and *Ceratobasidium cornigerum* (*Rhizoctonia solani*, Table 16.3), which tend to have broad host ranges and wide geographical distributions. Another example is that of *Sistotrema brinkmannii* (Table 16.3) in which the different intersterile groups can sometimes even be found on the same piece of wood. A number of examples occur in *Coprinus*, e.g. the six breeding groups close to *Coprinus lagopus* and the remarkable 18 additional groups recently identified by Kemp (1983) in the section *Lanatuli* (Table 16.3).

One explanation for such multiplicities of intersterile groups (i.e. 'sibling species' − see Brasier & Rayner, Chapter 25) could be that they were originally a product of allopatric speciation and that their ranges have subsequently overlapped, perhaps through man's influence (see below).

However, many cases such as those of the intersterile groups in *Sistotrema brinkmannii* and *Coprinus lagopus*, give an appearance of sympatry reinforced by total reproductive isolation. Hence, we may simply be failing to identify, through paucity of critical ecological studies, the subtlety of the niches they occupy or the subtle differences in their colonisation strategies. Certainly, the material influences on niche divergence already discussed are rather simple, when in reality ecological strategies are complex. Perhaps multiple temporal windows for colonisation occur as the physical and microbial status of the substratum changes. If such sibling species do not occupy distinct niches, then, where their reproductive isolation is complete, it is difficult to understand why the fittest does not eliminate the others (cf. Gause, 1932). A part of the answer to this problem may lie in the origin of the isolating mechanisms which separate them.

Reproductive isolation

Classic speciation theory requires that some form of reproductive isolation reinforces, sustains and in some cases promotes the divergence of an originally interbreeding population into sub-populations. Dobzhansky (1937) has classified the various processes of reproductive isolation, and a scheme along similar lines can be tailored to the special properties of the fungi, as outlined below.

Geographical isolation

Much primary speciation must have resulted from separation due to geographical barriers, e.g. *Auricularia auricula*, *Hirschioporus abietinus* and *Stereum hirsutum* (Table 16.1; Ainsworth, Chapter 19).

Prezygotic isolation

(i) Ecological isolation Where a population is diverging in response to different ecological niches (see above), if the majority of fertilisations are local the habitats themselves may tend to isolate the resulting sub-populations.

(ii) Gametic isolation Potentially, at least three forms of gametic isolation could occur in fungi. (a) *Temporal isolation*, where gamete release is asynchronous. An example is provided by the two closely related Chytridiomycete species *Coelomyces dodgei* and *C. punctatus* where the gametes are released over different time periods (Federici, 1972). A slightly different situation occurs in *Phytophthora*, where gametangia of different

heterothallic species often associate, but meiosis in the antheridium and oogonium may be asynchronous and fertilisation disrupted (E. R. Sansome & C. M. Brasier, unpublished). (b) *Behavioural isolation*, in which plasmogamy is often preceded by chemically mediated growth and recognition processes, such as interactive zygophore development in Zygomycetes, hormonally regulated gametangial development in Oomycetes, trichogyne-spermatial responses in Ascomycetes and homing in Basidiomycetes. These are the courtship rituals of the fungal world. Their breakdown would inevitably lead to reproductive isolation. Among possible examples are the absence or failure of mating responses between some members of the Mucorales (see Schipper, Chapter 17); and the cessation of gametangial development at the 'gametangial initial' stage in pairings between heterothallic *Phytophthora* species (C. M. Brasier unpublished). (c) *Post-contact isolation*, involves a breakdown of the next step in fungal mating: the fusion of gametangia, or of trichogyne with spermatium, or of hypha with hypha. This is probably the commonest mechanism of reproductive isolation in the fungi, particularly in the somatogamous Basidiomycetes where it frequently takes the form of a strong rejection response following fusion, as for example in lethal homing reactions in *Coprinus* (Kemp, 1975, 1977) or in barrage formation between monosporous mycelia of different *Armillaria* species (Korhonen, Chapter 20). There are probably many variations on this theme, some involving delayed rejection responses.

Post-zygotic isolation
Several forms of post-zygotic reproductive isolation are probably very important in the fungi.

(i) Zygotic abortion This includes meiotic failure in haploid based systems and failure of oospore formation in the diploid Oomycetes. Chromosome differences may often be a factor, as in *Phytophthora megakarya* and *P. palmivora*, which occur together on cocoa in West Africa, but have a very different chromosome size and number. Although the two species can mate, hybrid gametangia are probably unable to produce viable oospores (Sansome, Brasier & Sansome, 1979).

(ii) Genetic disharmony The immediate products of meiosis, such as ascospores, may be defective as in some *Neurospora intermedia* crosses (Perkins *et al*, 1976); or germlings produced by post-meiotic spores may lack vigour, reflecting serious genetic disturbances. Alternatively an F_1

product may be reasonably vigorous but nevertheless unfitted for survival in nature. The latter may result if the parent genomes carry special gene combinations which fit them to their respective ecological niches but which, when recombined, are disrupted, rendering the progeny poorly fitted for either niche. Thus, in *Ophiostoma ulmi*, while rare aggressive × non-aggressive crosses can be achieved in the laboratory the F₁ tend to be very weak pathogens and probably do not survive in nature (Brasier, 1977, 1983b, 1986a). Similarly progeny of crosses between sub-populations of *Aspergillus nidulans* in Britain (Croft & Jinks, 1977; Butcher, 1969; and see Croft, Chapter 21), of *Neurospora intermedia* in Asia (Perkins *et al.*, 1976) and of *Typhula ishikariensis* in the USA (Christen & Bruehl, 1979) all show reduced vigour and are thought to be selected against in nature.

(iii) Hybrid sterility Hybrid F₁ progeny may survive in nature, but be sexually sterile, e.g. in *Coelomyces dodgei* × *C. punctatus* crosses (see above) where the hybrid meiospores are motile and able to encyst on the secondary copepod host, but fail to produce a functional gametophyte (Federici, 1972). The sterile diploid form of *Phytophthora meadii* (see below) is probably another example of this phenomenon.

Points of general significance concerning isolating mechanisms

Although current information on reproductive isolation in the fungi is fragmentary, several points of interest or special significance emerge which deserve comment. First is the fact that within several Basidiomycete morphospecies including *Fomitopsis pinicola*, *Hirschioporus abietinus*, *Heterobasidion annosum* and *Stereum hirsutum* (Table 16.1; and see Korhonen, Chapter 20; Ainsworth, Chapter 19), an exception being *Auricularia auricula* (Table 16.1), what are apparently geographically isolated sub-populations tend to show less abrupt reproduction isolation than sympatrically isolated sub-populations. The above trend is consistent with an expectation that geographical isolation, being passive, is likely to lead through drift to a random degree of pre- or post-zygotic isolation, while sympatric divergence, being a more active process, is more likely to be reinforced by vigorous isolating mechanisms. Whether some general relationships between speciation processes and degree or type of associated reproductive isolation will emerge as further information accumulates remains to be seen.

Second, as emphasised by Burnett (1983), a consequence of the mycelial habit of fungi is that many reproductive isolating mechanisms probably arise through breakdown of processes associated with hyphal fusion i.e.

via post-contact isolation. A logical and attractive hypothesis following from this is that many such isolating mechanisms may result from the elevation of the existing vegetative incompatibility system operating between different individuals to the level of a barrier to gene flow between divergent sub-populations. Certainly, in entirely non-outcrossing or in imperfect fungi, vegetative incompatibility automatically becomes the primary isolating mechanism, as appears to be the case with the numerous vc or population groups in *Verticillium dahliae* (Table 16.3), *Fusarium oxysporum* (Puhalla, 1985) and *Stereum sanguinolentum* (Ainsworth, Chapter 19). Correspondingly, in most sexually outcrossing fungi, the potential disruption due to the expression of a vegetative incompatibility reaction between vegetatively incompatible but otherwise sexually compatible partners is probably suppressed either by the action of the mating type loci or through some property of the gametangia (Brasier, 1984; Rayner *et al.*, 1984). While the negation of this suppression system seems a likely route to reproductive isolation, the genetic basis of post-contact isolating mechanisms in fungi is, unfortunately, seriously unresearched. Moreover, the fact that genetic analysis will only be readily feasible in cases of partial reproductive isolation may lead to emphasis in such studies on cases of geographical or allopatric rather than of sympatric speciation (cf. above). To date, there is evidence for a separate five locus heterogenic incompatibility system operating between the S and P groups in *Heterobasidion annosum* (see Korhonen, Chapter 20), while Kemp (1977 & personal communication) has proposed a gene-for-gene model for sympatric speciation *via* lethal fusion reactions in two-spored basidiomycetes. In *Ophiostoma ulmi*, the partial fertility barrier between the EAN and NAN aggressive sub-groups appears to be polygenic (Brasier, 1984); whether the mechanism involved here includes the triggering of the vegetative incompatibility system at plasmogamy, or whether an entirely separate genetic incompatibility system has developed has yet to be fully determined. However, present evidence suggests the involvement of a separate system, as in *H. annosum*.

The third point is related to the previous one. Contemplating the sometimes remarkably large numbers of apparently sympatric reproductively isolated sub-groups or sibling species that occur within some fungal morphospecies (e.g. Table 16.3), Burnett (1983) and Kemp (1977, 1983) have emphasised that post-contact isolating mechanisms can probably arise very suddenly in fungi, e.g. as a result of a mutation blocking hyphal fusion. Burnett (1983) has taken this conclusion further by suggesting that in fungi sudden reproductive isolation may precede

Table 16.3. *Examples of sub-groups or sibling species with no evident niche*

Morphospecies, sub-group and other features

Thanatephorous cucumeris (Rhizoctonia solani) anastomosis groups

    AG 1–4   Each group is heterothallic, reproductively isolated, and tends to be polyphagous and of world-wide distribution (although AG 2 may specialise in crucifers). Trend for greater morphological and physiological homogeneity within rather than between groups (Parmeter, Sherwood & Platt, 1969).

Ceratobasidion cornigerum (R. solani) anastomosis groups

    CAG 1–7   Each group is heterothallic, reproductively isolated, sympatrically distributed within the USA, and is polyphagous except for CAG 1 which specialises in *Gramineae*. Some groups show cultural differences (Burpee *et al.*, 1980).

Sistotrema brinkmannii groups in North America, Europe and Australia

homothallic forms	No clear evidence for geographical barriers, or for ecological specificity between the intersterile groups. Different groups, including homothallics, can be obtained from the same wood pile, even from the same piece of wood (Ullrich, 1973).
bipolar forms: 5 intersterile groups	
tetrapolar forms: 2 intersterile groups	

Coprinus section *Lanatuli,* breeding groups

C. lagopus: 6 breeding groups in UK	All heterothallic, reproductively isolated, growing in woodland soils, Some show useful morphological differences (Kemp, 1983, 1985).
18 further breeding groups	Heterothallic, reproductively isolated. From various international locations. Fourteen of the groups are so far represented by only a single isolation (Kemp, 1985).

Verticillium dahliae populations P1–P16

P1	on cotton, pine, olive, rose, elm, maple, peanut, sesame	No sexual stage known. Each population is a separate vegetative compatibility group, and at least three of the four groups illustrated possess unique physiological and morphological characteristics, although sometimes widely geographically distributed (Puhalla, 1979; Puhalla & Hummel 1983)
P2	on cotton, tomato, grape, pistachio, olive, almond, apricot, eggplant	
P3	on cotton, potato, maple, eggplant, pea, spearmint, sumac, strawflower, cantaloupe	
P4	on potato, sugar beet, cantaloupe, horseradish, grape, mint	
etc.		

sympatric niche divergence rather than *vice versa*. However, while it is easy to accept rapid isolation as a process in itself, as already discussed, it is less easy to comprehend the initial or, indeed, continued survival of a genetically similar but reproductively isolated individual, or a population unit derived from it, within broadly the same niche without its having some form of selective advantage or spatial separation. Consolidation of this important point of theory must therefore await the acquisition of further ecological evidence.

Finding a solution to such intriguing and fundamental problems is important for an understanding of many speciation phenomena in fungi. The following speculative model is offered as another possible mechanism of speciation.

A theoretical model for rapid clonal speciation

In recent decades, the non-aggressive sub-group of the Dutch elm disease pathogen, *Ophiostoma ulmi*, has been in balance with its elm host in Europe. Local population samples of the non-aggressive are highly polymorphic for vegetative compatibility (vc) groups, and both A and B mating types occur in about equal frequency (Brasier, 1984). Hence, the fungus appears to be regularly outcrossing, to show basal population structure and to be subject to routine selection. The EAN and NAN aggressive sub-groups of *O. ulmi*, on the other hand, are responsible for the current massive epidemics of the disease in Europe. They are thus out of balance with their host, and they also show a rather different population structure. In samples from current epidemic front areas in Europe a single vc group, the vc 'super group', commonly predominates (about 50 − 98%), while the remainder of the same sample is heterogeneous for vc groups. At comparable post-epidemic sites the super-group has declined to about 10 − 20% of the population while the heterogeneous (presumably outcrossing) component has increased to around 80 − 90%. Hence, there appears to be a return towards basal population structure as the epidemics progress, a process that appears to have taken only a few years at one site in Portugal. One explanation for the occurrence of super-groups could be a simple founder effect, but evidence that the Λ mating type occurs at about 10 − 30% in the heterogeneous component, yet is virtually absent in the super-group component of a sample even at the post-epidemic sites where there must have been many intervening generations of outcrossing, suggests otherwise. An alternative explanation is that, in response to intense selection at the epidemic fronts, the population is partitioning into (i) a highly fitted, possibly clonal, component, the super group, which is

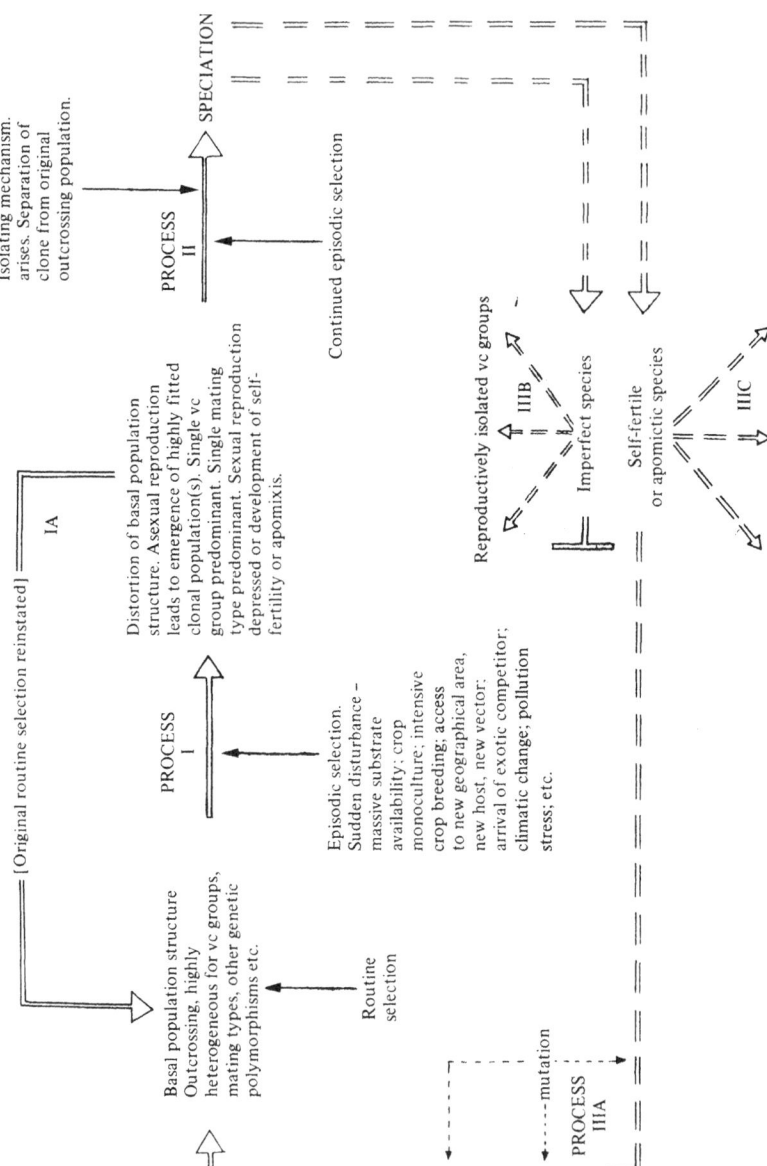

Fig. 16.3. A qualitative model for rapid clonal speciation. Open arrows represent consequences of a variety of selection and reproductive isolation events. See text for further details. (Based on Brasier (1986a) Figs 24 and 25.)

Table 16.4. *Possible examples of clone development in response to episodic selection*

Organism	Original (= basal) population structure?	Distorted population structure under episodic selection
Ophiostoma ulmi on elm	Chronic disease level (non-aggressive sub-group); heterogeneous for vc groups; A and B mating types present in equal frequency.	Epidemics (aggressive sub-groups); emergence of vc super-groups; loss of one mating type (Brasier, 1984, 1986*a* and see text).
Helminthosporium maydis on corn	Minor corn pathogen on genetically heterogeneous host. Both A and a mating types in equal frequency.	Epidemic form 'Race T' selected through exposure to *Tms* hybrid corn. Predominantly A mating type in northern corn belts (Nelson *et al.,* 1971; Leonard, 1973).
Phytophthora cinnamomi	At its putative origin (Taiwan) is morphologically heterogeneous, with A1 and A2 mating types in equal frequency.	At epidemic sites in Australia, A2 is predominant, and of only two isozyme genotypes. World-wide, A2 is predominant. Near clonal population reinforced through selfing of A2? (Brasier, 1978; Ko, Chang & Su, 1978; Old, Moran & Bell, 1984).
P. palmivora	Origin as chronic pathogen in Amazonia? Both A1 and A2 mating types prevalent world-wide, on a very wide total host range.	A characteristic morphological phenotype of the pathogen occurs on cocoa world-wide, which is of only A2 mating type. A clone, reinforced through selfing of A2? (Brasier & Griffin 1979).
Puccinia striiformis	Population structure? Sexual stage unknown.	World-wide, the wheat form of *P. striiformis (f. sp. tritici)* is genetically uniform, suggesting a clone (Newton, Caten & Johnson, 1985). Sexual stage lost?

mainly asexually disseminated, and may be losing the A mating type altogether; and (ii) a heterogeneous component tending to be recently sexually derived, i.e. closer to basal population structure (Brasier, 1986*a*, & unpublished observations).

The latter explanation would fit within a relatively sudden and major change in the selection pressures acting on the pathogen population sufficient to disturb the long term local ecological balance, i.e. episodic selection (see above). Under such circumstances some previously existing routine selection constraints could be released while others could become strongly directional. As shown in Fig. 16.3, rather than a gradual divergence into sub-populations, a common consequence of the sudden shift in composition of selection patterns under episodic selection might be the emergence and propagation of one or more correspondingly well fitted clonal populations. This could lead in turn to breakdown of the mating system and the development of self-fertility and apomixis or dependence on purely asexual modes of propagation (Fig. 16.3 process I) along the lines suggested by Thomas, Suryanarayana & Manavathu (1983) and Brasier (1986a).

Evidence for the emergence of clonal populations in response to severe ecological disturbance in other fungi is, as with *O. ulmi*, somewhat circumstantial, particularly because the structure of the presumed ancestral population is often uncertain. Nevertheless, Table 16.4 lists some highly suggestive cases each of which gives the impression of the narrowing of an originally variable fungal population towards a clone, accompanied either by reduction in or a loss of outcrossing ability (*O. ulmi*, *Helminthosporium maydis*) or a trend towards inbreeding or selfing (*Phytophthora cinnamomi* and *P. palmivora*) or loss of the sexual stage altogether (*Puccinia striiformis*).

If, in response to episodic selection, highly fitted clonal populations can emerge from basal populations, then a continuation of this process could lead to speciation (Fig. 16.3, top). The population might revert, as apparently does *O. ulmi* in Europe, to basal structure if something close to the original routine selection were reinstated (Fig. 16.3, process IA). However, if episodic selection continued, and if a reproductive isolating mechanism arose (e.g. by mutation or a chromosomal rearrangement) leading to the sudden genetic isolation of the clonal element from the remainder of the population (Fig. 16.3, process II) the clone would have rapidly speciated. This would result in an imperfect species if the clone had lost normal sexual reproductive capacity, and in a self-fertile (homothallic or apomictic) species if a switch to selfing had occurred; in either case the new species being fitted to the new disturbed environment. If, following what may sometimes be a chain of episodic selection events (Brasier, 1986a), the new set of environmental conditions were sustained, becoming the stable condition, then what was initiated by episodic selection would have become, for the fitted clone or new species, routine selection. If the niche was narrow and recurrently disturbed though widespread, then the fungus might remain a self-fertile, apomictic or imperfect organism. However, if at some point a more diverse environment or more flexible niche became available, a self-fertile or apomictic species could develop, *via* mutation, into a new regularly outcrossing organism with basal population structure (Fig. 16.3, process IIIA). The latter option would be barred to an imperfect species, but these could potentially give rise to further reproductively isolated clones simply through the development of new vegetative incompatibility types by mutation (Fig. 16.3, process IIIB); cf. *Verticillium dahliae* (Table 16.3) and *Fusarium oxysporum* (Puhalla, 1985). A self-fertile species could also diversify in this fashion (Fig. 16.3, process IIIC); cf. *Aspergillus nidulans*, Table 16.1 and Croft, Chapter 21, and *Stereum sanguinolentum* (Ainsworth, Chapter 19). Thus this qualitative model, though highly speculative, could account for a variety of speciation patterns in fungi.

Outcrossing versus non-outcrossing

The above model also highlights the question of the respective roles of outcrossing and non-outcrossing. The interplay of heterothallism and homothallism is a striking and recurrent theme in the fungi. Closely related but reproductively isolated homothallic and heterothallic subgroups are often identified within the same fungal morphospecies, as in *Sistotrema brinkmannii*, *Phomopsis oblonga*, *Calonectria rigidiscula*, *Nectria haematococca*, *Heterobasidion annosum*, *Stereum hirsutum* and *Didymium iridis* (Table 16.1, and Chapters 18, 19 and 20). Systems of homo-heterothallic or mating type switching have been observed in a number of fungi including yeast (see Carlile, Chapter 14), *Ophiostoma ulmi* (Brasier & Gibbs, 1975), *Stereum hirsutum* (Coates & Rayner, 1985) and *Coprinus* (P. J. Pukkila, personal communication). Secondarily homothallic forms can arise from heterothallics through the simple device of including two nuclei in the post-meiotic spores, as in *Neurospora* (Dodge, 1927) and *Agrocybe* (Sass, 1929) or through polysomy, as in *Phytophthora* (Mortimer, Shaw & Sansome, 1977). In heterothallic Oomycetes such as *Achlya* (Raper, 1966; Turner, Suryanarayana & Wee, 1984) and *Phytophthora* (Brasier, 1983a) the sexual mating system itself facilitates a subtle interplay between outbreeding and selfing. Moreover, a diploid Oomycete, or its dikaryotic Basidiomycete equivalent, has the added flexibility over a haploid Ascomycete of either concealing potential genetical variability in a heterozygote or of fixing a more uniform homozygous genotype through repeated selfing (see Brasier & Sansome,

1975). The latter feature is particularly pertinent to the propagation of a clonal population (see above).

If, as implied in Fig. 16.3, the ecological essence of the relationship between heterothallism and homothallism in fungi is also that between basal population structure and routine selection on the one hand and population structures distorted or derived from basal population structure on the other, then the relationship between heterothallism and homothallism in the fungi should be seen as broadly similar to that postulated by Stebbins (1962) and others for cross fertilisation and self-fertilisation in plants. Heterothallism is favoured by routine selection associated with ecological balance and moderate environmental fluctuation which allow genetic flexibility. Homothallism, apomixis and loss of sexual reproduction are the consequences of episodic selection and sudden ecological disturbance, or of routine selection associated with recurrent disturbance and environmental extremes which demand immediate genetic fitness and, in some cases, the secure production of spores. An illustrative example may be the data of Thomas *et al.* (1984) for *Achlya* species, where, in permanent ponds, both homothallic and heterothallic forms were found, while in a nearby temporary pond only homothallic forms were recorded, albeit in a small sample.

Secondary speciation and man

Since man has been increasingly responsible for the creation of many artificially uniform habitats where once there were more heterogeneous ecosystems, especially with regard to land management, manipulation of crops and the development of food stores, he may, *via* the above or similar processes, have unwittingly promoted much rapid speciation in fungi, particularly in plant pathogens.

Man's influence on speciation is clearly not confined to that resulting from alteration of habitats, but also involves his promotion of secondary speciation through bringing together previously geographically isolated sub-populations or species units. Two examples have recently come within the author's experience. The first case is that of *Phytophthora meadii*, a serious pathogen of rubber in Sri Lanka and southern India. A cytological investigation has shown that this heterothallic fungus exists as diploid and tetraploid forms. The diploid forms are sterile and suffer a breakdown of meiosis, suggesting that *P. meadii* is an allopolyploid or hybrid species (E. R. Sansome & C. M. Brasier, unpublished). *P. meadii* may be a product of hybridisation between an indigenous *Phytophthora* and *P. palmivora*, a species probably introduced into the area by man and itself a pathogen of rubber. The opportunity for this speciation event may have been presented by the introduction and intensive monoculture of rubber in the area in the past 200 years.

The second case concerns once again *Ophiostoma ulmi*, or rather the non-aggressive, EAN aggressive and NAN aggressive sub-groups within it, on the elms of Europe. The three sub-groups show varying degrees of genetic divergence and reproductive isolation (see Brasier, 1983b, 1986a). In the period 1920 − 1940 the non-aggressive sub-group was probably the only form of fungus present in Europe, causing the first epidemic of Dutch elm disease. It was probably introduced, by man, in the early 1900s. More recently, during the second epidemic of the disease, the EAN and NAN aggressive sub-groups have been replacing the non-aggressive. Again, man was probably responsible for their introduction. The near total reproductive isolation between the non-aggressive sub-group and the EAN or NAN has apparently prevented any aggressive × non-aggressive hybrids from emerging (Brasier, 1983b, 1986a). However, in areas where the EAN and NAN overlap, EAN/NAN hybrids have recently begun to appear, a consequence of their being only partially genetically isolated (Brasier, 1986b). Thus, the stage seems set for the emergence of yet another form of fungus in Europe. The original EAN and NAN sub-groups may disappear altogether.

These then are the 'three ages of *Ophiostoma ulmi*' in Europe. Whatever the morphospecies may be called, the organism studied as *O. ulmi* 50 years ago is quite different from the two forms, the EAN and NAN, predominating today. That studied in a few decades time seems likely to be quite different again.

Concluding comment

Fungal species are probably best viewed as potentially highly dynamic entities. Their populations can ebb and flow, split or recombine, rather like an amoeba, and especially under man's influence. A recognition of this potential and a deeper understanding, yet to come, of the processes governing speciation events is a necessary basis for the interpretation of fungal ecological behaviour, for the exploitation and manipulation of fungal populations, and for the development of a natural fungal taxonomy.

Acknowledgements I would like to thank Mark Anderson, Alan Rayner and Roger Kemp for helpful comments on the manuscript.

References

Adams, D. H. & Roth, L. F. (1967). Demarcation lines in paired cultures of *Fomes cajenderi* as a basis for detecting genetically distinct mycelia. *Canadian Journal of Botany*, **45**, 1583 − 1589.

Anderson, J. B. & Ullrich, R. C. (1979). Biological species of *Armillaria mellea* in North America. *Mycologia*, **71**, 402 − 414.

Andrews, J. H. & Rouse, D. I. (1982). Plant pathogens and the theory of *r*- and *K*-selection. *American Naturalist*, **120**, 283 − 296.

Betterley, D. A. & Collins, O. R. (1983). Reproductive systems, morphology and genetical diversity in *Didymium iridis* (Myxomycetes). *Mycologia*, **75**, 1044 − 1063.

Biggs, R. (1937). The species concept in *Corticium coronilla*. *Mycologia*, **29**, 686 − 706.

Boidin, J. (1951). Sur l'existence de races interstériles chez *Gloeocystidium tenue* (Pat.): Étude morphologique et comportement nucléaire de leurs cultures. *Bulletin de la Societé Mycologique de France*, **66**, 204 − 219.

Brasier, C. M. (1970). Variation in a natural population of *Schizophyllum commune*. *American Naturalist*, **104**, 191 − 204.

Brasier, C. M. (1977). Inheritance of pathogenicity and cultural characteristics in *Ceratocystis ulmi*. Hybridisation of protoperithecial and non-aggressive strains. *Transactions of the British Mycological Society*, **68**, 45 − 52.

Brasier, C. M. (1978a). Mites and reproduction in *Ceratocystis ulmi*. *Transactions of the British Mycological Society*, **70**, 81 − 89.

Brasier, C. M. (1978b). Stimulation of oospore formation in *Phytophthora* by antagonistic species of *Trichoderma* and its ecological implications. *Annals of Applied Biology*, **89**, 135 − 139.

Brasier, C. M. (1983a). Problems and prospects in Phytophthora research. In *Phytophthora: Its Biology, Taxonomy and Pathology*, ed. Erwin, D. C., Bartnicki-Garcia, S. & Tsao, P. H., pp 351 − 364. St. Paul, Minnesota: American Phytopathological Society.

Brasier, C. M. (1983b). The future of Dutch elm disease in Europe. In *Research on Dutch Elm Disease in Europe (Forestry Commission Bulletin vol. 60)*, ed. Burdekin, D. A., pp. 96 − 104. London: HMSO.

Brasier, C. M. (1984). Inter-mycelial recognition systems in *Ceratocystis ulmi*: their physiological properties and ecological importance. In *The Ecology and Physiology of the Fungal Mycelium*, British Mycological Society Symposium Volume 8, ed. Jennings, D. H. & Rayner, A. D. M., pp. 451 − 497. Cambridge University Press.

Brasier, C. M. (1986a). The population biology of Dutch elm disease: its principal features and some implications for other host-pathogen systems. *Advances in Plant Pathology*, **5**, 55 − 118.

Brasier, C. M. (1986b). Emergence of EAN/NAN hybrids of *Ophiostoma ulmi* in Europe. In *Report on Forest Research, 1986*, p. 37. London: HMSO.

Brasier, C. M. & Gibbs, J. N. (1975). Highly fertile form of the aggressive strain of *Ceratocystis ulmi*. *Nature*, **257**, 128 − 131.

Brasier, C. M. & Griffin, M. J. (1979). The taxonomy of *Phytophthora palmivora* on cocoa. *Transactions of the British Mycological Society*, **72**, 111 − 143.

Brasier, C. & Sansome, E. R. (1975). Diploidy and gametangial meiosis in *Phytophthora cinnamomi, P. infestans* and *P. drechsleri*. *Transactions of the British Mycological Society*, **65**, 49 − 65.

Brayford, D. (1983). *Phomopsis* as a biological control agent of Dutch elm disease. *Ph. D. thesis, University of Wales, Aberystwyth*,

Burdon, J. J. & Marshall, D. R. (1981). Isozyme variation between species and *formae speciales* of the genus *Puccinia*. *Canadian Journal of Botany*, **59**, 2628 − 2634.

Burnett, J. H. (1983). Speciation in fungi. *Transactions of the British Mycological Society*, **81**, 1 – 14.

Burnett, J. H. & Boulter, M. E. (1963). The mating systems of fungi, II. Mating systems of the Gasteromycetes *Mycocalia denudata* and *M. duriaeana*. *New Phytologist*, **62**, 217 – 236.

Burnett, J. H. & Partington, M. (1957). Spatial distribution of fungal mating type factors. *Proceedings of the Royal Physical Society of Edinburgh*, **26**, 61 – 68.

Burpee, L. L., Sanders, P. L., Cole, H. & Sherwood, R. T. (1980). Anastomosis groups among isolates of *Ceratobasidium cornigerum* and related fungi. *Mycologia*, **72**, 689 – 701.

Butcher, A. C. (1969). Non-allelic interactions and genetic isolation in wild populations of *Aspergillus nidulans*. *Heredity*, **24**, 621 – 631.

Chase, T. & Ullrich, R. C. (1984). Homothallic isolates of *Heterobasidion annosum*. *Mycologia*, **77**, 975 – 977.

Christen, A. A. & Bruehl, G. W. (1979). Hybridization of *Typhula ishikariensis* and *T. idahoensis*. *Phytopathology*, **69**, 263 – 266.

Coates, D. & Rayner, A. D. M. (1985). Evidence for a cytoplasmically transmissible factor affecting recognition and somato-sexual differentiation in the basidiomycete *Stereum hirsutum*. *Journal of General Microbiology*, **131**, 207 – 219.

Cooke, R. C. & Rayner, A. D. M. (1984). *Ecology of Saprotrophic Fungi*. London & New York: Longman.

Croft, J. H. & Jinks, J. L. (1977). Aspects of the population genetics of *Aspergillus nidulans*. In *Genetics and Physiology of Aspergillus*, British Mycological Society Symposium volume 1, ed. Smith, J. E. & Pateman, J. A., pp. 339 – 360. London: Academic Press.

Dawkins, R. (1976). *The Selfish Gene*. Oxford University Press.

Dobzhansky, T. (1937). *Genetics and the Origin of Species*. New York: Columbia University Press.

Dodge, B. O. (1927). Nuclear phenomena associated with heterothallism and homothallism in the Ascomycete *Neurospora*. *Journal of Agricultural Research*, **35**, 289 – 305.

Duncan, E. G. & MacDonald, J. A. (1967). Micro-evolution in *Auricularia auricula*. *Mycologia*, **59**, 803 – 818.

Eriksson, J. & Henning, E. (1896). *Die Getreideroste*. Stockholm: Norstedt.

Federici, B. A. (1982). Inviability of interspecific hybrids in the *Coelomyces dodgei* complex. *Mycologia*, **74**, 555 – 562.

Fischer, G. W. & Holton, C. S. (1957). *Biology and Control of Smut Fungi*. New York: The Ronald Press.

Ford, E. J., Burnett, J. A. & Snyder, W. C. (1967). Biologic speciation in *Calonectria* (Fusarium) *rigidiscula* in relation to green point gall of cocoa. *Phytopathology*, **57**, 710 – 712.

Fries, N. (1985). Intersterility groups in *Paxillus involutus*. *Mycotaxon*, **24**, 403 – 409.

Fries, N. & Mueller, G. (1984). Incompatibility systems, cultural features and species descriptions in the ectomycorrhizal genus *Laccaria* (Agaricales). *Mycologia*, **76**, 633 – 642.

Gause, G. F. (1932). *The Struggle for Existence*. Baltimore: Williams & Wilkins.

Gibbs, J. N. & Greig, B. W. (1986). Resin top caused by *Peridermium pini*. In *Report on Forest Research, 1986*, p. 37. London: HMSO.

Hansen, E. M., Brasier, C. M., Shaw, D. S. & Hamm, P. B. (1987). The taxonomic status of *Phytophthora megasperma*: evidence for emerging biological species groups. *Transactions of the British Mycological Society*, **87**, 557 – 573.

Hill, J. P. & Nelson, R. R. (1976). Ecological races of *Helminthosporium maydis* race T. *Phytopathology*, **66**, 873 – 876.

Hiura, U. (1962). Hybridization between varieties of *Erysiphe graminis*. *Phytopathology*, **52**, 664 − 666.

Hiura, U. (1978). Genetic basis of some formae speciales in *Erysiphe graminis*. In *The Powdery Mildews*, ed. Spencer, D. M., pp. 101 − 128. London: Academic Press.

Hsieh, W. H., Smith, S. N. & Snyder, W. C. (1979). Mating groups in *Fusarium moniliformis*. *Phytopathology*, **67**, 1041 − 1043.

Johnson, T. (1949). Intervarietal crosses in *Puccinia graminis*. *Canadian Journal of Research*, **27**, 45 − 65.

Kemp, R. F. O. (1975). Breeding biology of *Coprinus* species in the section *Lanatuli*. *Transactions of the British Mycological Society*, **65**, 375 − 388 (and see corrigendum, ibid., **66**, 567).

Kemp, R. F. O. (1977). Oidial homing and the taxonomy and speciation of Basidiomycetes with special reference to the genus *Coprinus*. In *The Species Concept in Hymenomycetes*, ed. Clemencon, H., pp. 259 − 273. Vaduz: Cramer.

Kemp, R. F. O. (1983). Incompatibility and speciation in the genus *Coprinus*. *Revista de Biologia*, **12**, 179 − 186.

Kemp, R. F. O. (1985). Do fungal species really exist? A study of basidiomycete species with special reference to those in *Coprinus* section *Lanatuli*. *Bulletin of the British Mycological Society*, **19**, 34 − 39.

Kimura, K. (1952). Studies on the sex of *Coprinus macrorhizus* Rea. *f. microsporus* Hongo. II. On the sexual strains. *Biological Journal of Okayama University*, **1**, 80 − 83.

Ko, W. H., Chang, H. S. & Su, H. J. (1978). Isolates of *Phytophthora cinnamomi* from Taiwan as evidence for an Asian origin of the species. *Transactions of the British Mycological Society*, **71**, 496 − 499.

Korhonen, K. (1978a). Intersterility and clonal size in the *Armillariella mellea* complex. *Karstenia*, **18**, 31 − 42.

Korhonen, K. (1978b). Intersterility groups of *Heterobasidion annosum*. *Communications Instituti Forestalis Fenniae*, **74**, 1 − 25.

Kuhlman, E. G. (1982). Varieties of *Gibberella fujikuroi* with anamorphs in *Fusarium* section *Liseola*. *Mycologia*, **74**, 759 − 768.

Lange, M. (1952). Species concept in the genus *Coprinus*. *Dansk Botanisk Arkiv*, **14**, 7 − 162.

Lemke, P. A. (1969). A reevaluation of homothallism, heterothallism and the species concept in *Sistotrema brinkmannii*. *Mycologia*, **61**, 57 − 76.

Leonard, K. J. (1973). Association of mating type and virulence in *Helminthosporium maydis* and observations on the origin of the race T population in the United States. *Phytopathology*, **63**, 112 − 115.

Macrae, R. (1967). Pairing incompatibility and other distinctions among *Hirschioporus* (Polyporus) *abietinus*, *H. fusco-violaceous* and *H. laricinus*. *Canadian Journal of Botany*, **45**, 1371 − 1398.

Marchal, E. (1902). De la spécialisation du parasitism chez l'Erysiphe graminis. *Comptes rendus hebdomadaires de l'Academie des Sciences, Paris*, **135**, 210 − 212.

Mather, K. (1953). The genetical structure of populations. *Symposia of the Society for Experimental Biology*, **7**, 66 − 95.

Mather, K. (1955). Polymorphism as an outcome of disruptive selection. *Evolution*, **9**, 52 − 61.

Matsumoto, N., Sato, T. & Arak, T. (1982). Biotype differentiation in the *Typhula ishikariensis* complex and their allopatry in Hokkaido. *Annals of the Phytopathological Society of Japan*, **48**, 275 − 280.

Matsuyama, N. & Kozaka, T. (1971). Comparative gel electrophoresis of soluble proteins and enzymes of rice blast fungus *Pyricularia oryzae* Cav. *Annals of the Phytopathological Society of Japan*, **37**, 259 − 265.

Matuo, T. & Snyder, A. (1973). Use of morphology and mating populations in the identification of formae speciales in *Fusarium solani*. *Phytopathology*, **6**, 562 − 565.

Mayr, E. (1942). *Systematics and the Origin of Species*. New York: Columbia University Press.

McKeen, C. V. (1952). A cultural and taxonomic study of three species of *Peniophora*. *Canadian Journal of Botany*, **30**, 764 − 787.

Minter, D. W. & Millar, C. S. (1980). Ecology and biology of three *Lophodermium* species on secondary needles of *Pinus sylvestris*. *European Journal of Forest Pathology*, **10**, 169 − 181.

Morrison, D. J., Thompson, A. J., Chu, D., Peet, F. G. & Sahota, T. S. (1985). Isozyme patterns of *Armillaria* intersterility groups occurring in British Columbia. *Canadian Journal of Microbiology*, **31**, 651 − 653.

Mortimer, A., Shaw, D. S. & Sansome, E. R. (1977). Genetical studies of secondary homothallism in *Phytophthora drechsleri*. *Archiv für Mikrobiologie*, **111**, 255 − 259.

Mounce, I. & Macrae, R. (1938). Interfertility phenomena in *Fomes pernicola*. *Canadian Journal of Research*, **16**, 354 − 376.

Mylyck, O. M. (1976). Heteromorphism for heterokaryon incompatibility genes in natural populations of *Neurospora crassa*. *Genetics*, **63**, 275 − 284.

Nelson, R. R., Ayers, J. E., Cole, H., Massie, L. B. & Forer, L. (1971). Distribution, race frequency, virulence and mating type of isolates of *Helminthosporium maydis* in the northeastern United States in 1970. *Plant Disease Reporter*, **55**, 495 − 498.

Newton, A. C., Caten, C. E. & Johnson, R. (1985). Variation for isozymes and double stranded RNA among isolates of *Puccinia striiformis* and two other cereal rusts. *Plant Pathology*, **34**, 235 − 247.

Old, K. M., Moran, G. F. & Bell, J. C. (1984). Isozyme variability among isolates of *Phytophthora cinnamomi* from Australia and Papua New Guinea. *Canadian Journal of Botany*, **62**, 2016 − 2022.

Parmeter, J. R., Sherwood, R. T. & Platt, W. D. (1969). Anastomosis groupings among isolates of *Thanatephorus cucumeris*. *Phytopathology*, **59**, 1270 − 1278.

Perkins, D. D., Turner, B. C. & Barry, E. G. C. (1976). Strains of *Neurospora* collected from nature. *Evolution*, **30**, 281 − 313.

Puhalla, J. E. (1979). Classification of isolates of *Verticillium dahliae* based on incompatibility. *Phytopathology*, **69**, 1186 − 1189.

Puhalla, J. E. (1985). Classification of strains of *Fusarium oxysporum* on the basis of vegetative incompatibility. *Canadian Journal of Botany*, **63**, 179 − 183.

Puhalla, J. E. & Hummel, M. (1983). Vegetative incompatibility groups within *Verticillium dahliae*. *Phytopathology*, **73**, 1305 − 1308.

Raper, J. R. (1940). Sexuality in *Achlya ambisexualis*. *Mycologia*, **32**, 710 − 727.

Raper, J. R., Krongelb, G. S. & Baxter, M. G. (1958). The number and distribution of incompatibility factors in *Schizophyllum commune*. *American Naturalist*, **92**, 221 − 234.

Rayner, A. D. M., Coates, D., Ainsworth, A. M., Adams, T. J. H., Williams, E. N. D. & Todd, N. K. (1984). The biological consequences of the individualistic mycelium. In *The Ecology and Physiology of the Fungal Mycelium*, British Mycological Society Symposium Volume 8, ed. Jennings, D. H. & Rayner, A. D. M., pp. 509 − 540. Cambridge University Press.

Sansome, E. R., Brasier, C. M. & Sansome, F. W. (1979). Further cytological studies on the 'L' and 'S' types of *Phytophthora* from cocoa. *Transactions of the British Mycological Society*, **73**, 293 – 302.

Sass, J. E. (1929). The cytological basis for homothallism and heterothallism in the Agaricaceae. *American Journal of Botany*, **16**, 663 – 701.

Schipper, M. A. A. (1975). On *Mucor mucedo*, *Mucor flavus* and related species. *Studies on Mycology*, **10**, 1 – 33.

Simchen, G. (1967). Independent evolution of a polygenic system in isolated populations of the fungus *Schizophyllum commune*. *Evolution*, **21**, 310 – 315.

Smith, M. J. (1966). Sympatric Speciation. *American Naturalist*, **100**, 637 – 650.

Spieth, P. T. (1975). Population genetics of an allozyme variation in *Neurospora intermedia*. *Genetics*, **80**, 785 – 805.

Stebbins, G. L. (1950). *Variation and Evolution in Plants*. New York: Columbia University Press.

Thomas, D. D. S., Suryanarayana, K. & Manavathu, E. K. (1983). Asexual Reproduction coupled with heterothallism; possible consequences for fungi. *Journal of Theoretical Biology*, **105**, 373 – 378.

Thomas, D. D. S., Suryanarayana, K. & Wee, M. L. (1984). Interthallic sexual reactions among *Achlya* isolations from natural populations. *Mycologia*, **76**, 601 – 607.

Todd, N. K. & Rayner, A. D. M. (1980). Fungal individualism. *Science Progress*, **66**, 331 – 334.

Ullrich, R. C. (1973). Sexuality, incompatibility and intersterility in the biology of *Sistotrema brinkmannii* aggregate. *Mycologia*, **65**, 1234 – 1249.

Verrall, A. F. (1937). Variation in *Fomes ignarius*. *Minnesota Technical Bulletin*, **117**, 1 – 41.

Vilgalys, R. & Miller, O. K. (1983). Biological species in the *Collybia dryophila* group in North America. *Mycologia*, **75**, 707 – 722.

Wahl, I., Eshed, N., Segal, A. & Sobel, Z. (1978). Significance of wild relatives of small grains and other wild grasses in cereal powdery mildews. In *The Powdery Mildews*, ed. Spencer, D. M., pp. 83 – 100. London: Academic Press.

Wang, C. G., Blanchette, R. A., Jackson, W. A. & Palmer, A. (1986). Differences in conidial morphology among isolates of *Sphaeoropsis sapinea*. *Plant Disease Reporter*, **69**, 838 – 841.

Webber, J. F., Brasier, C. M. & Mitchell, A. G. (1987). The role of the saprophytic phase in Dutch elm disease. In *Fungal Infestation of Plants: Establishment, Progress and Outcome of Infection*, British Mycological Society Symposium volume 13, ed. Pegg, G. F. & Ayres, P. G., in press. Cambridge University Press.

Webber, J. F. & Gibbs, J. N. (1982). Colonisation of elm bark by *Phomopsis oblonga*. *Transactions of the British Mycological Society*, **82**, 348 – 352.

Willetts, H. J. & Wong, J. A. (1980). The biology of *Sclerotinia sclerotiorum*, *S. trifoliorum* and *S. minor* with special emphases on specific nomenclature. *The Botanical Review*, **46**, 101 – 165.

Williams, E. N. D. & Todd, N. K. (1985). Numbers and distribution of individuals and mating type alleles in populations of *Coriolus versicolor*. *Genetical Research*, **46**, 251 – 262.

Williams, S., Verma, M. M., Jinks, J. L. & Brasier, C. M. (1976). Variation in a natural population of *Schizophyllum commune*. II. Variation within extreme isolates for growth rate. *Heredity*, **37**, 365 – 375.

17

Mating ability and the species concept in the Zygomycetes

M. A. A. SCHIPPER

Centraalbureau voor Schimmelcultures, PO Box 273, 3740 AG Baarn, The Netherlands

Introduction

Interest in the life cycles of Zygomycetes dates from the middle of the last century. After Tulasne (1855) demonstrated the occurrence of two phases in the life cycle of *Syzygites megalocarpus*, a search for the perfect stage in other Zygomycetes was started. Bainier (1883, 1884) was one of the most successful, finding a number of zygosporic colonies in his collections. However, the results of investigations into the conditions stimulating zygospore production in the laboratory were mostly disappointing, until Blakeslee (1904) showed that the majority of Zygomycetes were heterothallic and unable to produce zygospores unless two mating partners (+ and −) were present. Subsequently, Burgeff (1924), Plempel (1957) and others initiated the investigation of sex hormones in Zygomycetes, which has become a prominent field of research (for reviews see Jones, Williamson & Gooday, 1981; van den Ende, 1984). It was found that (+) and (−) mycelia produce mating type specific substances which, in mixed culture, act complementarily to produce pheromones, inducing zygophore formation and acting as chemotropic agents between mating partners. The pheromones have a promoting action on their own biosynthesis. The complete metabolic cycle is still unknown. Similar compounds have been found in a number of different species.

Zygospores and azygospores

The *initial stages* of the mating process depend on the mating type only and can occur in partners of quite different taxa. The *completion* of the process, starting with lysis of the fusion-wall, is regarded as species specific. Environmental and physiological conditions are of a secondary nature, although the conditions for mating are much narrower than those for growth, both with regard to medium and temperature (Schipper, 1973,

1975, 1976, 1978). Closely related species generally favour similar conditions for zygospore production. When, in addition, their zygospores are virtually identical in size, colour and ornamentation, interbreeding is often possible. This has been accepted as an indication that the partners are of the same species (Schipper, 1970, 1973, 1975, 1976). However, the yield of zygospores in various matings with one common partner, and under the same conditions, may differ greatly due to unknown physiological factors. For example, the interplay between weak partners may fail to result in zygote production. Thus, the absence of zygospores is not necessarily evidence of genetic distinction.

Another confusing element, in this respect, is the production of azygospores both in those legitimate matings which also yield normal zygospores and in interspecific pairings in which normal zygospores do not occur. Zygospores result from the interaction between different hyphae or parts of hyphae, each secreting its own sex-specific substance. In heterothallic species the ($+$) and ($-$) sites are located on hyphae from different thalli; in homothallic species the copulating hyphae are connected either in the substrate or on the aerial mycelium, but the ($+$) and ($-$) sites are distinctly separate. Azygospores are produced from one single site only, which, therefore, might be ($+/-$) in genetic constitution. Following a very careful study of mating processes of Mucorineae, Ling Young (1930) suggested that azygospore formation in intraspecific matings may result from weak sexual potency of the partners leading to anomalies of the fusion processes and the prevention of actual zygospore formation. Blakeslee and others accepted the occurrence of such azygospores in interspecific contrasts as anomalies caused by ' . . . strong sexual ability of the partners . . . '.

It seems that azygospores, whether formed in intraspecific or interspecific crosses are probably produced through similar processes. They are certainly not rare. In mating experiments with strains of 19 different species, Schipper (1978) found zygospore-like bodies in 150 interspecific pairings. These were often pale, small, sometimes slightly misshapen, and were produced rather late by comparison with most intraspecific matings. They were never really abundant and were generally produced within a wall of interwoven hyphae and incomplete conjugations. After careful observation of their development it was found that the fusion-wall did not disintegrate, and that only one suspensor was fully developed. Under the scanning electron microscope the azygospores in interspecific matings resembled normal zygospores, arrested at an early stage of development (Stalpers & Schipper, 1980).

Although most heterothallic strains react either as (+) or (−), both sexual potentialities are probably present (see Schipper & Stalpers, 1980). A strong impulse towards sexual reproduction which becomes blocked might result in the expression of the latent potency of one partner. The production of abundant sexual substances through the stimulatory activity of the partner, but not followed by conjugation, might cause aberrations of a natural developmental pathway, e.g. activation of the dormant counter-part.

Quite another problem is presented when strains differing in general morphology produce apparently normal zygospores between two seemingly active suspensors. This poses a basic taxonomic problem: is the supposition that strains which mate to form zygospores are of the same species incorrect, or if correct, do we have to accept extreme variability within a species? The following examples illustrate the point. Apparently normal zygospores were obtained in crosses between *Mucor indicus* and *M. inaequisporus* and *M. variosporus* (Schipper, 1978). While *M. inaequisporus* and *M. variosporus* are rather similar in morphology, *M. indicus* is quite distinct. *M. indicus* has sporangiophores that are repeatedly branched; sporangia 40 − 50 (− 80) μm dia; columellae applanate − subglobose; sporangiospores subglobose to cylindrical ellipsoidal, $5 \cdot 4 − 5 \cdot 7 \times 4 \cdot 4$ μm; development optimal at 30°. In contrast, *M. inaequisporus* has unbranched sporangiophores; sporangia 150(− 175) μm dia; columellae (largest) obovoid; sporangiospores extremely variable, ellipsoidal $5 − 7 \times 3 − 4$ μm, ellipsoidal to globose up to 30×23 μm, globose $12 − 18$ μm; no development at 30°. *M. inaequisporua* may be a giant form of *M. indicus*.

Zygospores of normal appearance were also obtained in *Rhizopus microsporus* $\times$ *R. rhizopodiformis* (Schipper & Stalpers, 1980). *R. microsporus* has columellae (largest) subglobose − conical; sporangiospores of max dia $7 \cdot 5$ to $8 − 9$ μm, angular and ornamented with ridges (under the SEM); and shows no growth at 50°. *R. rhizopodiformis* has columellae (largest) pyriform; sporangiospores of max dia 5 μm, globose, and ornamented with spines (SEM); and shows good growth at 50°. In such apparently positive matings, information on the occurrence or otherwise of a normal meiosis in the zygospore and studies on the phenotypes of the progeny might provide an explanation.

An example is the study of matings between *Rhizopus microsporus* and *R. rhizopodiformis* (Schipper, Gauger & van den Ende, 1985). Zygospores may germinate, after a period of dormancy, either via a germ-sporangium or via a germ-tube to produce a mycelium. The percentage germination is extremely variable and seems to depend on the degree of affinity of the parents. Single germ spores or hyphal tips are the basis for assessing colonies of the first generation (F_1) of a mating. Since all available strains of *R. rhizopodiformis* are of the (+) mating type, intraspecific matings in this species could not be studied. Zygospores of intraspecific *R. microsporus* matings germinated within four weeks and the resulting germ-sporangia produced mostly either (+) or (−) mycelia; a few colonies were azygosporic. In contrast, zygospores resulting from the mating of *Rhizopus microsporus* × *R. rhizopodiformis* required at least two months to germinate, and germination was mostly via germ-tubes. The great majority of F_1 colonies were of a morphology intermediate between the parent strains and produced abundant azygospores. These azygospores germinated without a lag period, i.e. they did not behave as normal zygospores, and the subsequent generations were of the same morphological appearance as the F_1. Recombination of characters was never observed. This indicates that in the zygospores resulting from the original mating there was neither true fusion of parental nuclei nor a subsequent meiosis. The reason for this failure to fuse completely might be a lack of homology between the chromosomes of the two parents, leading to abortive meioses, the products of which are either diploid or aneuploid. The phenotypic variability of the F_1 was limited (Schipper *et al.*, 1985).

The production of individuals with truly recombinant or new sets of characters is the result of mating between heterothallic partners. Indeed, heterothallism is the more common condition in the Zygomycetes. However, several species are homothallic. Compared to heterothallism, homothallism is rare, and often occurs in combination with the ability to produce azygospores. It is suggested that homothallism is usually derived from heterothallism (cf. Brasier, Chapter 16). For instance, the homothallic *Mucor genevensis*, which is occasionally encountered, might originate from the very common heterothallic *M. hiemalis*. Several other homothallic or azygosporic strains of usually heterothallic species have also been described (Schipper & Stalpers, 1980). The author has encountered a few homothallic strains of *Rhizomucor pusillus* and azygosporic specimens of *M. circinelloides*, *M. racemosus* and *M. indicus*, all four of which are common heterothallic species. Other evidence suggesting that homothallism is derived from heterothallism comes from observation of single isolates of normally heterothallic species expressing both (+) and (−) potentialities to some extent. Such mating ambivalence was reported by Nottebrock, Scholer & Wall (1974) in strains of *Absidia corymbifera*, while van den Ende & Stegwee (1971) and Werkmann-Hoogland (1977) reported the presence of both $(+)$ and $(-)$ substances in acceptedly heterothallic strains of *Blakeslea trispora* and *Mucor mucedo* respectively. The CBS strains of *Rhizopus pseudochinensis* var. *thermosus* and of *R. pygmaeus* were found to produce azygospores in ageing cultures under certain conditions. This also indicates the presence of both $(+)$ and $(-)$ factors in a single thallus of an otherwise normally heterothallic strain. It seems that under certain conditions either a dominance or a repression mechanism fails to operate completely, resulting in the expression of both the $(+)$ and $(-)$ potentialities, though with unequal vigour.

Breeding units and speciation

The impact of breeding units on speciation depends largely on the nature of the mating processes involved. Three possible levels of mating interaction in the Zygomycetes are illustrated by processes (a) to (c) in Table 17.1.

The process in (a) may include a normal meiotic exchange of characters, although whether zygospores are the products of a true mating, in the sense that karyogamy and meiosis occur, is, surprisingly, still largely a matter of speculation. Many workers, such as Brefeld (1908), Callen (1940), Cutter (1942), Sassen (1965) and Laane (1974), have suggested that there is no evidence for a truly sexual process. Gauger (1984) and others, however, accept the occurrence of meiosis based on their genetic analysis of progenies from crosses.

The process shown in (b) (Table 17.1) has been described above for crosses between *Rhizopus microsporus* and *R. rhizopodiformis*. Shortly after those experiments were concluded, an azygosporic *Rhizopus*, isolated from tempeh, was described as *R. azygosporus* (Yuan & Jong, 1984). Its description fits remarkably closely the morphology of the azygosporic strains obtained in the above mating. The mating and subsequent germination of the zygospores seemingly also occurred spontaneously. Another species morphologically intermediate between *R. microsporus* and *R. rhizopodiformis*, without azygospores, is *R. chinensis*. Of the seven available isolates of this species, three reacted as $(-)$ and four as $(+)$; one of the $(+)$ strains also producing azygospores in old slant cultures. The mating ability of all seven isolates was rather poor. Possibly, this was due to weakening of sexual activity and reversion from the $(+/-)$ condition to a situation of dominance or repression of one of these two potentialities, only the stronger being manifested.

The process in panel (c) of Table 17.1 usually stops at an earlier stage. The production of azygospores in interspecific pairings is a possibility, but

Table 17.1. *Three possible levels of mating process in the Zygomycetes*

	Mating processes		
	(a) Intraspecific	(b) Between related, non-identical populations	(c) Interspecific
Mating types	$(+) \times (-)$	$(+) \times (-)$	$(+) \times (-)$
Presumed nuclear behaviour	fusion	fusion (?)	no fusion
Product	zygospores	zygospores	azygospores induced on either $(+)$ or $(-)$ thallus
Phenotype of F1 or its equivalent	heterothallic $(+)$ or $(-)$; occasionally homothallic by accident	azygosporic; intermediate between parents	sterile; germination not observed

is certainly not the rule. Such induced azygospore production may be indicative of a breakdown of the block suppressing one of the (+) or (−) potentialities. To my knowledge, germination of such azygospores has never been achieved.

Conclusions

In the Mucorales, outcrossing or heterothallic strains prevail. Matings within a species yield viable zygospores, but the percentage germination of such zygospores is usually, at best, only at a low level, a feature which interferes with a proper investigation of the progeny. Secondary products of the interaction of mating partners are azygospores; borne on one suspensor from one thallus only. Azygospores are also the common product of seemingly unbalanced or incomplete reproductive processes (Table 17.1 b & c). Azygosporic strains of species that are normally heterothallic have a tendency to revert to vegetative colonies. Only by careful culturing and certain methods of long-term maintenance can the azygosporic condition be maintained.

Since the pheromones active in the initiation of conjugation processes are mating type specific, partners of opposite mating type but of quite different species may interact, though incompletely. However, there are also some additional barriers to mating in cases of interspecific partners. Firstly, mating conditions differ greatly. Some species favour temperatures of 30° and beyond, others only respond at temperatures of 15° or below. Species may be most reproductively prolific either on acid, neutral or alkaline media. The optimal media for different species also vary considerably in their nutrient composition. Secondly, the sites of sexual attraction differ between species. For example, *Mucor mucedo* produces its progametangia in the under portion of the colony, just above the medium. In contrast, *M. flavus* produces its organs in the upper, aerial, portion of the colony, which may be up to 60 mm above the agar surface. Nevertheless, the author has observed, in a pairing of *Mucor mucedo* and *M. flavus* (both growing well on neutral beerwort agar at 15°), tall zygophores of *M. flavus* bending to meet short zygophores of *M. mucedo*!

The Mucorales are very variable in general morphology within a species. Yet, this variability is difficult to explain by spontaneous matings and the subsequent germination of the ensuing zygospores. Zygospore-bearing isolates have been very uncommon among observations on hundreds of isolates from all over the world. On the other hand, as matings are most prolific in darkness, it is possible that some instructive material has been missed.

Acknowledgement I am grateful to Prof. Dr H. van den Ende, Department of Plant Physiology of the University of Amsterdam, for comments on my manuscript.

References

Bainier, G. (1883). Sur les zygospores des Mucorinées. *Annales des Sciences Naturelles*, **15**, 347 − 356.

Bainier, G. (1884). Nouvelles observations sur les zygospores des Mucorinées. *Annales des Sciences Naturelles*, **19**, 200 − 214.

Blakeslee, A. F. (1904). Sexual reproduction in the Mucorineae. *Proceedings of the American Academy of Arts and Sciences*, **40**, 205 − 319.

Brefeld, O. (1908). *Untersuchungen aus dem Gesamtgebiete der Mykologie 14, Die Kultur der Pilze*. Münster, Westphalia: Heinrich Schöningh.

Burgeff, H. (1924). Untersuchungen über Sexualität und Parasitismus bei Mucorineen. I. *Botanische Abhandlungen*, **4**, 1 − 135.

Callen, E. O. (1940). The morphology, cytology, and sexuality of the homothallic *Rhizopus sexualis* (Smith) Callen. *Annals of Botany*, **4**, 791 − 818.

Cutter, V. M. (1942). Nuclear behavior in the Mucorales. I. The *Mucor* pattern. *Bulletin of the Torrey Botanical Club*, **69**, 480 − 508.

Gauger, W. (1984). Genetic analysis utilizing single germ sporangia in two species of *Rhizopus*. *Journal of General and Applied Microbiology*, **30**, 337 − 345.

Jones, B. E., Williamson, J. P. & Gooday, G. W. (1981). Sex pheromones in *Mucor*. In *Sexual Interactions in Eukaryotic Microbes*, ed. O'Day, D. H. & Horgen, P. A., pp. 179 − 198. London & New York: Academic Press.

Laane, M. M. (1974). Nuclear behaviour during vegetative stage and zygospore formation in *Absidia glauca*. *Norwegian Journal of Botany*, **21**, 125 − 135.

Ling Yong (1930). Etude des phénomènes de la sexualité chez les Mucorinées. *Revue Générale de Botanique*, **42**, 567 − 768.

Nottebrock, H., Scholer, H. J. & Wall, M. (1974). Taxonomy and identification of Mucormycosis-causing fungi. I. Synonymity of *Absidia ramosa* with *A. corymbifera*. *Sabouraudia*, **12**, 64 − 74.

Plempel, M. (1957). Die Sexualstoffe der Mucoraceae. Ihre Abtrennung und die Erklärung ihrer Funktion. *Archiv für Mikrobiologie*, **26**, 151 − 174.

Sassen, M. M. A. (1965). Breakdown of the plant cell wall during the cell-fusion process. *Acta botanica neerlandica*, **14**, 165 − 196.

Schipper, M. A. A. (1970). Two species of *Mucor* with oval- and spherical spored strains. *Antonie van Leeuwenhoek*, **36**, 475 − 480.

Schipper, M. A. A. (1973). A study on variability in *Mucor hiemalis* and related species. *Studies in Mycology*, **4**, 1 − 40.

Schipper, M. A. A. (1975). On *Mucor mucedo*, *Mucor flavus* and related species. *Studies in Mycology*, **10**, 1 − 33.

Schipper, M. A. A. (1976). On *Mucor circinelloides*, *Mucor racemosus* and related species. *Studies in Mycology*, **12**, 1 − 40.

Schipper, M. A. A. (1978). On certain species of *Mucor* with a key to all accepted species. *Studies in Mycology*, **17**, 1 − 52.

Schipper, M. A. A., Gauger, W. & van den Ende, H. (1985). Hybridization of *Rhizopus* species. *Journal of General Microbiology*, **131**, 2359 − 2365.

Schipper, M. A. A. & Stalpers, J. A. (1980). Various aspects of the mating system in Mucorales. *Persoonia*, **11**, 53 − 63.

Stalpers, J. A. & Schipper, M. A. A. (1980). Comparison of zygospore ornamentation in intra- and interspecific matings in some related species of *Mucor* and *Backusella*. *Persoonia*, **11**, 39 − 52.

Tulasne, L. R. (1855). Note sur l'appareil reproducteur de quelques Mucedinées fongicoles. *Compte rendu de l'Academie des Sciences*, **41**, 615 − 618.

van den Ende, H. (1984). Sexual interaction in the lower filamentous fungi. In *Encyclopedia of Plant Physiology*, vol. 17, ed. Linskens, H. F. & Heslop-Harrison, J., pp. 333 − 349. Berlin & Heidelberg: Springer-Verlag.

van den Ende, H. & Stegwee, D. (1971). Physiology of sex in Mucorales. *Botanical Review*, **37**, 22 − 36.

Werkman-Hoogland, B. A. (1977). Sexual reproduction in Mucorales. *Thesis, University of Amsterdam*.

Yuan, G. F. & Jong, S. C. (1984). A new obligate azygosporic species of *Rhizopus*. *Mycotaxon*, **20**, 397 − 400.

18
Reproductive biology and speciation in Myxomycetes

O'NEIL RAY COLLINS

Department of Botany, University of California at Berkeley, California 94720, USA

Introduction

Myxomycetes (true slime moulds) are primitive holotrophs whose complex life cycle proceeds from spores to amoeboflagellates to plasmodia to fruiting bodies and back to spores again. They are, conceivably, direct descendants of the earliest sexual amoeboflagellates. Among extant groups, they probably occupy a phylogenetic position above protists but below animals and fungi. They are neither unicellular nor truly multicellular; their holotrophic mode of nutrition and motility are animal-like and their sessile fruiting bodies and meiospores are fungus-like. This combination of seemingly contradictory features gives Myxomycetes a reputation as biological curiosities which enhances their attractiveness as objects for evolutionary studies. So, I am delighted to discuss their breeding and speciation mechanisms in this Symposium, even if they are not first-class fungal citizens.

The amoeboflagellate stage

Myxomycetes are apparently monophyletic with structurally and functionally homologous life cycle stages. It is simplest to assume the amoeboflagellate stage derived from a common amoebal ancestor with a capacity to acquire reversible flagellate and microcyst states, depending on environmental conditions. Over time, integration of sex into the cycle led to evolution of a plasmodial and subsequently a sporulative stage (Collins, 1979).

In any case today's amoeba can still produce reversible flagellate and microcyst states and it is remarkably well adapted to its role in the reproductive biology of these apparently ancient organisms. Well-fed amoebae can undergo rapid vegetative reproduction through mitotic cell divisions

(a)

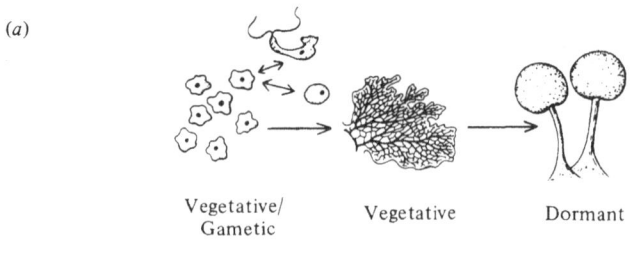

Vegetative/   Vegetative   Dormant
Gametic

(b)

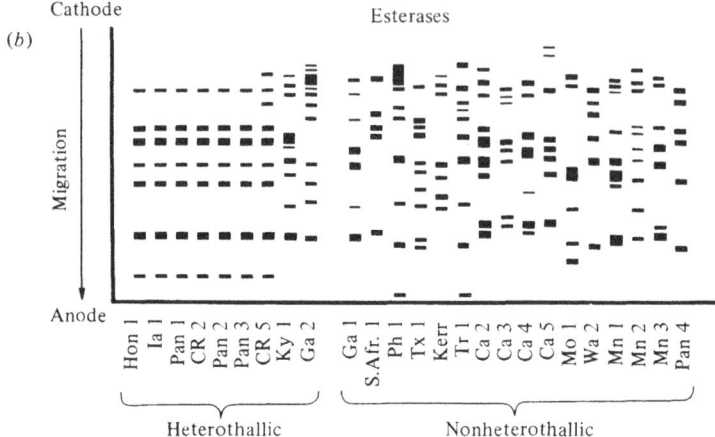

(c)

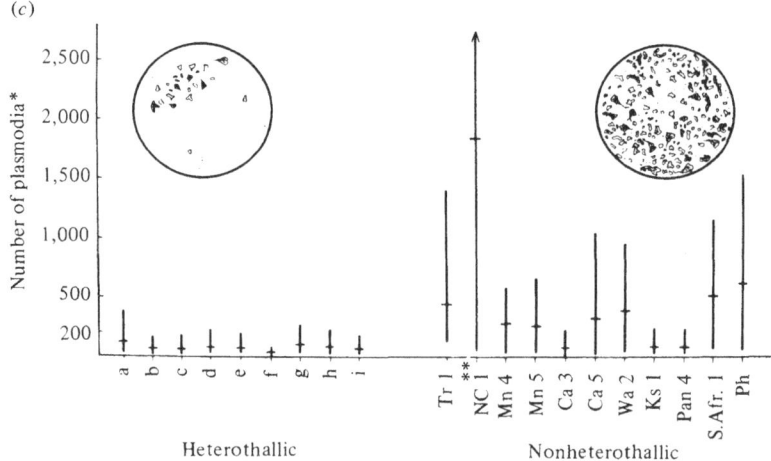

Fig. 18.1. (a), stages in the myxomycete life cycle. (b), isozyme profiles
for α-esterases from heterothallic and nonheterothallic isolates of *Didymium
iridis*. (c), relative fitness for plasmodial development of heterothallics and
nonheterothallics in cultures containing approximately equivalent
amoeboflagellate cell densities.

and can survive harsh environmental stress in the microcyst state. During a growing season many mitotic cycles are completed, providing many opportunities for natural selection. Each successful variant can give rise to new adaptive lines as a hedge against a changing environment. The amoeboflagellate stage is also essential for sexual reproduction, so that every cell has dual potential as either a vegetative unit, or a gamete (Fig. 18.1A), depending on circumstances.

Spore-to-spore reproductive cycles

Independent discovery of mating types in *Physarum polycephalum* (Dee, 1960) and *Didymium iridis* (Collins, 1961) gave new impetus to Myxomycete research. Following examination of additional isolates, it was shown that the two mating types identified in *D. iridis* were really a pair of alleles in a one-locus multiple allelic system (Collins, 1963), and Dee (1966) later reported a similar system in *P. polycephalum*. These discoveries meant that we could perpetuate clones of amoeboflagellates as vegetative entities, as well as use transfers from them for making crosses in a variety of mating type combinations. Further, it was shown (Collins, 1965, 1966) that some collections of *D. iridis* are not heterothallic such that individual clones are capable of completing the spore-to-spore cycle. In following years, *D. iridis* became prominent in the study of myxomycete life cycles, reproductive systems, and speciation, whilst genetical research on reproductive systems in *P. polycephalum* (Collins, 1975; Collins & Tang, 1977) and other species (Henney, 1967; Henney & Henney, 1968; Clark & Collins, 1976) broadened the information base.

Multiple allelic mating systems have so far been demonstrated only in Physarales (Collins, 1979) and Echinosteliales (McGuinness & Haskins, 1985), but a one-locus, two-allele system has been revealed in Stemonitales (Collins *et al.*, 1983) and as more specimens are studied a multiple allelic system may well be demonstrated. The relative difficulty of laboratory cultivation has restricted research on sex and reproductive systems in the remaining three orders.

As our survey of collections of *D. iridis* broadened, it became apparent that in nature perhaps nonheterothallics are more prevalent than heterothallics (32 out of 44 isolates examined were nonheterothallic) and several breeding sub-groups were detected (see below). These results led us to assemble information on 38 additional species of which nine contained both heterothallic and nonheterothallic forms while, among the remaining 29 species, seven were heterothallic only and 22 nonheterothallic only.

Table 18.1. *Distribution of heterothallism and nonheterothallism among nine myxomycete species*

Species	Number of isolates sampled
Both heterothallic and nonheterothallic	
Fuligo septica	7 (4 hetero- , 3 nonheterothallic)
Didymium iridis	44 (12 hetero- , 32 nonheterothallic)
Physarum pusillum	7 (3 hetero- , 4 nonheterothallic)
Heterothallic only	
Echinostelium coelocephalum	4
Physarum flaviocomum	5
Physarum polycephalum	7
Nonheterothallic only	
Didymium difforme	8
Echinostelium minutum	7
Stemonitis flavogenita	6*

* See text for laboratory derivation of heterothallism from one of these six isolates.

Data from those species represented by at least four isolates are shown in Table 18.1 from which it appears that in nature some species (or even higher taxa) may be solely heterothallic or nonheterothallic, whilst others may encompass both kinds of behaviour.

An evolutionary explanation for coexistence of the two different reproductive strategies in a single morphospecies has been a stimulus for our recent research. Initially, we assumed an ancestor − descendant relationship between the two systems, and current data on *D. iridis*, and to a lesser extent *Stemonitis flavogenita*, support this interpretation. Here, a key question has been whether nonheterothallics are homothallic (sexual) or apomictic (asexual). Because of difficulty in handling and counting Myxomycete chromosomes, our approach has been to compare quantities of nuclear DNA in amoebae and plasmodia (Collins & Therrien, 1976; Therrien, Bell & Collins, 1977). The rationale is that the DNA profiles for sexual forms should show an alternation of ploidy levels (n to 2n) between prezygotic (amoebal) and postzygotic (plasmodial) stages. On the other hand, apomicts should reveal the same nuclear DNA levels (n, 2n or Xn) in both amoebae and plasmodia. Although the data were not always an exact fit to these predictions, they consistently supported existence of either heterothallism or apomixis and not homothallism in *D. iridis*. Moreover, demonstration of interconvertibility between the two reproductive modes in the laboratory points to an explanation for intraspecific coexistence of the two systems in nature, not only in *D. iridis* and *S. flavogenita*, but also in other taxa.

Interconvertibility of reproductive cycles

Although Clark & Collins (1976) speculated on derivation of one kind of cycle from others, it was only later that Collins (1980) produced repeatable experimental evidence for derivation of heterothallism from apomixis. This resulted when a transfer of an isolate, designated Pan 4, which for 14 years had been known only as a vigorous nonheterothallic form, became sluggish in mass-spore derived cultures and nonplasmodial forming (self-sterile) in single-amoeba derived cultures. Nuclear DNA measurements revealed a haploid level in amoebae. Crosses between the nonplasmodia-forming clones yielded plasmodia in a pattern consistent with the segregation of a pair of mating alleles but with a skewed distribution. Subsequently, a second line from the Pan 4 isolate was examined and this one displayed the nonheterothallic traits of the original isolate, including vigour, self-fertility, and a 2n to 2n nuclear DNA state for amoebae and plasmodia. Recovery of two mating types in the heterothallic line where none had been known to exist previously could, hence, be explained if they had segregated following meiosis. Consistent with this view is that originally the apomict in nature was a derivative of a heterothallic ancestor. The convertant has been stable in subsequent generations, while transfers from the apomict can be induced to yield heterothallic descendants by appropriate laboratory manipulations (see below). Each conversion event always yields the same pair of mating types, as would be expected if the explanation for conversion is correct.

We do not yet understand what pressures select heterothallism over apomixis in the laboratory, but it is correlated with perpetuation of the amoebal stage through many mitoses on a nutrient-rich medium, at-least in the case of the Pan-4 isolate. In a recent experiment successive transfers of log-phase cells from clonal amoebal populations of a Pan-4 apomict were made in order to favour the amoebal over the plasmodial stage and haploid amoebae, with their faster division rate under high-nutrient regimes, over those of higher ploidy. After many generations, individual cells were recloned to test for possible self-sterility. The preliminary data indicate we have recovered self-sterile, cross-fertile clones and we are awaiting confirmation that our crosses produced 2n plasmodia and n F_1 meiospores, hence, demonstrating at least one mechanism, i.e. amoebal selection, for conversion from apomixis to heterothallism. This explanation is also consistent with data from earlier experiments involving mass-spore culturing of individual clones of Pan 4 followed by clonal analysis of the next spore generation. Results varied according to cultures, ranging from 100% converted, to partially converted, to 100% unconverted. Presence of both heterothallics and apomicts among F_1 progeny of some cultures suggests the parent plasmodium was a heterothallic-apomictic dikaryon, with each nuclear type yielding F_1 progeny according to its genotype.

In nature, where nutrients are probably frequently scarce, large amoebae should have a selective advantage over small ones because their greater speed of migration and their slower rate of division would provide exposure to larger feeding areas, as well as more feeding time. This could help to explain apparent stability of apomicts in nature. Also, although destabilisation of the Pan 4 apomict occurs, we have been unable to destabilise a second isolate, MO1, even though it has been subjected to many more laboratory manipulations than Pan 4. It, therefore, seems possible that convertibility is a variable and not a consistent feature of apomicts. This may be correlated with the amount of time the organism has existed in nature as an apomict and with the concomitant selection for apomictic existence which it has already undergone.

Although we have also demonstrated apomictic − heterothallic convertibility in *Stemonitis flavogenita* (Collins *et al.*, 1983) we know less about the process in this species. In *S. flavogenita* the process may be closely coupled with various levels of polyploidy, making interpretation more difficult. At present, we have examined six isolates (Table 18.1) and in only one have we demonstrated heterothallism. Our initial recovery of clones carrying mating types in one culture (in one generation) could not be repeated in many successive generations obtained from the original fertile crosses. However, after many attempts over a 10-year period, we obtained heterothallic clones which give heterothallism consistently from one generation to the next, although reversion to apomixis does occur among some F_1 progeny of some crosses. As with *D. iridis*, our recent heterothallic convertants carry the same mating types as the originals. Further, on the basis of our limited sample, *S. flavogenita* appears to be even more predominantly apomictic in nature than *D. iridis*.

Genetical divergence in *Didymium iridis*

Recently, I have chosen to call *D. iridis* a morphospecies because data increasingly suggest that this taxonomic entity is partitioned into several different subunits whose biological relationships probably range from local populations to subspecies, incipient species, and even perhaps sibling species. This accords with data from studies on both enzyme polymorphism and outcrossing experiments. As noted earlier, 44 isolates of *D. iridis* have been examined (Collins, 1976; Betterley, 1981; Betterley & Collins, 1983), 12 of which are heterothallic and 32 nonheterothallic. Nine of the 12 are fully sexually compatible. The remaining three fail to produce plasmodia in all inter-isolate crosses because reproductive barriers block either zygote production or development; we do not know which. Nonheterothallics do not interbreed either with one another or with any available heterothallic form.

Data from crosses agree with isozyme profiles of several different polymorphic enzymes (Betterley, 1981) as shown in Fig. 18.1*b*. Profiles for 25 of 44 isolates were obtained and among the nine heterothallics shown only the first seven are interfertile and, with the partial exception of CR5, they have virtually identical patterns. On the other hand, nonheterothallics show inconsistent patterns from one isolate to another, as would be expected if each is representative of separate ('clonal') noninterbreeding populations. Thus, the isozyme profiles indicate that genetic divergence has occurred among nonheterothallic lines as well as between non-interbreeding heterothallics. We have not tried to estimate the timing of the various levels of divergence. That may come later through the use of mitochondrial DNA comparisons. At present, we assume noninterfertile heterothallics diverged from their interbreeding counterparts while they existed as apomicts and this resulted in development of reproductive barriers.

In outcrossing experiments (Collins, Gong & Grantham, 1983; Collins & Gong, 1985) with our Pan 4 convertant we encountered various levels of sexual incompatibility, ranging from no plasmodial development to recovery of at least some viable F_1 progenies showing evidence for recombination. The 27:13 skewness of mating types recovered among the original sibling convertants (Collins, *et al.*, 1983) is of particular interest because ratios in F_1 progenies obtained from crosses of siblings were all skewed in the same direction (34:1, 41:1, 92:0 & 23:1). We first suspected a 'meiotic drive' phenomenon but later we obtained comparable skewness in the opposite direction. What is clear is that skewness and extra-ordinarily low viability are tightly correlated; both appear to reflect the generally reduced level of fitness in the convertant as compared with the progenitor apomict. It is proposed that meiotic dismantling of the well-adapted apomict genome had the effect of depressing fitness because genetic recombination exposed lethal and semi-lethal mutations not evident in the 2n to 2n asexual cycle.

Fig. 18.1*c* illustrates relative fitness of heterothallics as compared to nonheterothallics for plasmodial development as a function of size of amoebal populations. These data agree in general with similar results comparing heterothallics with nonheterothallics on the basis of plasmodial vigour, fruiting capacity, and spore viability.

Vegetative incompatibility in heterothallics and apomicts

Like amoebae, plasmodia migrate over solid substrates when food is scarce or absent. This behaviour has survival value since it provides potential for exposure to new grazing areas under appropriate circumstances. In the process of migrating, however, plasmodia may also come into contact with other plasmodia. Reactions displayed on contact are a function of genetical relationships (see also Carlile, Chapter 14). For example, identical plasmodia always fuse and become a single entity, which means that growth occurs both by anabolism and accretion, a situation which typifies sibling plasmodial interactions of apomicts (Betterley & Collins, 1984). On the other hand, nonidentical plasmodia display a range of behaviour depending on the level of genetical relatedness. Unrelated plasmodia, as for example those belonging to different species, are completely incompatible and do not fuse at all. Those belonging to the same species, however, may also be incompatible if there are differences at particular vegetative incompatibility loci in the polygenically controlled system (Collins, 1966; Collins & Clark, 1968; Carlile, 1974) in which phenotypic expression is a function of dominance or recessiveness at each locus. That is, offspring of heterothallics typically display interactions which range from compatible fusions, incompatible fusions, to no fusion at all (Ling & Clark, 1981; Collins, 1981). From our experience, heterothalically derived, randomly selected plasmodia are far less likely to display compatible fusions than incompatible reactions. This appears to mean that in each generation of siblings in nature (as in massspore derived plasmodial cultures) there is competition and selection for a particular dominant phenotypic class (Collins, 1966; Clark & Collins, 1973) which survives and kills its siblings with cytotoxins then uses them as food (Lane & Carlile, 1979; Clark & Hakim, 1980; Schrauwen, 1979, 1981, 1985).

While it is conceivable that selection for the most dominant plasmodial incompatibility phenotype in each generation of heterothallics may result in a long term advantage over apomicts, the short term advantage appears to favour apomicts (Fig. 18.2). Regardless, each apomictic isolate tends to have a different plasmodial incompatibility phenotype from that of other apomicts (see Table I in Betterley & Collins, 1984). This means that, in nature, a species such as *D. iridis* is not only partitioned into interbreeding heterothallics, reproductively isolated heterothallics, and reproductively isolated apomicts, but that it is also partitioned into numerous vegetatively isolated entities. Further, the vegetative incompatibility system apparently

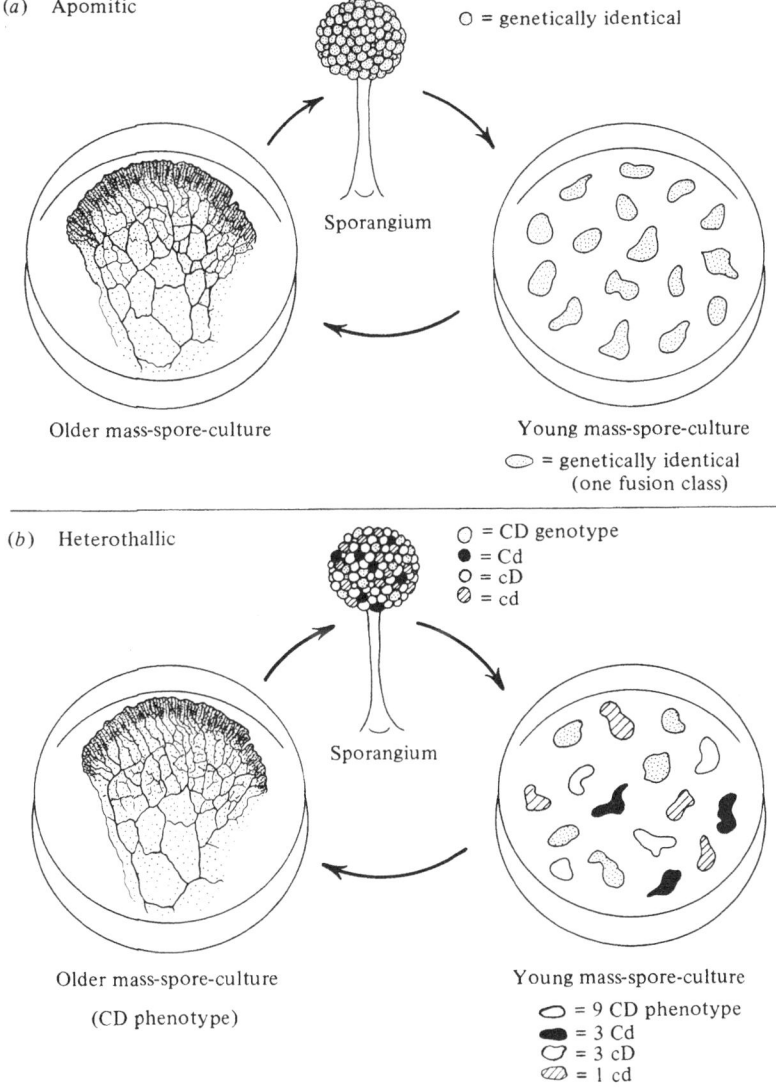

Fig. 18.2. Plasmodial (vegetative) incompatibility phenotypes in mass-spore cultures of *(a)* nonheterothallic (apomictic) and *(b)* heterothallic isolates.

serves to reinforce reproductive isolation, rather than serve as a bridge that might surmount barriers at the gametic level.

Evidence for variation within apomicts

Whilst mass-spore derived, as well as single-amoeba derived, plasmodia from a given apomictic isolate almost invariably fuse compatibly on contact, an exception is the CA 5 isolate, where only 70 out of 85 single-amoeba derived cultures were intercompatible (Betterley & Collins, 1984). The remaining 15 fell into three different, similarly sized, fusion classes. The simplest explanation is that the 15 exceptional plasmodia were recombinant, though apparently diploid. Evidence of such apparent recombinational events would ordinarily not be revealed by routine testing for apomixis or heterothallism. On the other hand, plasmodia in the exceptional classes might reflect chromosomal losses which permitted expression of recessive alleles not apparent at the higher ploidy level. We cannot currently distinguish between these alternatives as our cytological techniques would only detect gross polyploid variation, not small aneuploid differences (Collins, Therrien & Betterley, 1978; Mulleavy, 1979; Mulleavy & Collins, 1979).

A second example of genetical evidence of variation within an apomict was obtained very recently with the Staten Island isolate (S. I. 1), from which Lynch & Collins (unpublished) established 100 single-amoeba derived clones of which 98 were plasmodial and two did not advance past the amoebal stage. The latter are presumed to be heterothallic and of the same mating type, since they do not produce plasmodia when crossed with each other. In outcrosses to testers from all available heterothallic isolates, however, the two S. I. 1 heterothallic clones are sexually compatible with at least one tester from each isolate. We tentatively conclude that 2/100 clones were recovered either because of rare or partial meioses or because new genomic combinations came into existence by some other means, such as unequal mitotic divisions of polypoid nuclei.

Genetical data on apomictic − heterothallic convertibility and speciation

Preliminary data from outcrosses with the two variant S. I. 1 clones suggest we may be able to identify genes which determine whether a life cycle will be apomictic or heterothallic. For example, F_1 progeny from three different outcrosses with the variant clones are classifiable as (i) all heterothallic, (ii) all apomictic, and (iii) heterothallic and apomictic in about a 1:1 ratio. However, other outcrosses yield a range of other data and further substantiation is necessary before firm conclusions can be drawn. Nonetheless, even a clear genetical explanation for convertibility of the two life cycles would not be a description of how speciation occurs. It would instead help us better to understand the mechanism underlying a process which may lead to genetical divergence and subsequently to speciation. A major goal, therefore, is to identify specific genes and to understand their phenotypic expression at the cytological level. In this connection work carried out on apomictic strains of *Saccharomyces cerevisiae* (Klapholz & Esposito, 1980; Klapholz, Waddell & Esposito, 1985) may be especially relevant to our work.

Concluding remarks

Discovery of mating types in *Didymium iridis* and *Physarum polycephalum* in the early 1960's led to demonstration of a multiple allelic mating system. This system may have had a common origin from which it descended to all other Myxomycete taxa. Today it is coupled with vegetative reproductive modes as well as (at least in some cases) with an asexual (apomictic) mode. Further, separate sexual and vegetative (plasmodial) incompatibility systems contribute to the definition of an individual Myxomycete and to the speciation process.

Acknowledgements I acknowledge contributions of Faculty colleagues Dale Therrien, Edward Haskins and Thomas Gaither; former graduate students H. Ling, J. Clark, P. Mulleavy and D. Betterley and current graduate student Margaret Silliker; as well as research associates T. Tang, T. Gong and R. Lynch.

References

Betterley, D. (1981). Reproductive systems, enzyme polymorphism and speciation in the myxomycete *Didymium iridis*. *Ph. D. thesis, University of California, at Berkeley,*

Betterley, D. & Collins, O. R. (1983). Reproductive systems, morphology and genetical diversity in *Didymium iridis* (Myxomycetes). *Mycologia*, **75**, 1044 − 1063.

Betterley, D. & Collins, O. R. (1984). Vegetative incompatibility and myxomycete biology. *Mycologia*, **76**, 785 − 792.

Carlile, M. J. (1974). Incompatibility in the myxomycete *Badhamia utricularis*. *Transactions of the British Mycological Society*, **62**, 401 − 402.

Clark, J. & Collins, O. R. (1973). Directional cytotoxic reactions between incompatible plasmodia of *Didymium iridis*. *Genetics*, **73**, 247 − 257.

Clark, J. & Collins, O. R. (1976). The mating systems of eleven species of *Myxomycetes*. *American Journal of Botany*, **63**, 783 − 789.

Clark, J. & Hakim, R. (1980). Nuclear sieving of plasmodia. *Experimental Mycology*, **4**, 17 – 22.

Collins, O. R. (1961). Heterothallism and homothallism in two Myxomycetes. *American Journal of Botany*, **48**, 674 – 683.

Collins, O. R. (1963). Multiple alleles at the incompatibility locus in the myxomycete *Didymium iridis*. *American Journal of Botany*, **50**, 477 – 480.

Collins, O. R. (1965). Homothallic behavior in two Costa Rican isolates of the slime mold *Didymium iridis*. *American Journal of Botany*, **52**, 634 (abstract).

Collins, O. R. (1966). Plasmodial compatibility in heterothallic and homothallic isolates of *Didymium iridis*. *Mycologia*, **58**, 362 – 372.

Collins, O. R. (1975). Mating types in five isolates of *Physarum polycephalum*. *Mycologia*, **67**, 98 – 107.

Collins, O. R. (1976). Heterothallism and homothallism: a study of 27 isolates of *Didymium iridis*, a true slime mold. *American Journal of Botany*, **63**, 138 – 143.

Collins, O. R. (1979). Myxomycete biosystematics: some recent developments and future research opportunities. *Botanical Review*, **45**, 145 – 201.

Collins, O. R. (1980). Apomictic – heterothallic conversion in a myxomycete, *Didymium iridis*. *Mycologia*, **72**, 1109 – 1116.

Collins, O. R. (1981). Myxomycete genetics, 1960 – 1981. *Journal of the Elisha Mitchell Scientific Society*, **97**, 101 – 125.

Collins, O. R. & Clark, J. (1968). Genetics of plasmodial compatibility and heterokaryosis in *Didymium iridis*. *Mycologia*, **60**, 90 – 103.

Collins, O. R. & Gong, T. (1985). Genetical relatedness of a former apomict and a heterothallic isolate in *Didymium iridis* (Myxomycetes). *Mycologia*, **77**, 300 – 307.

Collins, O. R., Gong, T., Clark, J. & Tang, H. C. (1983). Apomixis and heterothallic in *Stemonitis flavogenita* (Myxomycetes, Stemitales). *Mycologia*, **75**, 614 – 622.

Collins, O. R., Gong, T. & Grantham, M. (1983). Genetical analyses of an apomictic – heterothallic convertant of *Didymium iridis*. *Mycologia*, **75**, 683 – 692.

Collins O. R. & Tang, H. C. (1977). New Mating types in *Physarum polycephalum*. *Mycologia*, **69**, 421 – 423.

Collins O. R. & Therrien, C. D. (1976). Cytophotometric measurement of nuclear DNA in seven heterothallic isolates of *Didymium iridis*, a Myxomycete. *American Journal of Botany*, **63**, 457 – 462.

Collins O. R., Therrien, C. D. & Betterley, D. A. (1978). Genetical and cytological evidence for chromosomal elimination in a true slime mold, *Didymium iridis*. *American Journal of Botany*, **65**, 660 – 670.

Dee, J. (1960). A mating-type system in an acellular slime mould. *Nature*, **185**, 780 – 781.

Dee, J. (1966). Multiple alleles and other factors affecting plasmodium formation in the true slime mould *Physarum polycephalum*. *Journal of Protozoology*, **13**, 610 – 616.

Henney, M. R. (1967). The mating type system of the myxomycete *Physarum flavicomum*. *Mycologia*, **59**, 637 – 652.

Henney, M. R. & Henney, H. R. (1968). The mating type systems of the Myxomycetes *Physarum rigidum* and *Physarum flavicomum*. *Journal of General Microbiology*, **53**, 321 – 332.

Klapholz, S. & Esposito, R. E. (1980). Isolation of SPO12-1 and SPO13-1 from a natural variant of yeast that undergoes a single meiotic division. *Genetics*, **96**, 567 – 588.

Klapholz, S., Waddell, C. S. & Esposito, R. E. (1985). The role of the SPO11 gene in meiotic recombination in yeast. *Genetics*, **110**, 187 – 216.

Lane, E. G. & Carlile, M. J. (1979). Post-fusion somatic incompatibility in plasmodia of *Physarum polycephalum*. *Journal of Cell Science*, **35**, 339 – 354.

Ling, H. & Clark, J. (1981). Somatic cell incompatibility in *Didymium iridis*: locus identification and function. *American Journal of Botany*, **68**, 1191 – 1199.

McGuinness, M. D. & Haskins, E. F. (1985). Multiple alleles at the mating type locus in *Echinostelium coelocephalum*. *Mycological Society of America Newsletter*, **36**, 33.

Mulleavy, P. (1979). Genetic and cytological studies in heterothallic and non-heterothallic isolates of the myxomycete *Didymium iridis*. *Ph. D. thesis, University of California at Berkeley*,

Mulleavy, P. & Collins, O. R. (1979). Development of apogamic amoebae from heterothallic lines of a myxomycete, *Didymium iridis*. *American Journal of Botany*, **66**, 1067 – 1073.

Schrauwen, J. A. M. (1979). Post-fusion incompatibility in *Physarum polycephalum*. *Archives of Microbiology*, **122**, 1 – 7.

Schrauwen, J. A. M. (1981). Post-fusion incompatibility in *Physarum polycephalum*. The involvement of DNA. *Archives of Microbiology*, **129**, 257 – 260.

Schrauwen, J. A. M. (1985). Post-fusion incompatibility in *Physarum polycephalum*: involvement of membranes. *Canadian Journal of Microbiology*, **31**, 782 – 785.

Therrien, C. D., Bell, W. R. & Collins, O. R. (1977). Nuclear DNA content of myxamoebae and plasmodia in six heterothallic isolates of a myxomycete, *Didymium iridis*. *American Journal of Botany*, **64**, 286 – 291.

19

Occurrence and interactions of outcrossing and non-outcrossing populations in *Stereum*, *Phanerochaete* and *Coniophora*

A. MARTYN AINSWORTH

School of Biological Sciences, University of Bath, Bath BA2 7AY, UK

Introduction

Many members of the genera *Stereum*, *Phanerochaete* and *Coniophora* (Basidiomycotina) exhibit holocoenocytic nuclear behaviour (Boidin, 1971), such that their hyphae have multinucleate compartments, often until basidium formation. Also, clamp connections often occur in pairs or whorls, even in a mycelium derived from a single basidiospore, i.e. a primary mycelium, a feature which gave rise to the notion that these fungi must be homothallic. However, recent studies have shown that in some populations a homogenic incompatibility system regulates the formation of distinct secondary mycelia from paired single basidiospore isolates in *Stereum hirsutum* (Coates, Rayner & Todd, 1981), *S. gausapatum* (Boddy & Rayner, 1982) and *S. rugosum* (Rayner & Turton, 1982). These studies have now been extended to include collections from different continents and a range of habitats. It is clear from the results of simple direct pairing between primary mycelia, from confirmatory interactions of subcultures from such pairings, and from progeny analyses, that two basic breeding strategies occur: outcrossing and non-outcrossing.

The non-outcrossing strategy

Here, fruit body tissue isolates and primary mycelia are morphologically indistinguishable if derived from the same wild fruit body. Interactions of sibling and some non-sibling primary mycelia result in a macroscopic intermingling response which is indistinguishable from control pairings between different subcultures of a single primary mycelium (Fig. 19.1a). Other non-sibling interactions result in a macroscopic rejection response at the mycelial confrontation zone (Fig. 19.1b).

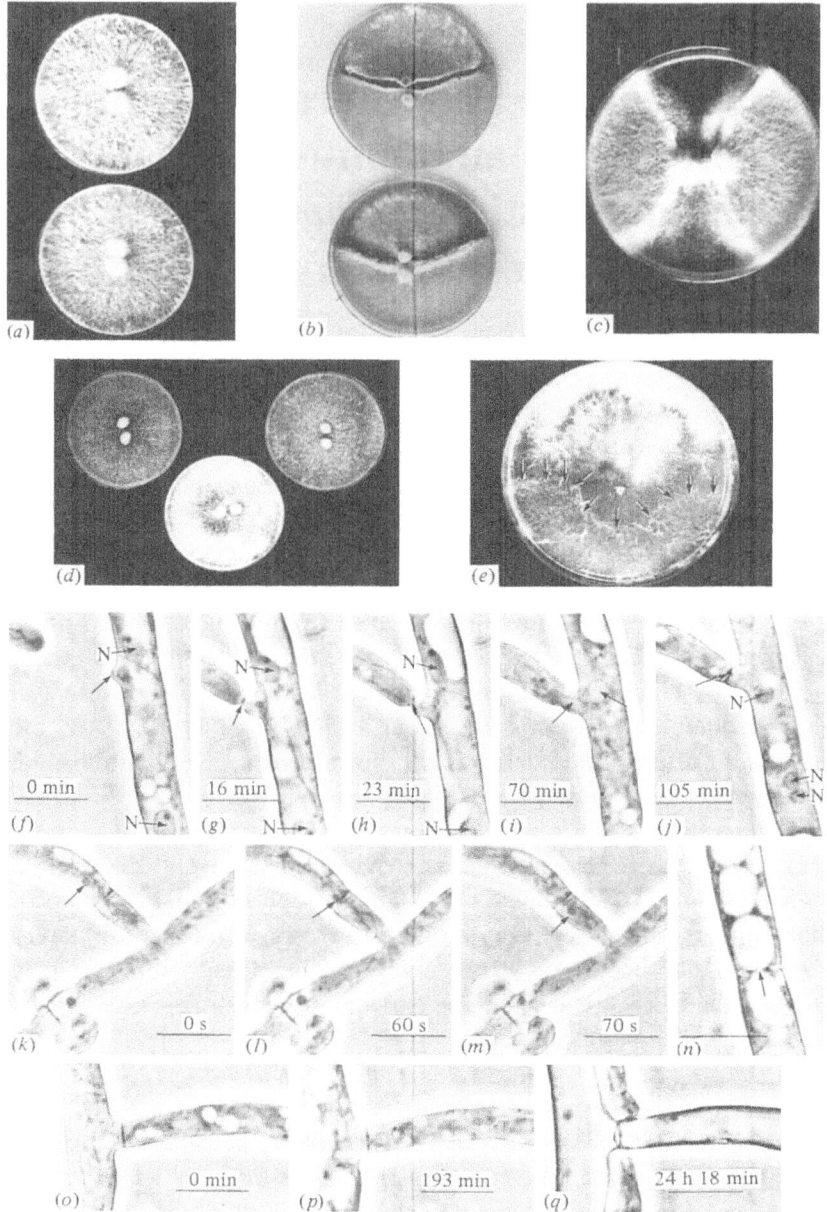

Fig. 19.1. Mycelial and hyphal interactions between primary mycelia associated with non-outcrossing and outcrossing strategies. (a) Intermingling between subcultures from the same mycelium (above) and sibs (below) of non-outcrossing *Stereum hirsutum*; (b) rejection between non-sibs

On this basis, primary mycelia can be assigned to different non-outcrossing interaction groups which can be used to investigate population structure in the field. However, whilst rejection is an indicator of genetic difference, intermingling may not necessarily signify genetic identity between non-sibs. This was illustrated by three British *S. sanguinolentum* interaction groups, between two of which rejection occurred, yet both intermingled with isolates belonging to a third group.

The outcrossing strategy

Progeny sets from fruit bodies of outcrossing populations exhibit morphological variation and a range of distinct recognition responses. As described by Rayner & Coates (Chapter 8), macroscopic recognition responses between primary mycelia are of three basic and superimposable types: rejection (Fig. 19.1b), inhibition/appression (Fig. 19.1c), and mating type compatibility resulting in either a generalised (Fig. 19.1d) or localised (Fig. 19.1e) establishment of stable secondary mycelium. In all the cases so far examined, secondary mycelial development has occurred according to a multiallelic unifactorial (i.e. bipolar) system; both localised and generalised secondary mycelial establishment occurred in some progeny sets, but not to any consistent pattern.

Some primary mycelia fruited in the laboratory and their basidiospore-derived progeny were morphologically indistinguishable and intermingled when paired, e.g. *S. insignitum* and *Phanerochaete laevis*. However,

of non-outcrossing *S. sanguinolentum*; (c) inhibition/appression in a bow-tie shaped region between mating type incompatible sibs of *S. hirsutum*; (d) two mating type compatible sibs of *Stereum* sp. (left and right dishes) which produce a generalised pattern of secondary mycelium establishment when paired (centre); (e) paired mating type compatible sibs of *Phanerochaete velutina* showing a localised pattern of secondary mycelium establishment in a region of rejection, the interface of this region (above) and remnant primary mycelium (below) is arrowed. (f)–(j) Nuclear division and septation during self fusion in *P. velutina*; arrows indicate (f) pre-fusion hyphal tip induction, (g) hyphal contact, (h) site of fusion pore opening, (i) pair of dividing nuclei and (j) dolipore septum across the fusion pore. Nuclei (N) are shown before (f–h) and after (j) division. (k)–(n) Septal erosion and nuclear migration; (k)–(m) show a nucleus (arrowed) passing through an incompletely eroded septum in *S. hirsutum*; (n) vacuole constricted during passage through an incompletely eroded septum (arrowed) in *P. velutina* (from Ainsworth & Rayner, 1986). (o)–(q) Sequential occurrence of septation (o), septal erosion (p) and hyphal rejection (q) after a mating type compatible reaction in *P. velutina*. Petri dishes are 9 cm diam in (a)–(d), 14 cm in (e). Scale markers represent 10 μm.

since all these mycelia were still capable of participating in secondary mycelial production, they must be assigned to the outcrossing breeding strategy.

Microscopic interactions

Each macroscopic interaction is the sum of a series of microscopic interhyphal responses initiated in varying proportions within the confrontation zone. Three basic types of interhyphal response have been identified in *Phanerochaete velutina* and *Stereum hirsutum* using the microculture chamber method of Aylmore & Todd (1984a).

1. Nuclear division and septation This is the only response seen after self-fusions (i.e. between different hyphae of a single primary or secondary mycelium) but can also occur after non-self fusion between certain primary mycelia. It involves aggregation and division of nuclei, usually derived from both participating hyphal compartments, in the fusion pore region, followed by septum formation across the pore itself (in *P. velutina*; Fig. 19.1f − j) or in close proximity to it (in *S. hirsutum*). This sequence differs from that in self-fusions in certain Basidiomycotina having strict uninucleate (monokaryotic) or binucleate (dikaryotic) life cycle stages, where a partial or complete nuclear replacement reaction occurs, e.g. *Coriolus versicolor* (Aylmore & Todd, 1984b), *Schizophyllum commune* (Todd & Aylmore, 1985) and *Chondrostereum purpureum* (A. M. Ainsworth, unpublished).

2. Septal erosion and initiation of nuclear migration These responses (Fig. 19.1k − n) can occur after mating type compatible non-self fusions and, in *S. hirsutum*, after certain mating type incompatible non-self fusions leading to macroscopic inhibition/appression. Septal erosion is often incomplete and, although septal repair has never been observed, complete septa may be synthesised in intercalary positions to produce new compartments.

3. Hyphal rejection This response involves a progressive vacuolation of the fusion compartment and up to several neighbouring compartments. Although an invariable response between secondary mycelia, hyphal rejection was only seen between mating type compatible primary mycelia of *P. velutina*. Such seemingly paradoxical rejection and acceptance phenomena associated with the action of complementary mating type factors have also been described in *Neurospora crassa* (Ascomycotina), although spatially restricted to vegetative (somatic) or sexual fusions respectively (see Perkins & Barry, 1977). In *P. velutina*, however, hyphal fusion may be regarded as somato-sexual (Rayner *et al.*, 1984) and complementary mating type factors may be involved in sequential expression of acceptance (e.g. septal erosion) and rejection reactions within the same compartments (Fig. 19.1o – q).

The distribution and interconversion of outcrossing and non-outcrossing strategies

Table 19.1 lists those taxonomic species in which outcrossing alone has been detected and Table 19.2 those in which both outcrossing and non-outcrossing occur. Whilst *S. sanguinolentum* and *S. hirsutum* were identified as such regardless of breeding strategy, the outcrossing collections of *S. subtomentosum* were provisionally assigned to this taxon on the basis of their cultural characteristics. *S. 'rameale'* is used throughout this chapter in the sense of European authors (e.g. Jahn, 1971), some of whom regard it as a synonym for *S. ochraceo-flavum* (e.g. Jülich & Stalpers, 1980). However, it should not be confused with *S. rameale* of American authors (e.g. Lentz, 1955; Chamuris, 1985) which is synonymous with *S. complicatum* and does not occur in Europe.

The presence of at least two breeding strategies within a species or group of morphologically similar species is documented several times in this volume (Brasier, Chapter 16; Schipper, Chapter 17; Collins, Chapter 18; Korhonen, Chapter 20). Within the Basidiomycotina, *Sistotrema brinkmanii* has probably received most attention. This taxon comprises groups with non-outcrossing, unifactorial and bifactorial strategies. In attempts to detect heterokaryon formation in forced matings within and between these groups, genetic recombination was detected only in some cases, and Lemke (1969) and Ullrich & Raper (1975) concluded that reproductive isolation was not total within the non-outcrossing group or between it and the unifactorial group. However, such conclusions have doubtful relevance to wild populations (cf. Caten & Jinks, 1966).

Within *Stereum*, outcrossing/non-outcrossing primary mycelial interactions usually resulted in a pigmented rejection response, but in *S. hirsutum*, British outcrossing forms were often replaced by non-outcrossing forms (Fig. 19.2a). However, where pairings between Russian outcrossing and other non-outcrossing forms were involved, development of a pigmented confrontation zone was accompanied by an adjacent inhibition/appression reaction, widest at its edges, in the outcrossing mycelium. This was superseded by emergence of aerial

Table 19.1. *Taxonomic species in which the outcrossing strategy alone has been detected*

Taxa and no. of breeding units identified	Geographical origin of samples	Substrata and no. of fruit bodies used for isolation of primary mycelia
Stereum hirsutum complex		
S. *gausapatum* (1)	Britain	*Quercus* (1); *Castanea* (1); *Fagus* (1); unknown angiosperm (1)
S. *complicatum* (1)	Eastern USA	*Quercus* (1)
	Eastern USA	*Betula* (2); *Fagus* (1) *Carpinus* (1)
Stereum striatum complex		
S. *striatum* (1)	Eastern USA	*Carpinus* (1)
Stereum ostrea complex		
S. *insignitum* (1)	France	*Fagus* (1); unknown (1)
Stereum rugosum (1)	Britain	*Alnus* (2); *Betula* (2); *Fagus* (1); *Corylus* (1) *Quercus* (1); *Prunus* (1); *Salix* (1); *Ilex* (1); *Rhododendron* (1)
Unidentified *Stereum* sp. (1)	Northern USA	Unknown (1)
Unidentified *Stereum* sp. (1)	Eastern USSR	*Quercus* (2)
Phanerochaete velutina (1)	Britain	*Fagus* (6); *Acer* (1)
P. *laevis* (1)	Britain	*Fagus* (2)
Coniophora puteana (3)	Britain	*Thuja* (1); unknown gymnosperm (1); *Fagus* (3); *Ulmus* (3); *Crataegus* (2); unknown angiosperm (1)
	Eastern USA	Unknown (1); *Picea* (1); *Abies* (1); *Prunus* (1)
	Norway	*Picea* (1)
	Estonia	*Picea* (2); *Populus* (1)

mycelium whose boundary distal to the confrontation zone also became pigmented (Fig. 19.2b). The interactive behaviour and cultural characteristics of mycelium subcultured from the region of altered morphology differed both from those of either precursive primary mycelium or secondary mycelium. Nevertheless, its localised establishment in a region of inhibition/appression paralleled that of secondary mycelium production in several other *Stereum* species, e.g. *S. insignitum* and *S. striatum*. This suggests that a genetic factor had been transmitted from the non-

Table 19.2. *Taxonomic species with both outcrossing and non-outcrossing strategies*

Taxa and no. of identified outcrossing breeding units or non-outcrossing interaction groups	Geographical origin of samples	Substrata and no. of fruit bodies used for isolation of primary mycelia
Stereum hirsutum complex		
S. hirsutum		
Outcrossing forms (1)	Western USA	*Umbellularia* (1); *?Umbellularia* (1); unknown (1)
	Britain	*Quercus* (1); *Fraxinus* (1)
	Caucasus Mountains USSR	*Carpinus* (1)
	Ukraine	*Populus* (1)
	Western Australia	*Casuarina* (1); *Eucalyptus* (5)
Non-outcrossing forms (11)	Finland	*Betula* (7)
	Norway	*Sorbus* (1)
	Eastern USSR	*Quercus* (1); *Betula* (1)
	Sweden	*Betula* (2); *Quercus* (1)
S. subtomentosum		
Outcrossing forms (1)	Eastern USSR	*Alnus* (1); *Betula* (1)
Non-outcrossing forms (7)	Eastern USA	Unknown (1)
	Britain	*Fagus* (1); unknown (2)
	Finland	*Corylus* (1)
	Norway	*Alnus* (1)
	Estonia	*Betula* (1)
Stereum striatum complex		
S. ochraceo-flavum		
Outcrossing forms (1)	Western USA	*Umbellularia* (1)
S. 'rameale'		
Non-outcrossing forms (7)	Eastern USSR	*Alnus* (2); *Quercus* (1)
	Britain	*Betula* (2); *Quercus* (16); *Castanea* (1) *Prunus* (4); *Acer* (1)
S. sanguinolentum		
Outcrossing forms (1)	Eastern Canada	*Abies* (1)
Non-outcrossing forms (33)	Britain	*Larix* (50); *Picea* (16); *Pinus* (24) *Pseudotsuga* (1); unknown gymnosperm (2)
	Finland	*Picea* (4); *Pinus* (1)

outcrossing to a portion of the outcrossing mycelium. In non-outcrossing forms of both *Stereum* species and *Sistotrema brinkmanii* the morphology of the primary mycelia is similar to that of secondary mycelia of outcrossing forms within the same (or similar) species. Furthermore, intraspecific pairings between the various non-outcrossing interaction groups and outcrossing mating type representatives failed to result in distinct secondary mycelial development from subcultures of the non-outcrossing resident. Such an inability to accept nuclei is also a feature of secondary mycelia and of those primary mycelia of bifactorial *Schizophyllum commune* which have mutations at both mating type factors rendering them 'heterokaryon mimics' (Raper, 1966). In contrast to the interactions between secondary and primary mycelia, those between non-outcrossing and outcrossing primary mycelia never resulted in conversion of the latter to heterokaryons by the Buller Phenomenon such as has been demonstrated and analysed in *Stereum hirsutum* by Coates & Rayner (1985a). Hence, the non-outcrossing primary mycelia are probably best regarded as functionally, but not truly, heterokaryotic.

Similar functional heterokaryosis has been conferred on British outcrossing primary mycelia of *S. hirsutum* by a mobile regulatory element (Coates & Rayner, 1985b; Rayner & Coates, Chapter 8) which may be relevant to the origin of a non-outcrossing from an outcrossing form. Furthermore, the presence of this element was initially indicated by

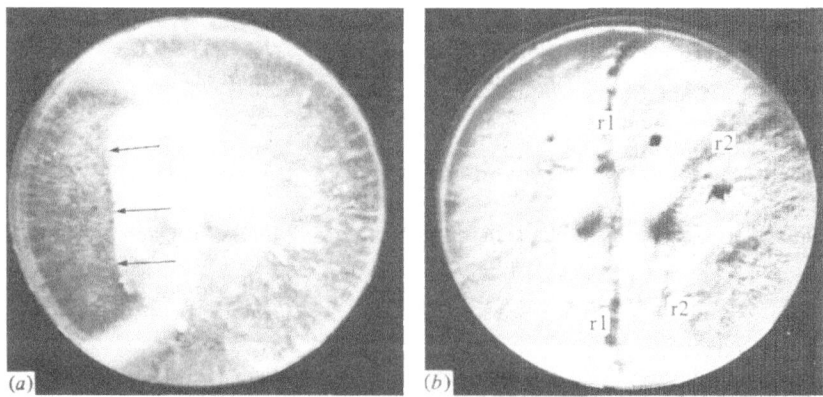

Fig. 19.2. (a) Replacement of a British outcrossing isolate of *S. hirsutum* by a mycelial front (arrowed) of a Finnish non-outcrossing isolate. (b) Interaction between a Finnish non-outcrossing isolate (left) and an outcrossing isolate (right) from the Caucasus Mountains (USSR), containing a central distinctive region bounded by pigmented rejection responses (r1, r2).

spontaneous conversion of an outcrossing primary mycelium during cold
storage. Since all collections from a particular area have exhibited only
one or the other breeding strategy, it is suggested that such conversion
might initially have occurred under stress conditions operating at the edge
of the geographical range of an ancestral outcrossing form. Subsequent
spread of the non-outcrossing form could then be facilitated by clonal
dissemination. On this basis, the large and widespread interaction groups
in *Stereum sanguinolentum* and *S. 'rameale'* might have resulted from
woodland colonisation by a few founding non-outcrossing forms. The
small, geographically restricted, interaction groups could then be derived
from the founder groups by mutation.

Outcrossing and intraspecific ecological behaviour

The possibility of a correlation between ecological behaviour and
impairment or absence of secondary mycelium formation was examined in
British strains of *Stereum rugosum*, *S. gausapatum* and *Coniophora
puteana* (Table 19.1). Pairings between mating type representatives from
eleven fruit bodies of *S. rugosum* from nine woody genera indicated that
the samples contained 21 different mating type factors with one repeat. *S.
gausapatum* shows a high degree of selectivity for oaks (*Quercus*),
nevertheless, fruit bodies were obtained from other members of the
Fagaceae, namely sweet chestnut (*Castanea*) and beech (*Fagus*). All
pairings of mating type representative primary mycelia resulted in
secondary mycelium formation indicating the presence of eight mating
type factors. *C. puteana* fruit bodies occur on both gymnospermous and
angiospermous substrata and the interactions between mating type
representatives from the eleven British samples (Table 19.1) indicated the
presence of two intersterile groups or breeding units. However, these units
were not correlated with obvious ecological specialisation

Outcrossing and geographical (allopatric) isolation

Whilst only two breeding units of *Coniophora puteana* were
identified in British isolates, at the international level, mating type representative primary mycelia fell into three breeding units (Table 19.1) such
that secondary mycelial establishment only occurred within a unit,
whereas rejection occurred between units (Fig. 19.3a & b). Rejection also
occurred during certain intra-unit interactions involving Norwegian or
Estonian isolates. On the basis of secondary mycelial formation, the
outcrossing collections of *S. hirsutum* were assigned to a single breeding
unit (Table 19.2). However, in international pairings involving an

Australian isolate, secondary mycelial formation followed an inhibition/appression reaction which often affected the entire Australian mycelium. This contrasted with mating type compatible interactions between Australian primary mycelia in which inhibition/appression was only visible around the periphery of the mated cultures. It is suggested that the reaction of Australian isolates in international pairings was due to a shift in the proportion of hyphal fusion fates favouring

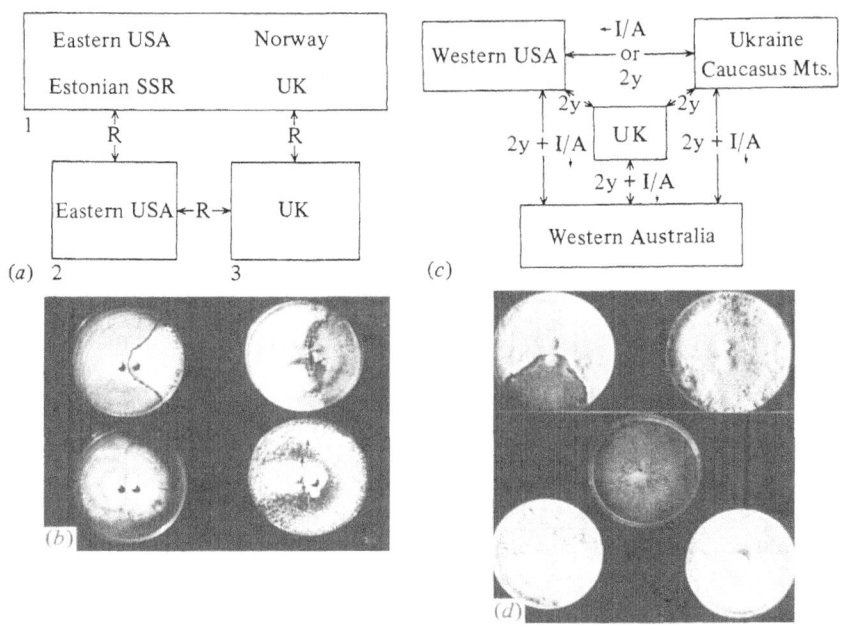

Fig. 19.3. Breeding units in *Coniophora puteana* (a, b) and *Stereum hirsutum* (c, d). (a) Boxes 1 – 3 enclose those interactions yielding secondary mycelium. Rejection responses (R) occurred between members of different boxes and also during certain interactions within box 1 (data not shown); (b) rejection between isolates from boxes 2 and 3 (upper left), between boxes 1 and 3 (upper right – British isolates; lower left – British and Estonian isolates). Secondary mycelial emergence (arrowed) in a mating type compatible interaction of a box 1 isolate from Britain and an Estonian isolate is shown in the lower right plate. (c) Secondary mycelium (2y) was established in all intracontinental pairings (within boxes) and most intercontinental pairings. Inhibition/appression interactions (I/A) occurred unilaterally in the direction shown. (d) Unilateral inhibition/appression (upper left) and mating type compatible (upper right) interactions between Ukrainian (above) and American (below) isolates. A subcultured region of inhibition/appression produced during the interaction of an American isolate (lower right) with one from the Caucasus Mountains, USSR (lower left) is shown in the centre plate.

inhibition/appression morphology compared with that occurring between
Australian isolates. However, whether this is the cause or consequence of
reduced gene flow between Australian *S. hirsutum* and the others is
unclear.

Secondary mycelium was also produced in pairings between
American/British, Russian/British, and some Ukrainian SSR/American
isolates (Fig. 19.3c&d). The remaining Russian/American pairings failed
to establish secondary mycelia. Hence, an interesting 'A-B-C' relationship
emerged between those isolates originating in the USA, Britain and from
near the Caucasus Mountains in the USSR (cf. *Hirschioporus abietinus*,
Macrae, 1967). When a secondary mycelium was not produced, an
inhibition/appression reaction occurred throughout the American
mycelium (Fig. 19.3d) and also, rarely, in a narrow band within that of the
Russian isolates adjacent to the confrontation zone. Subcultures of the
affected regions produced putative heterokaryons or heteroplasmons
which were pigmented, slowly extending, lacking in aerial growth and
generally of a 'senescent' morphology (Fig. 19.3d; cf. Rayner & Coates,
Chapter 8). Such mycelia showed no macroscopic interaction when
paired, although repeated subculturing sometimes resulted in irregular
patches of normal *S. hirsutum* primary mycelial morphology. Subcultures
of the latter failed to intermingle with either precursive primary mycelial
type thereby resembling mycelia produced during certain pairings of *S.
hirsutum* outcrossing and non-outcrossing isolates (see above).

A single breeding unit was also detected in *S. gausapatum* from Britain
and the USA (Table 19.1), although several international pairings failed to
result in secondary mycelial establishment. In the latter case, an
inhibition/appression reaction occurred either throughout both mycelia or
only in those of British origin. Subcultures from such appressed regions
were of a similarly 'senescent' morphology to those just described for
international pairings of *S. hirsutum*.

Outcrossing and intrageneric interactions

Mating type representative primary mycelia from each geographical area were used in interspecific pairings to see whether current
taxonomic relationships could be supported.

Phanerochaete laevis and *P. velutina* produced a pigmented rejection
response followed by overarching of the latter by mycelial cords of the
former. Regardless of breeding strategy, similar rejection responses,
without an accompanying mode change, were seen in all interactions of
Stereum species from different species complexes (Fig. 19.4a) except for

those between *Stereum subtomentosum*, *S. ochraceo-flavum* and *S. 'rameale'*. These latter exceptional reactions involved attenuation and vacuolation of hyphal apices of both primary mycelia when they were still about 1 cm apart (Fig. 19.4b).

Within species complexes, *Stereum* inhibition/appression reactions occurred throughout mycelia of *S. hirsutum* paired against those of *S. complicatum* (Fig. 19.4c) and in bow-tie shaped regions between interacting mycelia of *S. striatum* and *S. ochraceo-flavum*. Subcultures from these regions were usually 'senescent', as described previously for other interactions within the genus. However, subcultures from some *S. hirsutum/S. complicatum* inhibition/appression reactions − where the former isolate was Russian or British − morphologically resembled those of the inoculated *S. complicatum* isolate with which they intermingled when paired. This indicates mycelial replacement by regeneration from the appressed regions as has been demonstrated in the bow-tie reaction of mating type incompatible sibs of *S. hirsutum* (Coates & Rayner, 1985c) and supports a close taxonomic relationship between *S. hirsutum* and *S. complicatum*, and between *S. striatum* and *S. ochraceo-flavum*. It is, therefore, of interest that Welden (1971) placed the former species pair in the *S. hirsutum* complex whilst the latter were recognised as varieties of *S. striatum* (Tables 19.1 & 19.2).

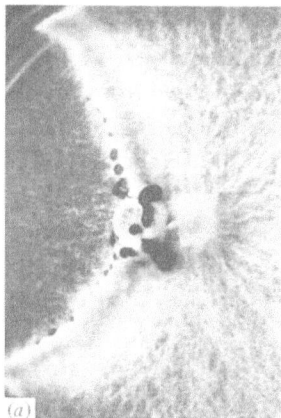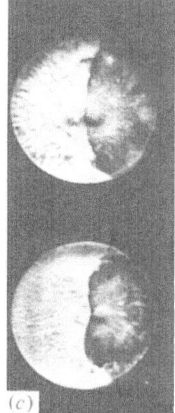

Fig. 19.4. Intrageneric interaction types. (a) Rejection with pigmented droplet exudation between *Stereum rugosum* (left) and *S. gausapatum*. (b) non-fusion of outcrossing and non-outcrossing isolates of *S. ochraceo-flavum/S. 'rameale'* in the upper half of each dish and *S. subtomentosum* in the corresponding lower halves. (c) Unilateral inhibition/appression reactions seen in *S. hirsutum* isolates (right) when paired against those of *S. complicatum* (left).

Concluding comments

Within several taxonomic species of *Stereum*, non-outcrossing forms can be assigned to groups which are genetically isolated from each other and from putatively ancestral outcrossing forms. From an evolutionary standpoint, a shift from outcrossing to non-outcrossing may be associated with a short term selective advantage, as in the formation of myxomycete plasmodia (Collins, Chapter 18; and see Brasier, Chapter 16). This would suit adoption of a ruderal ecological strategy or adjustment to climatic stresses at the edge of the geographic range (Cooke & Rayner, 1984). However, since this shift also reduces the capacity to generate variation within a clonal line, it will, in the long term, be selectively disadvantageous in an unstable environment.

Several mating type compatible interactions between collections of *Stereum* species showed impaired secondary mycelial establishment, often accompanied by inhibition/appression and/or rejection responses. Although based on results from a small number of collections, this is nevertheless in accordance with gradual speciation associated either with geographical (allopatric) isolation, e.g. in *S. hirsutum* and *S. gausapatum*, or with ecological specialisation, e.g. of the species pair *S. ochraceoflavum/S. striatum*, the latter is usually found on hornbeams (*Carpinus*). By contrast, there was little evidence of genetic exchange between breeding units of *Coniophora puteana* within a single geographical area (sympatric), even when this was a single woodland, which suggests that a more abrupt form of isolation has occurred within this species.

Acknowledgements I am grateful to the Science and Engineering Research Council for financial support, to Dr Alan Rayner for much valuable discussion and, together with Dr C. M. Brasier, for critically reading this manuscript; to Dr Robert Aylmore and the late Dr Norman Todd for assistance with techniques and to all those mycologists who have collected and identified material used in these studies.

References

Ainsworth, A. M. & Rayner, A. D. M. (1986). Responses of living hyphae associated with self and non-self fusions in the basidiomycete *Phanerochaete velutina*. *Journal of General Microbiology*, **132**, 191 – 201.
Aylmore, R. C. & Todd, N. K. (1984a). A microculture chamber and improved method for combined light and electron microscopy of filamentous fungi. *Journal of Microbiological Methods*, **2**, 317 – 322.

Aylmore, R. C. & Todd, N. K. (1984b). Hyphal fusion in *Coriolus versicolor*. In *The Ecology and Physiology of the Fungal Mycelium*, British Mycological Society Symposium volume 8, ed. Jennings, D. H. & Rayner, A. D. M., pp. 103 − 125. Cambridge University Press.

Boddy, L. & Rayner, A. D. M. (1982). Population structure, inter-mycelial interactions and infection biology of *Stereum gausapatum*. *Transactions of the British Mycological Society*, **78**, 337 − 351.

Boidin, J. (1971). Nuclear behaviour in the mycelium and the evolution of the basidiomycetes. In *Evolution in the Higher Basidiomycetes*, ed. Petersen, R. H., pp. 129 − 148. Knoxville, Tennessee: University of Tennessee Press.

Caten, C. E. & Jinks, J. L. (1966). Heterokaryosis: its significance in wild homothallic ascomycetes and fungi imperfecti. *Transactions of the British Mycological Society*, **49**, 81 − 93.

Chamuris, G. P. (1985). On distinguishing *Stereum gausapatum* from the '*S. hirsutum* complex'. *Mycotaxon*, **22**, 1 − 12.

Coates, D. & Rayner, A. D. M. (1985a). Heterokaryon − homokaryon interactions in *Stereum hirsutum*. *Transactions of the British Mycological Society*, **84**, 637 − 645.

Coates, D. & Rayner, A. D. M. (1985b). Evidence for a cytoplasmically transmissible factor affecting recognition and somato-sexual differentiation in the basidiomycete *Stereum hirsutum*. *Journal of General Microbiology*, **131**, 207 − 219.

Coates, D. & Rayner, A. D. M. (1985c). Genetic control and variation in expression of the 'bow-tie' reaction between homokaryons of *Stereum hirsutum*. *Transactions of the British Mycological Society*, **84**, 191 − 205.

Coates, D., Rayner, A. D. M. & Todd, N. K. (1981). Mating behaviour, mycelial antagonism and the establishment of individuals in *Stereum hirsutum*. *Transactions of the British Mycological Society*, **76**, 41 − 51.

Cooke, R. C. & Rayner, A. D. M. (1984). *Ecology of Saprotrophic Fungi*. London: Longman.

Jahn, H. (1971). Stereoide pilze in Europa (Stereaceae Pil. emend. Parm. u.a., Hymenochaete). *Westfälische Pilzbriefe*, **8**, 69 − 176.

Jülich, W. & Stalpers, J. A. (1980). *The Resupinate Non-poroid Aphyllophorales of the Temperate Northern Hemisphere*. Amsterdam: North-Holland Publishing Co.

Lemke, P. A. (1969). A reevaluation of homothallism, heterothallism and the species concept in *Sistotrema brinkmanii*. *Mycologia*, **61**, 57 − 76.

Lentz, P. L. (1955). *Stereum and Allied Genera of Fungi in the Upper Mississippi Valley*. Agriculture Monograph No. 24. United States Department of Agriculture.

Macrae, R. (1967). Pairing incompatibility and other distinctions among *Hirschioporus* (*Polyporus*) *abietinus*, *H. fusco-violaceus*, and *H. laricinus*. *Canadian Journal of Botany*, **45**, 1371 − 1398.

Perkins, D. D. & Barry, E. G. (1977). The cytogenetics of *Neurospora*. *Advances in Genetics*, **19**, 133 − 285.

Raper, J. R. (1966). *Genetics of Sexuality in Higher Fungi*. New York: The Ronald Press.

Rayner, A. D. M., Coates, D., Ainsworth, A. M., Adams, T. J. H., Williams, E. N. D. & Todd, N. K. (1984). The biological consequences of the individualistic mycelium. In *The Ecology and Physiology of the Fungai Mycelium*, British Mycological Society Symposium volume 8, ed. Jennings, D. H. & Rayner, A. D. M., pp. 509 − 540. Cambridge University Press.

Rayner, A. D. M. & Turton, M. N. (1982). Mycelial interactions and population structure in the genus *Stereum*: *S. rugosum*, *S. sanguinolentum* and *S. rameale*. *Transactions of the British Mycological Society*, **78**, 483 − 493.

Todd, N. K. & Aylmore, R. C. (1985). Cytology of hyphal interactions and reactions in *Schizophyllum commune*. In *Developmental Biology of Higher Fungi*, British Mycological Society Symposium volume 10, ed. Moore, D., Casselton, L. A., Wood, D. A. & Frankland, J. C., pp. 231 − 248. Cambridge University Press.

Ullrich, R. C. & Raper, J. R. (1975). Primary homothallism − relation to heterothallism in the regulation of sexual morphogenesis in *Sistotrema*. *Genetics*, **80**, 311 − 321.

Welden, A. L. (1971). An essay on *Stereum*. *Mycologia*, **63**, 790 − 799.

20
Breeding units in the forest pathogens *Armillaria* and *Heterobasidion*

KARI KORHONEN

Finnish Forest Research Institute, PO Box 18, SF-01301 Vantaa, Finland

Introduction

The agaric genus *Armillaria* includes a number of important root-rot fungi of woody plants. Another prominent root-rotter is the polypore *Heterobasidion annosum*. Although not closely related, these fungi share several characteristics. They are both saprophytes, but also show various degrees of pathogenicity. They both have a wide host range, a very wide geographical distribution and are of great economic importance.

Until the end of the 1970s, almost all the fungi belonging to the genus *Armillaria* were called *A. mellea* in the forest pathology literature. Although a number of *Armillaria* species had been described by taxonomists, most of them were not adopted for general use owing to difficulties in identification. *H. annosum*, in spite of its considerable variability, is still regarded as being a single taxonomic species today.

The unravelling the mating systems of *Armillaria* and *H. annosum* in the mid-1970s made it possible to investigate their breeding behaviour and to identify natural or biological species within them. This knowledge has laid the basis for a new taxonomy. In forest pathology the new species concepts have, in several cases, provided a reasonable explanation for the seemingly enigmatic ecological behaviour of these pathogens. Moreover, recent cytological and genetical investigations of these fungi have revealed new information about nuclear behaviour and intersterility mechanisms in Hymenomycetes. These studies are reviewed in this chapter.

Armillaria

There have been many different opinions and much confusion regarding the nomenclature and taxonomy of *Armillaria* (Watling, Kile & Gregory, 1982). Altogether, more than 50 species have been described.

Most species within the genus have an annulus and this group is often called the '*A. mellea* complex' (Singer, 1975). The group of exannulate species has only a few members, and includes *A. tabescens*, a wood-inhabiting fungus found in the Northern Hemisphere, and *A. ectypa*, a rare European species growing in *Sphagnum* bogs and ecologically quite different from the other species of the genus. In broader classification terms the genus *Armillaria* is traditionally included in Tricholomatales, and sometimes even combined with the genus *Clitocybe*. These links have recently been questioned or rejected (Peabody & Motta, 1979; Bennell, Watling & Kile, 1985).

Two prominent characteristics separate the members of *Armillaria* from other Hymenomycetes: rhizomorphs and diploidy in the vegetative stage. The rhizomorphs of *Armillaria* are fast-growing, highly differentiated, hyphal associations which promote effective vegetative spread. The occurrence of vegetative diploidy was discovered only recently.

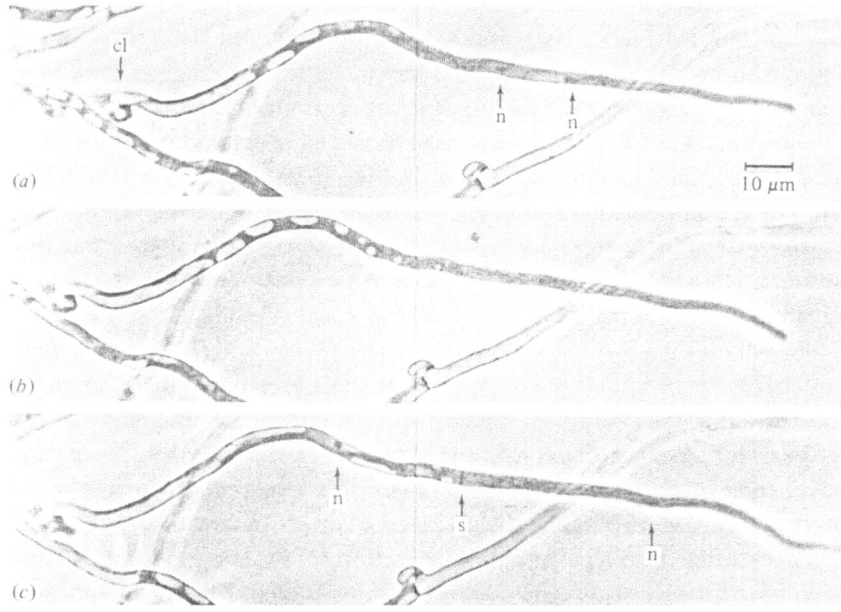

Fig. 20.1. Somatic diploidisation in the dikaryotic tip cell of *Armillaria cepistipes*, as seen by phase contrast microscopy. (a) Before division, showing two haploid nuclei close to each other; (b) 20 minutes later the same nuclei have become invisible as they undergo karyogamy and mitosis; (c) after 60 minutes, two uninucleate diploid cells have been formed. cl = clamp connection; n = nucleus; s = septum; arrows indicate the nucleoli. The nucleoplasm can be seen as a light area around the nucleolus.

Nuclear cycle and sexuality

The apparent absence of clamp connections in the mycelium of *Armillaria* is probably the main reason why the sexual mechanism remained unknown until Hintikka (1973), on the basis of the external appearance of the mycelia, identified a distinct tetrapolar mating pattern. Since compatible matings produced mycelia with uninucleate tip cells, he suspected that their nuclei were diploid. Subsequent cytological investigations revealed a dikaryotic phase with clamp connections in compatible matings (Fig. 20.1). However, the two nuclei of the dikaryon soon fused in the tip cells, giving rise to monokaryotic diploid hyphae (Korhonen & Hintikka, 1974). Supporting genetical evidence for diploidy was reported by Ullrich & Anderson (1978). Diploidy has also been confirmed by DNA measurements (Franklin, Filion & Anderson, 1983; Peabody & Peabody, 1984).

Another stage where clamp connections can be found is the fruit body. The basidia of most species have a basal clamp connection, whereas the basidia of other species are clampless. This small cytological difference seems to relate to a difference in the nuclear cycle (Fig. 20.2). Clampless basidia develop from uninucleate subhymenial cells with diploid nuclei, whereas clamped basidia arise from binucleate cells with haploid nuclei, karyogamy occurring in the basidium (Korhonen, 1980; Peabody & Peabody, 1984; Lamoure & Guillaumin, 1985). The origin of these haploid nuclei is an enigma: either some haploid nuclei persist after

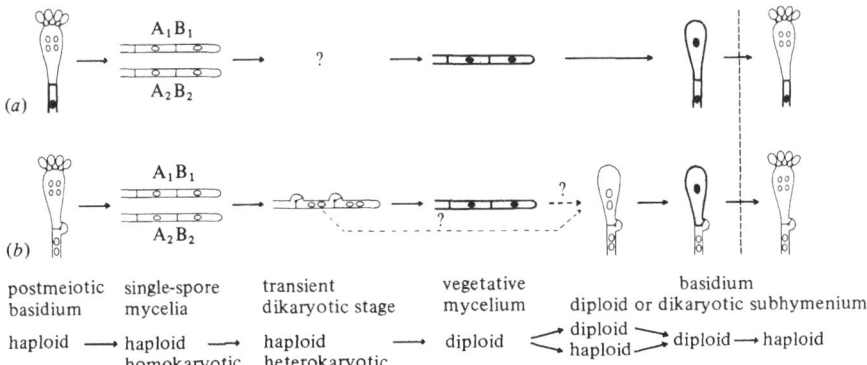

Fig. 20.2. The nuclear cycle of *Armillaria* species which have a diploid subhymenium (a), and those which have a dikaryotic subhymenium (b). The former cycle includes one diploidisation and one haploidisation; the latter, possibly two of each of these events. The diploid stages are shown in heavy outline.

mating, or haploidisation of diploid nuclei occurs before fruiting. According to Tommerup & Broadbent (1975), haploidisation takes place in the gill initials of fruit body primordia. Recently, Peabody & Peabody (1985) have reported the widespread occurrence of haploidy among nuclei in the fruit body, which suggests that haploidisation is completed at some earlier stage of fruit body development. It is interesting to note that those species which possess a dikaryotic subhymenium, also form a distinct dikaryophase on mating. Such a dikaryophase has never been observed in matings of *A. mellea*, which has a diploid subhymenium (Fig. 20.2).

In the vegetative mycelium, haploidisation may occasionally occur in pairings between diploid and haploid mycelia, as indicated by the recombination of incompatibility factors (Guillaumin, 1986). Haploidisation via aneuploidy can also be induced artificially with the aid of benomyl (Anderson, Petsche & Franklin, 1985).

Breeding units

Mating experiments have revealed seven intersterility groups within *Armillaria* in Europe. Five of them belong to the *A. mellea* complex, and two, *A. tabescens* and *A. ectypa*, are exannulate. Most of these intersterility groups have fitted quite easily into those species described earlier by taxonomists, whilst others have caused considerable difficulties (for review see Roll-Hansen, 1985; Guillaumin, 1986). As regards their pathological properties, *A. mellea* (*sensu stricto*) is the species which causes most damage to broadleaved trees, especially in fruit orchards and vineyards. *A. obscura* (syn. *A. ostoyae*) is a pathogen of conifers; while *A. bulbosa*, *A. cepistipes* and *A. borealis* are weak pathogens only, the two last named being northerly in distribution (Rishbeth, 1985a & b; Gregory & Watling, 1985; Guillaumin *et al.*, 1985; Roll-Hansen, 1985). In Europe, *A. tabescens* appears to be almost non-pathogenic except towards some exotics (Guillaumin, 1986).

Ten intersterility groups have so far been reported within *Armillaria* in North America: nine annulate groups (Anderson & Ullrich, 1979; Guillaumin *et al.*, 1985; Morrison, Chu & Johnson, 1985) and the exannulate *A. tabescens* (Anderson, 1982). The European species *A. bulbosa*, *A. mellea* and *A. obscura* have their counterparts in North America, and within each of these species interfertility across the Atlantic appears to be practically complete. The breeding relationship between the European *A. cepistipes* and some North American groups is still unclear. *A. borealis* has not been found in North America and at least three unnamed North American annulate groups apparently have no counterparts in Europe (Anderson, Ullrich & Korhonen, 1980; Morrison *et al.*, 1985; Wargo & Shaw, 1985). The extent of interfertility between European and North American strains of *A. tabescens* is unknown.

The taxonomic species described from Australia (Kile & Watling, 1983) and New Zealand seem to correspond with the intersterility groups. Five annulate species have been reported from this area. At least four of them are true biological species and all four have proved to be intersterile with the various European and North American groups (Guillaumin, 1986). Until now, the breeding relationships of *Armillaria* isolates from other parts of the world, such as Asia, Africa and South America, have been investigated only sporadically. Consequently, our knowledge of the number and distribution of the different *Armillaria* species in the world is still far from complete.

All *Armillaria* species so far investigated have a tetrapolar incompatibility system. However, the European *A. ectypa* may be an exception; preliminary observations suggest that it is homothallic (Guillaumin, 1986). Whether it is diploid or haploid is unknown.

Heterobasidion annosum

The genus *Heterobasidion* is often included in the heterogeneous family Poriaceae, but according to recent investigations it is most closely related to Bondarzewiaceae (Gluchoff-Fiasson, David & Dequatre, 1983). Two prominent characteristics of the genus are the occurrence of asperulate and amyloid basidiospores and the ability to produce asexual spores, or conidia. Besides *H. annosum* the genus also includes *H. insulare* which is found in Southern and Eastern Asia and is very similar to *H. annosum* but non-pathogenic (Hood, 1985).

Nuclear cycle and sexuality

The cells in the hyphae of *H. annosum* are multinucleate and most septa are simple. Clamp connections can be found in varying frequency, depending on the strain. The presence of clamp connections appears to be a reliable indication of heterokaryosis in both European and North American strains in which the mating system has been shown to be bipolar (Fig. 20.3; Korhonen, 1978; Chase & Ullrich, 1983; Holt, Gockel & Hüttermann, 1983). Strains from Australia, New Zealand and Fiji are different: mycelia of single-spore origin have clamp connections and fruit readily. Those isolated from the same basidiome appear morphologically identical and when paired, do not show mating reactions. This indicates a homothallic sexual system (Chase, Ullrich & Korhonen, 1985).

Breeding units and pathological behaviour

The pathological behaviour of *H. annosum* in Europe is sometimes strikingly variable and this has long been a cause for suspecting the
existence of different races within the fungus. Thus, in South-East
Finland, *H. annosum* is a serious pathogen of Scots pine, yet it does not
generally attack this tree species in the western part of the country, where
it commonly causes butt rot of Norway spruce. In mating experiments, the
south-eastern form proved to be genetically isolated from the dominant
form in the spruce forests of western Finland (Korhonen, 1978). Based on
their principal hosts, these two groups of *H. annosum* have been called,
preliminarily, the P (pine) and S (spruce) groups. The ecological
differences between them are so significant that they must have important
consequences for practical disease control. The S group causes butt rot of
Norway spruce and may kill some Scots pine saplings growing near
infected spruce stumps, but is rare on other native tree species in Finland.
Group P has a broader host range, including pine of all ages, juniper,
spruce, birch and sometimes other broadleaved trees. In Finland it attacks
spruce to a much lesser extent than group S (Korhonen, 1978).

Thus, in North Europe at least, *H. annosum* consists of two biological
species. Small morphological differences have also been found between
them, suggesting that they may possibly be regarded as representing
different taxonomic species. However, the situation may not be that simple
in other parts of the world. In Southern Europe, for instance, there may be
another breeding group of *H. annosum*, close to the S group but only
partially interfertile with it (F. Moriondo, personal communication).

The S and P groups have also been identified from North America and

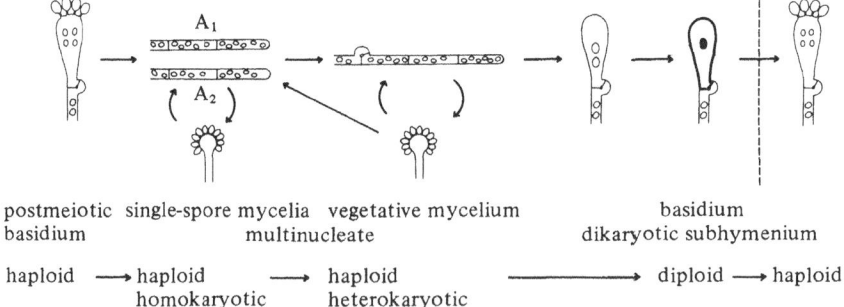

postmeiotic single-spore mycelia vegetative mycelium   basidium
basidium   multinucleate   dikaryotic subhymenium

haploid ⟶ haploid   ⟶ haploid   ⟶ diploid ⟶ haploid
            homokaryotic   heterokaryotic

Fig. 20.3. The nuclear cycle of heterothallic strains of *Heterobasidion
annosum*.

recent investigations by Chase (1986) have thrown light on the mechanism which regulates interfertility between them. Working with North European and North American isolates, Chase has identified five specific 'intersterility' genes, each with two alleles. Interfertility occurs when two isolates being paired are homoallelic for the + allele of at least one of the five loci. That the North European S and P isolates are not homoallelic at any locus accounts for the near complete intersterility between them. In North America, various combinations of intersterility alleles are found and a high degree of interfertility exists between S and P groups.

Current knowledge of the world-wide distribution of these intersterility groups is weak. Both S and P groups occur in central Europe and the P group also appears to be dominant in Norway spruce stands there (C. Delatour, R. Siepmann, personal communications). Pure cultures, most probably belonging to the S group, have been identified from India and Japan (Korhonen, 1978). The homothallic isolates from Australia, New Zealand and Fiji are intersterile with the European and North American strains. This intersterility, together with the occurrence of homothallism, differences in morphology (Hood, 1985), and weak pathogenicity suggest that this fungus is taxonomically separate from *H. annosum*. However, hybrid heterokaryons between homothallic and heterothallic strains can be produced by nutritional forcing (Chase *et al.*, 1985). This may indicate a recent derivation of one group from the other. The interfertility relationships of *H. insulare* are unknown.

Evolutionary considerations

It seems that speciation is currently occurring within the genus *Heterobasidion*. The genetic mechanism regulating interfertility between *H. annosum* subunits, as revealed by Chase (1986), may provide new insights into speciation processes among the Basidiomycetes.

The genus *Armillaria* appears to be an exceptionally interesting group of fungi from the point of view of evolution. Two fascinating features of *Armillaria* are the occurrence of diploidy in the vegetative stage and the existence of two types of nuclear cycle, one with a dikaryotic and the other with a diploid subhymenium. Whether the life cycle of those species with a dikaryotic subhymenium includes two haploidisations is an open question at present. There is some evidence for the occurrence of haploidisation in fruit body primordia, although the mechanism is unknown and the genetical significance of such a system is hard to understand. If this putative haploidisation does not occur, then some haploid nuclear pairs must be maintained in the mycelium after mating and the diploid hyphae would be, effectively, a genetic *cul de sac*. Haploidisation may, however, occur at a low rate in the diploid mycelium.

The ancestors of *Armillaria* had, presumably, a dikaryotic vegetative mycelium. Their present diploidy may represent an extended diploid phase of a parasexual cycle. Most *Armillaria* species in Europe still have a visible dikaryotic phase both at the beginning and end of the vegetative diploid phase. By comparison with these types, *A. mellea sensu stricto* may represent a more advanced type where the dikaryophase has disappeared completely from the fruit bodies (Guillaumin, 1986). However, the fact that *A. obscura* has the 'more advanced' cycle in the laboratory and the 'primitive' cycle under natural conditions casts some doubt over such an evolutionary sequence.

Among Hymenomycetes, *Armillaria* is so far the only group which has been shown to have a diploid mycelium in nature. Other stable diploids in the Hymenomycetes have been obtained only in the laboratory. Among them, the diploid *dik⁻* genotype of *Schizophyllum commune* closely resembles *Armillaria* (Koltin & Raper, 1968). Anderson & Ullrich (1982) have suggested that similar diploids may arise in natural populations of *S. commune*, but that they are selected against in competition with dikaryons.

Why, therefore, has diploidy been so selectively advantageous in *Armillaria*? Is it necessary for the development of highly differentiated vegetative organs like rhizomorphs? How old a characteristic is it on the time scale of evolution? Does this evolutionary experiment, the diploid Hymenomycetes, represent a declining, static or expanding group of fungi? Among Agaricales, the genus *Armillaria* seems to be quite solitary, without any closely related groups. Thus, the genus must be relatively old. As *Armillaria* has cutinised rhizomorphs, one would expect their preservation as fossils. In fact, rhizomorph-like fossils have been found, the oldest from the Carboniferous period, but their true origin seems to be very uncertain (Pirozynski, 1976).

Certainly, from an evolutionary standpoint, *Armillaria* can presently be regarded as being a very successful group of fungi. Members of the genus have spread to most parts of the world. As wood destroyers they are leaders among the Agaricales, and in many places their rhizomorphs form an important component of the forest floor. There are many closely related species with different nuclear cycles within the genus, suggesting a recently active period of evolution. In contrast to *Heterobasidion*, however, genetical isolation between intersterility groups of *Armillaria* within continents seems to be complete, even between those groups which, externally, are still almost impossible to distinguish from each other.

There are a number of open questions concerning *Armillaria*. While future investigators will provide answers to many of them, one is likely to remain unanswered: could the ecological success of *Armillaria* be a starting point for a larger group of diploid agarics in the distant future?

Acknowledgements I wish to thank Teuvo Ahti, Clive Brasier, Thomas Chase, Veikko Hintikka, Ian Hood, Lalli Laine, Francesco Moriondo and Anne Sairanen for their help during preparation of this chapter.

References

Anderson, J. B. (1982). Bifactorial heterothallism and vegetative diploidy in *Clitocybe tabescens*. *Mycologia*, **74**, 911 – 916.

Anderson, J. B., Petsche, D. M. & Franklin, A. L. (1985). Nuclear DNA content of benomyl-induced segregants of diploid strains of the phytopathogenic fungus *Armillaria mellea*. *Canadian Journal of Genetics and Cytology*, **27**, 47 – 50.

Anderson, J. B. & Ullrich, R. C. (1979). Biological species of *Armillaria mellea* in North America. *Mycologia*, **71**, 402 – 414.

Anderson, J. B. & Ullrich, R. C. (1982). Diploids of *Armillaria mellea*: synthesis, stability, and mating behavior. *Canadian Journal of Botany*, **60**, 432 – 439.

Anderson, J. B., Ullrich, R. C. & Korhonen, K. (1980). Relationships between European and North American biological species of *Armillaria mellea*. *Experimental Mycology*, **4**, 87 – 95.

Bennell, A. P., Watling, R. & Kile, G. (1985). Spore ornamentation in *Armillaria* (Agaricales). *Transactions of the British Mycological Society*, **84**, 447 – 455.

Chase, T. E. (1986). Genetics of sexuality and speciation in the fungal forest pathogen *Heterobasidion annosum*. *Ph. D. thesis, University of Vermont*.

Chase, T. E. & Ullrich, R. C. (1983). Sexuality, distribution and dispersal of *Heterobasidion annosum* in pine plantations of Vermont. *Mycologia*, **75**, 825 – 831.

Chase, T. E., Ullrich, R. C. & Korhonen, K. (1985). Homothallic isolates of *Heterobasidion annosum*. *Mycologia*, **77**, 975 – 977.

Franklin, A. L., Filion, W. G. & Anderson, J. B. (1983). Determination of nuclear DNA content in fungi using mithramycin: vegetative diploidy in *Armillaria mellea* confirmed. *Canadian Journal of Microbiology*, **29**, 1179 – 1183.

Gluchoff-Fiasson, K., David, A. & Dequatre, B. (1983). Contribution à l'étude des affinités entre *Heterobasidion annosum* (Fr.) Bres. et les Bondarzewiaceae. *Cryptogamie, Mycologie*, **4**, 135 – 143.

Gregory, S. C. & Watling, R. (1985). Occurrence of *Armillaria borealis* in Britain. *Transactions of the British Mycological Society*, **84**, 47 – 55.

Guillaumin, J. J., Lung, B., Romagnesi, H., Marxmüller, H., Lamoure, D., Durrieu, G., Berthelay, S. & Mohammed, C. (1985). Systematique des Armillaires du groupe Mellea. Consequences phytopathologiques. *European Journal of Forest Pathology*, **15**, 268 – 277.

Guillaumin, J. J. (1986). Contribution à l'étude des Armillaires phytopathogenes, en particulier du groupe Mellea: cycle caryologique, notion d'espèce, role biologique des espèces. *Doctoral thesis, Université Claude Bernard – Lyon I*.

Hintikka, V. (1973). A note on the polarity of *Armillariella mellea*. *Karstenia*, **13**, 32 – 39.

Holt, C. E., Gockel, H. & Hüttermann, A. (1983). The mating system of *Fomes annosus* (Heterobasidion annosum). *European Journal of Forest Pathology*, **13**, 174 – 181.

Hood, I. A. (1985). Pore width in *Heterobasidion annosum* (Fries) Brefeld. *New Zealand Journal of Botany*, **23**, 495 – 498.

Kile, G. A. & Watling, R. (1983). *Armillaria* species from south-eastern Australia. *Transactions of the British Mycological Society*, **81**, 129 – 140.

Koltin, Y. & Raper, J. R. (1968). Dikaryosis: genetic determination in *Schizophyllum*. *Science*, **160**, 85 – 86.

Korhonen, K. (1978). Intersterility groups of *Heterobasidion annosum*. *Communicationes Instituti Forestalis Fenniae*, **94 (6)**, 1 – 25.

Korhonen, K. (1980). The origin of clamped and clampless basidia in *Armillariella ostoyae*. *Karstenia*, **20**, 23 – 27.

Korhonen, K. & Hintikka, V. (1974). Cytological evidence for somatic diploidization in dikaryotic cells of *Armillariella mellea*. *Archiv für Mikrobiologie*, **95**, 187 – 192.

Lamoure, D. & Guillaumin, J. J. (1985). Le cycle caryologique des Armillaires du groupe mellea. *European Journal of Forest Pathology*, **15**, 288 – 293.

Morrison, D. J., Chu, D. & Johnson, A. L. S. (1985). Species of *Armillaria* in British Columbia. *Canadian Journal of Plant Pathology*, **7**, 242 – 246.

Peabody, D. C. & Motta, J. J. (1979). The ultrastructure of nuclear division in *Armillaria mellea*: meiosis I. *Canadian Journal of Botany*, **57**, 1860 – 1872.

Peabody, D. C. & Peabody, R. B. (1984). Microspectrophotometric nuclear cycle analyses of *Armillaria mellea*. *Experimental Mycology*, **8**, 161 – 169.

Peabody, D. C. & Peabody, R. B. (1985). Widespread haploidy in monokaryotic cells of mature basidiocarps of *Armillaria bulbosa*, a member of the *Armillaria mellea* complex. *Experimental Mycology*, **9**, 212 – 220.

Pirozynski, K. A. (1976). Fossil fungi. *Annual Review of Plant Pathology*, **14**, 237 – 246.

Rishbeth, J. (1985a). Infection cycle of *Armillaria* and host response. *European Journal of Forest Pathology*, **15**, 332 – 341.

Rishbeth, J. (1985b). *Armillaria*: resources and hosts. In *Developmental Biology of Higher Fungi*, British Mycological Society Symposium Volume 10, ed. Moore, D., Casselton, L. C., Wood, D. A. & Frankland, J. C., pp. 87 – 101. Cambridge University Press.

Roll-Hansen, F. (1985). The *Armillaria* species in Europe. *European Journal of Forest Pathology*, **15**, 22 – 31.

Singer, R. (1975). *Agaricales in Modern Taxonomy*. 3rd edition. Vaduz: J. Cramer.

Tommerup, I. C. & Broadbent, D. (1975). Nuclear fusion, meiosis and the origin of dikaryotic hyphae in *Armillariella mellea*. *Archiv für Mikrobiologie*, **103**, 279 – 282.

Ullrich, R. C. & Anderson, J. B. (1978). Sex and diploidy in *Armillaria mellea*. *Experimental Mycology*, **2**, 119 – 129.

Wargo, P. M. & Shaw, C. G. (1985). *Armillaria* root rot: the puzzle is being solved. *Phytopathology*, **69**, 826 – 832.

Watling, R., Kile, G. A. & Gregory, N. M. (1982). The genus *Armillaria* – nomenclature, typification, the identity of *Armillaria mellea* and species differentiation. *Transactions of the British Mycological Society*, **78**, 271 – 285.

21

Genetic variation and evolution in *Aspergillus*

J. H. CROFT

Department of Genetics, University of Birmingham, PO Box 363, Birmingham B15 2TT, UK

Introduction

The study of evolution depends to a large extent upon a knowledge of the degree of phylogenetic relationship between the various strains or species under study. The study of taxonomy, on the other hand, being essentially one of classification, does not of necessity require that the degree of genetic relationship between the strains is understood. Nevertheless, most taxonomists would probably express the hope that any system of classification would have an underlying genetic basis.

The genus *Aspergillus* has long had one of the better taxonomic descriptions found among the fungi (Raper & Fennell, 1965) and it is inevitable that such a taxonomic structure should give rise to hypotheses concerning the evolution of the genus. The taxonomic structure proposed by Raper and Fennell implies a divergent evolution from a common ancestor into genetically isolated lines giving some 18 main species groups, with the latter having further diverged to give rise to the currently described 200 or more species or varieties. On this evolutionary model, the members of each species group would be expected to be closely related to each other, possibly as sibling species. Evidence for further differentiation within each species to give genetically distinct heterokaryon compatibility groups suggests that divergent evolution is still progressing (Croft & Jinks, 1977). However, such an essentially hierarchical model (see Fig. 1 in Croft & Dales, 1984) may not adequately reflect evolution in the genus. For example, genetic exchange between what on present grounds might appear to be quite unrelated gene pools (that is, between individuals from different species groups) may have occurred resulting in a more complex network of genetic relationships.

Underlying this problem is the question of how accurately characters chosen for taxonomic utility reflect the true genetic relationship between strains. For example, a spore colour difference used to separate two species groups could be due to a single mutation. Such a fundamental problem could be circumvented by ensuring that the characters under study reflect underlying genetic relationships. This can be best achieved either by the study of polymorphisms in the genetic material itself, or the study of characters whose genetic control can be clearly demonstrated. In the following account the possible microevolutionary relationships within and between taxa in the 'nidulans' species group of *Aspergillus* will be assessed from the standpoint of such a genetical and molecular approach.

Genetical methods
Intraspecific variation
There is considerable natural genetic variation within individual species of *Aspergillus*. This has been most thoroughly investigated in samples of the British population of *A. nidulans*, where studies of characters in wild type isolates under different environmental conditions (Butcher, Croft & Grindle, 1972) and of the segregation of many single locus and polygenically controlled characters among sexual progenies have been conducted (see Croft & Jinks, 1977, for review). One of the principal results of these analyses has been the demonstration that *A. nidulans* comprises a number (at least 19) of separate heterokaryon compatibility groups. Heterokaryon incompatibility is under heterogenic control with an allelic difference at any one of at least eight specific *het* genes being sufficient to prevent heterokaryon formation between a pair of strains (Grindle, 1963a & b; Croft, 1985). Analysis of genetic variation for other characters shows that the members of any one heterokaryon compatibility group are so similar that it is clear that they are clonally related and probably are asexual derivatives of a common ancestor which have become widely dispersed, presumably *via* conidia (Croft & Jinks, 1977). The various combinations of alleles at the *het* loci found in the naturally occurring heterokaryon compatibility groups clearly suggest that the population has been derived from the progeny of a number of sexual crosses between pre-existing strains. To determine if this occurred early in the evolution of the species, or is a continuing process, would require further sampling on a much wider geographical basis and also a re-sampling of the British population. It is of interest to note that the structure of the population of some imperfect species of *Aspergillus* appears to be similar to that of the sexual *A. nidulans* (Caten, 1971) but there is a need for further data in this area. Nevertheless, the differentiation of the

population of a single species of *Aspergillus* into genetically distinct, sexually interfertile, but heterokaryon incompatible groups of strains efficiently dispersed by asexual propagules could clearly lead to their continued divergence into genetically isolated sibling species (see also Brasier, Chapter 16).

Another observation which may be of significance is that approximately 10% of wild isolates of *A. nidulans* are diploid (Upshall, 1981; Caten & Howell, unpublished). Most of these diploids appear to be fully homozygous, but one was found to be heterozygous for a temperature sensitive mutation. Haploid strains carrying this allele were also found among wild isolates. This suggests that parasexual events may be of importance in nature, at least within a heterokaryon compatibility group.

Interspecific genetic analysis

Although sexual hybridisation is possible between some pairs of species from the 'nidulans' group, the crosses are of very low viability and give rise to progenies consisting of aneuploids and allodiploids (Tyc, 1968; Croft, unpublished). Another feature of such crosses is that intrachromosomal recombination has not been detected among the progenies. As far as the nuclear genome is concerned, the allodiploid represents a full hybrid between the two species and, as a diploid strain, is amenable to parasexual analysis following haploidisation either spontaneously or by treatment with haploidising agents such as benomyl. Although allodiploids may be produced by sexual hybridisation of some pairs of closely related species, the most consistent way to obtain them between the wider range of species combinations within the 'nidulans' group is by protoplast fusion (see Croft & Dales, 1983), though it should be pointed out that not all combinations of 'nidulans' group species will produce hybrids by the use of this technique (Kevei & Peberdy, 1984). Subsequent parasexual studies of the allodiploids produced in this way have permitted considerable analysis of the genetic differences between the pairs of species (Croft & Dales, 1983). Thus, for example, in the case of the *A. nidulans* plus *A. nidulans* var. *echinulatus* hybridisation, the presence of complex translocations in *A. nidulans* var. *echinulatus* involving the chromosomes equivalent to the *A. nidulans* linkage groups III, V and VIII has been demonstrated. More detailed genetic analysis has revealed other structural differences between the two species and has also shown that they differ at one *het* gene on each of the chromosome pairs equivalent to the *A. nidulans* linkage groups II, VI and VII and at least at one *het* gene on the III − V − VIII complex.

Molecular methods

Though the above methods of genetic analysis have produced considerable information about phylogenetic relationships both within and between species of the 'nidulans' group, the use of these methods appears to be possible only where the genetic relationship is very close. If the genetic distance between a pair of species is beyond a certain level, no such general genetic analysis is possible, though recent gene cloning and gene manipulation methods (Ballance & Turner, 1985; Tilburn *et al.*, 1984) will allow the study of individual genes and their controlling sequences in different genetic backgrounds. This should contribute greatly to the understanding of the evolution of individual genes, or of groups of genes of related function. However, one approach which can yield considerable information concerning phylogenetic relationships is the study of DNA

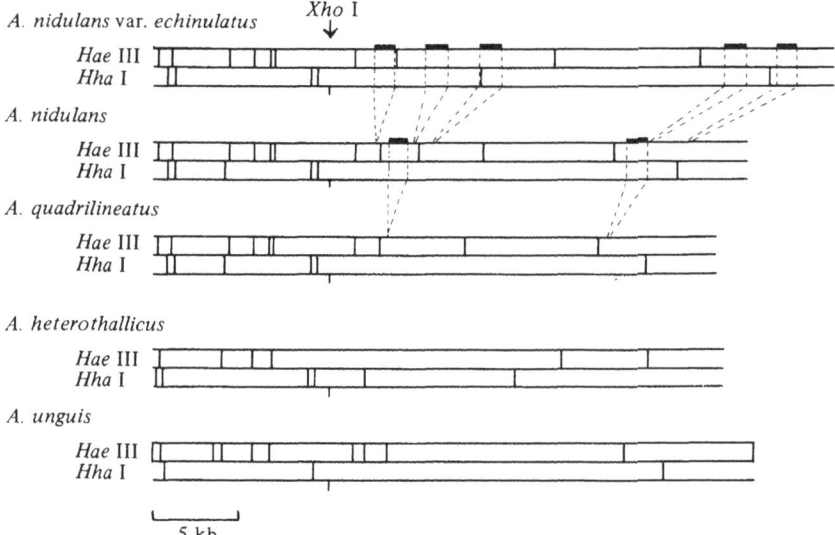

Fig. 21.1. Linearised restriction maps of the circular mitochondrial genomes of *Aspergillus nidulans* var. *echinulatus*, *A. nidulans*, *A. quadrilineatus*, *A. heterothallicus* and *A. unguis*. For simplicity, results for only two restriction enzymes, *Hae* III and *Hha* I or *Cfo* I, are given. In all cases the maps are drawn in relation to a unique *Xho* I site found in all of these species in a gene for tRNA^{phe}. The first three species are very closely related with few restriction site polymorphisms. They differ in size because of the presence of absence of several optional introns, the positions of which are indicated by solid bars connected by vertical broken lines. The other two species are less closely related to *A. nidulans* and to each other and they contain many more restriction site polymorphisms. The intron content of these species is not known.

sequence variation, and in particular the analysis of restriction fragment length polymorphisms (RFLPs) in both mitochondrial (mt) and nuclear DNA (and see Barrett, Chapter 6).

Mitochondrial DNA

Most information concerning DNA sequence variation in *Aspergillus* is available from studies of mtDNA. The mitochondrial genome of *A. nidulans* has been sequenced almost completely (Waring, Davies *et al.*, 1981; Köchel *et al.*, 1981; Grisi *et al.*, 1982; Netzker *et al.*, 1982; Brown, Davies *et al.*, 1983; Waring, Brown *et al.*, 1984; Brown, Waring *et al.*, 1985) providing a useful comparison with the genomes of other *Aspergillus* species and a potentially valuable contribution to the study of phylogeny at the microevolutionary level.

To date, no intraspecific sequence variation has been detected in the mt genome of *Aspergillus*. Thus, in isolates of *A. nidulans* collected between 1939 and 1981 and from Britain, Europe and North America, ten strains so far investigated have yielded identical restriction maps for up to eight restriction enzymes. Similar results have been found in *A. nidulans* var. *echinulatus*. This result is perhaps unexpected and is certainly in contrast to the amount of polymorphism found in *Neurospora* Taylor, Smolich & May, 1986).

The mt genomes of the different taxonomic species in the 'nidulans' group each has a quite distinct restriction site map. Detailed analyses of these differences show several causes for this variation and it is clear that the 'nidulans' group contains several very closely related species and others where the relationship is more remote (Fig. 21.1).

Closely related species Many pairs of species in the 'nidulans' group will hybridise and produce viable allodiploid products following protoplast fusion (Kevei & Peberdy, 1984). It is also possible to transfer and to recombine mt genomes between these same species (Earl *et al.*, 1981; Croft & Dales, 1984). The mt genomes of these species vary in size from about 30 to 40 kilobase pairs (kb) and the majority of the restriction sites are common to all species, there being just a few species-specific sites in most cases. The major difference between them, and cause of the size variation, is the presence or absence of optional introns in certain regions of the genome (Fig. 21.2). The significance of this variation in intron content is not clear. Introns may be able to move about in the genome (Colleaux *et al.*, 1986) or possibly they are a feature of ancestral DNA and gradually became lost in different combinations during evolution (see

Scazzocchio, Chapter 4). This problem is further compounded by the interesting observation that the third intron of the cytochrome oxidase I gene in *A. nidulans* shows a very high sequence homology with the second intron of the equivalent gene in *Schizosaccharomyces pombe* (Lang, 1984; Waring *et al.*, 1984); further, both introns are located at exactly equivalent positions within the respective genes. In addition, the coding sequences around this intron show a lower homology between the two species than do the intronic sequences. This is an unusual result, since fungal introns do not normally share high sequence homology except for two conserved decapeptide regions (Scazzocchio, Chapter 4) and it is tempting to use this observation as evidence for the horizontal transfer of DNA sequences between the mt genomes of the two fungi.

It has also been suggested that, despite the control of somatic fusion events by the *het* system, rare somatic fusion events may be involved in the evolution of these closely related species (Croft & Dales, 1984). This suggestion is based on the observation that some recombinant mt DNA molecules obtained in the laboratory by protoplast fusion between pairs of species resemble the mt genomes of certain species in nature. As yet there is no independent experimental evidence to support this suggestion.

More distantly related species Several species within the 'nidulans' group have failed to produce any viable hybrid products with *A. nidulans* either at nuclear or mitochondrial level following protoplast fusion experiments (Kevei & Peberdy, 1984). Physical maps of the mt genomes of these species (*A. unguis*, *A. stellatus* and *A. heterothallicus*) show considerable variation (Jadayel, 1986). Estimates can be made from these maps of the

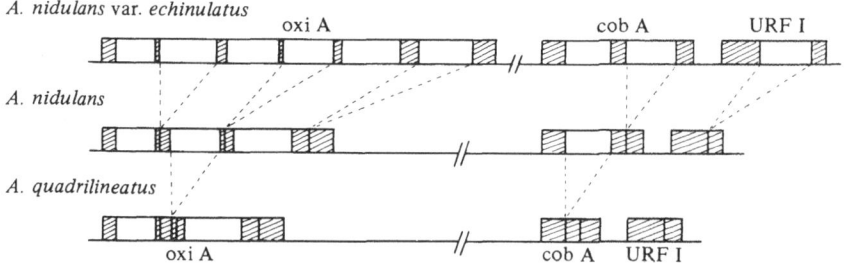

Fig. 21.2. The complex mosaic structure of *oxiA*, *cobA* and URF1 (the latter codes for a component of the respiratory chain NADH dehydrogenase) in *Aspergillus nidulans* var. *echinulatus*, *A. nidulans* and *A. quadrilineatus*. The exons, or coding sequences, are shown cross-hatched and the introns are unshaded. The relative positions of the nine introns are shown by the connected dotted lines.

Table 21.1. *Nucleotide substitutions in mitochondrial genomes: comparisons between* Aspergillus nidulans *and four other 'nidulans' group species*

Species compared with *Aspergillus nidulans*	Estimates (%) from restriction digests		Estimates (%) from sequence data
	4 bp enzymes	6 bp enzymes	
A. nidulans var. echinulatus	1.4 to 6		0.17
A. unguis	11.3	12.9	
A. heterothallicus	13.6	14.4	12.0
A. stellatus	16.6	14.4	

The estimates of % nucleotide substitutions were made from restriction digests involving enzymes which recognise 4 or 6 bp sequences as indicated; data obtained from direct sequencing of short regions (see text) are given for comparison.

proportion of nucleotide sites which are polymorphic in pairwise combinations of species (Table 21.1). However, these methods require many assumptions and carry many pitfalls (Upholt, 1977; Engels, 1981; and for review see Birley & Croft, 1986), not least of which is the likelihood that many restriction sites will be carried in optional introns, the presence or absence of which has yet to be determined in these species

A small amount of sequence data is available which allows direct comparisons of base substitution frequencies (Table 21.1). In the case of *A. nidulans* versus *A. nidulans* var. *echinulatus*, only five base substitutions were found ($0 \cdot 17\%$) in a sequence of 665 base pairs (bp) covering the second exon of the gene coding for *apo*-cytochrome *b* (*cobA*) and most of the common *cobA* intron (Spooner, 1984). This is a low frequency when compared with the estimates for the whole mt genome obtained from restriction map data. In the case of *A. nidulans* versus *A. heterothallicus*, a 124 bp sequence derived from *A. heterothallicus* has been compared to the equivalent region in *A. nidulans* (Jadayel, 1986). This sequence is located in the fourth exon of the *oxiA* gene of *A. nidulans*. Although there is a very high amino acid homology in the product of this region, there are 15 base substitutions (12%); a frequency in good agreement with estimates obtained from restriction map data. Eleven of these base substitutions occur in the third-base position of the codon and consequently only four of the 41 amino acids coded by this region are substituted.

Finally, in most cases where the physical order of the genes on the mt genome has been determined, it has been found to be similar to that in *A. nidulans*. The only striking example of a genome rearrangement occurs in *A. heterothallicus*, where the order of *oxiA*, *oxiB*, *cobA* and URFs 1 and 4 has been changed (Fig. 21.3).

Nuclear DNA

Some preliminary experiments have been carried out to test the possible use of cloned DNA probes for the determination of phylogenetic relationships in the 'nidulans' group of species using Southern blotting (Southern, 1975, 1979; and see Pukkila, Chapter 5; Barrett, Chapter 6). In our experiments, the probes were chosen at random from a genomic library of one isolate of *A. nidulans* cloned in the λ phage vector, EMBL3. The fragments cloned in this vector will be of the order of 17 to 23 kb in size. A small number of these clones were then individually labelled and used as probes against total nuclear DNA from a range of strains representing different species, mainly from the 'nidulans' group, digested with various restriction enzymes.

The random probes have proven to be very diagnostic at the level of the different species within the 'nidulans' group (Chapman & Croft, unpublished). Thus, for most probes and for most digests no RFLP is seen within a single species. In some combinations of probes and digests a small

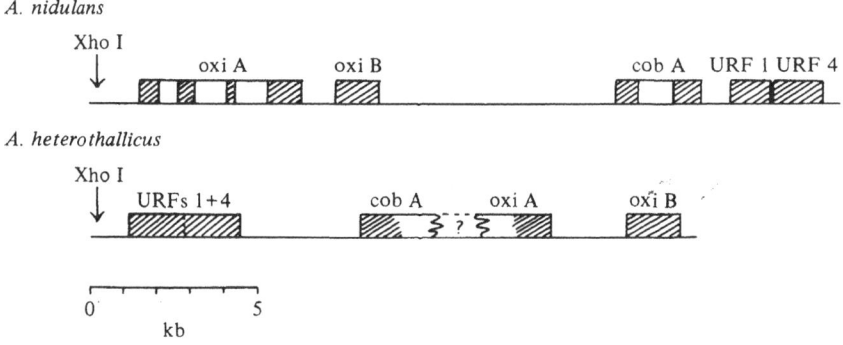

Fig. 21.3. A comparison of parts of the mitochondrial genomes of *Aspergillus nidulans* and *A. heterothallicus*, showing the structural rearrangements involved. The maps are aligned at the unique *Xho* I site located within the tRNA[phe] gene. The introns within the *A. nidulans oxiA* and *cobA* genes are shown unshaded. The intron content of these genes in *A. heterothallicus* is unknown but the amount of space available between the detected parts of these genes suggests that few or no introns may be present.

amount of polymorphism is seen and, as would be predicted from the genetic data, this is limited to differences between heterokaryon compatibility groups. No polymorphism was detected within a heterokaryon compatibility group (Fig. 21.4).

When the *A. nidulans* random clones were used to probe genomic DNA from other species of the 'nidulans' group, clear polymorphisms were revealed between species which, from the protoplast fusion and mitochondrial studies, were considered to be distinct but closely related. However, where the protoplast fusion and mitochondrial studies had indicated a distant relationship, such as between *A. nidulans* and *A. unguis*, the random probes used failed to hybridise at all under the conditions of stringency used in these experiments. The suspected distant relationship was thus confirmed, even though *A. unguis* is classified in the 'nidulans' group. The *A. nidulans* probes also failed to hybridise to DNA from *A. niger*. This is not unexpected since *A. niger* is not classified in the 'nidulans' group. However, by contrast, strong hybridisation was detected to DNA from *A. terreus* which also is not classified in the 'nidulans' group

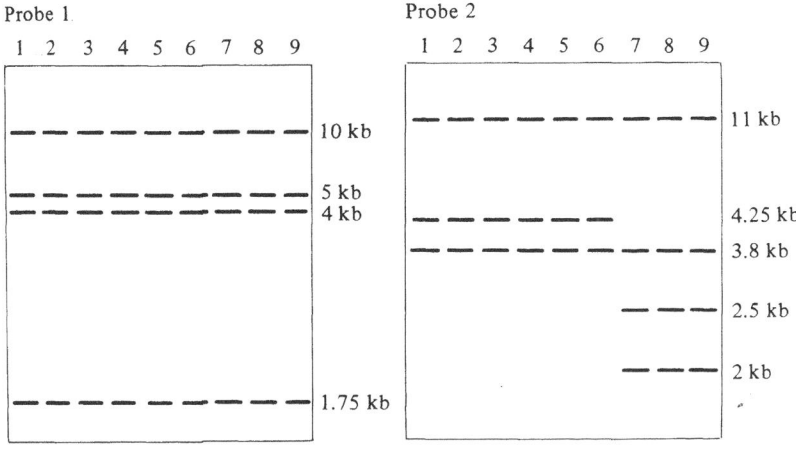

Fig. 21.4. Diagrammatic representations of the autoradiographs obtained by the hybridisation of two labelled random genomic clones (probes 1 and 2) to electrophoresed *Eco* RI digests of the total nuclear DNA of 9 wild isolates of *Aspergillus nidulans* representing three heterokaryon compatibility (h−c) groups. The strains used were Birmingham isolates 3, 28 and 38 (lanes 1 to 3) representing h−c group A, isolates 1, 26 and 36 (lanes 4 to 6) representing h−c group B, and isolates 33, 37 and 74 (lanes 7 to 9) representing h−c group C. No polymorphism was detected using probe 1, but there was an extra *Eco* RI site in the h−c group C isolates in the region homologous to probe 2. An approximate size scale for the fragments is indicated for each diagram.

(Fig. 21.5), thus showing that *A. nidulans* and *A. terreus* are much more closely related than the current taxonomy indicates.

Discussion

As this brief account has shown, the use of both genetic and molecular methods is providing a considerable amount of information concerning phylogenetic relationships both within and between *Aspergillus* species. The analyses of genetic crosses within a species and, using the technique of protoplast fusion, between closely related species have largely upheld the model of divergent evolution for *Aspergillus* put forward by Croft & Jinks (1977). In this model it is proposed that each species consists of a number of distinct lines (the heterokaryon compatibility groups), each maintained and dispersed by asexual propagation, which become genetically isolated from each other, thus evolving into closely related sibling species. However, results from protoplast fusion experiments and, particularly, from molecular analysis of both mitochondrial and nuclear DNA, show that this model is an oversimplification

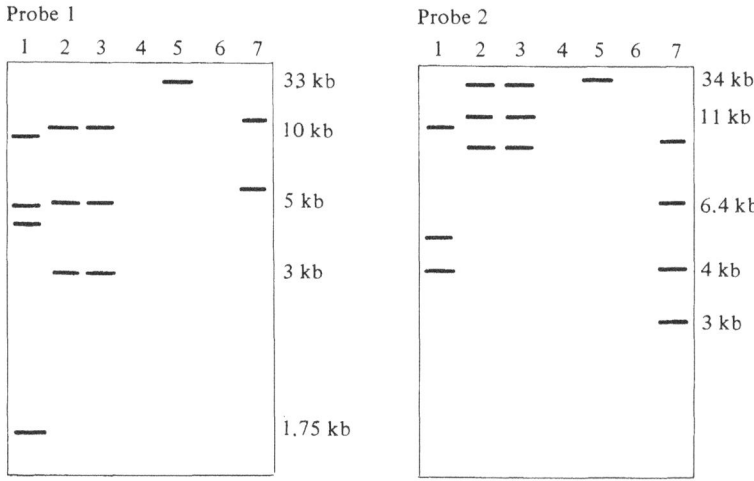

Fig. 21.5. Diagrammatic representations of autoradiographs obtained by the hybridisation of the two labelled random clones (probes 1 and 2) to electrophoresed *Eco* RI digests of the total nuclear DNA of seven species of *Aspergillus*. The strains used were: lane 1, *A. nidulans* 258 (recent USA isolate); lanes 2 & 3, *A. nidulans* var. *echinulatus* 25 and 209; lane 4, *A. unguis;* lane 5, *A. terreus;* lane 6, *A. niger;* and lane 7, *A. quadrilineatus* 12. No hybridisation was detected with either probe to *A. unguis* or *A. niger.* Approximate size scales for the fragments are indicated.

since it is clear that a wide range of genetic variation is represented within the 'nidulans' group such that taxa with no detectable relationship with *A. nidulans* are included. By contrast, it has been shown that *A. terreus*, which is classified in a separate species group, is closely related to *A. nidulans*. These results may be due to misclassification of these species, but also it seems that a much more complex network of interrelationships exists within this genus than was proposed in the simple hierarchical model of divergent evolution.

The use of molecular methods clearly has considerable potential for the study of evolution in the genus *Aspergillus* as a whole. The study of mtDNA may not be the most useful approach, particularly because variation in the presence or absence of optional introns, even in closely related species, greatly affects the estimates of restriction site polymorphism. This would be further confounded by the occurrence of genome rearrangements of the sort demonstrated in *A. heterothallicus* (Fig. 21.3). Such features would remain undetected without detailed study. Nevertheless, data reported here and by Kozlowski & Stepién (1982) suggest that the mt genome could provide efficient information concerning interspecific phylogenetic relationships in *Aspergillus*. In the 'nidulans' group, a detailed description of the mt genome of at least one species, *A. nidulans*, is now available and the introns themselves are the main feature of the differences between closely related species. This variation and the constancy of the mt genome within a species provides a very sensitive test of species identity. One of the most striking features of mtDNA is its extraordinary structural variability and *Aspergillus* seems likely to provide valuable material for microevolutionary study of the mt genome itself.

The RFLPs in nuclear DNA revealed by the use of random genomic clones clearly demonstrates that this method will be of considerable use for the determination of phylogenetic relationships in *Aspergillus*. This approach could be refined by the use of random cDNA clones, or other known cloned genes as probes, and if genes showing a range of degrees of sequence conservation were used, the method could be applied to the study of phylogenetic relationships over a range of genetic distances. The combination of all these methods has provided the basis for a detailed understanding of the evolutionary relationships between a number of taxa belonging to one *Aspergillus* species group. Continued and structured use of these methods seems likely to provide information which will eventually allow a comprehensive assessment of evolution within this genus as a whole.

322   J. H. Croft

References

Ballance, D. J. & Turner, G. (1985). Development of a high-frequency transforming vector for *Aspergillus nidulans*. *Gene*, **36**, 321 − 331.

Birley, A. J. & Croft, J. H. (1986). Mitochondrial DNAs and phylogenetic relationships. In *DNA Systematics*, vol I Evolution, ed. Dutta, S. K., pp. 107 − 137. Boca Raton, Florida: CRC Press.

Brown, T. A., Davies, R. W., Ray, J. A., Waring, R. B. & Scazzocchio, C. (1983). The mitochondrial genome of *Aspergillus nidulans* contains reading frames homologous to human URFs1 and 4. *EMBO Journal*, **2**, 427 − 435.

Brown, T. A., Waring, R. B., Scazzocchio, C. & Davies, R. W. (1985). The *Aspergillus nidulans* mitochondrial genome. *Current Genetics*, **9**, 113 − 117.

Butcher, A. C., Croft, J. H. & Grindle, M. (1972). Use of genotype-environmental interaction analysis in the study of natural populations of *Aspergillus nidulans*. *Heredity*, **29**, 263 − 283.

Caten, C. E. (1971). Heterokaryon incompatibility in imperfect species of *Aspergillus*. *Heredity*, **26**, 299 − 283.

Colleaux, L., d'Auriol, L., Betermier, M., Cottarel, G., Jacquier, A., Galibert, F. & Dujon, B. (1986). Universal code equivalent of a yeast mitochondrial intron reading frame is expressed into *E. coli* as a specific double strand endonuclease. *Cell*, **44**, 521 − 533.

Croft, J. H. (1985). Protoplast fusion and incompatibility in *Aspergillus*. In *Fungal Protoplasts, Applications in Biochemistry and Genetics*, ed. Peberdy, J. F. & Ferenczy, L., pp. 255 − 240. New York: Marcel Dekker.

Croft, J. H. & Dales, R. B. G. (1983). Interspecific somatic hybridisation in *Aspergillus*. In *Protoplasts 1983*, Proceedings of the 6th International Protoplast Symposium, ed. Potrykus, I., Harms, C. T., Hinnen, A., Hütter, R., King, P. J. & Shillito, R. D., pp. 179 − 186. Basel, Switzerland: Birkhäuser Verlag.

Croft, J. H. & Dales, R. B. G. (1984). Mycelial interactions and mitochondrial inheritance in *Aspergillus*. In *The Ecology and Physiology of the Fungal Mycelium*, British Mycological Society Symposium volume 8, ed. Jennings, D. H. & Rayner, A. D. M., pp. 433 − 450. Cambridge University Press.

Croft, J. H. & Jinks, J. L. (1977). Aspects of the population genetics of *Aspergillus nidulans*. In *Genetics and Physiology of Aspergillus*, British Mycological Society Symposium volume 1, ed. Smith, J. E. & Pateman, J. A., pp. 339 − 360. London: Academic Press.

Earl, A. J., Turner, G., Croft, J. H., Dales, R. B. G., Lazarus, C. M., Lunsdorf, H. & Kuntzel, H. (1981). High frequency transfer of species specific mitochondrial DNA sequences between members of the Aspergillaceae. *Current Genetics*, **3**, 221 − 228.

Engels, W. R. (1981). Estimating genetic divergence and genetic variability with restriction endonucleases. *Proceedings of the National Academy of Sciences, USA*, **78**, 6329 − 6333.

Grindle, M. (1963a). Heterokaryon compatibility of unrelated strains in the *Aspergillus nidulans* group. *Heredity*, **18**, 191 − 204.

Grindle, M. (1963b). Heterokaryon compatibility of closely related wild isolates of *Aspergillus nidulans*. *Heredity*, **18**, 397 − 405.

Grisi, E., Brown, T. A., Waring, R. B., Scazzocchio, C. & Davies, R. W. (1982). Nucleotide sequence of a region of the mitochondrial genome of *Aspergillus nidulans* including the gene of ATPase subunit 6. *Nucleic Acids Research*, **10**, 3531 − 3539.

Jadayel, D. M. (1986). Variation in the organisation and structure of the mitochondrial DNA of species of *Aspergillus*. *Ph.D. thesis, University of Birmingham*.

Kevei, F. & Peberdy, J. F. (1984). Further studies on protoplast fusion and interspecific hybridisation within the *Aspergillus nidulans* group. *Journal of General Microbiology*, **130**, 2229 − 2236.

Köchel, H. G., Lazarus, C. M., Basak, N. & Kuntzel, H. (1981). Mitochondrial tRNA gene clusters in *Aspergillus nidulans*: organization and nucleotide sequence. *Cell*, **23**, 625 − 633.

Kozlowski, M., Bartnik, E. & Stepién, P. P. (1982). Restriction enzyme analysis of mitochondrial DNA of members of the genus *Aspergillus* as an aid in taxonomy. *Journal of General Microbiology*, **128**, 471 − 476.

Lang, F. B. (1984). The mitochondrial genome of the fission yeast *Schizosaccharomyces pombe*: highly homol ogous introns are inserted at the same position of the otherwise less conserved cox1 genes in *Schizosaccharomyces pombe* and *Aspergillus nidulans*. *EMBO Journal*, **3**, 2129 − 2136.

Netzker, R., Köchel, G., Basak, N. & Kuntzel, H. (1982). Nucleotide sequence of *Aspergillus nidulans* mitochondrial genes coding for ATPase subunit 6, cytochrome oxidase subunit 3, seven unidentified proteins, four tRNA's and L-rRNA. *Nucleic Acids Research*, **10**, 4783 − 4794.

Raper, K. B. & Fennell, D. I. (1965). *The Genus Aspergillus*. Baltimore: Williams & Wilkins.

Southern, E. M. (1979). Gel electrophoresis of restriction fragments. *Methods in Enzymology*, **68**, 152 − 176.

Southern, E. M. (1975). Detection of specific sequences among DNA fragments separated by gel electrophoresis. *Journal of Molecular Biology*, **98**, 503 − 517.

Spooner, R. A. (1984). An investigation of the physical nature and recombination behaviour of introns of the mitochondrial genomes of *Aspergillus* species. *Ph. D. thesis, University of Bristol*.

Taylor, J. W., Smolich, B. D. & May, G. (1986). Evolution and mitochondrial DNA in *Neurospora crassa*. Evolution **40**, 716 − 739.

Tilburn, J., Scazzocchio, C., Taylor, G. T. & Zabicky-Zissman, J. H. (1984). Transformation by integration in *Aspergillus nidulans*. *Gene*, **26**, 205 − 221.

Tyc, M. (1968). An attempt to produce interspecific hybrids between *Aspergillus nidulans* and *Aspergillus rugulosus*. *Aspergillus Newsletter*, **9**, 20 − 21.

Upholt, W. B. (1977). Estimation of DNA sequence divergence from comparison of restriction endonuclease digests. *Nucleic Acids Research*, **4**, 1257 − 1265.

Upshall, A. (1981). Naturally occurring diploid isolates of *Aspergillus nidulans*. *Journal of General Microbiology*, **122**, 7 − 10.

Waring, R. B., Brown, T. A., Ray, J. A., Scazzocchio, C. & Davies, R. W. (1984). Three variant introns in the same general class in the mitochondrial gene for cytochrome oxidase subunit 1 in *Aspergillus nidulans*. *EMBO Journal*, **3**, 2121 − 2128.

Waring, R. B., Davies, R. W., Lee, S., Grisi, E., McPhail Berks, M. & Scazzocchio, C. (1981). The mosaic organization of the apocytochrome b gene of *Aspergillus nidulans* revealed by DNA sequencing. *Cell*, **27**, 4 − 11.

22

Speciation in *Phytophthora:* evidence from the *Phytophthora megasperma* complex

EVERETT M. HANSEN

Department of Botany and Plant Pathology, Oregon State University, Corvallis, Oregon 97331, USA

Introduction

The species of *Phytophthora* are plant pathogens. This creates interesting opportunities for speciation in the genus, while making the outcome of the process of special importance to agriculture and crop protection. Most species of *Phytophthora* are soil borne, with limited natural means of long distance dispersal. Agricultural commerce has introduced the fungi to many new environments and provided additional opportunities for reproductive isolation, a prerequisite to speciation, through cropping practices. The result appears to be a period of rapid evolution in many species of the genus, and a challenge for plant pathologists.

Progress in understanding the genetics, let alone speciation in this group has come slowly. Sansome published her dramatic cytological evidence for diploidy in 1965 (Sansome, 1965), but supporting genetical evidence was in dispute for some time (Caten & Day, 1977). Real progress in the genetic analysis of key features, such as the inheritance of mating type or pathogenicity, has come only recently (e.g. Shattock, Tooley & Fry, 1986).

There are few morphological features on which to base a taxonomy of *Phytophthora*. Sporangia and oospores are single celled and basically similar in many species. Chlamydospores and hyphal swellings are distinctive in a few cases, but *Phytophthora* has earned the reputation as a difficult genus taxonomically, despite abundant variation in behaviour in the environment, especially in pathogenicity. Although more than 60 species have been described, estimates of the number of 'good' morphological species range as low as three (Leonian, 1934; Waterhouse, Newhook & Stamps, 1983).

The accepted taxonomic treatment by Waterhouse (1963) divides 43 taxa into six groups, based primarily on features of the asexual sporangia, the occurrence of heterothallism or homothallism, and the attachment of antheridia to oogonia. While Waterhouse stated that the groups were ' . . . not necessarily meant to imply a natural classification . . . ' her scheme has become, *de facto*, the basis for speculation about relationships within the genus. Waterhouse and others have stressed the need for additional non-morphological characters to aid in establishing species limits and affinities (Waterhouse *et al.*, 1983; Brasier, 1983). Certainly, any investigations of speciation in the genus will have to rely heavily on the study of a wide range of morphological, physiological and biochemical traits in large numbers of isolates, in order to detect patterns at and below the species level.

Phytophthora megasperma

Phytophthora megasperma provides perhaps the most completely studied example of apparent speciation in progress in this genus. The evidence comes from a variety of sources, and points to changes in host specific pathogenicity and karyotype as the principal isolating mechanisms. I will develop this example in some detail because it illustrates: (i) the problems inherent in *Phytophthora* taxonomy; (ii) an integrative approach to their solution that relies on morphological, as well as biochemical methods; and (iii) the consequences of the action of diverse isolating mechanisms on a fungal population.

Phytophthora megasperma is an economically important pathogen of many woody and herbaceous plants in Northern temperate to subtropical regions. *P. megasperma* is reasonably distinctive in the Waterhouse (1963) classification, containing two varieties distinguished by oospore size. Almost from its description (Drechsler, 1931), however, there has been controversy about sub-specific groupings of isolates based on host range. The soy bean pathogen with small oospores, *P. megasperma* var. *sojae*, was straightforward enough, but what was to be done with the isolates described subsequently from alfalfa or clover? Waterhouse included them in var. *sojae* on the basis of their small oospore size, but these isolates had specific pathogenicity to their host of origin and pathologists needed some nomenclatural recognition of that important practical feature. A system of *formae speciales* descriptions was, therefore, developed for these host specific isolates. Hence, *P. megasperma* f. sp. *glycinae*, f. sp. *medicaginis*, and f. sp. *trifolii* (Kuan & Erwin, 1980; Pratt, 1981).

Meanwhile, more isolates from a diverse array of woody hosts were being examined. It soon became evident that oospore size was a variable character, both among the single zoospore progeny of a single isolate (Hamm & Hansen, 1982), and within the general population of *P. megasperma* from many hosts (Fig. 22.1; Kuan & Erwin, 1980). Additional difficulties arose as large-spored isolates, pathogenic to alfalfa, were found (Barr, 1980; Hansen & Hamm, 1983) and isolates aggressive towards Douglas fir also proved to be pathogenic to soybean under standard test conditions (Hamm & Hansen, 1981).

To clarify the situation, we examined isolates from alfalfa, soybean, Douglas fir, and a range of other, mostly woody, hosts in a single comparison (Hansen & Hamm, 1983). Morphological and growth variables were measured and pathogenicity to the respective hosts was tested. Oogonial sizes varied continuously among the isolates, although soybean isolates were always relatively small. Some alfalfa and some Douglas fir isolates had small spores and some had large spores, while

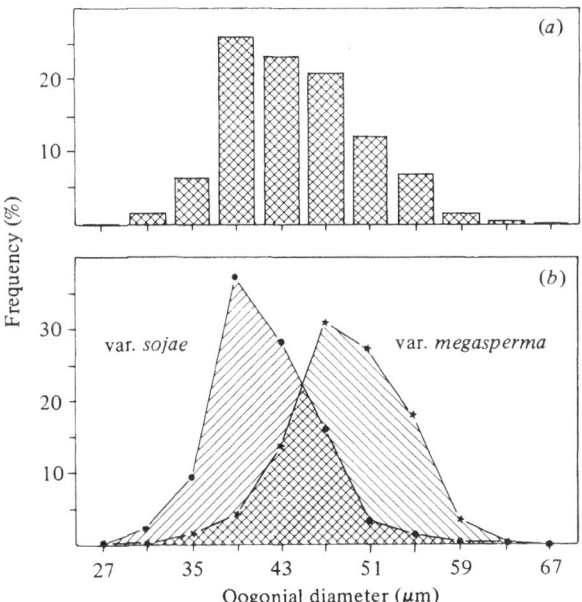

Fig. 22.1. Oogonial diameters for single zoospore isolates of *Phytophthora megasperma*. Based on measurements of ten oogonia from each of eight single spore isolations from 35 parental isolates. Data are shown plotted for all isolates *(a)* (above) and separately for isolates considered to be var. *megasperma* and var. *sojae (b)* (below). (Data from Hamm & Hansen (1982).)

clover isolates tended to be intermediate. Isolates from other hosts had oogonia that ranged from small to large. Only when additional features were considered did order begin to emerge. If oogonial diameter and sporangial length were plotted on opposing axes, the various host groups clustered together. When 14 morphological and growth variables were combined by cluster analysis, five statistically distinct groups were identified (Fig. 22.2). These corresponded with the host of origin, except that isolates from alfalfa and Douglas fir appeared in three of the groups. In pathogenicity tests, some feeder root necrosis was evident with most host-pathogen combinations, but tap root lesions and plant death were confined to specific interactions. The soybean and clover groups, distinguished by morphological and growth characters, killed only soybean and clover plants respectively. Alfalfa isolates pathogenic only to alfalfa and Douglas fir isolates pathogenic to both fir and soybeans also formed distinct groups. The last group was composed of the remaining alfalfa, Douglas fir, and woody host isolates. These seemed to have a broad host range and were weakly pathogenic to both alfalfa and Douglas fir in our tests. We hypothesised that ' . . . morphological differentiation among *Phytophthora* species is a consequence of the isolation of populations by agricultural practice and host specialization . . . '.

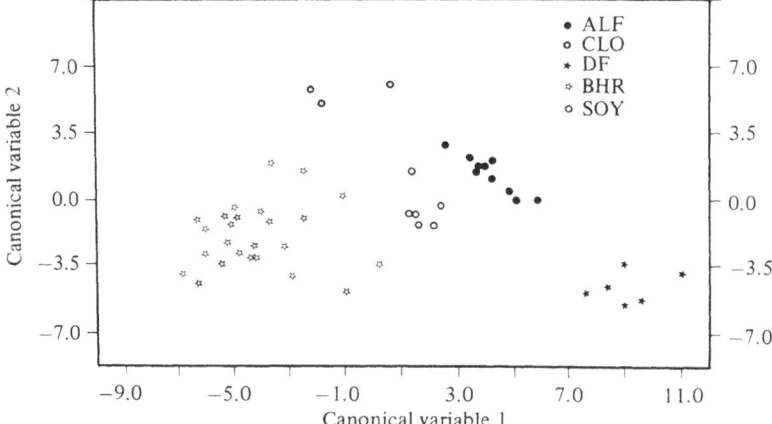

Fig. 22.2. Separation of 51 isolates of *Phytophthora megasperma* into main host groups by compressing 14 morphological and growth characteristics into two canonical variables (Hansen & Hamm, 1983).

Table 22.1. *Average similarity coefficients (%) for the six protein groups of* Phytophthora megasperma

Protein group	ALF	SOY	CLO	DF	AC	BHR
ALF	97	47	60	52	39	43
SOY		76	50	50	31	34
CLO			89	53	36	38
DF				91	37	35
AC					88	68
BHR						93

The data presented are based on comparisons between three isolates each from SOY, CLO, AC, and BHR protein groups, two ALF isolates and four DF isolates. From Hansen *et al.* (1987).

Proteins and DNA

While the conclusion was attractive in several respects, it failed to address two key points. First, that the diverse assemblage of isolates from various hosts is difficult to characterise on pathogenic ability. Second, it ignored the important observation by Sansome & Brasier (1974) that some large-spored isolates had roughly twice the chromosome number of small-spored isolates. To remedy these deficiencies, chromosome number, DNA content, and protein profiles as revealed by polyacrylamide gel electrophoresis, were compared.

The results confirmed and expanded our concept of host groups within *P. megasperma* (Hansen *et al.*, 1987). Six distinct protein patterns were recognised. Five corresponded to the host groups previously recognised and the sixth was a subgroup of the group from various hosts. The protein groups were designated: ALF = alfalfa; SOY = soybean; CLO = clover; DF = Douglas fir; and AC = apple-cherry, a subgroup of the BHR = broad host range group. Similarity coefficients calculated between protein patterns of all pairs of isolates indicated that SOY, CLO, and ALF groups, all with relatively small oospores, were related, as were the AC and BHR groups (Table 22.1). DF isolates were more similar to the legume groups than to the BHR group. The latter group, although from many hosts and encompassing isolates with small, intermediate, and large oogonia, had a common protein pattern with only minor differences between isolates. Large-spored isolates from alfalfa and Douglas fir had the BHR protein pattern. Finally, the six protein groups of *P. megasperma* were more similar to each other than to other species of *Phytophthora*. Others (Irwin

Table 22.2. *Characteristics of the subgroups of* Phytophthora megasperma

Protein group	Mean growth rate (mm day^{-1})	Mean oogonial diameter (μm)	Chromosome number	Mean relative DNA (% of *P. infestans*)
ALF	4.7±0.2	37±2.5	12−15	20
CLO	3.3±1.2	43±0.1	11−15	ND
SOY	4.9±0.8	39±2.0	12−15	21
DF	10.0±0.3	43±1.6	17−24	19
AC	6.1±0.7	36±4.0	13−20	19
BHR-KI	6.4±0.5	43±3.9	12−17	21
BHR-KII	5.9±0.4	43±3.1	15−23	ND
BHR-KIII	6.0±1.1	46±8.2	22−28	ND
BHR-KIV	6.9±0.8	50±3.4	26−34	37

ND = not determined. Data from Hansen *et al.* (1987).

& Dale, 1982; Elliott & Maxwell, 1984) reached a similar conclusion in a comparison of isolates now referred to the ALF and SOY groups.

Although exact chromosome numbers could not be counted with certainty, at least four karyotypes were distinguished (Table 22.2). The ALF, SOY, and CLO protein groups had a base of n = 12 − 15. DF isolates were n = 17 − 24, and BHR isolates ranged from n = 12 − 17 to n = 26 − 34. Five overlapping ranges of chromosome number were recorded in the latter group: AC, n = 13 − 20; BHR Karyotype I, n = 12 − 17; BHR KII, n = 15 − 23; BHR KIII, n = 22 − 28 and BHR KIV, n = 26 − 34. This range in chromosome numbers within a single protein group was both unexpected and exciting. Chromosome number correlated positively with oogonial size within this group. Fluorometric measurements of DNA in individual zoospore nuclei stained with DAPI (4,6-diamido-2-phenyl-indole) confirmed that isolates with high chromosome counts had roughly twice the amount of DNA of those with low counts (Table 22.2). Rutherford & Ward (1985) used the DAPI technique to show that nine physiologic races of *P. megasperma* f. sp. *glycinea* (n = 12 − 15) had similar DNA content, supporting our observations, at least at the low end of the range of counts.

Evidence for biological species

In summary, five main protein groups were distinguished within *P. megasperma* and the BHR group contains at least four karyotypes.

These must represent multiples of a common genome, either a polyploid series or heteroploids with aneuploids and chromosome duplications. Do they represent biological or sibling species, that is, are the populations reproductively isolated one from another? There is no direct genetic evidence, but three lines of indirect evidence support the assertion. First, the protein pattern of the various groups are uniform worldwide. Thus, ALF isolates from the East and West coasts of North America, South Africa, and Japan are indistinguishable. Similarly, BHR isolates from three hosts and three continents are as similar as multiple isolates from the same hosts in the same locale. Second, differences in protein pattern are correlated with morphological and behavioural differences that allow identification of the groups without either pathogenicity testing, electrophoresis, or chromosome counting. Third, the groups appear to remain distinct in all these respects, even when growing alongside each other (sympatrically). The close association of different groups has been documented in Ontario, New York, Oregon, and Japan (E. M. Hansen, unpublished data; Barr, 1980; Matsumoto & Sato, 1985). Three of the protein groups are found on alfalfa (ALF, DF, and BHR), and both DF and BHR groups are found in Douglas fir nurseries.

Stability of the karyotype groups within the BHR group is less certain. This population may represent a constantly changing mix of chromosome numbers as in some myxomycetes (Collins, Therrien & Betterley, 1978), or it may be a polyploid series with, for example, n = 10, 15, 20, 25, and 30, and be more stable. There are qualitative as well as quantitative differences between the various BHR karyotypes, including differences in staining intensity with aceto-orcein, apparent chromosome size, and occurrence of chromosome complexes. Some differences are such that it is difficult to imagine successful matching of chromosomes during meiosis. The apparent uniformity of the protein pattern within the BHR group could be accounted for in several ways: (i) the various karyotypes are not reproductively isolated; (ii) they are isolated, but are of such recent origin that differences have not accumulated; or (iii) they are isolated, but protein similarity is maintained through asexual reproduction (zoospores and chlamydospores) and barriers to heterokaryon formation.

Phytophthora megasperma regularly produces oospores following meiosis, but it is homothallic and a presumed inbreeder. If heterokaryon formation were inhibited through a vegetative incompatibility system, then genotypes might be very stable. Abundant asexual reproduction through zoospores is a feature of the genus and identical protein patterns from opposite sides of the world attest to the vegetative propagation and transfer of the fungus through agricultural commerce (Hansen *et al.*, 1987).

We hypothesise that *P. megasperma* is actively speciating with differences in host specificity and karyotype providing the requisite isolation between populations (Fig. 22.3). The responsiveness of the fungal genotype to host genotype is apparent in many plant pathogens, including *Phytophthora* species. This is regularly seen in the selection of virulence genes in a fungal population to match new host resistance genes. It seems reasonable that other genes imparting fitness on a particular host would be

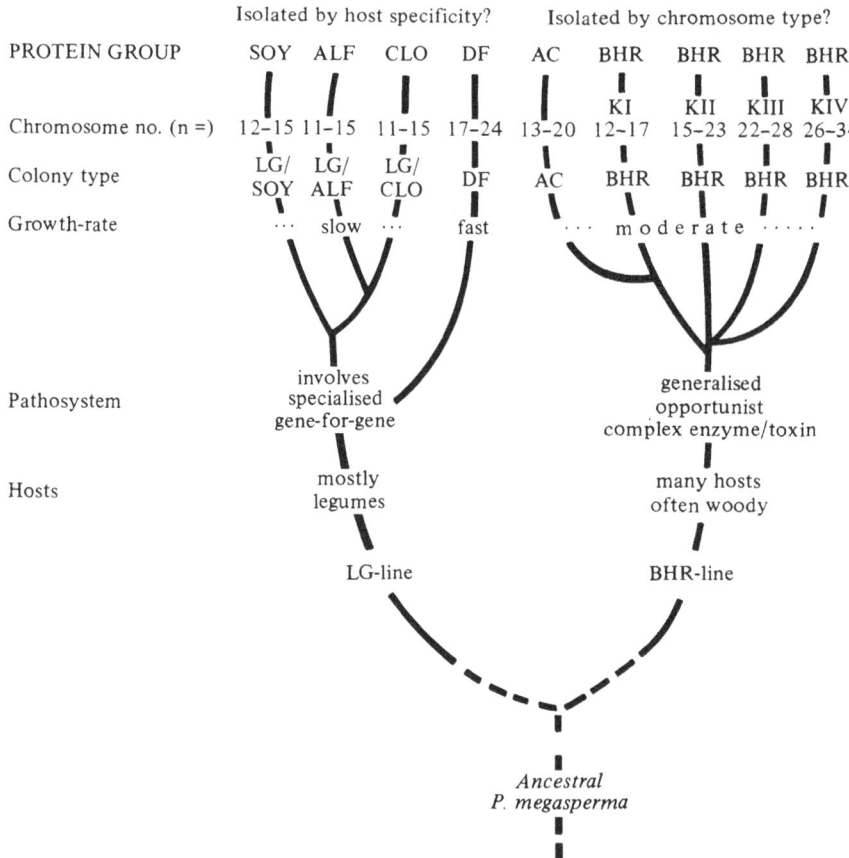

Fig. 22.3. An interpretation of the phylogenetic structure of *Phytophthora megasperma.*Branch ends represent emerging biological species isolated initially by host specificity or karyotype. It is hypothesised that a split between generalist and specialist modes of pathogenesis provided the initial impetus for speciation within the ancestral population (Hansen *et al.*, 1987).

selected as well, resulting, in time, in physiologically and sometimes morphologically distinct populations. Other species of *Phytophthora* show evidence of active speciation under similar influences.

Evidence of speciation in other *Phytophthora* **species**

Phytophthora cinnamomi is a heterothallic species with two mating types, A1 and A2. The predominance of the A2 type throughout the world, the ability of the A2 to self fertilise in response to *Trichoderma* and the apparently different host ranges of A1 and A2 has led to the suggestion (Brasier, 1978) that A1 and A2 are becoming reproductively isolated in response to the combined pressures of host specialisation and geographical dispersal by man. Recent evidence for different isozyme patterns in A1 and A2 isolates from Australia and the South Pacific islands (Old, Moran & Bell, 1984) supports this view and suggests that sexual recombination between A1 and A2 is rare in these areas.

A study of protein electrophoretic patterns among isolates of *P. pseudotsugae* (Hamm & Hansen, 1983) and another with *P. nicotianae* and *P. citricola* isolates (Erselius & de Vallavielle, 1984), both revealed nearly identical patterns within each species, even when isolates had originated on different hosts and in the case of *P. nicotianae*, even between isolates from France, the Ivory Coast, and the United States. In contrast, other species have more variable protein patterns suggesting that sub-specific groupings similar to those in *P. megasperma* might be revealed if enough isolates were examined. *Phytophthora cactorum* (Hamm & Hansen, 1983) and *P. citricolor* (Erselius & de Vallavielle, 1984; Zentmeyer *et al.*, 1974) are particularly variable species.

A vegetative incompatibility system has not been demonstrated from any species of *Phytophthora*, but limited evidence suggests that barriers to heterokaryon formation exist in *P. megasperma* and *P. infestans*. Long & Keen (1977) successfully forced heterokaryons between auxotrophic mutants of a single isolate, but failed in all attempts made between different isolates (N. H. Keen, personal communication). The author's attempts to force heterokaryons between genetically marked strains have so far been unsuccessful. Poedinok & Dyakov (1981) reported greatly reduced heterokaryon formation in inter-isolate crosses of *P. infestans* compared to intra-isolate crosses, although some of their strains incorporated cytoplasmic, not nuclear, markers and their results are more properly interpreted in terms of success of anastomosis than hetero-karyosis. Malcomson (1970) presented evidence suggesting hetero-karyosis in *P. infestans* based on new combinations of pathogenic races of the fungus from zoospore mixtures; some mixtures did not yield new combinations, suggesting incompatibility.

Phytophthora palmivora, like *P. megasperma* was known as a surprisingly variable species from shortly after its description. Only when a large group of isolates from cocoa was gathered and compared for morphology, growth behaviour and chromosome number did a different picture emerge (Brasier & Griffin, 1979). Three distinct groups are now recognised at the species level: *P. palmivora*, *P. capsici*, and a new species *P. megakarya*. Once the pattern had been recognised, the groups were found to be distinguishable by combinations of spore characters, physiological characters, chromosome number and colony habit, and to have different geographical distributions. Moreover, within *P. palmivora sensu stricto*, cocoa isolates from throughout the world are remarkably uniform and distinctive in terms of colony form, have smaller oogonia and are all of A2 mating type, while isolates from other hosts, such as rubber and coconut, tend to have a different colony pattern, larger oogonia and are often A1 type (Brasier & Griffin, 1979). This has led to the suggestion that separate, possibly reproductively isolated forms are emerging on cocoa and rubber and perhaps also on coconut (Brasier, Griffin & Maddison, 1981; C. M. Brasier, personal communication).

Phytophthora infestans, in contrast, has long been considered a 'good' species. It is heterothallic, with both A1 and A2 mating types found in Mexico, but until recently only the A1 type was found elsewhere in the world. Isozyme heterozygosity is greater in the interbreeding Mexican population than in the essentially clonal European population (Tooley, Fry & Villarreal Gonzalez, 1985). Recent observations suggest that this classic species is in reality more complex and perhaps, is in the process of speciation. Chromosome number is not constant within the species; isolates with diploid, tetraploid, and intermediate chromosome numbers have been described (Sansome, 1977; P. W. Tooley, personal communication). The A2 mating type now appears to be fairly widespread in Europe, possibly of long standing or perhaps, introduced from Egypt (Hohl & Iselin, 1984; Shaw *et al.*, 1985). Finally, isolates morphologically almost identical to *P. infestans*, but pathogenic to different hosts and differing in physiological properties have been reported (H. Hohl, personal communication).

Conclusions

Species are stirring in *Phytophthora*, but the action is barely visible at the level of classical taxonomic characters. In a genus of plant pathogens, host preference is an important character, especially as it isolates populations one from another and allows other differences to accumulate. Similarly, karyotype is a potentially powerful isolating mechanism. Growth patterns and growth rates, temperature effects, response to fungicides and antibiotics, nutritional requirements and a myriad of other possible stimuli can provide valuable clues to emerging species. Many of these responses may be controlled by single genes and consequently, be of very limited significance, just as any single morphological feature must be interpreted with caution until its genetic basis is understood. Collectively, however, they greatly extend our ability to distinguish population units. Protein electrophoresis is an especially powerful tool because it integrates so many features; the banding pattern of soluble proteins on a polyacrylamide gel represents the diverse products of 50 or more genes (Hansen *et al.*, 1987; and see Barrett, Chapter 6). Although the bands on such a gel cannot be equated with morphological features, groups of isolates identified by protein banding pattern similarity have regularly turned out to have similar morphology and behaviour. In *P. megasperma*, for example, the ALF protein group is uniquely pathogenic to alfalfa and has distinctive morphology and growth, but cannot be distinguished from other legume and Douglas-fir isolates by classical taxonomic means. As illustrated in Figs. 22.1 to 22.3, the addition of new characters to the identification matrix refines the picture. Confidence that we are looking at genuine population units comes when additional characters correlate with established groups, rather than suggesting new ones (Table 22.2).

What patterns of relationship within the genus will be revealed as additional features are examined? Certainly, more species will be recognised, but the grouping of species in the future will be quite different from those proposed by Waterhouse (1963). They were, of necessity, grounded on the International Code of Botanical Nomenclature with its severely limiting 'type' concept and restriction to morphologically based species. *Phytophthora* and other fungi exist as dynamic populations and new morphological species probably seldom arise *de novo*, with a unique spore shape for example. Brasier (1983) highlighted many of these issues in a perceptive paper at the Phytophthora Symposium in 1981.

Speciation in *Phytophthora* and perhaps in other genera of plant pathogens as well, seems to be accelerated by the reproductive isolation afforded by specialisation on hosts of agricultural importance. If this is true, then changes in agricultural practice will, in future, continue to stimulate speciation in this important genus. If taxonomy is to reflect reality and remain relevant to disease control strategies, it must adopt a corresponding flexibility.

References

Barr, D. J. S. (1980). Heterothallic-like reaction in the large-oospore forms of *Phytophthora megasperma*. *Canadian Journal of Plant Pathology*, **2**, 116 – 118.

Brasier, C. M. (1978). Stimulation of oospore formation in *Phytophthora* by antagonistic species of *Trichoderma* and its ecological implications. *Annals of Applied Biology*, **89**, 135 – 139

Brasier, C. M. (1983). Problems and prospects in *Phytophthora* research. In *Phytophthora: Its Biology, Taxonomy and Pathology*, ed. Erwin, D. C., Bartnicki-Garcia, S. & Tsao, P. H., pp. 351 – 364. St Paul, Minnesota: American Phytopathological Society.

Brasier, C. M. & Griffin, M. J. (1979). Taxonomy of 'Phytophthora palmivora' on cocoa. *Transactions of the British Mycological Society*, **72**, 111 – 143.

Brasier, C. M., Griffin, M. J. & Maddison, A. C. (1981). The cocoa black pod Phytophthoras. In *Epidemiology of Phytophthora on Cocoa in Nigeria*, ed. Gregory, P. H. & Maddison, A. C., pp. 18 – 30, Phytopathological Paper no. 25. Kew, UK: Commonwealth Mycological Institute.

Caten, C. E. & Day, A. W. (1977). Diploidy in plant pathogenic fungi. *Annual Review of Phytopathology*, **15**, 295 – 318.

Collins, O. R., Therrien, C. D. & Betterley, D. A. (1978). Genetical and cytological evidence for chromosomal elimination in a true slime mold, *Didymium iridis*. *American Journal of Botany*, **65**, 660 – 670.

Drechsler, C. (1931). A crown rot of hollyhocks caused by *Phytophthora megasperma* n. sp. *Journal of the Washington Academy of Science*, **21**, 513 – 526.

Elliott, C. K. & Maxwell, D. P. (1984). Evaluation of isozymes for the identification of isolates of *Phytophthora megasperma*. *Phytopathology*, **74**, 866 (abstract).

Erselius, L. J. & De Vallavieille, C. (1984). Variation in protein profiles of *Phytophthora*: comparison of six species. *Transactions of the British Mycological Society*, **83**, 463 – 472.

Hamm, P. B. & Hansen, E. M. (1981). Host specificity of *Phytophthora megasperma* from Douglas-fir, soybean and alfalfa. *Phytopathology*, **71**, 65 – 68.

Hamm, P. B. & Hansen, E. M. (1982). Single-spore isolate variation: the effect on varietal designation in *Phytophthora megasperma*. *Canadian Journal of Botany*, **660**, 2931 – 2938.

Hamm, P. B. & Hansen, E. M. (1983). *Phytophthora pseudotsugae*, a new species causing root rot of Douglas-fir. *Canadian Journal of Botany*, **61**, 2626 – 2631.

Hansen, E. M., Brasier, C. M., Shaw, D. S. & Hamm, P. B. (1987). The taxonomic structure of *Phytophthora megasperma*: evidence for emerging biological species groups. *Transactions of the British Mycological Society*, **87**, 557 – 573.

Hansen, E. M. & Hamm, P. B. (1983). Morphological differentiation of host-specialized groups of *Phytophthora megasperma*. *Phytopathology*, **73**, 129 – 134.

Hohl, H. R. & Iselin, K. (1984). Strains of *Phytophthora infestans* with A2 mating type behaviour. *Transactions of the British Mycological Society*, **83**, 529 – 530.

Irwin, J. A. G. & Dale, J. L. (1982). Relationships between *Phytophthora megasperma* isolates from chickpea, lucerne, and soybean. *Australian Journal of Botany*, **30**, 199 − 210.

Kuan, T. L. & Erwin, D. C. (1980). Formae speciales differentiation of *Phytophthora megasperma* isolates from soybean and alfalfa. *Phytopathology*, **70**, 333 − 338.

Leonian, L. H. (1934). Identification of Phytophthora species. *West Virginia Agricultural Experiment Station Bulletin*, **262**, 2 − 36.

Long, M. & Keen, N. T. (1977). Evidence for heterokaryosis in *Phytophthora megasperma* var. *sojae*. *Phytopathology*, **67**, 670 − 674.

Malcolmson, J. F. (1970). Vegetative hybridity in *Phytophthora infestans*. *Nature*, **225**, 971 − 972.

Matsumoto, N. & Sato, T. (1985). Host non-specific *Phytophthora megasperma* with small oogonia from alfalfa-field soil. *Research Bulletin of the Hokkaido National Agricultural Experiment Station*, **143**, 75 − 84.

Old, K. M., Moran, G. F. & Bell, C. J. (1984). Isozyme variability among isolates of *Phytophthora cinnamomi* from Australia and Papua New Guinea. *Canadian Journal of Botany*, **62**, 2016 − 2022.

Poedinok, N. L. & Dyakov, Y. T. (1981). Observation of vegetative incompatibility in *Phytophthora infestans*. *Mikologiya i Fitopatologiya*, **15**, 275 − 279.

Pratt, R. G. (1981). Morphology, pathogenicity, and host range of *Phytophthora megasperma*, *P. erythroseptica*, and *P. parasitica* from arrowleaf clover. *Phytopathology*, **71**, 276 − 282.

Rutherford, F. S. & Ward, E. W. B. (1985). Estimation of relative DNA content in nuclei of races of *Phytophthora megasperma* f. sp. *glycinea* by quantitative fluorescence microscopy. *Canadian Journal of Genetics and Cytology*, **27**, 614 − 616.

Sansome, E. R. (1965). Meiosis in diploid and polyploid sex organs of *Phytophthora* and *Achlya*. *Cytologia*, **30**, 103 − 117.

Sansome, E. R. (1977). Polyploidy and induced gametangial formation in British isolates of *Phytophthora infestans*. *Journal of General Microbiology*, **99**, 311 − 316.

Sansome, E. R. & Brasier, C. M. (1974). Polyploidy associated with varietal differentiation in the megasperma complex of *Phytophthora*. *Transactions of the British Mycological Society*, **63**, 461 − 467.

Shattock, R. C., Tooley, P. W. & Fry, W. E. (1986). *Phytophthora infestans* genetics: determinations of recombination, segregation and selfing by isozyme analysis. *Phytopathology*, **76**, 410 − 413.

Shaw, D. S., Fyfe, A. M., Hibberd, P. G. & Abtel-Sattar, M. A. (1985). Occurrence of the rare A2 mating type of *Phytophthora infestans* on imported Egyptian potatoes and the production of sexual progeny with A1 mating types from the UK. *Plant Pathology*, **34**, 552 − 556.

Tooley, P. W., Fry, W. E. & Villarreal Gonzalez, M. J. (1985). Isozyme characterization of sexual and asexual *Phytophthora infestans* populations. *Journal of Heredity*, **76**, 431 − 435.

Waterhouse, G. M. (1963). Key to the species of *Phytophthora* de Bary. *Mycological Papers of the Commonwealth Mycological Institute*, **92**, 1 − 22.

Waterhouse, G. M., Newhook, F. J. & Stamps, D. J. (1983). Present criteria for classification of *Phytophthora*. In *Phytophthora: Its Biology, Taxonomy and Pathology*, ed. Erwin, D. C., Bartnicki-Garcia, S. & Tsao, P. H., pp. 139 − 147. St Paul, Minnesota: American Phytopathological Society.

Zentmeyer, G. A., Jefferson, L., Hickman, C. J. & Chang-Ho, Y. (1974). Studies of *Phytophthora citricola* isolated from *Persea americana*. *Mycologia*, **66**, 830 − 845.

23

The origin of Fungi and pseudofungi

T. CAVALIER-SMITH

Department of Biophysics, Cell and Molecular Biology, King's College, University of London, 26-29 Drury Lane, London WC2B 5RL, UK

Introduction

Ultrastructural and molecular data confirm earlier speculations that the organisms studied by mycologists are polyphyletic. Setting aside the 'slime moulds', of four disparate groups belonging in the kingdom Protozoa (Cavalier-Smith, 1981a, 1983), we have two convergent 'fungoid' groups: the kingdom Fungi *sensu stricto*, which (like most animals) have non-discoidal plate-like mitochondrial cristae (Cavalier-Smith, 1981a, 1983), and the Heterokontimycotina (Dick, 1976) with tubular cristae. Like Kreisel (1969) and Shaffer (1975), I exclude the latter from the kingdom Fungi, and have placed them as the subdivision Pseudofungi in the phylum Heterokonta, of the kingdom Chromista (Cavalier-Smith, 1981a, 1986a). The Pseudofungi, comprising the Oomycetes, Hyphochytrea, labyrinthulids and thraustochytrids, are probably monophyletic and evolved from a heterokont alga that lost chloroplasts (for details see Cavalier-Smith, 1986a). I think the kingdom Fungi *sensu* Cavalier-Smith (1981a, 1983; see Table 23.1) is monophyletic; and that the Entomophorales evolved from a chytridiomycete by loss of cilia [and I urge mycologists to use 'cilia' instead of 'flagella' as it is briefer, less confusing (Cavalier-Smith, 1986b) and based on sound precedent (de Bary, 1887)], and gave rise to all other Eufungi. To substantiate this view, which treats chytridiomycetes as the most primitive Fungi, one needs to show how all major eufungal taxa could have evolved via an entomophoralean from a chytridiomycete ancestor. This paper seeks to do this and to clarify the origin of Fungi by answering three interrelated questions: (i) Which chyridiomycetes are most primitive? (ii) From what did they evolve? (iii) How did this occur?

Table 23.1. *Classification of the kingdom Fungi*

Phylum 1. Archemycota divisio nova
  Subphylum 1. Chytriomycotina subdivisio nova
    Class 1. Spizomycetes classis nova (includes Spizellomycetales)
    Class 2. Rumpomycetes classis nova (includes Chytridiales;
        Monoblepharidales)
    Class 3. Allomycetes classis nova (includes Blastocladiales)
  Subphylum 2. Zygomycotina
    Class 1. Zygomycetes
    Class 2. Trichomycetes

Phylum 2. Ascomycota emend.
  Subphylum 1. Euascomycotina emend. (includes most lichen fungi and most
        fungi imperfecti)
    Class 1. Plectomycetes
    Class 2. Loculoascomycetes
    Class 3. Pyrenomycetes
    Class 4. Discomycetes
  Subphylum 2. Laboulbeniomycotina subdivisio nova
    Class 1. Laboulbeniomycetes (includes Laboulbeniales, Spathulosporales)

*Phylum 3. Endomycota divisio nova**
    Class 1. Saccharomycetes classis nova (includes Endomycetales)
    Class 2. Taphrinomycetes classis nova (includes Taphrinales,
        Schizosaccharomyces, Protomycetales)

Phylum 4. Basidiomycota
  Subphylum 1. Uredomycotina subdivisio nova
    Class 1. Uredomycetes classis nova (includes Uredinales)
  Subphylum 2. Orthomycotina subdivisio nova
    Class 1. Septomycetes classis nova (includes Septobasidiales, Exobasidiales,
        Tilletiales, Brachybasidiales)
    Class 2. Ustomycetes *sensu* von Arx, van der Walt & Liebenberg (1982)
        (includes Ustilaginales, Sporidiales)
    Class 3. Gelimycetes classis nova (includes Tulasnellales, Dacrymycetales,
        most Tremellales, saprophytic Auriculariales)
    Class 4. Holobasidiomycetes (also includes hymenolichens)

* See note on p. 339.

The origin of the Fungi

Chytridiomycetes have two key features that narrow the search
for their ancestors: flattened, non-discoidal cristae, and a single posterior
cilium on motile cells. Only two other eukaryote groups have both
characters: metazoan animals and 'choanoflagellate' protozoa (or choanociliates, as I call them). Fungi resemble animals more than plants in four
other respects: (i) both commonly have chitinous exoskeletons (ii) both

store glycogen not starch (iii) both lack chloroplasts, and (iv) unlike plants, in their mitochondrial code, UGA codes for tryptophan, not chain termination. I suggest that Fungi and animals had a common ancestor with all six of these properties, and which was a choanociliate protozoan.

It has been long thought that sponges, the most primitive animals, evolved from choanociliates (Leadbeater, 1983). To convert a choanociliate into a chytridiomycete would be even easier than to convert it into a sponge: it would need only to acquire a chitinous wall and thereby lose phagotrophy (Fig. 23.1).

The most likely ancestral chytridiomycetes are the Spizellomycetales: their zoospores seem primitive in two major and two minor respects: (i) they are the only ones with ribosomes dispersed in the cytoplasm as in most protists (other chytridiomycetes have a nuclear cap of ribosomes and therefore derived, not ancestral, (iii) they are more amoeboid, (iv) in some, the non-ciliated centriole is about 90° to the ciliated one, as in choanociliates and animals.

Spizellomycetales also have the most diverse ciliary roots: as these include variants similar to the two entirely different patterns in the Blastocladiales and Chytridiales, they probably gave rise independently to both orders; Monoblepharidales probably evolved from Chytridiales (Barr, 1983). More importantly, the ciliary roots of certain Spizellomycetales (Barr, 1981) closely resemble those of choanociliates (Hibberd, 1975; Leadbeater & Morton, 1974): in both, fan-shaped arrays of microtubules diverge from arc-like dense bodies beside the kinetosome and the narrow end of the fan is cross-striated.

Diversity and classification of the ciliated Fungi

Chytridiomycete zoospores are ultrastructurally far more diverse than in any other class of protists. Blastocladiales differ from other Chytridiomycetes in having single Golgi cisternae not stacked as dictyosomes; their ciliary roots (a cone-shaped array of microtubules surrounding the nucleus and ribosomal cap) are totally unlike Chytridiales and Monoblepharidales; they have a distinctive complex of lipid drops, microbodies, and mitochondrion, and are mitotically distinctive (Heath, 1986); I therefore separate them as a new class: Allomycetes. The distinctiveness of the Spizellomycetales also merits a new class, the Spizomycetes. For Chytridiales and Monoblepharidales (including Harpochytriales) I propose a new class, Rumpomycetes, characterised by a rumposome (a unique tubular derivative of the endoplasmic reticulum attached to the lipid drop in the microbody/lipid/mitochondrion complex), nuclear cap, and microtubular roots not enclosing the nucleus.

I group Rumpomycetes, Allomycetes, and Spizomycetes in the sub-division Chytriomycotina, characterised by the posteriorly ciliated zoospore (vernacular 'chytriomycete'). Treating chytriomycetes as a

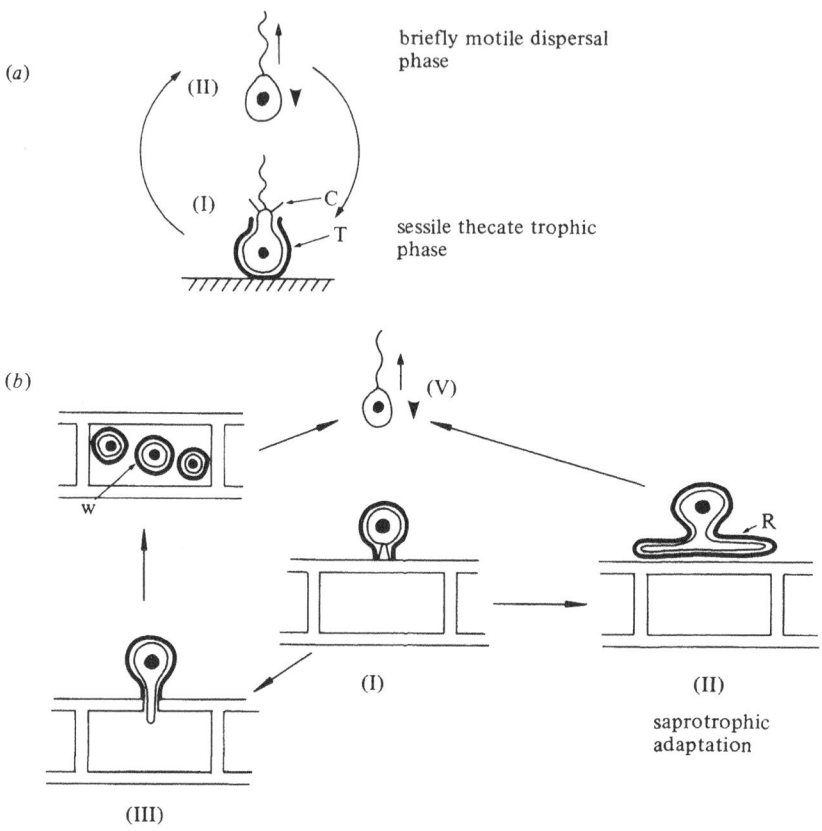

Fig. 23.1. Origin of the first fungus from a thecate choanociliate. (a) the life cycle of a thecate choanociliate (family Salpingoecidae). It phagocytoses bacteria wafted by its cilium on to its collar, C; the small arrow shows the direction of ciliary waves and water currents, and the broad arrow-head the direction of swimming. The chemistry of the theca, T, is unclear; presumably, it was chitinous in species ancestral of fungi. (b) An epiphytic salpingoecid undergoes a mutation sticking its apical, rather than basal, end to the algal wall, so that thecal secretion prevents it feeding phagotrophically (I); survival would depend on either saprotrophism (II) and evolution of rhizoids (R), or secretion of digestive enzymes allowing penetration of the wall and parasitism of the algal cell (III); with loss of collar and cilium, the theca would become a complete wall (W). The posteriorly uniciliate dispersal phase of the choanociliate (V) is retained.

division (Copeland, 1956; Whittaker, 1969; Margulis, 1974; Corliss, 1984) separates them too sharply from Zygomycotina. 5S rRNA sequences (Walker, 1984a), and the morphological arguments below, support their placement in a single phylum, which I name Archemycota (Table 23.1 & Fig. 23.2).

The origin of the Zygomycotina

Absence of cilia, not possession of zygospores, is what distinguishes Zygomycotina from Chytriomycotina, for the latter do have zygospores, though by tradition rather than logic they are seldom so called. That the ancestral zygomycete lost cilia is strongly suggested by the presence in *Basidiobolus* (McKerracher & Heath, 1985) of a single procentriole consisting of a ring of 11 − 12 singlet microtubules; it even has microtubular roots radiating from it as in many chytriomycetes. In *Chlamydomonas* a single mutation converts 9-triplet centrioles into 9-singlet centrioles (Goodenough & St Clair, 1975); 9-singlet centrioles are stages in triplet basal body assembly (Cavalier-Smith, 1974; Dippell, 1968) and exist in mature centrioles in a few other protists. As tubule numbers other than 9 are known (Schrevel & Besse, 1975), there is little doubt that the *Basidiobolus* procentriole is a degenerate kinetosome, supporting a chytriomycete ancestry for Zygomycotina.

Of the chytriomycete classes (Table 23.1), the Allomycetes are the most likely ancestors, being typically syncytial and lacking dictyosomes; their closed mitosis with centrioles outside the nucleus (Heath, 1986) probably preadapted them to ciliary loss, which would be difficult in Spizomycetes or Rumpomycetes which have polar fenestrae with centrioles directly at spindle poles. Chytriomycete centrioles are integral to spindle poles, so loss of cilia and centrioles would affect centrosomal structure; if this took different directions as the Entomophorales rapidly diversified it would explain why the Entomophthorales are more diverse in mitotic ultrastructure than other fungi (Heath, 1986). If derived taxa (e.g. Mucorales and the three eufungal phyla) each evolved from a different entomophoralean ancestor, they would inherit a different ancestral centrosomal pattern, yet individually be mitotically quite uniform.

An entomophoralean could also have given rise, via the Kickxellales (not necessarily monophyletically; Moss & Young, 1978) to the Trichomycetes.

'Ascomycete' origins

It is commonly thought that euascomycetes evolved from the

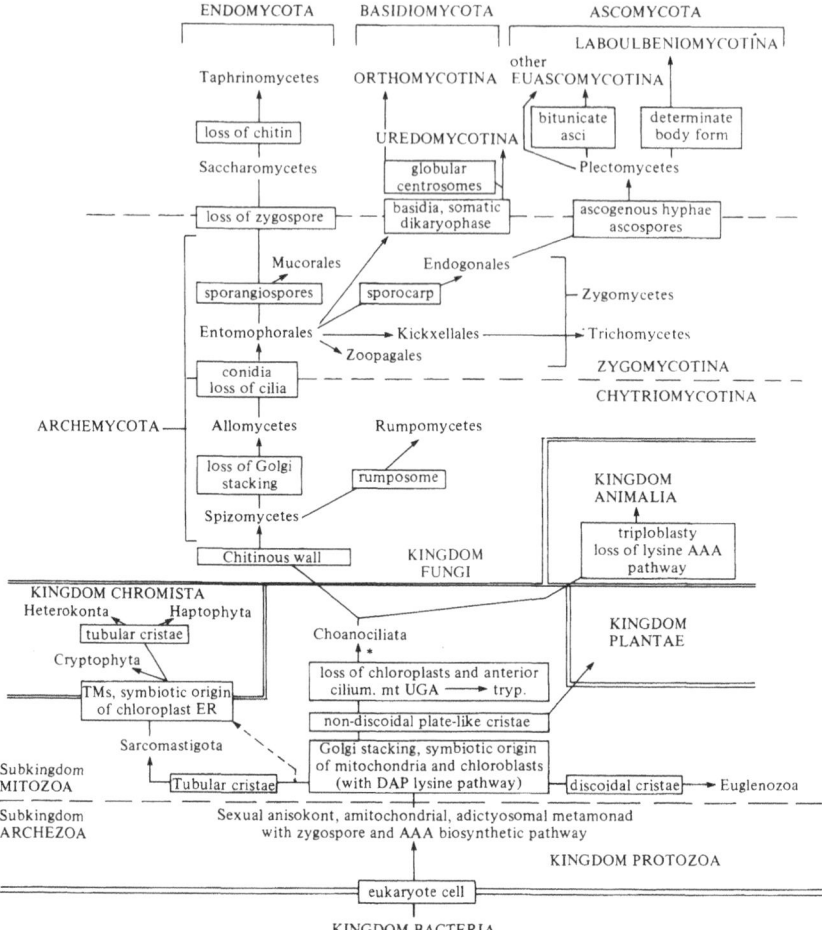

Fig. 23.2. Phylogenetic relationships of the Fungi. The subkingdom Archezoa contains the four protozoan phyla totally without mitochondria or chloroplasts (Cavalier-Smith, 1983). The α-aminoadipic acid (AAA) lysine biosynthetic pathway was retained by Fungi and euglenoids, but replaced in Chromista and Plantae by the diaminopimelic acid (DAP) pathway from the cyanobacterial symbiont that evolved into chloroplasts (Cavalier-Smith, 1981b). The DAP pathway in the Pseudofungi, their mastigonemes (TMs) on the anterior cilium, and their tubular cristae establish their origin from a photosynthetic heterokont (Cavalier-Smith, 1986a). Fungi, Choanociliata and Animalia together form a clade, the Opisthokonta, whose common ancestor (asterisk) had a single, posterior, cilium, but which (as shown by the presence of the second non-ciliated centriole) itself evolved from a biciliated anisokont by losing the anterior cilium. For discussion of the origin of eukaryotes see Cavalier-Smith (1981b, 1987).

simpler hemiascomycetes, but some authors have argued that hemiascomycetes evolved by simplification from euascomycetes. I suggest that neither view is correct, and that there is no direct phyletic link between them and they evolved independently from different zygomycetes. Traditionally, the two groups have been considered a monophyletic unit because unlike other fungi both form endospores. However, presence of endospores is insufficient evidence for monophyly. I argue that hemiascomycete endospores and euascomycete ascospores are products of parallel evolution.

The epigenetic programme producing zygospores could be converted to one producing endospores by two mutations: one delaying secretion of resistant wall materials until after meiosis, and one causing presumptive plasma membrane vesicles to fuse endogenously with each other around the post-meiotic nuclei. Because of the delay in wall secretion this would force the resistant wall materials to be secreted endogenously to form endospores. In euascomycetes the presumptive plasma membrane forms initially as an ascus vesicle around all four meiotic products; it is only secondarily divided up around individual ascospores. In hemiascomycetes the presumptive plasma membrane forms from the start around each endospore (Curry, 1985). Each mode of development seems equally functional and is totally conserved within each taxon. As I see no reason why either should ever evolve into the other, I suggest that both are inherited without change from two different mutations that independently produced endospores in the ancestor of each group. The selective advantage of the change would not have been the formation of endospores per se but the delay of spore formation until after meiosis, yielding a larger number and greater genetic diversity of spores than in zygospore formation.

As hemiascomycetes also lack ascocarps and ascogenous hyphae, there is no good reason to place them in the Ascomycota or to call their spores ascospores; I therefore restrict the Ascomycota to the Euascomycetes, and erect a new phylum (Endomycota) for the hemiascomycetes (Table 23.1 & Fig. 23.2).

Origin of the Ascomycota

The Ascomycota as just redefined can be simply derived from the zygomycete group Endogonales. Only two significant modifications are needed to convert the subterranean endogonalean sporocarp into a plectomycete ascocarp: (i) evolution of endospores from the zygospore, as just discussed; (ii) evolution of ascogenous hyphae.

Development of an ascogenous hypha and of *Endogone* zygospores are similar. In both, the end of a hooked heterokaryotic coenocytic cell is cut off by a septum so as to contain two haploid nuclei that later fuse to form a diploid nucleus that then undergoes meiosis. The difference is that in *Endogone* the hooked cell is produced directly by gametangial copulation, whereas in Ascomycota cell multiplication occurs between copulation and nuclear fusion in each ascogenous hypha; its function is clearly to increase the number of zygotes produced from each gametangial fusion. This will further increase the number and genetic variety of spores produced from each copulation and thus, stems from the same selective forces that favoured the shift from zygospore to ascospores. The mechanism also is similar, i.e. a mutation delaying a developmental process: here the delay of nuclear fusion, allowing growth in the dikaryotic product of syngamy to multiply the dikaryotic cells capable of nuclear fusion.

Three other ascomycete classes probably evolved independently of each other from an early plectomycete by mutations that spread because they favoured aerial ascospore dispersal: active discharge involving bitunicate asci in Loculoascomycetes; and open perithecia or apothecia in Pyrenomycetes and Discomycetes. The loss of mycelial character in Laboulbeniales may adapt them to ectoparasitism on small discrete insect hosts in contrast to ramifying plants. The marine Spathulosporales, which parasitise red algae, are fundamentally similar; the main differences are their uninucleate, rather than binucleate ascospores, and greater development of sterile perithecial tissue. The determinate development of Laboulbeniomycetes differs so strongly from mycelial ascomycetes that I propose a new subphylum, Laboulbeniomycotina, for them; the subphylum Euascomycotina is thus restricted to the four mycelial classes: Plectomycetes, Loculoascomycetes, Pyrenomycetes, Discomycetes (Table 23.1).

My view of Spathulosporales as a specialised terminal group contrasts with Kohlmeyer's (1975, 1979) version of the red algal theory of ascomycete origins, which sees them as a link between red algae and Euascomycotina. There are six major objections to a red algal origin for ascomycetes: (i) it requires total loss of their plastids, yet all parasitic, non-photosynthetic red algae studied by electron microscopy retain leucoplasts (Goff, 1982), implying that red algal plastids play an indispensable biochemical role other than photosynthesis; (ii) the red algal Golgi apparatus consists of dictyosomes, i.e. stacks of cisternae, whereas ascomycete Golgi cisternae are unstacked, as in Zygomycetes; (iii) Florideophycidae (the red algal subclass proposed as ancestral to ascomycetes) have a well-developed diplophase, absent in Ascomycota (though present in some Endomycota); (iv) chitin is unknown in red algae; (v) lysine biosynthesis in red algae is by the DAP pathway, not the AAA pathway found in all Fungi; (vi) parasitic red algal biology is very different from that of fungi; the algal parasite cells become cytoplasmically continuous with those of the host and their organelles mingle (Goff & Coleman, 1985), which may be why parasites are always close relatives of their hosts.

There is no specific evidence for a direct link between fungi and red algae. A transition from an *Endogone*-like zygomycete to a plectomycete involves far fewer and less drastic changes than from a red alga to *Spathulospora*; their selective advantages are also clear, which is not so for the loss of leucoplasts and diploidy, shift from cellulose to chitin, and from DAP to AAA lysine biosynthesis − all necessary on the red algal hypothesis, which should be put to rest.

The origin of the Endomycota

Endomycota differ from ascomycetes in endospore development and in lacking ascocarps and ascogenous hyphae, as already discussed, and also in having little or no chitin in their walls. I propose two distinct classes (Table 23.1): (i) Saccharomycetes: which comprise Endomycetales, i.e. the budding, endospore-forming yeasts (e.g. *Saccharomyces*) and their filamentous relatives (*Dipodascus & Eremascus*); (ii) Taphrinomycetes: including Taphrinales and the fission yeast *Schizosaccharomyces* (both totally lack chitin), and the Protomycetales (wall chemistry unknown). By analogy with *Mucor*, and with Basidiomycota where eight different orders have given rise to yeasts by simplification (Oberwinkler, 1982), I think in both classes yeasts are derived and the filamentous forms ancestral; molecular sequences are needed from filamentous endomycetes to test this.

Because filamentous Endomycota are structurally so simple, morphology does not reveal a likely zygomycete ancestor. If ancestral *Taphrina* were parasitic on ferns, not flowering plants, an obvious possibility would be an entomophoralean fern-parasite like *Completoria*. Like the entomophoralean desmid-parasite *Ancylistes* it is as somatically reduced as the filamentous Endomycota, so their common ancestor might have become similarly reduced as a cellular endoparasite of lower green plants. But if Saccharomycetes evolved from a taphrinomycete, this must have preceded their total loss of chitin. A more plausible ancestral endomycote would be a filamentous saccharomycete with chitinous walls and a large and indefinite number of endospores, e.g. *Dipodascus*. *Dipodascus* copulates by gametangia, like zygomycetes; many zygomycetes produce numerous post-meiotic endospores (i.e. the sporangiospores); in some, e.g. Mortierellaceae, the zygospore is commonly suppressed. The so-called 'ascus' of *Dipodascus* may, therefore, be a non-stalked, non-columellate, zygomycete-type of sporangium. The gap between *Dipodascus* and such zygomycetes is so small that one could almost put the endomycetes as a zygomycete subclass; the only change, beyond suppression of the zygospore, would be to shorten the life cycle so that sporangia develop immediately after meiosis. All Endomycota have clearly been selected for rapid growth and concomitant somatic simplification, which also accounts for the usual presence of only eight or four endospores. Such an origin by reducing the number of sporangiospores explains why they develop differently from ascospores, which originally were four in number and secondarily increased to eight in many species.

Loss, or restriction to the bud scar, of chitin probably occurred during the independent origin from a filamentous ancestor like *Eremascus* of the budding and fission endomycete yeasts. *Taphrina*, I suggest, evolved from a chitin-free *Schizosaccharomyces*-like ancestor that became an endoparasite on ferns; its unique life cycle with diploid, haploid and dikaryotic phases seems a terminal development unrelated to the dikaryophase in Basidiomycota and Ascomycota where diploid cells do not normally divide.

The fact that mitochondrial DNA of the ascomycetes *Aspergillus* and *Neurospora* codes for about 14 polypeptides (Sederoff, 1984) and animal mtDNA for 13, suggests this was also true of the ancestral opisthokont and that in endomycetes (where it codes for only about eight polypeptides) several mitochondrial genes were transferred to the nucleus. Like their wall chemistry, this suggests that Endomycota were not ancestors of Ascomycota. Study of mitochondrial DNA of *Dipodascus*, *Eremascus* and a variety of zygomycetes would test the present thesis, and perhaps show from which zygomycete group Endomycota evolved.*

*An inserted uracil base in 5S RNA, between positions 52 and 53, which is unique to Endomycetales and Ascomycota, suggests that the former may have evolved from the latter and be unrelated to Taphrinomycetes (Erdmann & Wolters, 1986) which, alone, may have evolved as suggested here. *Note added in proof:* For this and other reasons I would now place the Taphrinomycetes in the Zygomycontina and the Saccharomycetes as a third subphylum in the Ascomycota.

The origin of the Basidiomycota

Basidiomycota differ from ascomycetes in two fundamental respects: their post-meiotic spores (basidiospores) are exospores, not endospores, and they have a vegetative heterokaryophase. Past attempts to homologise basidia and asci have been wishful thinking. They are epigenetic alternatives to the same end that work equally well, so there would be no value in changing over from one to another. Developmentally it is easier to envisage an origin of basidiospores from entomophoralean conidia than from ascospores. Basidiospores are formed by conidia-like budding of the presumptive spores and their separation from the metabasidium by sterigmal septa.

Entomophorales typically have actively discharged conidia formed in basically this way; such a zygomycete parasitic on a fern is a more suitable ancestor for rusts than the *Taphrina* fern-parasite suggested by Savile (1968, 1976). Apart from the improbable switch from endospores to exospores, *Taphrina* has no chitin (Petit & Schneider, 1983) and so could not be ancestral to Basidiomycota. Two key changes are necessary to form a basidiomycete from an entomophoralean: (i) evolution of the 'drop' spore discharge mechanism which differs from any of the four methods used by Entomophorales (Tucker, 1981); (ii) suppression of nuclear fusion in the zygospore and its connection instead to conidia formation – this could in principle be achieved by a single transposition, placing a master nuclear fusion gene promoter under the control of conidiation controlling elements. In one step, this would convert a ballistosporous conidiophore into a basidium and create a vegetative dikaryophase.

Thus, the ancestral basidiomycete would have had zygospores as well as basidiospores – analogous to many 'teliomycetes' where the teliospore, like a zygospore, is a premeiotic structure. But teliospores differ from zygospores in not being immediate products of cell of hyphal fusion. However, the very transposition that temporally shifted nuclear fusion from syngamy to preconidiation possibly did the same for zygospore formation, which would have been epigenetically linked to it; if so the teliospore developmental programme may be directly derived from the zygospore programme. Even if the teliospore in Uredinales is derived, it may be a reversion to an ancestral state, rather than entirely new. The transverse septation of teliomycete metabasidia supports their derivation from vegetative conidiophores; the similar metabasidia of the Auriculariales and Septobasidiales have long suggested that they are derived from teliomycetes. But the fundamental basidiomycete cleavage is not between teliomycetes and others, nor between hetero- and homobasidiomycetes; it is (Savile, 1976) between rusts (here put in their own subdivision Uredomycotina) and the other basidiomycetes (here designated Orthomycotina) which must have diverged in the very early centrosome diversification phase of eufungal evolution (Fig. 23.3). The five orders grouped here as Septomycetes may seem a mixed bag — indeed they are, but this is, I think, because they are the most ancient orthomycete group within which much diversification occurred (e.g. origin of dolipore septa) and from which

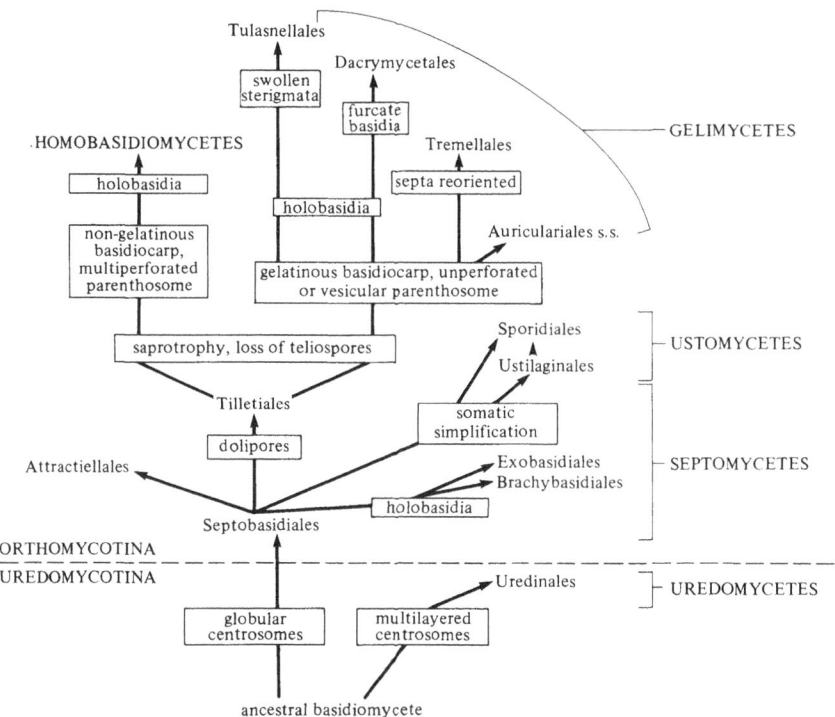

Fig. 23.3 Proposed phylogeny for Basidiomycota. This provides the rationale for subdivision into five classes and their grouping into Orthomycotina and Uredomycotina. 5S RNA sequences (Walker & Doolittle, 1983; Walker, 1984b) support a link between Exobasidiales and Tilletiales and suggest that Septomycetes are intermediate between Ustomycetes and Gelimycetes/Homobasidiomycetes. Septal diversity (Khan & Kimbrough, 1982) suggests that some Tremellales and many 'Auriculariales' belong in the Septomycetes.

many intermediates are extinct. From these parasitic septomycetes evolved the Ustomycetes by somatic simplification, and the basically saprophytic Homobasidiomycetes and Gelimycetes with their macroscopic basidiocarps. Large basidiocarps with dense hymenia create a problem for the liberation of lower basidiospores from a transversely septate metabasidium, as they will tend to get trapped among other basidia. This problem was solved in three different ways: in Auriculariales by elongating the lower sterigmata to place all basidiospores on the same level; in Tremellales by reorienting the septa to make them longitudinal; in 'Holobasidiomycetes' by losing the septa and reorienting the sterigmata. On this view of basidiomycete divergence (Fig. 23.3) Phragmobasidiomycetes (septate basidia) and the Holobasidiomycetes (non-septate basidia) are artificial groups; the holobasidia of the Tulasnellales, Dacrymycetales, Exobasidiales, and Homobasidiomycetes all seem convergent responses to the same selective force.

The classical ascomycete theory of basidiomycete origin converts the transient reproductive ascomycete dikaryophase into the long-lived vegetative dikaryophase of basidiomycetes, but this would require drastic changes with no evident selective advantage. Evolution of a dikaryophase merely involves a delay in nuclear fusion, and so could readily evolve convergently. On the present zygomycete theory the vegetative dikaryophase is merely the inevitable byproduct of the mutation that converted ballistosporous conidiophores into basidia; its selective advantage was the vast increase in genetic diversity of spores from a single copulation. On this view the quite different life cycles and spore development of the three phyla of higher fungi evolved not because their actual forms were directly selected, but because they were necessary epigenetic consequences of mutations increasing spore diversity.

Acknowledgements I thank numerous colleagues for preprints or reprints, the Royal Society for a Travelling Fellowship, the Australian National University for a Visiting Fellowship, Des Clark-Walker for stimulating discussions, and David Moore for presenting my paper at the symposium.

References

Barr, D. J. S. (1981). The phylogenetic and taxonomic implications of flagellar rootlet morphology among zoosporic fungi. *BioSystems*, **14**, 359 − 370.

Barr, D. J. S. (1983). The zoosporic grouping of plant pathogens: entity or non-entity? In *Zoosporic Plant Pathogens*, ed. Buczacki, S. J., pp. 43 − 83. London: Academic Press.

Cavalier-Smith, T. (1974). Basal body and flagella development during the vegetative cell cycle and the sexual cycle of *Chlamydomonas reinhardii*. *Journal of Cell Science*, **16**, 529 – 556.

Cavalier-Smith, T. (1981a). Eukaryote kingdoms: seven or nine? *BioSystems*, **14**, 461 – 481.

Cavalier-Smith, T. (1981b). The origin and early evolution of the eukaryote cell. In *Molecular and Cellular Aspects of Microbial Evolution*, Society for General Microbiology Symposium 32, ed. Carlile, M. J., Collins, J. F. & Moseley, B. E. B., pp. 33 – 84. Cambridge University Press.

Cavalier-Smith, T. (1982). The origins of plastids. *Biological Journal of the Linnean Society*, **17**, 289 – 306.

Cavalier-Smith, T. (1983). A 6-kingdom classification and a unified phylogeny. In *Endocytobiology II*, ed. Schwemmler, W. & Schenk, H. E. A., pp. 1027 – 1034. Berlin: de Gruyter.

Cavalier-Smith, T. (1986a). The kingdom Chromista: origin and systematics. In *Progress in Phycological Research*, vol. 4, ed. Round, F. E. & Chapman, D. J., 309 – 347. Bristol: Biopress.

Cavalier-Smith, T. (1986b). Cilia versus undulipodia. *BioScience*, **36**, 293.

Cavalier-Smith, T. (1987). The origin of eukaryote and archaebacterial cells. *Annals of the New York Academy of Sciences*, in press.

Copeland, H. F. (1956). *The Classification of the Lower Organisms*. Palo Alto, California: Pacific Books.

Corliss, J. O. (1984). The kingdom Protista and its 45 phyla. *BioSystems*, **17**, 87 – 126.

Curry, K. J. (1985). Ascosporogenesis in *Dipodascopsis tothii* (Hemiascomycetidae). *Mycologia*, **77**, 401 – 411.

De Bary, A. (1887). *Comparative Morphology of Fungi, Mycetozoa and Bacteria*. Oxford: Clarendon Press.

Dick, M. W. (1976). The ecology of aquatic Phycomycetes. In *Recent Advances in Aquatic Mycology*, ed. Jones, E. B. G., pp. 513 – 542. London: Elek.

Dippell, R. V. (1968). Development of basal bodies in *Paramecium*. *Proceedings of the National Academy of Sciences, USA*, **61**, 461 – 468.

Erdmann, V. A. & Wolters, J. (1986). Collection of published 5S, 5·8S and 4·5S ribosomal RNA sequences. *Nucleic Acids Research*, **14**, r1 – r59.

Goff, L. J. (1982). The biology of parasitic red algae. In *Progress in Phycological Research*, vol. 1, ed. Round, F. E. & Chapman, D. J., pp. 289 – 369. Amsterdam: Elsevier Biomedical Press.

Goff, L. J. & Coleman, A. W. (1985). The role of secondary pit connections in red algal parasitism. *Journal of Phycology*, **21**, 483 – 508.

Goodenough, U. W. & St. Clair, H. S. (1975). Bald-2: a mutation affecting the formation of doublet and triplet sets of microtubules in *Chlamydomonas reinhardtii*. *Journal of Cell Biology*, **66**, 480 – 491.

Heath, I. B. (1986). Nuclear division: a marker for protist phylogeny? In *Progress in Protistology*, vol. 1, ed. Corliss, J. O. & Patterson, D. J., in press. Bristol: Biopress.

Hibberd, D. J. (1975). Observations on the ultrastructure of the choanoflagellate *Codosiga botrytis* (Ehr.) Saville-Kent, with special reference to the flagellar apparatus. *Journal of Cell Science*, **17**, 191 – 219.

Khan, S. R. P. & Kimbrough, J. W. (1982). A reevaluation of the basidiomycetes based upon septation and basidial structures. *Mycotaxon*, **15**, 103 – 120.

Kohlmeyer, J. (1975). New clues to the possible origins of Ascomycetes. *BioScience*, **25**, 86 – 93.

Kohlmeyer, J. & Kohlmeyer, E. (1979). *Marine Mycology: the Higher Fungi*. New

York: Academic Press.

Kreisel, H. (1969). *Grundzüge eines natürlichen Systems der Pilze.* Lehre: Cramer.

Leadbeater, B. S. C. (1983). Observations on the life history and ultrastructure of the marine choanoflagellate *Proterospongia choanojuncta. Journal of the Marine Biological Association, UK*, **63**, 135 − 160.

Leadbeater, B. S. C. & Morton, C. (1974). A microscopical study of a marine species of *Codosiga* James Clark (Choanoflagellata) with special reference to the ingestion of bacteria. *Biological Journal of the Linnean Society*, **6**, 337 − 347.

Margulis, L. (1974). Five kingdom classification and the origin and evolution of cells. *Evolutionary Biology*, **7**, 45 − 78.

McKerracher, L. J. & Heath, I. B. (1985). The structure and cycle of the nucleus-associated organelle in two species of *Basidiobolus. Mycologia*, **77**, 412 − 417.

Moss, S. T. & Young, T. W. K. (1978). Phyletic considerations of the Harpellales and Asellariales (Trichomycetes, Zygomycotina) and the Kickxellales (Zygomycetes, Zygomycotina). *Mycologia*, **70**, 944 − 963.

Oberwinkler, F. (1982). The significance of the morphology of the basidium in the phylogeny of basidiomycetes. In *Basidium and Basidiocarp*, ed. Wells, K. & Wells, E. K., pp. 9 − 35. New York: Springer-Verlag.

Petit, M. & Schneider, A. (1983). Chemical analysis of the wall of the yeast form of *Taphrina deformans. Archives of Microbiology*, **135**, 141 − 146.

Powell, M. J. (1978). Phylogenetic implications of the microbody-lipid globule complex in zoosporic fungi. *BioSystems*, **10**, 167 − 180.

Savile, D. B. O. (1968). Possible interrelationships between fungal groups. In *The Fungi, an Advanced Treatise*, vol. III, ed. Ainsworth, G. C. & Sussman, A. S., pp. 649 − 675. New York: Academic Press.

Savile, D. B. O. (1976). Evolution of the rust fungi (Uredinales) as reflected by their ecological problems. *Evolutionary Biology*, **9**, 137 − 207.

Schrevel, J. & Besse, C. (1975). Un type flagellaire fonctionnel de base 6+0. *Journal of Cell Biology*, **66**, 492 − 507.

Sederoff, R. R. (1984). Structural variation in mitochondrial DNA. *Advances in Genetics*, **22**, 1 − 108.

Shaffer, R. L. (1975). The major groups of Basidiomycetes. *Mycologia*, **67**, 1 − 18.

Tucker, B. E. (1981). A review of the non-entomogenous Entomophorales. *Mycotaxon*, **13**, 481 − 505.

von Arx, J. A., van der Walt, J. P. & Liebenberg, N. V. D. M. (1982). The classification of *Taphrina* and other fungi with yeast-like cultural states. *Mycologia*, **74**, 285 − 296.

Walker, W. F. (1984a). 5S ribosomal RNA sequences from Zygomycotina and evolutionary implications. *Systematic and Applied Microbiology*, **5**, 448 − 456.

Walker, W. F. (1984b). 5S rRNA sequences from Attractiellales, and basidiomycetous yeasts and fungi imperfecti. *Systematic and Applied Microbiology*, **5**, 352 − 359.

Walker, W. F. & Doolittle, W. F. (1983). 5S rRNA sequences from eight basidiomycetes and fungi imperfecti. *Nucleic Acids Research*, **11**, 7625 − 7630.

Whittaker, R. H. (1969). New concepts of kingdoms of organisms. *Science*, **163**, 150 − 160.

24

Yeasts and anastomoses: their occurrence and implications for the phylogeny of Eumycota

H . PRILLINGER

Botanical Institute, University of Regensburg, D-8400 Regensburg, FRG

Introduction

The advent during the last thirty years of vast amounts of new ultrastructural, biochemical, and genetic data forces fungal phylogeneticists to reconsider the significance of yeasts and sexuality in the evolution of chitinous Eumycota (i.e. the Chytridio-, Zygo-, Asco-, and Basidiomycetes). In this chapter these neglected but fundamental issues will be discussed first with respect to concepts of the evolution of sexuality, secondly to the role of sexual and vegetative anastomoses in the evolution of heterothallism, and thirdly to the relevance of yeast-stages in speciation processes and phylogeny.

Concepts of sexuality in Eumycota

Two markedly different concepts of sexuality in Eumycota have arisen, depending on the primary event(s) or mechanism believed to be involved (Prillinger, 1982, 1984). According to the view favoured here, the primary mechanism is 'sexual differentiation'; this being the process which leads to karyogamy and meiosis either within the same strain (homothallism) or after the crossing of two different mating types (heterothallism). According to the other view, the primary mechanism is the polarity resulting from homogenic incompatibility, a phenomenon which suppresses karyogamy and meiosis in monoecious organisms. Adoption of one or other of these viewpoints profoundly affects evolutionary interpretations (Table 24.1).

To explain the evolution of sexuality via sexual differentiation in Eumycota, two separate steps seem to be essential: (i) polyphyletic evolution from mitotic to meiotic life cycles in polykaryotic and coenocytic homothallic organisms; (ii) polyphyletic origin of heterothallism in

Table 24.1. *Homogenic incompatibility and sexual differentiation; two different concepts of sexuality in Eumycota and their implications for evolutionary interpretations*

Issue	Concept	Interpretation
Occurrence of mating types	Homogenic incompatibility	Arise by genotypically determined suppression of karyogamy and meiosis within a monoecious organism; not caused by sterility defects of gametes or nuclei (Esser, 1962, 1971; Esser & Kuenen, 1965)
	Sexual differentiation	Genetically differentiated elements of polyphyletic origin which control the process of sexual reproduction (cf. Kniep, 1922; Burgeff, 1924; Bauch, 1930; Prillinger, 1982)
Sexual dimorphism	Homogenic incompatibility	Primitive
	Sexual differentiation	Advanced (spermatia derived from yeasts; cf. Deml *et al.*, 1982a & b)
Polykaryotic (multinucleate cells, hyphae or hyphal compartments)	Homogenic incompatibility	Derived (Pascher, 1931; Zickler, 1952)
	Sexual differentiation	Primitive (cf. Batko, 1974)
Occurrence of yeasts	Homogenic incompatibility	Derived taxa of fungi, extremely reduced Ascomycetes (de Bary, 1884)
	Sexual differentiation	Primitive mode of morphogenesis
Occurrence of Ascomycetes and Basidiomycetes	Homogenic incompatibility	Derived from autotrophic algae (Rhodophyceae, Demoulin, 1974) or heterotrophic Oomycetes (e.g. Savile, 1955)
	Sexual differentiation	Closely related to Zygomycetes and Chytridiomycetes (von Wettstein, 1921; Bartnicki-Garcia, 1970; Vogel *et al.*, 1970; Fuller, 1976)

different groups of fungi (Fig. 24.1). During the first step, sexuality, in the sense of genetic recombination between different nuclei, is selected by deprivation of nutrients, especially nitrogen, after a period of vegetative multiplication. Possible ultrastructural evidence for polyphyletic evolution from mitosis to meiosis comes from studies of synaptonemal complexes in meiotic prophase-I nuclei. Despite claims for the universal occurrence of synaptonemal complexes in eukaryotic organisms displaying four-strand crossing over (von Wettstein, Rasmussen & Holm, 1984), typical examples have not been detected in *Schizosaccharomyces pombe* (Olson, *et al.*, 1978; Hirata & Tanaka, 1982), *Aspergillus nidulans* (Egel-Mitani, Olson & Egel, 1982), *Microbotryum (Ustilago) violaceum* (Prillinger, 1984), *Phleogena faginea* and *Agaricostilbum pulcherrimum* (R. Bauer, personal communication). Moreover, although linear rather than typical tripartite structures do appear in *S. pombe*, no structures resembling synaptonemal complexes have yet been detected in *A. nidulans*.

The data on synaptic meiosis are indicated in Fig. 24.1, where the broad black arrow indicates progression towards synaptic meiosis, and the thinner arrows signify the possibly different situations in *A. nidulans* and *S. pombe*. Currently, there are no data suggesting evolution of synaptic meiosis within a homothallic fungus before heterothallism was established in the Eumycota. So, how could heterothallism evolve via sexual differentiation? In chitinous fungi, this question is inextricably connected with the occurrence of sexual and vegetative anastomoses.

Sexual and vegetative anastomoses in chitinous fungi

Van Tieghem (1875) was, apparently, the first to record the absence of vegetative anastomoses in the polykaryotic Zygomycete genera *Pilobolus*, *Pilaira*, *Mucor*, *Phycomyces*, *Spinellus*, *Syzygites* (*Sporodinia*), *Chaetostylum*, *Helicostylum*, *Thamnidium*, *Chaetocladium*, *Rhizopus* and *Circinella*. However, he did observe vegetative anastomoses in *Mortierella*, *Syncephalis* and *Piptocephalis*. These data have partially been confirmed by other authors and to the list without vegetative anastomoses may be added *Cunninghamella*, and several members of the Mycotyphaceae, including *Mycotypha africana*, *M. microspora*, *Benjaminiella multispora* and *B. poitrasii* (Prillinger, 1984; Forst & Prillinger, unpublished). Moreover, from the literature on parasitic Zygomycetes, the Entomophthorales and especially the Zoopagaceae (Zoopagales), there is no indication of the occurrence of vegetative anastomoses, except for a few nematode-trapping fungi (Drechsler, 1933).

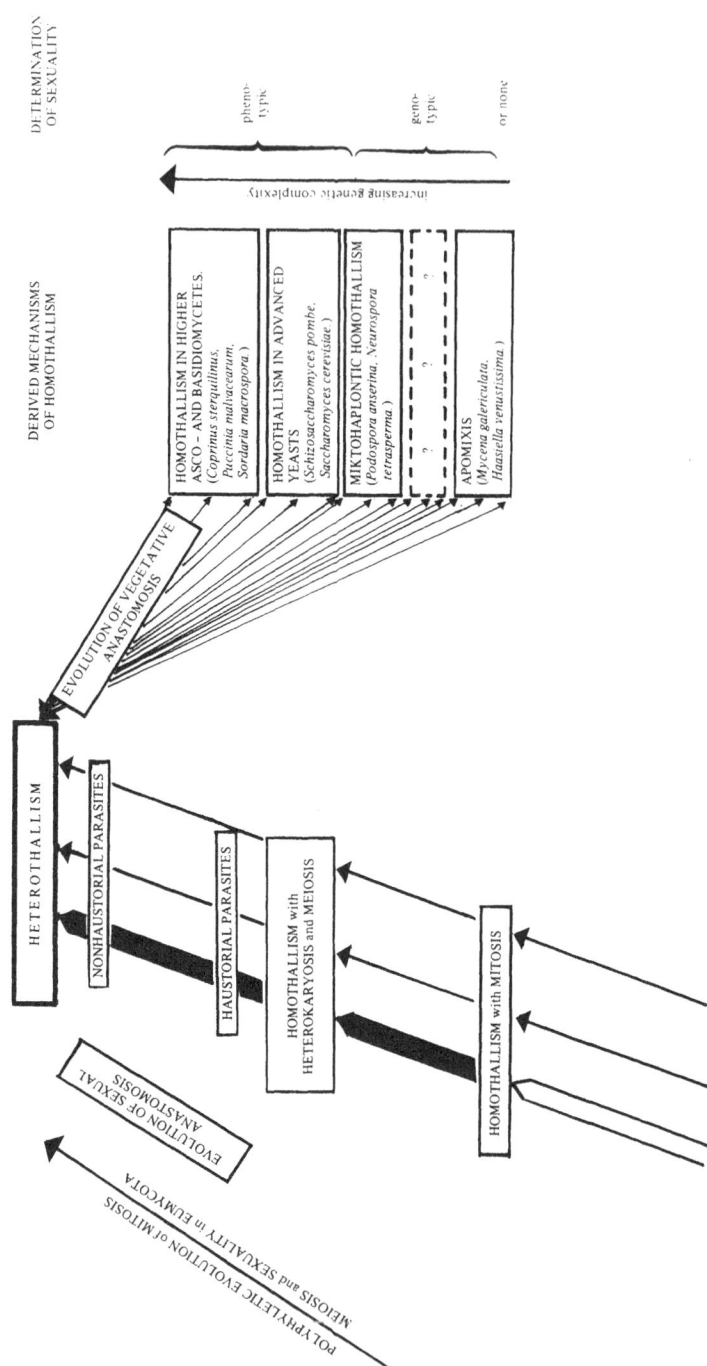

Fig. 24.1. Evolution of sexuality in Eumycota.

In the flagellate Chytridiomycetes vegetative anastomosis has only been reported in the genus *Coelomyces* (Keilin, 1921; Couch, 1945). However, the polykaryotic hyphae of these fungi occur parasitically in the fat body of mosquito larvae and do not possess cell walls, so they may be a special case. Also, the original report of vegetative anastomoses has been questioned (Umphlett, 1962; Martin, 1969; Bland & Couch, 1985).

Vegetative anastomoses do, however, commonly appear in advanced groups of Zygomycetes with septate, oligokaryotic mycelium, e.g. in *Martensiomyces pterosporus* and *Linderina pennispora* of the Kickxellaceae (Prillinger, 1984). According to Zycha, Siepmann & Linneman (1969), vegetative anastomoses also occur in the Endogonaceae.

Although vegetative anastomoses are abundant in Ascomycetes and Basidiomycetes there have, until recently, been few specific studies subsequent to the classical work of Buller (1931, 1933). Consequently, it is not yet clear whether vegetative anastomoses are lacking in some primitive groups, at least in early ontogenetic stages. However, vegetative anastomoses are absent from the pseudomycelium of various homobasidiomycetous yeasts (see below), and from the siphonal (coenocytic) mycelial front of certain *Phlebia* species (Boddy & Rayner, 1983).

Considering the proposed main line of evolution in Eumycota from polykaryotic to dikaryotic hyphae (Prillinger, 1984), clamp connections seem to represent the most advanced type of hyphal anastomoses, ensuring a definite number of nuclei (commonly two) in neighbouring hyphal compartments. They may also indicate that it needs a long time to establish synchrony between cell and nuclear divisions during truly mycelial morphogenesis. According to the interpretation in Fig. 24.1, clamp connections may have evolved within a derived homothallic organism, the whorled clamps found in rather primitive groups of Basidiomycetes (Kemper, 1937; Thielke, 1984; Boidin *et al.*, 1985; Ainsworth, Chapter 19) being the phylogenetic precursor of this line. Clamps as a well-established indicator of sexuality in Basidiomycetes form an intermediate stage between sexual and vegetative anastomoses.

Sexual anastomoses occur in a significantly broader taxonomic range of Eumycota than vegetative anastomoses, notably in the Zoopagales, Entomophthorales and the Mucorales. This suggests that sexual anastomoses are phylogenetically more primitive than vegetative anastomoses which possibly arose via repeated alternations of heterothallism and derived mechanisms of homothallism (Fig. 24.1).

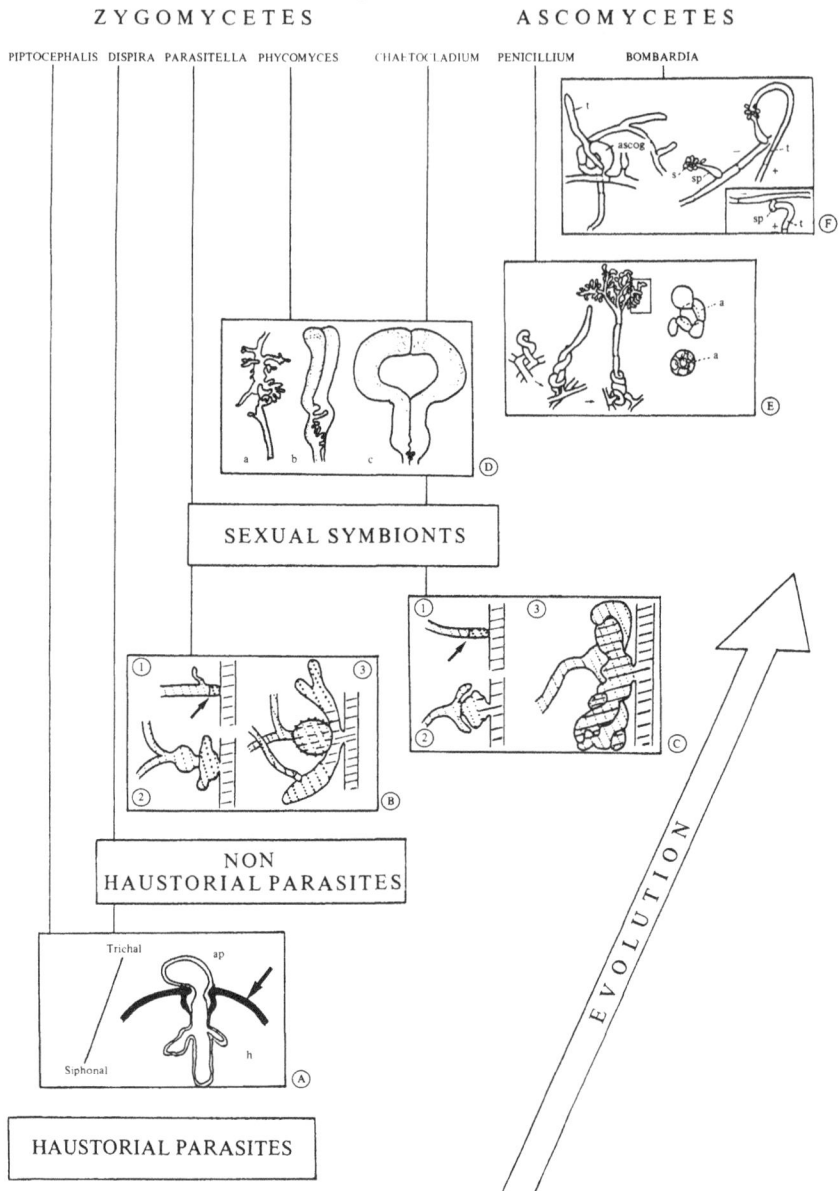

Fig. 24.2. Evolution of heterothallism in Eumycota. (A) Haustorial parasites (after Jeffries & Young, 1976), ap = appressorium, h = haustorium, arrow shows cell wall of fungal host. (B) & (C) Non-haustorial parasites (after Burgeff, 1920, 1924; Satina & Blakeslee, 1926), hatched = nuclei &

A scheme for the evolution of sexual anastomoses appears in Fig. 24.2. The following points are salient to the development of this scheme. The coming together of genetically different nuclei (heterokaryosis) is greatly enhanced if two polykaryotic organisms integrate their cytoplasmic contents, as actually occurs in the parasitism of *Chaetocladium brefeldii* on Zygomycetes (Fig. 24.2C; and see below). As suggested by Savile (1955), parasitism is a primitive trait, at least in aflagellate chitinous fungi, this being confirmed for various Basidiomycetes, especially the Auriculariales, by ultrastructural data (Oberwinkler & Bandoni, 1984; Oberwinkler, 1986). Fungal phylogeny may, therefore, be traced back to the concept of coevolution of parasites and their hosts (cf. Savile, 1976, 1979). Evolution from haustorial parasites to a mutualistic fungal-algal interaction, involving hyphal sheaths or non-haustorial globular envelopes was recently reported by Oberwinkler (1984) in the Basidiolichenes, and correlated well with morphological differentiation of the thallus. Early ontogenetic stages in the formation of sexual anastomoses in some Zygomycetes are reminiscent of an originally parasitic hyphal interaction (Fig. 24.2D).

For the evolution of heterothallism three steps are postulated based on comparative morphology of extant organisms. A good example of *haustorial parasitism* is provided by extant Zoopagales, especially the Piptocephalidaceae. Members of this family are generally obligately parasitic on the Mucorales (*sensu* Benjamin, 1979). However, remarkable exceptions are *Syncephalis wynneae* on the ascomycete *Wynnea macrotus* (Thaxter, 1897) and *Piptocephalis xenophila* on the ascomycetes *Byssochlamys fulva*, *Acremonium* species, *Diplodina* species, *Neurospora tetrasperma*, *Venturia inaequalis* and different species of *Penicillium* (cf. Fig. 24.2E) and *Aspergillus* (Dobbs & English, 1954). Unlike *Piptocephalis*, species of *Syncephalis* produce intracellular hyphae with unlimited growth, prevent the sporulation of their hosts (Zycha, *et al.*, 1969), commonly have a broader host range and can be cultivated, with

cytoplasm of the host, stippled = nuclei and cytoplasm of the parasite, hatched and stippled = heterokaryosis, the arrow indicates septum formation in the parasite. (D) Early successive stages in the gametangiogamy of *Phycomyces nitens* (Orban, 1919; Burgeff, 1924). (E) Successive stages of gametangiogamy (arrows) in *Penicillium stipitatum* (after Emmons, 1935), note the budding asci (a). (F) Spermatia–trichogyne fertilisation in *Bombardia lunata* (after Zickler, 1937, 1952), s = spermatia; insert shows a spermatogonium (sp) fusing with a trichogyne (t).

difficulty, in pure culture (Ellis, 1966). These features seem likely to be primitive. Furthermore, the bladder-like outgrowth after the characteristic entwining of sexual hyphae in *Syncephalis*, as beautifully demonstrated by Thaxter (1897), renders a '*Syncephalis*-like organism' of particular interest in the possible evolution of sexuality and fruit bodies in Asco-mycetes (cf. Fig. 24.2E). The septate Dimargaritales exhibit a host range and pattern of development of haustoria (with a slightly swollen appres-sorium and a more or less restricted, branched haustorium) similar to *Piptocephalis* but several species have been grown in axenic culture, so that their parasitism is not obligate (Benjamin, 1979). They occur mostly on mucoralean hosts, the ascomycetes *Chaetomium* and *Monascus* being interesting exceptions (Ayers, 1935; Brunk & Barnett, 1966; Mandelbrot & Erb, 1972).

The morphological structures and physiological reactions involved in *non-haustorial parasitism* are so similar to some processes of sexual anastomosis as to strongly suggest that the evolution of heterothallism could have taken this route. The process was described by Burgeff (1920, 1924) in *Parasitella* and *Chaetocladium*, although he interpreted the sex-limited parasitism of *P. simplex* on the heterothallic *Absidia glauca* as a possible route for evolution of this type of parasitism *from* a sexual reaction.

The host range of *Parasitella* and *Chaetocladium* within the Mucorales is smaller than that of the haustorial parasites *Syncephalis*, *Piptocephalis* and *Dispira* (Burgeff, 1924, Ayers, 1935). The parasitic interactions start with septum formation in that part of the parasitic hypha touching the host (Fig. 24.2B[1] & C[1]; cf. gametangium formation). The cell in contact fuses with the hypha of the host (Fig. 24.2B[2] & C[2]) and becomes a gall. Whereas nuclei and cytoplasm of the parasite and host within the gall remain distinct in *Parasitella*, integration of the different nuclei and cytoplasms does occur in *Chaetocladium*. Moreover, though *Parasitella* produces only a zygospore-like resting spore in connection with the gall (Fig. 24.2B[3]), *Chaetocladium* grows out over and surrounding the gall and envelops it with spirally twisted hyphae (Fig. 24.2C[3]). Two further observations by Satina & Blakeslee (1926) are of special interest here. Unlike Burgeff, they found stages in the parasitism of *Parasitella* on certain *Mucor* species, where the intervening septum wall became dissolved and nuclei entered the resting spore. Satina & Blakeslee (1926) also noticed that when grown together saprotrophically, the opposite sexes of *Parasitella* formed numerous zygospores and some perfect galls.

I regard the non-haustorial parasitism of *Chaetocladium* as a land-mark on the road to heterothallism in Ascomycetes. Fig. 24.2E shows an ascocarp growing out from the twisted hyphae in the probably derived homothallic *Penicillium stipitatum*. Similarly, mitotic sporangiophores can grow out from the gall of *Chaetocladium* (Burgeff, 1920, 1924). The processes in *Parasitella*, on the other hand, are on the route to sexual anastomosis in the Zygomycetes (Fig. 24.2D). In agreement with Savile (1968), fertilisation by means of spermatia and a trichogyne has developed by convergent evolution in Ascomycetes, Basidiomycetes and Rhodophyceae. A transition from gametangiogamy to spermatia-trichogyne fertilisation is shown in Fig. 24.2F in *Bombardia lunata* (after Zickler, 1937, 1952).

To conclude this section it should briefly be mentioned that, at least in the primitive heterobasidiomycetous 'bunt' fungi, sexual anastomoses are not always species specific (and similarly, for Zygomycetes see Schipper, Chapter 17). Thus, Kniep (1926) obtained fusions either between any two species of dicotyledonous bunts with reticulate spores or between any two species of monocotyledonous bunts with smooth or punctate spores. The promycelia of *Ustilago nuda* and *U. tritici*, which do not produce yeast stages, fused with yeast stages of *U. hordei* and *U. bromivora*. No fusions occurred between monocotyledonous and dicotyledonous bunts. This finding is in accord with modern phylogenetic concepts of bunt taxonomy (Fig. 24.3; Blanz & Gottschalk, 1984; Deml, 1986; Deml, Oberwinkler & Bauer, 1985; Gottschalk & Blanz, 1985) and the above concept on the evolution of heterothallism.

Yeasts and the phylogeny of Eumycota

Only three concepts on the phylogeny of yeasts were well established in the mycological literature, those of de Bary (1884), Brefeld (1891) and Corner (1971), until a fourth was recently introduced by Oberwinkler (1977, 1982, 1986).

The concept of de Bary was based on the homology of sexual organs and a special type of spore development called 'free cell formation' (Harper, 1897). Accordingly, yeasts were interpreted as extremely reduced Ascomycetes, whose homology with Euascomycetes is only apparent during spore formation. However, this view is not supported by the following more recent data: (i) ultrastructural data on nuclear division in the yeasts *Saccharomyces cerevisiae* (Moens & Rapoort, 1971, Peterson & Ris, 1976), *Wickerhamia fluorescens* (Rooney & Moens, 1973), and *Cephaloascus fragrans* (Ashton & Moens, 1979) are more in accord with a protozoan ancestry than relatedness to Euascomycetes (Kubai, 1978;

364   H. Prillinger

Raikov, 1982); (ii) ascospore formation in Euascomycetes deviates ultra-structurally, in the formation of an ascus vesicle, from the process known in yeasts, indeed ascospore formation may differ between yeasts as is especially evident between *S. cerevisiae* and *Taphrina* (Syrop & Beckett, 1972; Ashton & Moens, 1979; Beckett, 1981); (iii) endospores are not restricted to ascomycetous yeasts, but also occur in basidiomycetous

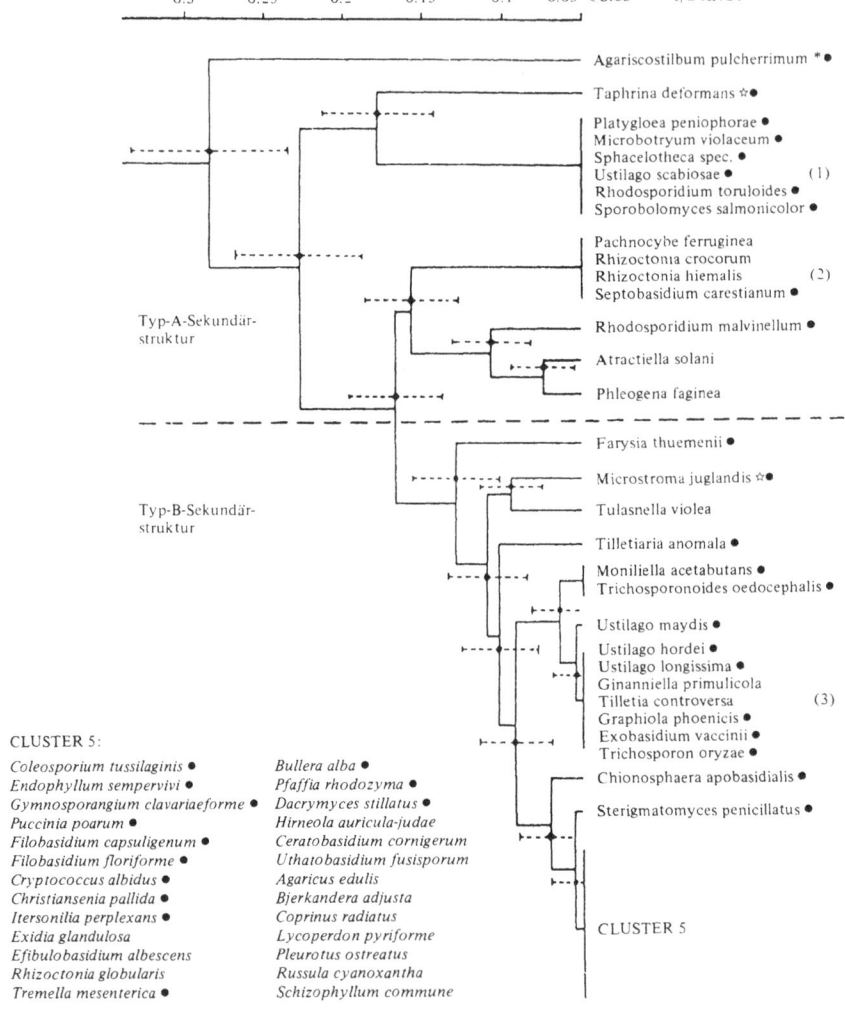

CLUSTER 5:

Coleosporium tussilaginis ●   Bullera alba ●
Endophyllum sempervivi ●   Pfaffia rhodozyma ●
Gymnosporangium clavariaeforme ●   Dacrymyces stillatus ●
Puccinia poarum ●   Hirneola auricula-judae
Filobasidium capsuligenum ●   Ceratobasidium cornigerum
Filobasidium floriforme ●   Uthatobasidium fusisporum
Cryptococcus albidus ●   Agaricus edulis
Christiansenia pallida ●   Bjerkandera adjusta
Itersonilia perplexans ●   Coprinus radiatus
Exidia glandulosa   Lycoperdon pyriforme
Efibulobasidium albescens   Pleurotus ostreatus
Rhizoctonia globularis   Russula cyanoxantha
Tremella mesenterica ●   Schizophyllum commune

Fig. 24.3. Phylogenetic dendrogram deduced from the comparison of 5S rRNA nucleotide sequences of 57 Basidiomycetes. Species in which yeast stages are known are marked with a closed circle (from Gottschalk & Blanz, 1985; reproduced by permission of Einhorn-Verlag).

yeasts (e.g. *Rhodosporidium* (Fell, Hunter & Tallman, 1973), *Trichosporon* (do Carmo, 1970), *Cystofilobasidium* (Oberwinkler *et al.*, 1983), and *Agaricostilbum* (Oberwinkler, personal communication)); (iv) all data from molecular taxonomy, especially coenzyme Q (Yamada & Kondo, 1972; Yamada, 1983; Barnett, Payne & Yarrow, 1983), sequences of cytochrome *c* (Heller & Smith, 1966) and 5S rRNA (Piechulla *et al.*, 1981; Küntzel, Heidrich & Piechulla, 1981; Gottschalk & Blanz, 1985; and see Fig. 24.3); (v) data on the ontogeny of primitive Heterobasidiomycetes (Brefeld, 1891; Kobayashi & Tubaki, 1965; Oberwinkler, 1977, 1978a, 1982, 1986).

Von Arx (1979) recently revived the concept of Brefeld (1891) to the effect that ' . . . in general the yeast cell can be considered to be a conidium . . . '. However, although Brefeld clearly recognised the morphologically primitive trait of yeasts in the ontogeny of Basidiomycetes, he strictly refuted the occurrence of yeasts in the Zygomycetes. In his opinion yeasts evolved from unispored sporangiola of the '*Chaetocladium*' type, which Benjamin (1979) would call today a sporangiolumspore. Brefeld detected yeasts or microconidia during basidiospore germination in the Tremellales (Fig. 24.4c), Auriculariales, Dacrymycetales, Ustilaginales, Tilletiales, Exobasidiales, *Microstroma*, and he was the only author to illustrate a yeast stage in the Entomophthorales (Fig. 24.4a), which is salient to Oberwinkler's concept which is discussed below. There are objections to Brefeld's concept. An ultrastructural investigation of the process of budding in *Rhizopus (Mucor) rouxii* (Lara & Bartnicki-Garcia, 1974) revealed no significant differences from the same process in sporulating yeasts. The fine structure of the process of spore formation within a sporangiolum (Fletcher, 1972, 1973) is similar to the process within a sporangium, but is significantly different from the process of ascospore formation in yeasts (Beckett, 1981). Unispored sporangiola of the '*Chaetocladium*' type represent an advanced and highly specialised situation in the Zygomycetes, contradicting Savile's phylogenetic principles that new major groups should always be derived from unspecialised lower groups of great genetic plasticity (Savile, 1955). Humber (1985) has recently given evidence that saprotrophic Mucorales may have evolved from parasitic polykaryotic Entomophthorales.

The concept of Corner (1971), which regards the yeasts as polyphyletic, reduced, end-products does not explain the occurrence of yeast stages during basidiospore or sporangiospore ontogeny, or recent experiments on speciation (Prillinger, 1986). It was, however, supported recently by Jülich (1982).

366   *H. Prillinger*

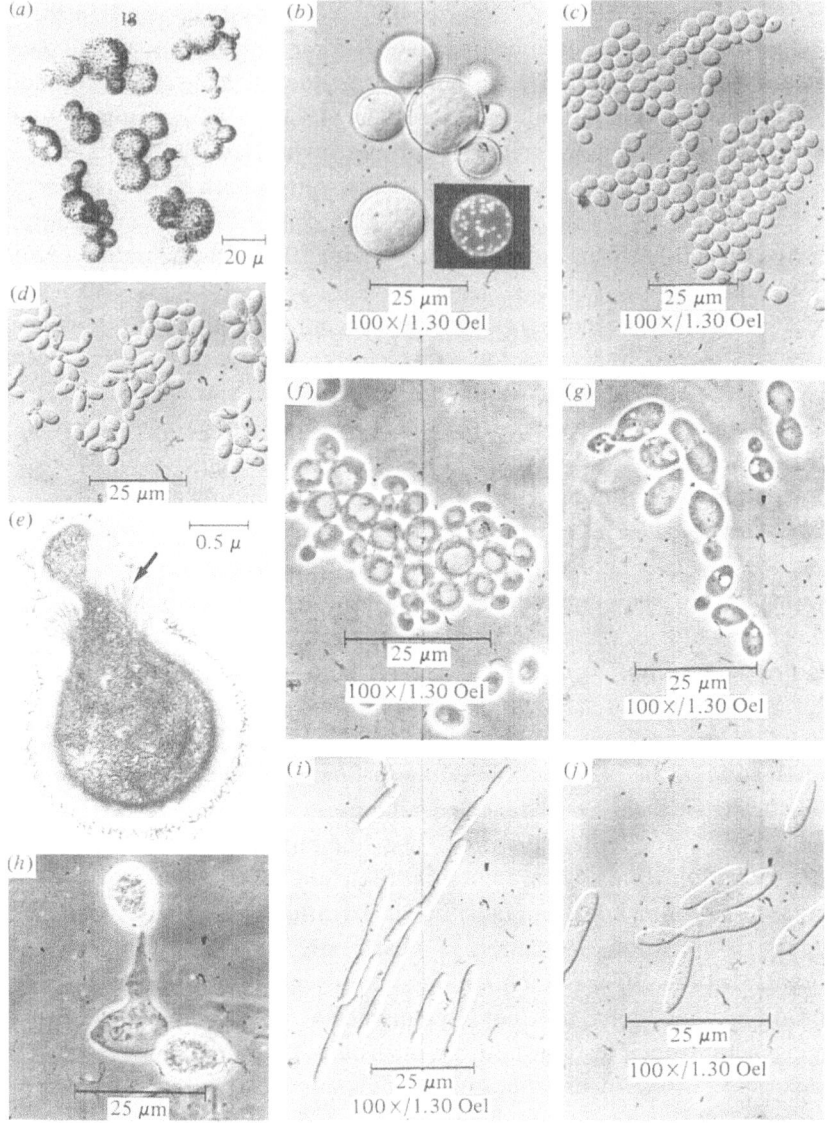

Fig. 24.4. Morphology of yeast cells from Zygomycetes and Basidiomycetes.
(a) *Entomophthora (Empusa) muscae* (after Brefeld, 1871). (b) *Mycotypha microspora* observed with Nomarski interference contrast optics, insert shows the polykaryotic yeast *M. africana* (CBS 122.64) observed after DAPI staining (Hooley *et al.*, 1982). (c) *Tremella foliacea*. (d) *Taphrina populina*. (e) *Taphrina californica*, electronmicrograph with characteristic bud scars arrowed (kindly provided by Dr I.B. Heath; reproduced by permission of the

Oberwinkler (1977, 1982, 1986, personal communication) has introduced an interpretation of yeasts or yeast stages as primitive morphological states in Zygomycetes, Ascomycetes and Basidiomycetes, thus uniting the yeasts of Zygomycete affinity with those of other fungi and pointing to a common ancestry. The similarity between presumed archaic conditions and the environments which commonly induce yeast-like growth in *Rhizopus* (*Amylomyces*), *Actinomucor* and different species of *Mucor* (Bartnicki-Garcia, 1963, 1973; Benjamin, 1979), make these yeasts of special interest in the phylogeny of Eumycota. However, the sparse data on Zygomycetous yeasts in parasitic groups (only Brefeld's observations on *Entomophthora muscae* and Benjamin's (1979) on *Basidiobolus ranarum*) need further corroboration. According to Oberwinkler (1982), in the primitive Heterobasidiomycetes are included those species producing basidiospores which are capable of forming secondary spores (Fig. 24.4h) and/or yeast-like cells (cf. Figs. 24.3 & 24.4). Here it is significant that Bauer (1986), Deml (1986) and Oberwinkler (1986) have recently demonstrated that the process of budding or sterigma formation in ballistospores (cf. Fig. 24.4) at the beginning of the ontogeny of different Heterobasidiomycetes could not be ultrastructurally differentiated from the process, occurring in the meiosporangium, which completes the life cycle. These structures may, therefore, have evolved in a unicellular organism. Oberwinkler's group has also been able to disprove that pycniospores only function as male sexual organs, a dogma established after the demonstration of the pycniospore-trichogyne fertilisation (Craigie, 1927). Thus, Deml, Bauer & Oberwinkler (1982a & b) successfully isolated yeasts from pycnidia of *Coleosporium tussilaginis* (Fig. 24.4g) and obtained products of copulation *in vitro* and on the natural host. Bauer (in Gottschalk & Blanz, 1984) extended this observation to isolation of yeasts from pycnidia of *Gymnosporangium clavariaeforme*, *Puccinia poarum* and *Endophyllum sempervivi*, and I have confirmed these results with *Gymnosporangium sabinae* (Fig. 24.4f).

Oberwinkler's concept has recently been further substantiated in Basidiomycetes by a comparative analysis of the 5S rRNA nucleotide sequences of 57 species by Gottschalk & Blanz (1985; and see Fig. 24.3).

National Research Council of Canada). (f) *Gymnosporangium sabinae*. (g) *Coleosporium tussilaginis*. (h) *Gymnosporangium clavariaeforme*, mitotic ballistospores = secondary spores. (i) *Polyporus lepideus*, yeast stage. (j) *Ganoderma adspersum*, yeast stage.

Although the dendrogram only represents relationships based on a single character, two features are striking: organisms in which yeast stages have been described occur in all sequence clusters and a similar pattern could be obtained for the occurrence of parasitism.

Interestingly, the dendrogram of Gottschalk & Blanz (1985) includes *Taphrina deformans*, an organism so far classified in the Ascomycetes (von Arx, van der Walt & Liebenberg, 1982). This is, however, far less surprising if one accepts Savile's interpretation of the evolution of rust fungi (Savile, 1955) and further anatomical and physiological characters discussed by Oberwinkler (1978b, 1986) and Heath *et al.* (1982). From my own investigations on *Taphrina epiphylla*, I regard the so-called ascus of this fungus as a 'siphonal' (cf. Fig. 24.6) germination stage of a chlamydospore.

Yeasts and speciation in Homobasidiomycetes

The idea that yeasts represent the basic morphogenetic building block from which chitinous aflagellate Eumycota are constructed raises the possibility that expression of this phase in groups where it is normally absent could result in genetic isolation and the generation of new breeding units. This possibility has been reinforced by observations, which still require complete substantiation, of yeasts in Homobasidiomycetes which can be detected experimentally by prolonged inbreeding. These experiments were first prompted by the observation of a yeast stage in the mycoparasitic fungus *Asterophora lycoperdoides* which occasionally developed from the chlamydospores of this fungus which normally germinate into a homothallic, dikaryotic mycelium (Jahrmann & Prillinger, 1983; Prillinger, 1986). However, although we have been able to obtain fruit bodies more than once from such chlamydospores germinating to form yeasts, we have not been able to confirm conspecificity of the mycelial and yeast stage using molecular methods (Laaser, personal communication). Further work is in progress to determine whether these yeasts truly belong to *Asterophora* or an unknown tremellaceous hyperparasite. In recent experiments, a fertile resupinate *Polyporus lepideus* strain was obtained from an isogenised strain after two years of vegetative propagation. This strain can be crossed with its parents and the character segregates as a monogenic mutant (Prillinger, 1983, 1986). A postulated yeast strain was obtained from a cross of isogenised ramarioid *Polyporus* strains (Fig. 24.4i; Fig. 24.5a & b) which, although not yet confirmed by genetic crossing or DNA reassociation, appears to belong to *P. lepideus*. This conclusion is based on a number of observations: (i) a trichogyne-like

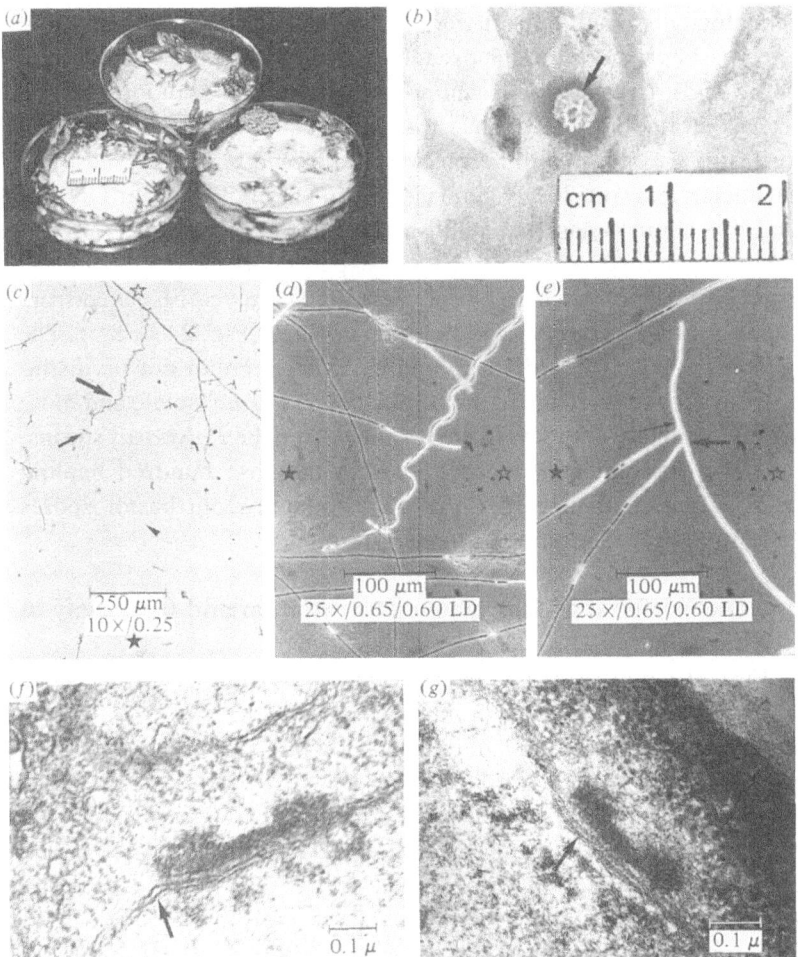

Fig. 24.5. Homobasidiomycetous yeasts in *Polyporus lepideus* (a − e) and *Ganoderma adspersum* (f & g). (a) Isogenised ramarioid inbreeding strains with non-sporulating fruit bodies; bottom = haploid parents, top = dikaryotic cross (see Prillinger, 1986). (b) A yeast colony (arrow) which appeared in the mycelium of the sterile dikaryotic cross. (c) & (d) Trichogyne-like approach of a hypha (large arrow) from the mycelial culture towards the yeast cells (small arrow) produced on the pseudomycelium by the yeast strain, open star = site of inoculum of mycelial strain, closed star = site of yeast strain inoculum, triangle = empty hyphae of the pseudomycelium. (e) Vegetative anastomosis, thin arrow = hyphal annealing, thick arrow = vegetative anastomosis. (f) & (g) Spindle pole body (SPB) of the nucleus in the hypha (f) and the yeast (g) of *Ganoderma adspersum*, the nuclear envelope is arrowed.

approach, albeit not resulting in fusion, of a hypha coming from a mycelial culture to yeasts produced on the pseudomycelium from the yeast culture (Fig. 24.5c & d); (ii) random anastomoses between the pseudomycelium from a yeast strain and true hyphae from a mycelial strain (Fig. 24.5e); (iii) close similarity between interphase spindle pole bodies of yeast and mycelial nuclei; and (iv) closely similar GC content of the nuclear DNA of yeast and mycelium (Laaser, unpublished). Recently, a yeast strain has been detected twice during germination experiments with *Ganoderma adspersum* basidiospores which closely resembled the yeast isolated from *Polyporus* cultures (Laaser & Luschka, unpublished; Fig. 24.4i & j). The spindle pole body (Fig. 24.5f & g) and GC values could not be distinguished from a mycelial strain of the original *Ganoderma* isolate and more surprisingly, neither were they distinguishable from the *Polyporus* strains. A yeast strain has recently appeared within the first hundred haploid cultures (i.e. sooner than with *Polyporus*) isolated from basidiospores after vegetative propagation of *Lentinus crinitus*.

Conclusions: morphological differentiation and phylogeny of Eumycota

Fig. 24.6 depicts a phylogenetic scheme accounting for the range of morphological organisation in Eumycota. It is based on a scheme proposed by Pascher (1931), but unlike the latter envisages evolution from polykaryotic, via oligokaryotic to mono- or dikaryotic systems. The yeast form, denoted by the term 'coccal' (i.e. a unicellular organism having a rigid wall outside its plasma membrane) occupies a basal position, just above monadal (flagellate) and rhizopodial forms. The ballistospore (cf. Fig. 24.4h) is a specific adaptation of unicellular morphological differentiation to terrestrial life, evolving polyphyletically in chitinous fungi including Entomophthorales (Benjamin, 1979; Tucker, 1981) and Basidiomycetes (McLaughlin, Beckett & Yoon, 1985; Oberwinkler, 1986). Pseudoparenchyma, with approximately isodiametric cells and synchronous cell and nuclear divisions (as in the aecidia of rust fungi, in the Laboulbeniales, and in fruiting structures of diverse Ascomycetes and Basidiomycetes), differs from the true parenchyma of higher plants by the absence of a phragmoplast. The term 'pseudotrichal' (pseudoseptate) should not be confused with the term 'trichal' (truly septate). The pseudotrichal pattern is characterised by the absence of vegetative anastomoses and it could often be well differentiated by the existence of complete septa (Gull & Trinci, 1975; Garrison *et al.*, 1975) and a protoplast which creeps inside the apical compartments of pseudohyphae retaining an essentially

unicellular condition. The route from pseudotrichal to siphonal (coenocytic) needs further confirmation. Our comparative investigations with
basidiomycetous yeasts so far suggest that there is no direct route from
coccal to the advanced filamentous stages, the association of millions of
unicellular organisms within loose or firm colonies always being an
essential prerequisite.

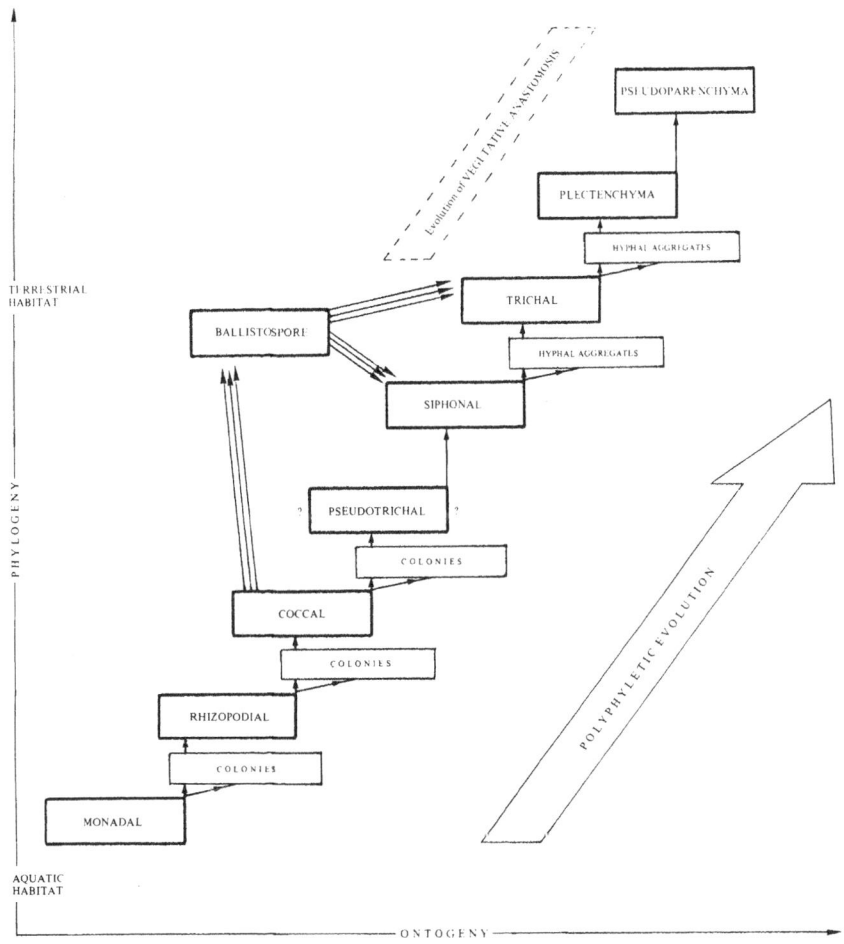

Fig. 24.6. Evolutionary scheme for morphological differentiation in Eumycota
(modified from Jahrmann & Prillinger, 1983; Prillinger, 1984).

Acknowledgements For assistance with experimental work and supply of cultures and data I thank Prof. F. Oberwinkler, Drs R. Bauer and G. Deml, Mr G. Laaser, Ms Ch. Dörfler and Mr Th. Forst. Mrs S. Korcz, Ms M. Jacobsen and my wife helped in preparation of the manuscript. I thank Prof. H. P. Molitoris, Prof. J. Webster and, especially, Dr A. D. M. Rayner for their critical review and revision of the text. Experimental work described here was supported by the Deutsche Forschungsgemein schaft.

References

Ashton, M. L. & Moens, P. B. (1979). Ultrastructure of sporulation in the Hemiascomycetes *Ascoidea corymbosa*, *A. rubescens*, *Cephaloascus fragrans*, and *Saccharomycopsis capsularis*. *Canadian Journal of Botany*, **57**, 1259 – 1284.

Ayers, T. T. (1935). Parasitism of *Dispira cornuta*. *Mycologia*, **27**, 235 – 261.

Barnett, J. A., Payne, R. W. & Yarrow, D. (1983). *Yeasts: Characteristics and Identification*. Cambridge University Press.

Bartnicki-Garcia, S. (1963). Mold – yeast dimorphism of *Mucor*. *Bacteriological Reviews*, **27**, 293 – 304.

Bartnicki-Garcia, S. (1970). Cell wall composition and other biochemical markers in fungal taxonomy. In *Phytochemical Phylogeny*, ed. Harborne, J. B., pp. 81 – 103. New York: Academic Press.

Bartnicki-Garcia, S. (1973). Fundamental aspects of hyphal morphogenesis. *Symposia of the Society for General Microbiology*, **23**, 245 – 267.

Batko, A. (1974). Filogeneza a struktury taksonomiczne Entomophthoraceae. In *Ewolucja Biologiczna: Szkice Teoretyczne i Metodologiczne*, ed. Nowinskiego, C., pp. 209 – 305. Wroclaw: Polska Akademia Nauk, Instytut Filozofii i Socjologii.

Bauch, R. (1930). Über multipolare Sexualität bei *Ustilago longissima*. *Archiv für Protistenkunde*, **70**, 417 – 466.

Bauer, R. (1986). Basidiosporenentwicklung und -keimung bei Heterobasidiomyceten. *Berichte der Deutschen Botanischen Gesellschaft*, **99**, 67 – 81.

Beckett, A. (1981). Ascospore formation. In *The Fungal Spore: Morphogenetic Controls*, ed. Turian, G. & Hohl, H. R., pp. 107 – 129. New York: Academic Press.

Benjamin, R. K. (1979). Zygomycetes and their spores. In *The Whole Fungus*, ed. Kendrick, W. B., pp. 573 – 616. Ottawa: National Museums of Canada.

Bland, C. E. & Couch, J. N. (1985). Structure and development. In *The Genus Coelomyces*, ed. Couch, J. N. & Bland, C. E., pp. 23 – 80. Orlando: Academic Press.

Blanz, P. A. & Gottschalk, M. (1984). Comparison of 5S ribosomal RNA sequences from smut fungi. *Systematic and Applied Microbiology*, **5**, 518 – 526.

Boddy, L. & Rayner, A. D. M. (1983). Mycelial interactions, morphogenesis and ecology of *Phlebia radiata* and *P. rufa* from oak. *Transactions of the British Mycological Society*, **80**, 437 – 448.

Boidin, J., Lanquetin, P., Gilles, G., Candoussau, F. & Hugueney, R. (1985). Contribution à la connaisance des Aleurodiscoideae a spores amyloides (Basidiomycotina, Corticiaceae). *Bulletin de la Société Mycologique de France*, **101**, 333 – 367.

Brefeld, O. (1871). Untersuchungen über die Entwicklung der *Empusa muscae* und *Empusa* (Entomophthora) *radicans* und die durch sie verusachten Epidemien der Stubenfliegen und Raupen. *Abhandlungen der naturforschenden Gesellschaft in Halle*, **12**, 1 −50.

Brefeld, O. (1891). Ascomyceten II. In *Untersuchungen aus dem Gesammtgebie der Mykologie*, Xth ed. pp. 157 − 378. Münster: Commissions-Verlag H. Schöningh.

Brunk, M. & Barnett, H. L. (1966). Mycoparasitism of *Dispira simplex* and *D. parvispora*. *Mycologia*, **58**, 518 − 523.

Buller, A. H. R. (1931). *Researches on Fungi*. vol. IV. London: Longmans, Green & Co..

Buller, A. H. R. (1933). *Researches on Fungi*. vol. VI. London: Longmans, Green & Co..

Burgeff, H. (1920). Sexualität und Parasitismus bei den Mucorineen. *Berichte Deutsche Botanische Gesellschaft*, **38**, 318 − 327.

Burgeff, H. (1924). Untersuchungen über Sexualität und Parasitismus bei Mucorineen. In *Botanische Abhandlungen*, ed. Goebel, K., pp. 1 − 135. Jena: G. Fischer.

Corner, E. J. H. (1971). *Das Leben der Pflanzen*. Lausanne: Editions Recontre.

Couch, J. N. (1945). Revision of the genus *Coelomyces*, parasitic in insect larvae. *Journal of the Elisha Mitchell Scientific Society*, **61**, 124 − 136.

Craigie, J. H. (1927). Discovery of the function of the pycnia of the rust fungi. *Nature*, **120**, 765 − 767.

de Bary, A. (1884). *Vergleichende Morphologie und Biologie der Pilze, Mycetozoen und Bacterien*. Leipzig: Wilhelm Engelmann.

Deml, G. (1986). Keimung phragmobasidialer Brandpilze. *Berichte Deutsche Botanische Gesellschaft*, **99**, 83 − 88.

Deml, G., Bauer, R. & Oberwinkler, F. (1982a). Studies in Heterobasidiomycetes. Part 9. Axenic cultures of *Coleosporium tussilaginis (Uredinales). I. Isolation, identification and characterization of the cultures. Phytopathologische Zeitschrift*, **104**, 39 − 45.

Deml, G., Bauer, R. & Oberwinkler, F. (1982b). Untersuchungen an Heterobasidiomyceten , Teil 16 Axenische Kultur von *Coleosporium tussilaginis* (Pers.) Lév. (Uredinales) II. Kreuzungsversuche mit monokaryotischen Stämmen.. *Phytopathologische Zeitschrift*, **103**, 149 − 155.

Deml, G., Oberwinkler, F. & Bauer, R. (1985). Studies in Heterobasidiomycetes. Part 38 *Sphacelotheca polygoni-persicariae G. Deml & Oberw. spec. nov.. Phytopathologische Zeitschrift*, **113**, 231 − 242.

Demoulin, V. (1974). The origin of Ascomycetes and Basidiomycetes: the case for a red algal ancestry. *Botanical Reviews*, **40**, 315 − 345.

Dobbs, C. G. & English, M. P. (1954). *Piptocephalis xenophila* sp. nov. parasitic on non-mucorine hosts. *Transactions of the British Mycological Society*, **37**, 375 − 389.

do Carmo Sousa, L. (1970). *Trichosporon* Behrend. In *The Yeasts, a Taxonomic Study*, ed. Lodder, J., pp. 1309 − 1352. Amsterdam: North Holland Publishing Company.

Drechsler, C. (1933). Morphological diversity among fungi capturing and destroying nematodes. *Journal of the Washington Academy of Sciences*, **23**, 138 − 141.

Egel-Mitani, M., Olson, L. W. & Egel, R. (1982). Meiosis in *Aspergillus nidulans*: another example for lacking synaptonemal complexes in the absence of crossover interference. *Hereditas*, **97**, 179 − 187.

Ellis, J. J. (1966). On growing *Syncephalis* in pure culture. *Mycologia*, **58**, 465 − 469.

Emmons, C. W. (1935). The ascocarps in species of *Penicillium*. *Mycologia*, **27**, 128 – 150.

Esser, K. (1962). Die Genetik der sexuellen Fortpflanzung bei Pilzen. *Biologisches Zentralblatt*, **81**, 161 – 172.

Esser, K. (1971). Breeding systems in fungi and their significance for genetic recombination. *Molecular and General Genetics*, **110**, 86 – 100.

Esser, K. & Kuenen, R. (1965). *Genetik der Pilze*. Berlin: Springer-Verlag.

Fell, J. W., Hunter, I. L. & Tallman, A. S. (1973). Marine basidiomycetous yeasts (Rhodosporidium spp. n.) with tetrapolar and multiple allelic bipolar mating systems. *Canadian Journal of Microbiology*, **19**, 643 – 657.

Fletcher, J. (1972). Fine structure of developing merosporangia and sporangiospores of *Syncephalastrum racemosum*. *Archiv für Mikrobiologie*, **87**, 269 – 284.

Fletcher, J. (1973). Ultrastructural changes associated with spore formation in sporangia and sporangiola of *Thamnidium elegans* Link. *Annals of Botany*, **37**, 963 – 971.

Fuller, M. S. (1976). Mitosis in fungi. *International Review of Cytology*, **45**, 113 – 153.

Garrison, R. G., Mariat, F., Boyd, K. S. & Tally, J. F. (1975). Ultrastructural and electron cytochemical studies of *Entomophthora coronata*. *Annales de Microbiolgie*, **126b**, 149 – 173.

Gottschalk, M. & Blanz, P. A. (1984). Highly conserved 5S ribosomal RNA sequences in four rust fungi and atypical 5S rRNA secondary structure in *Microstroma juglandis*. *Nucleic Acids Research*, **12**, 3951 – 3958.

Gottschalk, M. & Blanz, P. A. (1985). Untersuchungen an 5S ribosomalen Ribonukleinsäuren als Beitrag zur Klärung von Systematik und Phylogenie der Basidiomyceten. *Zeitschrift für Mykologie*, **51**, 205 – 243.

Gull, K. & Trinci, A. P. J. (1975). Septal ultrastructure in *Basidiobolus ranarum*. *Sabouraudia*, **13**, 49 – 51.

Harper, R. A. (1897). Kerntheilung und freie Zellbildung im Ascus. *Jahrbuch für wissenschaftliche Botanik*, **30**, 249 – 284.

Heath, I. B., Ashton, M. L., Rethoret, K. & Heath, M. C. (1982). Mitosis and the phylogeny of *Taphrina*. *Canadian Journal of Botany*, **60**, 1696 – 1725.

Heller, J. & Smith, E. L. (1966). *Neurospora crassa* cytochrome *c*. *Journal of Biological Chemistry*, **241**, 3158 – 3180.

Hirata, A. & Tanaka, K. (1982). Nuclear behaviour during conjugation and meiosis in the fission yeast *Schizosaccharomyces pombe*. *Journal of General and Applied Microbiology*, **28**, 263 – 274.

Hooley, P., Fyfe, A. M., Maltese, C. E. & Shaw, D. S. (1982). Duplication cycle in nuclei of germinating zoospores of *Phytophthora drechsleri* as revealed by DAPI staining. *Transactions of the British Mycological Society*, **79**, 563 – 566.

Humber, R. A. (1985). Phylogenetic position of the Entomophthorales within the Zygomycetes. *Mycological Society of America Newsletter*, **36**, 29.

Jahrmann, H. J. & Prillinger, H. (1983). Das Vorkommen eines ''Hefe'' -Stadiums bei dem Homobasidiomyceten *Asterophora* (Nyctalis) *lycoperdoides* (Bull.) Ditm. ex S. F. Gray und seine Bedeutung für die Phylogenese der Basidiomyceten. *Zeitschrift für Mykologie*, **49**, 195 – 235.

Jeffries, P. & Young, T. W. K. (1976). Ultrastructure of infection of *Cokeromyces recurvatus* by *Piptocephalis unispora* (Mucorales). *Archives of Microbiology*, **109**, 277 – 288.

Jülich, W. (1982). *Higher taxa of Basidiomycetes*. Bibliotheca Mycologica vol 85. Vaduz: A.R. Gantner Verlag.

Karling, J. S. (1977). *Chytridiomycetarum Iconographia*. Vaduz: J. Cramer.
Keilin, D. (1921). On a new type of fungus: *Coelomyces stegomyiae* n. g., n. sp. parasitic in the body cavity of the larva of *Stegomyia scutellaris* Walker (Diptera, Nematocera, Culicidae). *Parasitology*, **13**, 225 – 234.
Kemper, W. (1937). Zur Morphologie und Cytologie der Gattung *Coniophora*, insbesondere des sogenannten Kellerschwammes. *Zentralblatt fuer Bakteriologie Parasitenkunde Infektionskrankheiten und Hygiene Abteilung II*, **97**, 100 – 124.
Kniep, H. (1922). Über Geschlechtsbestimmung und Reduktionsteilung. *Verhandlungen der physikalisch-medizinischen Gesellschaft Würzburg*, **47**, 1 – 29.
Kniep, H. (1926). Über Artkreuzungen bei Brandpilzen. *Zeitschrift für Pilzkunde*, **10**, 217 – 251.
Kobayashi, Y. & Tubaki, K. (1965). Studies on cultural characters and asexual reproduction of Heterobasidiomycetes. I. *Transactions of the Mycological Society of Japan*, **6**, 29 – 36.
Kubai, D.F. (1978). Mitosis and fungal phylogeny. In *Nuclear Division in the Fungi*, ed. Heath, I. B., pp. 177 – 229. New York: Academic Press.
Küntzel, H., Heidrich, M. & Piechulla, B. (1981). Phylogenetic tree derived from bacterial, cytosol and organelle 5S rRNA sequences. *Nucleic Acids Research*, **9**, 1451 – 1461.
Lara, S. L. & Bartnicki-Garcia, S. (1974). Cytology of budding in *Mucor rouxii*: wall ontogeny. *Archiv für Mikrobiologie*, **97**, 1 – 16.
Mandelbrot, A. K. & Erb, K. (1972). Host spectrum of the mycoparasite *Dimargaris verticillata*. *Mycologia*, **64**, 1124 – 1129.
Martin, W.W. (1969). A morphological and cytological study of development in *Coelomyces punctatus* parasitic in *Anopheles quadrimaculatus*. *Journal of the Elisha Mitchell Scientific Society*, **85**, 59 – 72.
McLaughlin, D. J., Beckett, A. S., Yoon, K. S. (1985). Ultrastructure and evolution of ballistosporic basidiospores. *Botanical Journal of the Linnean Society*, **91**, 253 – 271.
Moens, P. B. & Rapport, E. (1971). Spindles, spindle plaques, and meiosis in the yeast *Saccharomyces cerevisiae* (Hansen). *Journal of Cell Biology*, **50**, 344 – 361.
Oberwinkler, F. (1977). Das neue System der Basidiomyceten. In *Beiträge zur Biologie der niederen Pflanzen*, ed. Frey, W., Hurka, H. & Oberwinkler, F., pp. 59 – 105. Stuttgart: Fischer-Verlag.
Oberwinkler, F. (1978a). Was ist ein Basidiomycet?. *Zeitschrift für Mykologie*, **44**, 13 – 29.
Oberwinkler, F. (1978b). Taxonomy of Basidiomycetes with ontogenetic yeast phases. *Abstracts, 12th International Congress of Microbiology (Munich)*, 94, C38.
Oberwinkler, F. (1982). The significance of the morphology of the basidium in the phylogeny of Basidiomycetes. In *Basidium and Basidiocarp: Evolution, Cytology, Function and Development*, ed. Wells, K. & Wells, E. K., pp. 9 – 35. New York: Springer-Verlag.
Oberwinkler, F. (1984). Fungus-alga interactions in Basidiolichens. *Beiheft 79 Zur Nova Hedwigia (Festschrift J. Poelt)*, 739 – 774.
Oberwinkler, F. (1986). Anmerkungen zur Evolution und Systematik der Basidiomyceten. *Botanische Jahrbücher für Systematik Pflanzengeschichte und Pflanzengeographie*, **107**, 541 – 580.
Oberwinkler, F. & Bandoni, R. (1984). *Herpobasidium* and allied genera. *Transactions of the British Mycological Society*, **83**, 639 – 658.
Oberwinkler, F., Bandoni, R., Blanz, P. & Kisimova-Horovitz, L. (1983). *Cystofilobasidium*: a new genus in the Filobasidiaceae. *Systematic and Applied*

Microbiology, **4**, 114 — 122.

Olson, L. W., Edén, U., Egel-Mitani, M. & Egel, R. (1978). Asynaptic meiosis in fission yeast?. *Hereditas*, **89**, 189 — 199.

Orban, G. (1919). Untersuchungen über die Sexualität von *Phycomyces nitens*. *Beihefte zum Botanischen Centralblatt I*, **36**, 1 — 59.

Pascher, A. (1931). Systematische Übersicht über die mit Flagellaten in Zusammenhang stehenden Algenreihen und Versuch einer Einreihung dieser Algenstämme in die Stämme des Pflanzenreiches. *Beihefte zum Botanischen Centralblatt II*, **48**, 317 — 332.

Peterson, J. B. & Ris, H. (1976). Electron-microscopic study of the spindle and chromosome movement in the yeast *Saccharomyces cerevisiae*. *Journal of Cell Science*, **22**, 219 — 242.

Piechulla, B., Hahn, U., McLaughlin, L. C. & Küntzel, H. (1981). Nucleotide sequence of 5S ribosomal RNA from *Aspergillus nidulans* and *Neurospora crassa*. *Nucleic Acids Research*, **9**, 1445 — 1450.

Prillinger, H. (1982). Zur genetischen Kontrolle und Evolution der sexuellen Fortpflanzung und Heterothalliie bei Chitinpilzen. *Zeitschrift für Mykologie*, **48**, 297 — 324.

Prillinger, H. (1983). The phenomenon of haploid fruiting in *Polyporus ciliatus* and its significance for the phylogeny of basidiomycetes. *Abstracts, 3rd International Mycological Congress (Tokyo)*, 246.

Prillinger, H. (1984). Zur Evolution von Mitose, Meiose und Kernphasenwechsel bei Chitinpilzen. *Zeitschrift für Mykologie*, **50**, 267 — 352.

Prillinger, H. (1986). Morphologische Atavismen bei Homobasidiomyceten durch natürliche und künstliche Inzucht und ihre Bedeutung für die Systematik. *Berichte der Deutschen Botanischen Gesellschaft*, **99**, 31 — 42.

Raikov, I. B. (1982). *The Protozoan Nucleus: Morphology and Evolution*. Cell Biology Monographs, no. 9. New York: Academic Press.

Rooney, L. & Moens, P. B. (1973). Nuclear Division and meiosis in the ascomycetous yeast *Wickerhamia fluorescens*. *Canadian Journal of Microbiology*, **19**, 1383 — 1387.

Satina, S. & Blakeslee, A. F. (1926). The *Mucor* parasite *Parasitella* in relation to sex. *Proceedings of the National Academy of Sciences, USA*, **12**, 202 — 207.

Savile, D. B. O. (1955). A phylogeny of the Basidiomycetes. *Canadian Journal of Botany*, **33**, 60 — 104.

Savile, D. B. O. (1968). Possible interrelationships between fungal groups. In *The Fungi, An Advanced Treatise*, vol. III, ed. Ainsworth, G. C. & Sussman, A. S., pp. 649 — 675. New York: Academic Press.

Savile, D. B. O. (1976). Evolution of the rust fungi (Uredinales) as reflected by their ecological problems. *Evolutionary Biology*, **9**, 137 — 207.

Savile, D. B. O. (1979). Fungi as aids in higher plant classification. *Botanical Review*, **45**, 377 — 503.

Syrop, M. J. & Beckett, A. (1972). The origin of ascospore-delimiting membranes in *Taphrina deformans*. *Archiv für Mikrobiologie*, **86**, 185 — 191.

Thaxter, R. (1897). New or peculiar Zygomycetes. 2. *Syncephalastrum* and *Syncephalis*. *Botanical Gazette*, **24**, 1 — 15.

Thielke, C. (1984). Wirtelschnallen bei homo- und heterothallischen *Coprinus*-Arten (Agaricales). *Plant Systematics and Evolution*, **148**, 35 — 49.

Tucker, B. E. (1981). A review of the nonentomogenous Entomophthorales. *Mycotaxon*, **13**, 481 — 505.

Umphlett, C. J. (1962). Morphological and cytological observations on the mycelium of

Coelomyces. Mycologia, **54**, 540 − 554.

van Tieghem, M. P. (1875). Nouvelles recherches sur les Mucorinées. *Annales des Sciences Naturelles Botanique, Sér. 6*, **1**, 5 − 175.

Vogel, H. J., Thompson, J. S. & Shockman, C. D. (1970). Characteristic metabolic patterns of prokaryotes and eukaryotes. *Symposia of the Society for General Microbiology*, **20**, 107 − 119.

von Arx, J. A. (1979). Propagation in the yeasts and yeastlike fungi. In *The Whole Fungus*, ed. Kendrick, W. B., pp. 555 − 571. Ottawa: National Museums of Canada.

von Arx, J. A., van der Walt, J. P. & Liebenberg, N. V. D. M. (1982). The classification of *Taphrina* and other fungi with yeastlike cultural states. *Mycologia*, **74**, 285 − 296.

von Wettstein, F. (1921). Das Vorkommen von Chitin und seine Verwertung als systematisch-phylogenetisches Merkmal im Pflanzenreich. *Sitzungsberichte der österreichischen Akademie der Wissenschaften Mathematisch-Naturwissenschaftliche Klasse, Abt. I*, **130**, 3 − 20.

von Wettstein, D., Rasmussen, S. W. & Holm, P. B. (1984). The synaptonemal complex in genetic segregation. *Annual Review of Genetics*, **18**, 331 − 413.

Yamada, Y. (1983). Coenzyme Q system in yeasts and yeast-like fungi. *Abstracts, 3rd International Mycological Congress (Tokyo)*, 350.

Yamada, Y. & Kondô, K. (1972). Taxonomic significance of coenzyme Q system in yeasts and yeast-like fungi. In *Yeasts: Models in Science and Technics*, Proceedings of the first Specialized International Symposium on Yeasts, Smolenice, 1971, pp. 363 − 373. Bratislava: Slovak Academy of Sciences.

Zickler, H. (1937). Die Spermatienbefruchtung bei *Bombardia lunata. Berichte der Deutschen Botanischen Gesellschaft*, **55**, 114 − 119.

Zickler, H. (1952). Zur Entwicklungsgeschichte des Askomyceten *Bombardia lunata* Zckl. *Archiv für Protistenkunde*, **98**, 1 − 70.

Zycha, H., Siepmann, R. & Linnemann, G. (1969). *Mucorales*. Lehre: J. Cramer.

25

Whither terminology below the species level in the fungi?

C. M. BRASIER AND A. D. M. RAYNER*

*Forest Research Station, Alice Holt Lodge, Farnham, Surrey GU10 4LH, UK, and *School of Biological Sciences, University of Bath, Claverton Down, Bath BA2 7AY, UK*

Introduction

As will be apparent from the foregoing chapters modern myco-logical research is steadily bringing to light population structures below the species level, some simple some complex, which cannot readily be fitted into current systems of taxonomic rank and terminology. It also seems likely that the arrangement of these population structures and the delimitation of the sub-groups involved will become of increasing economic and ecological relevance, especially in plant pathology (e.g. Hansen, Chapter 22). However, this may cause problems since the way in which patterns of variation within a species are identified, interpreted and formally classified, has always presented some difficulties to mycologists. Moreover, with the presently growing interest in fungal population biology, and the increasing application of new (molecular) or neglected (classical genetical and biochemical) approaches to determining relation-ships, there is a risk that these difficulties will be aggravated. At the same time, the increasing output of new data, concepts and criteria presents a real opportunity for the formulation of approaches to terminology and classification which will be of long-term value to mycologists of all persuasions. Now may be the time to take stock of the situation, and to attempt to capitalise on the opportunities it presents, rather than to allow the status quo to continue. At stake may be the extent to which mycologists with different interests will be able to communicate in the future. Similar problems have been identified by those working with plants (Styles, 1986). This chapter takes up the case for the fungi.

Some root causes: past and present approaches to terminology

To begin with it is necessary to outline what, in our view, are some of the historical reasons for difficulties attending infra-specific fungal classification and why these difficulties might be compounded rather than resolved by newly acquired information. We suggest that the problem may be traced to four main roots.

Need for a more population orientated approach

A tendency to neglect population studies with fungi has, historically, been coupled with a dependence on a nomenclatural approach associated with the collection, preservation and deposition of a single and sometimes fortuitously selected type specimen. Such a herbarium specimen, however carefully chosen, will be more useful in the application and conservation of morphological than of physiological criteria. Moreover, whilst providing a baseline for nomenclatural discussion, there is a danger that in the longer term such an approach may lead to a poor understanding of the vitality and plasticity of fungal populations and of the need to define homogeneous genetic units within what may be a wide spectrum of variation, of the existence of reproductive barriers, and of the effects of ecological specialisation and geographical isolation (Brasier, 1983). Hence, the delimitation of continuous and discontinuous variation within and between populations, a necessary ingredient of a predictively valuable taxonomy, has often been lacking, and when it has been attempted, the existing nomenclatural terminology appears in some cases to have been unhelpful (cf. some of the foregoing chapters).

Dependence on botanical nomenclatural practice

The rules of fungal taxonomy are currently laid down by the International Code of Botanical Nomenclature (see Hawksworth, 1974). Notwithstanding the validity or problems of applying the Code's rules to plant nomenclature (Styles, 1986), fungi are not plants (Whittaker, 1969). Concepts valid in plant nomenclature may not, therefore, always be either useful or relevant to fungal nomenclature and *vice versa*. This is evident, for example, in the special case made for the mycological concept of 'special forms' (*formae speciales*). Moreover, although the Code provides a set of terms for the purpose of infra-specific classification (Fig. 25.1), it is clear that these terms are somewhat limited in scope, particularly with respect to population criteria, are presumably intended for morphologically distinguishable categories (Ainsworth, 1962), are in some cases imprecise, and generally too rigidly hierarchical.

Hierarchical inflexibility

The formal aim of most classifications is to rank organisms into taxa which are usually arranged into a strict, non-overlapping hierarchy, with each taxon being regarded as the sum of its subordinate taxa. Such hierarchical systems, which are applied just as much above as below the species level, have the attractions of apparent precision and order, but if

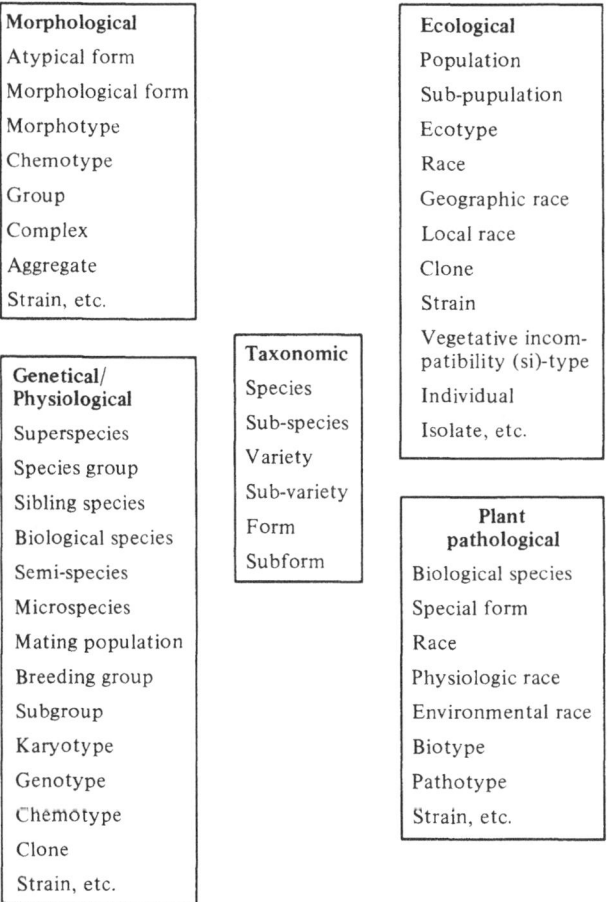

Fig. 25.1. Terms currently in use in infraspecific classification in relation to four dominant special interest areas.

applied rigidly may create as many problems as they solve. Firstly, whatever the criteria chosen (e.g. from within morphology, physiology, ontogeny, interfertility, and host-range), in natural populations the limit of nomenclatural units can, and often do, overlap. Secondly, it is sometimes desirable for practical purposes to group organisms according to one or more selected common characteristic(s) without regard to their overall similarities or dissimilarities. Thirdly, the existence of a strictly hierarchical system largely depends on there being sufficient information available about variation and population structure to make a valid judgement about where to place a newly recognised grouping. Although in the vast majority of cases this evidence may be lacking, the constraints of the hierarchical system often prompt a premature decision on ranking. The latter may then become fixed, even though in the longer term it may prove erroneous as further information accumulates. Historically, this difficulty has been reflected in the frequent re-assortments which have accompanied attempts to weight taxonomic criteria on an *a priori* basis.

Divisions between mycological special interest areas

In mycology, the acquisition of knowledge has, as in other disciplines, inevitably led to the diversification of mycology itself and thereby to the emergence of distinct special interest areas — such as morphology, ecology, plant pathology, genetics and physiology — by a process somewhat reminiscent of evolution itself. These special interest areas have tended to acquire independent goals and needs, each respectively attempting to resolve questions of pattern and rank within a particular complex of morphological forms, habitat relationships, host type and parasitic relationships, or of genetical and physiological diversity. Inevitably, the devotees of a particular area have tended to give greater emphasis to diagnostic features close to their own interests, leading to the emergence of several independent, amorphous and often largely unofficial terminologies (Fig. 25.1). In practice, these approaches often tend to be interconnected, as is reflected in the overlap between the different terminologies, and in the common practice of using the same term to denote very different levels of relationship both within and between the different disciplines. For example, the term 'race' or 'physiological race' is currently used by plant pathologists in at least four different senses (Caten, 1987) though perhaps used most commonly to signify possession of the same set of virulence loci, whereas it is used by ecological geneticists and by population biologists at large to mean discrete, often major, polymorphic sub-populations within a species (e.g. Dobzhansky *et al.*, 1977).

Similarly, the term 'biological species' has been used synonymously with 'special form' on the one hand (e.g. in plant pathology, Gaümann, 1950) and with reproductively isolated breeding units on the other (e.g. in genetics; see Korhonen, Chapter 20). The ubiquitous term 'strain' has been used to define anything from an isolate to a specific genotype, or from a minor to a major species sub-unit.

How should we progress?

As the fungi become increasingly studied from a population and evolutionary standpoint, and as our understanding of fungal speciation processes develops, a rather different set of species and infra-specific concepts seems likely to emerge. If this is the case, should the current systems of official and unofficial classification and diverse terminologies be the sole basis on which these further developments are likely to proceed? In the absence of a directing force combined with a broad mycological consensus, history indicates that it is unlikely that a more standard and simplified set of terms will emerge automatically (see du Rietz, 1930). Rather, the future may see the continued emergence and evolutionary refinement of several rather different sets of terms and concepts, along the lines of the four topic areas depicted in Fig. 25.1 with perhaps, at best, only occasional points of cross reference. Such an outcome would be divisive, impeding mycological communication. We would prefer to encourage the development of a more practical, simpli-fied, population-based system of standard terminology, of meaning and value to, and ideally promoting communication between, all mycologists. Thus, as for plants (Styles, 1986), there may be a strong case for a wide ranging re-evaluation of the utility of current subspecific terminology, and for a consideration of the possibility of devising a system encompassing features appropriate to the fungi and to fungal life styles, governed by the apparent *processes* of fungal speciation.

To illustrate how such a unified system might be devised, we present in Fig. 25.2 a draft system which is intended to simplify current terminology while being sufficiently flexible (i) to accommodate the new wave of information, (ii) to allow for increased availability of information in a particular case and (iii) to cater for wide variations in the state of knowledge between individual cases. We have attempted to incorporate the following features.

1. The system combines a simplified hierarchical ranking of terms for denoting field population units, the top four of which should only be used when sufficient evidence has accrued, together with an ancillary non-

hierarchical 'bio-type' nomenclature. The latter would be used as a means of grouping samples with a common characteristic, signified by pre-fixing the term 'type'. Conceptually, this use of bio-type is therefore more collective than that of Johannsen (1909), du Rietz (1930), and Gaümann (1950), and much closer to the type system in use in the Bacteriological Code. This biotype system is also conceptually similar to the 'deme' terminology advocated by some plant taxonomists (Gilmour & Heslop-Harrison, 1954), but involves a less radical departure from existing terminology.

2. The problem of delimiting fungal individuals, which often arises as a consequence of indeterminate growth in fungi, is met by the use of the terms *genet* and *ramet*, adapted from the terms originally developed for vegetatively reproducing higher plants (Kays & Harper, 1974; Harper, 1977). By slightly modifying the use of these terms for the fungi, genets can be taken to mean genetically discrete units or assemblages, whilst ramets are offshoots derived from genets by asexual or vegetative propagation, and which in most higher fungi can readily fuse *via* hyphal anastomosis. Generally, any *sample* of a fungus from the field can be regarded as a ramet of a genet. Single genets may be very widely dispersed in extensively asexually propagated populations, or they may occupy discrete domains in outcrossing populations. In many higher fungi individual genets may often, but need not necessarily, correspond to vegetative (vc) or somatic (si) incompatibility types (see below).

3. Terms equivalent to ramet and genet are used to describe *laboratory* populations. Thus, an *isolate* is the first culture of a fungus derived from the field, and this becomes a *strain* when it has been asexually propagated and maintained in culture (for definition of isolate and strain see Ainsworth, 1961, 1962). We would emphasise, however, the difference between strain and genet in the sense that the former is derived from a lower subunit (isolate) whereas the latter gives rise to the lower unit (ramet).

4. Several traditionally-established but otherwise ambiguous, ill-defined or problematical terms would be dispensed with. These include variety, sub-variety, form and sub-form, biological species, special form and clone. Physiologic race would be replaced by patho-type *sensu* Robinson (1969).

5. The system allows an *upward* progression through the hierarchy as information accumulates. Starting with individual ramets, isolates or strains, these may first be grouped into biotypes on the basis of common characteristics. Where single common characteristics are concerned they

are to be specified by the relevant prefix. Where several characteristics are shared, then biotype itself may be used as a general term. However, at this stage, the sharing of several characteristics, if also associated with marked discontinuity from other types, may be sufficient evidence for assignment either to an individual genet, if there is evidence of genetic uniformity, or to a race or higher unit (see below), if an appropriately large sample is involved.

6. Five higher units (more could be inserted) from race upwards are delimited as follows. Between genet and species are two widely used ranks in population biology and taxonomy, race and sub-species, to cover distinct, divergent sub-populations which have not, in general, reached total reproductive isolation. Such populations would commonly show an array of correlated differences for a variety of physiological or morphological characters. As a guide, with *races* the similarities exhibited by the sub-populations for important characters would broadly outweigh their differences; with *sub-species* their differences would broadly outweigh their similarities (cf. Brasier, 1982). Partial reproductive isolation might or might not be associated with either rank. Species would represent populations exhibiting several or more sharp, genetically based, discontinuities in their variation patterns and total reproductive isolation, whether geographical, ecological or genetic (i.e. *sensu* Dobzhansky, Mayr, Huxley, Stebbins and others; and see Hawksworth, 1974). Sibling species (e.g. Dobzhansky *et al.*, 1977) would be clusters of completely, or virtually completely, reproductively isolated populations which are otherwise very similar on morphological or physiological grounds. Such populations appear to be common in fungi (Brasier, Chapter 16).

Usually, only some of the above ranks would apply in any one case, and the theoretical distance between ranks, e.g. lying between species and sub-species or sub-species and race, could be large or small depending on the natural properties of the system under study. The scheme would, in particular, allow for the complex arrays of totally or partially reproductively isolated population units at higher infraspecific levels that are increasingly being found in fungi.

As an example of how such a scheme could operate, the particular case of *Ophiostoma* (*Ceratocystis*) *ulmi* can be used. From an initial sample of ramets in diseased twigs, laboratory isolates were prepared from which individual strains were propagated. Amongst a range of strains it was discovered that these could be grouped into distinct patho-, morpho- and physiotypes. Moreover, since pathogenicity, morphology and physiology were then found to be correlated, at first two and subsequently, three general biotypes could be identified, two aggressive and one non-aggressive. Initially these biotypes were designated 'strains', a reflection of the difficulties faced at the time with infra-specific nomenclature (Brasier, 1982). Further work showed that these biotypes were also distinct gamotypes, and as a result of a world-wide sample, it is now considered that they comprise two sub-species, one aggressive and the other non-aggressive, with the aggressive sub-species containing two races, a Eurasian and a North American race (cf. Brasier: Chapter 16).

At the same time as the larger population units, now recognised as sub-species, were being progressively defined, it was also established that within each sub-species and race, strains could be grouped into vegetative incompatibility types, or vc-biotypes. Many distinct vc-types could be discriminated in a given *O. ulmi* sub-species or race. Since these also tend to be distinct genotypes they could, for operational purposes, be considered to be genets. Of particular importance for the classification of the population structure of the EAN and NAN races was the subsequent discovery of the widespread geographical occurrence of a single predominant vc-type − the vc 'super-group' − which sometimes accounted for as much as 98% of a regional population sample (cf. Brasier, Chapter 16). Such extensive vc-biotypes could also, operationally, be regarded as large genets comprised of numerous ramets, although the extent of their genetic homogeneity needs to be further investigated.

Concluding comments

The purpose of this chapter is to draw attention to some of the problems attending infra-specific terminology and classification in fungi and to illustrate how an approach to their solution might be made. While we do not propose an immediate departure from existing procedures, nor consider the above scheme (Fig. 25.2) as necessarily the best way forward, we hope that this chapter, in part if not in its entirety, may − by stimulating discussion − act as a catalyst to progress. That further developments should be based upon a population approach is, we believe, of paramount importance. It seems appropriate to end with an echo from the past: ' . . . Probably most taxonomists think that just their way of using the terms concerned should be considered the right one − but why, probably very few of them would be able to tell. The situation has grown still worse through the suggestion of various geneticists to replace the old terms of taxonomy with a lot of new ones better defined. And certainly it is no exaggeration to say, that the attaining of a stable and generally accepted system of well defined terms for the designation of the various

fundamental units is one of the most urgent needs in present taxonomy
. . . ' (du Rietz, 1930).

Acknowledgement We thank Professor D. L. Hawksworth for helpful
discussions.

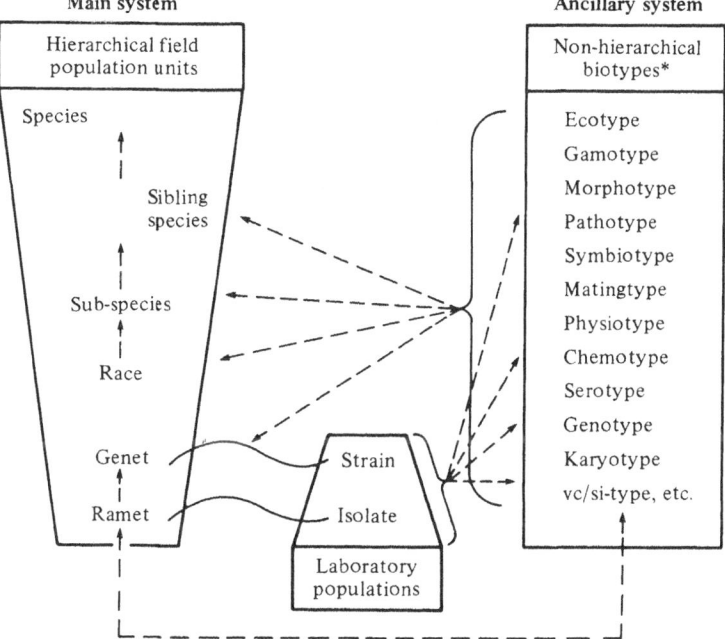

Fig. 25.2. A dual system for infraspecific terminology. Arrows indicate possible progression as information accumulates about population structure and
variation. *Prefixes indicate characteristics in common as follows: ECO,
habitat; GAMO, interbreeding ability (independent of mating type); MOR-
PHO, morphology; PATHO, pathogenic ability towards the host; SYMBIO,
host specificity; MATING, mating type incompatibility factors; CHEMO,
chemical moiety relationships (cf. Pitt & Hawksworth, 1986); SERO,
serological (antigen) profile; PHYSIO, physiological property relationship;
KARYO, chromosome character; GENO, one or more genes (cf. genet);
VC/SI, vegetative or somatic incompatibility genes (may or may not belong
to the same genet).

References

Ainsworth, G. C. (1961). *Dictionary of the Fungi*. 5th Edition. Kew: Commonwealth Mycological Institute.

Ainsworth, G. C. (1962). Pathogenicity and the taxonomy of fungi. *Symposia of the Society for General Microbiology*, **12**, 249 − 269.

Brasier, C. M. (1982). Genetics of pathogenicity in *Ceratocystis ulmi* and its significance for elm breeding. In *Resistance to Diseases and Pests in Forest Trees*, ed. Heybroek, H. M., Stephan, B. R. & von Weissenberg, K., pp. 224 − 235. Wageningen, The Netherlands: Pudoc.

Brasier, C. M. (1983). Problems and prospects in *Phytophthora* research. In *Phytophthora: its Biology, Taxonomy, Ecology and Pathology*, ed. Bartnicki-Garcia, S. & Tsao, P. H., pp. 351 − 364. St Paul, Minnesota: American Phytopathological Society.

Caten, C. E. (1987). The concept of race in plant pathology. In *Populations of Plant Pathogens*, ed. Wolfe, M. S. & Caten, C. E., in press. Oxford: Blackwell Scientific Publications.

Dobzhansky, T., Ayala, F. J., Stebbins, G. L. & Valentine, J. W. (1977). *Evolution*. San Francisco: W. H. Freeman & Company.

du Rietz, G. E. (1930). The fundamental units of biological taxonomy. *Svensk Botansk Tidskrijft*, **24**, 333 − 428.

Gaümann, E. (1950). *Principles of Plant Infection*. London: Crosby Lockwood & Son.

Gilmour, J. S. L. & Heslop-Harrison, J. (1954). The deme terminology and the units of microevolutionary change. *Genetica*, **27**, 147 − 161.

Harper, J. L. (1977). *Population Biology of Plants*. London: Academic Press.

Hawksworth, D. L. (1974). *Mycologists Handbook*. Kew: Commonwealth Mycological Institute.

Johannsen, W. (1909). *Elements der exacten Erblichkeitslehre*. 1st ed. Jena: Fischer.

Kays, S. & Harper, J. L. (1974). The regulation of plant and tiller density in a grass sward. *Journal of Ecology*, **62**, 97 − 105.

Pitt, J. I. & Hawksworth, D. L. (1986). The naming of chemical variants in Penicillium and Aspergillus. In *Advances in Penicillium Systematics*, ed. Samson, R. A. & Pitt, J. I., pp. 89 − 91. New York & London: Plenum Press.

Robinson, R. A. (1969). Disease resistance terminology. *Review of Applied Mycology*, **48**, 593 − 606.

Styles, B. T. (1986). *Infraspecific classification of wild and cultivated plants*. Oxford University Press.

Whittaker, R. H. (1969). New concepts of kingdoms of organisms. *Science*, **163**, 150 − 160.

26

The cell wall: a crucial structure in fungal evolution

SALOMON BARTNICKI-GARCIA

Department of Plant Pathology, University of California, Riverside, CA 92521, USA

Introduction − the significance of cell walls

The cell wall is the structure that gives fungi most of their unique features, but its role in the evolution of these and, indeed other organisms possessing walls, has not received widespread attention. Nevertheless, this role must, for a variety of reasons have been pivotal in the evolution of four of the five biological kingdoms, viz., Monera, Protista, Plantae and Fungi (Bartnicki-Garcia, 1984).

Firstly, by safely containing turgor pressure, the wall has been crucial to the survival and evolution of those organisms, namely bacteria and fungi, whose nutrition depends on absorption of organic matter by passive diffusion. The protoplasts of these organisms could consequently afford to generate a high concentration of intracellular metabolites and hence, sustain correspondingly increased rates of metabolism and growth. This increased growth potential was probably the primordial reason for the success of walled organisms (Bartnicki-Garcia, 1984).

However, the importance of the wall extends beyond provision of protection against plasmoptysis in that it enables cells to assume a wide variety of shapes suited to correspondingly diverse ecological settings. Thus, in fungi, the plasticity of the *growing* cell wall permitted the creation of a rich repertoire of vegetative and particularly, reproductive structures. This is reflected today by the seemingly innumerable fungal species whose natural relationships it is our challenge to decipher. Furthermore, the evolutionary success of the fungi may be attributed basically to the spreading and penetrating power of the mycelium (see Rayner & Coates, Chapter 8), which has allowed them effectively to invade and exploit even the most recalcitrant forms of natural organic matter. The mycelial habit made fungi champions of one of the three main evolutionary lines in

(b)

MUREIN SACCULUS

CHITIN MICROFIBRIL

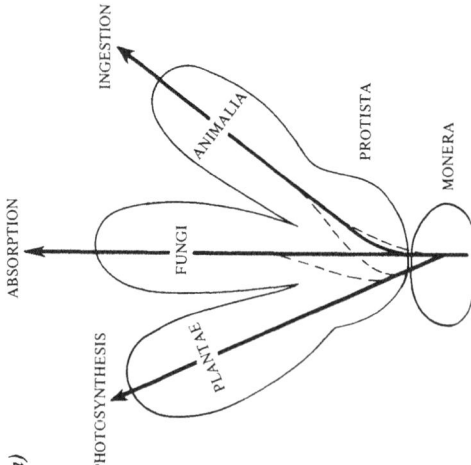

(a)

INGESTION

ABSORPTION

PHOTOSYNTHESIS

ANIMALIA

PROTISTA

MONERA

FUNGI

PLANTAE

biology, the one based on nutrition by absorption of preformed organic matter (Fig. 26.1a).

Origin of the fungal kingdom

Estimates for the oldest fossils of eukaryotes vary from about $1 \cdot 4$ to $0 \cdot 85$ billion years ago (Schopf, 1970; Schopf & Oehler, 1976), but whether any of these belonged to a veritable fungus remains unknown. For now, in the absence of definitive fossil evidence, we can therefore only speculate on a probable evolutionary sequence that led to the acquisition of those main features which distinguish fungi from other living creatures, namely eukaryotic protoplasts, microfibrillar cell walls, and mycelial morphology.

It is outside the scope of this chapter to dwell on the origin of the eukaryotic cell, except to reiterate that whilst much has been written about the evolution of eukaryotic organelles (e.g. Margulis, 1970; Cavalier-Smith, 1978; Fredrick, 1981), one major organelle, the cell wall, is generally overlooked.

Presumptive stages in the evolution of fungal cell walls

The sequence of events in Fig. 26.2 is a modified version of a conjectural exercise (Bartnicki-Garcia, 1984) that attempted to trace the evolutionary path of cell walls, and to find answers to three ultimate questions: when, how and why cell walls came into being. Five distinctive phases were recognised.

A wall-less world Inasmuch as walls are not essential to living cells, we must consider them to have been absent from the primeval biota. These primordial cells (of which no fossil record is available) must have possessed three main attributes: (i) specific and efficient catalysts to make their own cell components, − proteins; (ii) a central depository of information needed to make the catalysts, − nucleic acids; and (iii) a

Fig. 26.1. (a) Relationship of Whittaker's five kingdoms (Whittaker, 1969) to the three major evolutionary lines based on adaptation to three distinct modes of nutrition (Bartnicki-Garcia, 1984). (b) Schematic comparison of the basic structural elements of prokaryotic and eukaryotic cell walls. G = N-acetyl-D-glucosamine, M = acetyl muramic acid, PP = peptide. (c) Microfibrillar structure of a cyst wall of *Phytophthora palmivora* (Tokunaga & Bartnicki-Garcia, 1971). Bar = 1 μm. (d) Isolated chitosomes from yeast cells of *Mucor rouxii* in a negatively stained preparation (Bartnicki-Garcia *et al., 1976). Bar = 100 nm.*

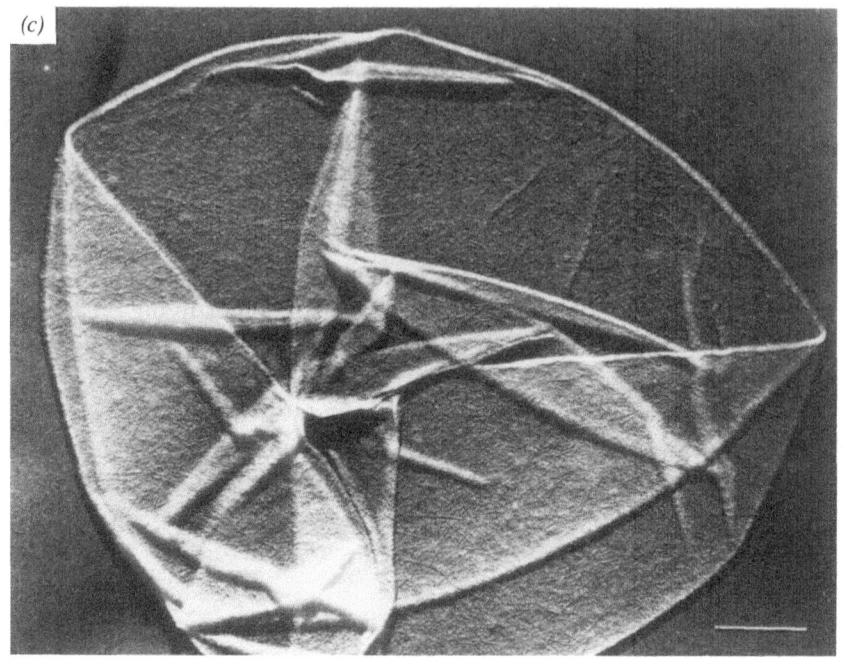

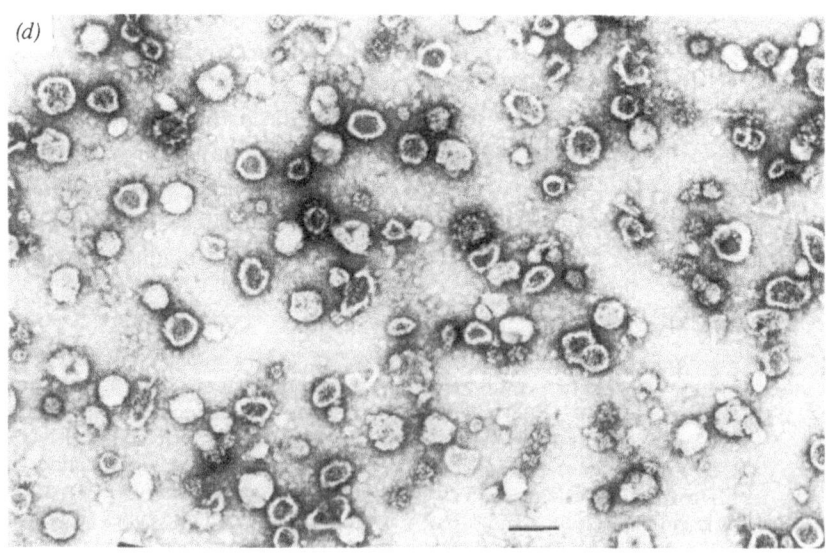

barrier to shield catalysts and metabolites from the external milieu — phospholipid/protein membrane.

Conceivably, these pioneering cells did not need a protective envelope outside their plasma membrane, either because the environment was hypertonic or because they operated under low turgor.

Invention of the cell wall Recent findings show cell walls to be present in the oldest known microfossils; bacterium-like organisms found in sediments 3·5 billion years old (Awramik, Schopf & Walter, 1983; Schopf & Walter, 1983). Seemingly, cell walls evolved rather 'rapidly' after the appearance of the first living cells about 3·5 to 4 billion years ago.

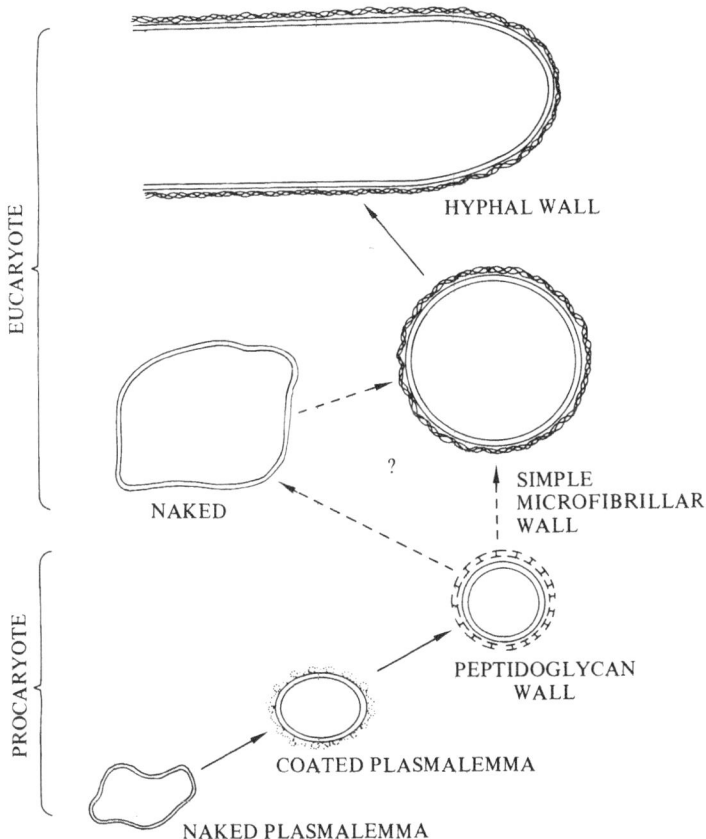

Fig. 26.2. Plausible stages in the evolution of fungal cell walls (modified from Bartnicki-Garcia, 1984).

Presumably, the origin of cell walls dates back to the time when some prokaryotes strengthened their plasma membranes with cytoplasmic polymers (proteins, glycoproteins or polysaccharides) which either leached out of the cytoplasm or were somehow actively discharged and remained anchored to the cell surface (Fig. 26.2). The resultant reinforced plasmalemma permitted the cells to function at a greater turgor commensurate with the increased rates of metabolism and growth which gave them a decisive selective advantage over their naked neighbours.

The prokaryotic wall Through the driving forces of mutation and natural selection, the presumed haphazard excretion of polymeric materials evolved into an orderly biochemical machinery capable of producing polymers specifically suited to give the cell envelope a high degree of tensile strength. For nearly 2×10^9 y the world flora contained only prokaryotic cells but we do not know when or how the wall structure of contemporary prokaryotes, namely an envelope based on a sacculus of peptidoglycan or murein (Fig. 26.1b; Rogers & Perkins, 1968), evolved. There is no reliable chemical evidence to tell whether these wall polymers were present in microfossils of ancestral prokaryotes.

Once the wall was invented as a protective jacket, cells turned it to other uses, particularly morphogenesis. Limited evidence of the resultant morphological plasticity is seen in prokaryotes, but its full splendour was attained only later, by eukaryotes.

The eukaryotic wall With few exceptions, the walls of eukaryotes have a *microfibrillar* structure (Aronson, 1965; Preston, 1974) and have, therefore, a drastically different cell wall organisation from that of prokaryotes. From the lowest to the highest members of the plant or fungal kingdoms, the same basic architecture prevails: a skeleton of interwoven microfibrils, conferring tensile strength, embedded in an amorphous matrix. The microfibril is a highly crystalline, parallel array of long, tightly packed, polysaccharide chains (Fig. 26.1b), the two most common types being cellulose (plants) and chitin (fungi).

Evidently, during the progression to eukaryotic cell organisation the prokaryotic wall architecture was discarded, but we do not know whether the microfibrillar wall arose *de novo* on the surface of some ancestral, wall-less eukaryote, or gradually from the prokaryotic wall (Fig. 26.2). Indeed, the origin of eukaryotic cell walls must be ranked as one of the great unsolved mysteries of biology. Although it has been suggested that eukaryotes might have evolved chimerically, with contributions from the two phylogenetic lines of the prokaryotes − Archaebacteriales and

Eubacteriales (Fox *et al.*, 1980) – neither line has cell walls comparable to those of extant eukaryotes (Kandler, 1982).

The drastic revision in wall architecture between prokaryote and eukaryote invites some fundamental questions. First, why was the prokaryotic cell wall (murein-network) discarded in favour of the micro-fibrillar wall? Was the prokaryotic wall unsuitable to support the range of morphogenetic expression exhibited by eukaryotes? If so, were its deficiencies in tensile strength, rigidity, malleability, plasticity, resistance to predatory enzymic digestion, or all of these? Since no critical com-parative data concerning these properties exist, the prevalence of micro-fibrillar walls, from the simplest to the most complex plants and fungi, must be taken as *prima facie* evidence for the superiority of the micro-fibrillar architecture. Second, by rephrasing the preceding line of questioning, another perspective emerges. If the microfibrillar wall architecture was so superior, why didn't prokaryotes develop and retain it? It was certainly not for lack of time (they had 2 to 3 $\times$ 10^9 y!) or for lack of enzymic capacity – cellulose can be synthesised by bacteria (Colvin & Beer, 1960) though, significantly, not as a wall component.

I believe the key evolutionary impediment to the construction of micro-fibrillar walls was not the lack of specific enzymes (glycosyl transferases and hydrolases) but the absence of subcellular compartments and ancillary structures needed for the spatial organisation and regulation of these enzymes. The magnitude of the topological obstacle is evident from examination of a wall of utmost simplicity such as the cyst wall of *Phytophthora palmivora* (Fig. 26.1c). The zoospore was fully equipped with all the necessary enzymatic paraphernalia to assemble this wall *de novo* from soluble precursors, most likely from UDP-Glc, in a matter of two minutes (Bartnicki-Garcia & Wang, 1983). But how could the micro-fibrils be woven into such a remarkably uniform envelope in both thickness and texture? Could such cell walls be spontaneously generated by a battery of stationary enzymes evenly spread over the cell surface? Is the topology of microfibril assembly resolved by mobile synthesising complexes operating on, or in, the plasma membrane? Is there a need for internal (cytoplasmic) guiding components during fibrillogenesis? Although no definitive answers to these questions are available, I have suggested that the construction of a microfibrillar wall requires the intracellular sophistication of the eukaryotic cell not only to generate the microfibrillar network and other wall polymers, but also to make the wall grow in surface area and to mould it into a diversity of shapes (Bartnicki-Garcia, 1984).

Fig. 26.1c also illustrates an instance where a non-extending microfibrillar wall is merely deposited on the surface of an existing protoplast whose shape it adopts. This type of wall synthesis has been described as morphogenetically passive (Bartnicki-Garcia, 1973a) and probably represents a primitive pattern that was, perhaps, a phylogenetic preamble to the more advanced type usually called *extension growth*. During extension growth, the total surface area of the wall increases through the insertion of new polymers without any significant reduction in wall thickness. As this morphogenetically active wall grows, the size and often the shape of the cell changes.

Mycelial walls: apical growth   The origin of the fungal kingdom might be traced to the time when some primordial walled eukaryotes, the progenitors of the fungi, discovered apical growth; i.e. how to mould their walls into long, branching tubes − hyphae − that extended continuously at their tips. Thus, the most fundamental shape of a fungus, the mycelium, was created.

In general, many of the various cellular shapes in the life cycle of a fungus can be understood in terms of modulation of two basic wall growth patterns. Hyphal development is basically a polarised version of extension growth. A cylindrical tube is generated through a steep radial gradient of wall formation centred at the zenith of the hyphal tip (Bartnicki-Garcia & Lipmann, 1969). This gradient is believed to result from a polarised pattern in the discharge of wall-forming vesicles (Bartnicki-Garcia, 1973b; Gooday & Trinci, 1980). By contrast, non-polarised (i.e. generalised) wall growth gives rise to spherical cells: yeast and yeast-like buds, conidia, sporangia, etc (Bartnicki-Garcia, 1973b; Bartnicki-Garcia & Wang, 1983; cf. Rayner & Coates, Chapter 8).

Requisites for the construction of microfibrillar cell walls

The construction of morphogenetically active microfibrillar walls requires complex interplay of the following different subcellular structures and processes; presumably, only the eukaryotic cell can satisfy these complicated demands (Bartnicki-Garcia, 1984).

An exocytotic system of wall-destined vesicles   This constitutes the essence of the wall-making process of eukaryotes. Cell wall formation in fungi depends on the existence of a vesicular secretory system serving as the vehicle to carry to the cell surface the enzymes and precursors needed for wall growth. The packaging of wall precursors and products into small discrete units or vesicles together with a mechanism to orchestrate their exocytotic movement, provides an effective solution to the problem of spatial regulation of wall synthesis and hence, morphogenesis. On the basis of size, fungi contain at least two types of vesicle involved in wall formation (Girbardt, 1969; Grove & Bracker, 1970). The large vesicles have been given various names, the small ones are usually called microvesicles.

Our studies of chitin formation in fungi led to the discovery of a sharp functional specialisation in wall-destined vesicles. The delivery of microfibril-synthesising enzyme, chitin synthetase, is delegated to microvesicles (Fig. 26.1d) called *chitosomes* (Bracker, Ruiz-Herrera & Bartnicki-Garcia, 1976; Bartnicki-Garcia & Bracker, 1984). Seemingly, the sole function of chitosomes is to deliver chitin synthetase, in zymogen form, to the surface of fungal cells (Bracker *et al.*, 1976; Bartnicki-Garcia, Bracker & Ruiz-Herrera., 1978; Bartnicki-Garcia, Ruiz-Herrera & Bracker, 1979; Hanseler, Nyhlen & Rast, 1983). The larger vesicles are involved in major secretion processes including the delivery of the non-fibrillar or matrix wall components as well as new membrane for extension of the plasmalemma (Grove & Bracker, 1970; Gooday & Trinci, 1980).

A cytoskeleton to guide the movement of the vesicles The establishment of a method to orchestrate the polarised movement of wall-destined vesicles needed for apical growth must be regarded as a crucial evolutionary landmark for the fungi. Presently, however, the directing mechanisms (microtubules/microfilaments/electric fields?) await elucidation (Bartnicki-Garcia, 1973b). We are only beginning to learn about the interaction of vesicles with components of the cytoskeleton that might function in guiding vesicles to their destination (Hoch & Howard, 1980; Adams & Pringle, 1984).

A site for the deployment of microfibril synthetases Whereas the non-fibrillar components of the wall appear to be synthesised in the cytoplasm, all available evidence indicates that microfibrils are made *in situ* (Bartnicki-Garcia & Wang, 1983). The plasma membrane has been proposed as the final site of action of chitin synthetase (Duran, Bowers & Cabib, 1975), but the exact location and mode of microfibril assembly has not been fully resolved (Bartnicki-Garcia & Bracker, 1984).

A mechanism for temporal regulation of microfibril synthesising enzymes Chitosomal chitin synthetase exists in a zymogenic state that requires limited proteolysis by surface (periplasmic?) proteases for its activation (Bartnicki-Garcia *et al.*, 1978; Cabib & Farkas, 1971). Presumably, zymogenicity evolved as an effective means of preventing premature operation of the enzyme *en route* to the cell surface.

Integrated assembly of microfibrils and matrix components The most abundant components of fungal walls are not microfibrils but amorphous matrix materials (Aronson, 1965; Bartnicki-Garcia, 1968). Whether these two kinds of wall ingredient are interdependent remains open, but the observation that secretion of amorphous materials precedes microfibril assembly (Sing & Bartnicki-Garcia, 1975) suggests that the former may play an important role in the latter.

Coordination between lytic and synthetic processes It is commonly believed that cell wall growth requires a coordinated balance between the synthesis of new wall polymers and the lysis of existing ones (Bartnicki-Garcia, 1973b; Rosenberger, 1979). Consequently, the crucial problem in microfibrillar wall biogenesis may not be the assembly of microfibrils *per se*, but the need for a mechanism to coordinate synthetic and lytic processes so that the wall attains an exact measure of controlled plasticity necessary for cell wall extension. As suggested earlier (Bartnicki-Garcia, 1973b), this may be established through coordinated discharge of vesicles carrying different ingredients for wall extension.

Wall chemistry and the phylogeny of extant fungi

During evolution, fungi acquired the ability to synthesise a wide variety of cell wall polysaccharides (Aronson, 1965; Bartnicki-Garcia, 1968; Wessels & Sietsma, 1981). The nature of these polymers in any particular fungus is not capricious, but is related to its taxonomic position and thus reflects its evolutionary history. Previously, I suggested that dual combinations of major wall polysaccharides can provide a basis for taxonomic (Bartnicki-Garcia, 1968) and phylogenetic (Bartnicki-Garcia, 1970) conclusions (Fig. 26.3).

Given the diversity of wall polymers, the question arises as to whether fungi are polyphyletic and thus, whether the sequence of cell wall evolution outlined above occurred more than once. There are strong biochemical arguments (Vogel, 1964; Klein & Cronquist, 1967; Bartnicki-Garcia, 1970) suggesting that the Kingdom Fungi is at least diphyletic with two principal but separate lines of evolution: a major line of chitinous fungi and a minor one of cellulosic fungi (Fig. 26.3). The chitinous line comprises organisms that synthesise L-lysine by the unique

aminoadipic acid pathway (Vogel, 1964). The cellulosic line has fungi
which make lysine by the common diaminopimelic acid pathway. Other
biochemical and cytological features support this dichotomy of the fungal
kingdom (Bartnicki-Garcia, 1970). Hence, it seems certain that the fungal
mycelium was invented independently by the cellulosic and the chitinous
fungi.

The scheme in Fig. 26.3 showing the relationship between taxonomic
groupings and the occurrence of dual combinations of major structural cell
wall polysaccharides (chitin, β-1,3-1,6-glucan, cellulose, chitosan, α- and
β-mannans) remains valid and needs only minor adjustments. Significantly, the modifications have resulted either in a revision of taxonomic
views or the realisation that the chemical distinction applies not to an entire
fungal class, but to some selected subgrouping(s) within the class. The

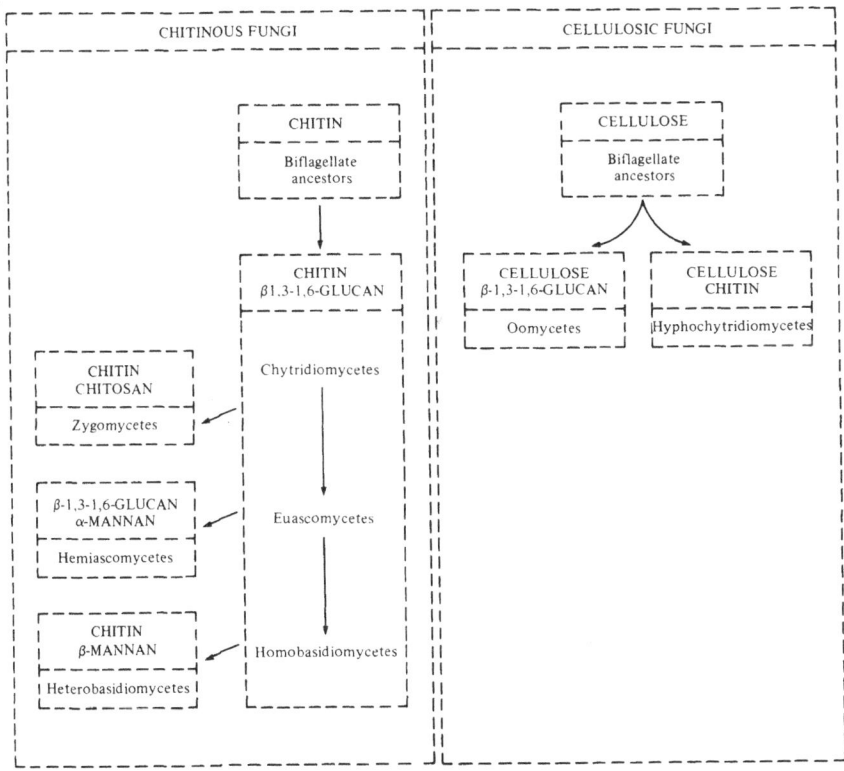

Fig. 26.3. Correlation of fungal phylogeny with the presence of dual combinations of structural cell wall polysaccharides. Modified from Bartnicki-
Garcia (1970).

following are a few salient adjustments to the original scheme (Bartnicki-Garcia, 1970). (i) The chitosan+chitin wall type characteristic of most Zygomycetes (Mucorales) does not extend to the Entomophthorales. The latter appear to lack chitosan, having instead the chitin + β-glucan wall type common to most fungi (Hoddinott & Olsen, 1972). (ii) Substantial amounts of chitin have been found in cell walls of members of the family Leptomitaceae, Order Leptomitales, Class Oomycetes (Lin, Sicher & Aronson, 1976; Bertke & Aronson, 1980). The presence of chitin does not change the fact that these fungi have typical components of oomycetous walls (β-1,3-1,6-glucan and cellulose). Since chitin is absent from members of the other family of the Leptomitales, the Rhipidiaceae (Bertke & Aronson, 1985), a sharp phylogenetic dichotomy is suggested within this Order (cf. Beakes, Chapter 27). (iii) Among organisms originally classified as Oomycetes, the Thraustochytriaceae is the only group lacking the two β-glucans characteristic of Oomycetes. Instead, these peculiar fungi have walls of unusual ultrastructure made largely of L- and D-galactose polymers (Darley, Porter & Fuller, 1973; see Beakes, Chapter 27). Accordingly, most taxonomists have agreed to remove these exceptional organisms from the Oomycetes. (iv) The unusual wall composition found in several genera of the Myxomycetes, with a predominance of galactosamine polymers (McCormick, Blomquist & Rusch, 1970; Chapman, Nelson & Orlowski, 1983; Raub, 1984), strengthens the belief that these organisms are not closely related to mycelial fungi.

Conclusions

The appearance (about 10^9 years ago) of eukaryotic cells with a superior intracellular organisation vastly increased the morphogenetic potential of living cells. One crucial but often overlooked advantage of most eukaryotes over prokaryotes is their ability to construct cell walls with a more advanced, *microfibrillar* architecture. Presumably, the capacity to weave and mould microfibrillar envelopes of increasing complexity and diversity was a decisive factor in fungal evolution. Organisms thus evolved capable of assuming the multitude of cellular shapes characteristic of fungi. Within the microfibrillar wall architecture, fungi evolved a diversity of cell wall chemical composition. Differences in wall polysaccharides can thus be used to help trace the course of fungal phylogeny.

References

Adams, A. E. M. & Pringle, J. R. (1984). Relationship of actin and tubulin distribution to bud growth in wild type and morphogenetic mutant *Saccharomyces cerevisiae*. *Journal of Cell Biology*, **98**, 934 − 945.

Aronson, J. M. (1965). The cell wall. In *The Fungi, An Advanced Treatise*, vol. I, ed. Ainsworth, G. C. & Sussman, A. S., pp. 49 − 76. New York: Academic Press.

Awramik, S. M., Schopf, J. W. & Walter, M. R. (1983). Filamentous fossil bacteria from the archean of Western Australia. *Precambrian Research*, **20**, 357 − 374.

Bartnicki-Garcia, S. (1968). Cell wall chemistry, morphogenesis, and taxonomy of fungi. *Annual Review of Microbiology*, **22**, 87 − 108.

Bartnicki-Garcia, S. (1970). Cell wall composition and other biochemical markers in fungal phylogeny. In *Phytochemical Phylogeny*, ed. Harborne, J. B., pp. 81 − 103. London: Academic Press.

Bartnicki-Garcia, S. (1973a). Cell wall genesis in a natural protoplast: the zoospore of *Phytophthora palmivora*. In *Yeast, Mould and Plant Protoplasts*, ed. Villanueva, J. R., Garcia-Acha, I., Gascon, S. & Uruburu, F., pp. 77 − 91. London: Academic Press.

Bartnicki-Garcia, S. (1973b). Fundamental aspects of hyphal morphogenesis. In *Microbial Differentiation*, ed. Ashworth, J. M. & Smith, J. E., pp. 245 − 267. Cambridge University Press.

Bartnicki-Garcia, S. (1984). Kingdoms with walls. In *Structure, Function and Biosynthesis of Plant Cell Walls*, ed. Dugger, W. M. & Bartnicki-Garcia, S., pp. 1 − 18. Rockville, USA: American Society of Plant Physiologists.

Bartnicki-Garcia, S. & Bracker, C. E. (1984). Unique properties of chitosomes. In *Microbial Cell Wall Synthesis and Autolysis*, ed. Nombela, C., pp. 101 − 112. Amsterdam, The Netherlands: Elsevier Science Publishers.

Bartnicki-Garcia, S., Bracker, C. E., Reyes, E. & Ruiz-Herrera, J. (1978). Isolation of chitosomes from taxonomically diverse fungi and synthesis of chitin microfibrils in vitro. *Experimental Mycology*, **2**, 173 − 192.

Bartnicki-Garcia, S., Bracker, C. E. & Ruiz-Herrera, J. (1978). Synthesis of chitin microfibrils in vitro by chitin synthetase particles, chitosomes, isolated from *Mucor rouxii*. In *Proceedings, 1st International Conference on Chitin/Chitosan*, ed. Muzzarelli, R. A. A. & Pariser, E. R., pp. 450 − 463. Cambridge, USA: Massachusetts Institute of Technology.

Bartnicki-Garcia, S. & Lippman, E. (1969). Fungal morphogenesis: cell wall construction in *Mucor rouxii*. *Science*, **165**, 302 − 304.

Bartnicki-Garcia, S., Ruiz-Herrera, J. & Bracker, C. E. (1979). Chitosomes and chitin synthesis. In *Fungal Walls and Hyphal Growth*, British Mycological Society Symposium volume 8, ed. Burnett, J. H. & Trinci, A. P. J., pp. 149 − 168. Cambridge University Press.

Bartnicki-Garcia, S. & Wang, M. C. (1983). Biochemical aspects of morphogenesis in *Phytophthora*. In *Phytophthora, Its Biology, Taxonomy, Ecology and Pathology*, ed. Erwin, D. C., Bartnicki-Garcia, S. & Tsao, P. H., pp. 121 − 137. St. Paul, Minnesota: American Phytopathological Society.

Bertke, C. C. & Aronson, J. M. (1980). Hyphal wall composition in *Apodachlyella completa*. *Current Microbiology*, **4**, 235 − 238.

Bertke, C. C. & Aronson, J. M. (1985). Hyphal wall composition of *Mindeniella spinospora* and *Araiospora* sp. *American Journal of Botany* **72**, 467 − 471.

Bracker, C. E., Ruiz-Herrera, J. & Bartnicki-Garcia, S. (1976). Structure and transformation of chitin synthetase particles (chitosomes) during microfibril synthesis *in vitro*. *Proceedings of the National Academy of Sciences, USA*, **73**, 4570 − 4574.

Cabib, E. & Farkas, V. (1971). The control of morphogenesis: an enzymatic mechanism for the initiation of septum formation in yeast. *Proceedings of the National Academy of Sciences, USA*, **68**, 2052 − 2056.

Cavalier-Smith, T. (1978). The evolutionary origin and phylogeny of microtubules, mitotic spindles and eukaryote flagella. *BioSystems*, **10**, 93 − 114.

Chapman, C. P., Nelson, R. K. & Orlowski, M. (1983). Chemical composition of the spore case of the acellular slime mold *Fuligo septica*. *Experimental Mycology*, **7**, 57 − 65.

Colvin, J. R. & Beer, M. (1960). The formation of cellulose microfibrils in suspensions of *Acetobacter xylinum*. *Canadian Journal of Microbiology*, **6**, 631 − 637.

Darley, W. M., Porter, D. & Fuller, M. S. (1973). Cell wall composition and synthesis via golgi-directed scale formation in the marine eucaryote, *Schizochytrium aggregatum*, with a note on *Thraustochytrium* sp. *Archiv für Mikrobiologie*, **90**, 89 − 106.

Duran, A., Bowers, B. & Cabib, E. (1975). Chitin synthetase zymogen is attached to the yeast plasma membrane. *Proceedings of the National Academy of Sciences, USA*, **72**, 3952 − 3955.

Fox, G. E., Stackebrandt, E., Hespell, R. B., Gibson, J., Maniloff, J., Dyer, T. A., Wolfe, R. S., Balch, W. E., Tanner, R. S., Magrum, L. J., Zablen, L. B., Blakemore, R., Gupta, R., Bonen, L., Lewis, B. J., Stahl, D. A., Luehrsen, K. R., Chen, K. N. & Woese, C. R. (1980). The phylogeny of prokaryotes. *Science*, **209**, 457 − 463.

Fredrick, J. F. (1981). *Origins and Evolution of Eukaryotic Intracellular Organelles*. New York: New York Academy of Sciences.

Girbardt, M. (1969). Die ultrastruktur der Apikalregion von Pilzhyphen. *Protoplasma*, **67**, 413 − 441.

Gooday, G. W. & Trinci, A. P. J. (1980). Wall structure and biosynthesis in fungi. In *The Eukaryotic Microbial Cell*, Society for General Microbiology Symposium 30, ed. Gooday, G. W., Lloyd, D. & Trinci, A. P. J., pp. 207 − 251. Cambridge University Press.

Grove, S. N. & Bracker, C. E. (1970). Protoplasmic organization of hyphal tips among fungi: vesicles and spitzenkorper. *Journal of Bacteriology*, **104**, 989 − 1009.

Hanseler, E., Nyhlen, L. E. & Rast, D. M. (1983). Isolation and properties of chitin synthetase from *Agaricus bisporus* mycelium. *Experimental Mycology*, **7**, 17 − 30.

Hoch, H. C. & Howard, R. J. (1980). Ultrastructure of freeze-substituted hyphae of the basidiomycete *Laetisaria arvalis*. *Protoplasma*, **103**, 281 − 297.

Hoddinott, J. & Olsen, O. A. (1972). A study of the carbohydrates in the cell walls of some species of the Entomophthorales. *Canadian Journal of Botany*, **50**, 1675 − 1679.

Kandler, O. (1982). Cell wall structures and their phylogenetic implications. *Zentralblatt für Bakteriologie, Mikrobiologie und Hygiene. I. Abteilung Originale C, Allgemeine, Angewandte und Oekologische Mikrobiologie*, **3**, 149 − 160.

Klein, R. M. & Cronquist, A. (1967). A consideration of the evolutionary and taxonomic significance of some biochemical, micromorphological, and physiological characters in the Thallophytes. *Quarterly Review of Biology*, **42**, 105 − 296.

Lin, C. C., Sicher, R. C. & Aronson, J. M. (1976). Hyphal wall chemistry in
   Apodachlya. *Archives of Microbiology*, **108**, 85 – 91.
Margulis, L. (1970). *Origin of Eukaryotic Cells*. New Haven, Connecticut: Yale
   University Press.
McCormick, J. J., Blomquist, J. C. & Rusch, H. P. (1970). Isolation and
   characterization of a galactosamine wall from spores and spherules of
   Physarum polycephalum. *Journal of Bacteriology*, **104**, 1119 – 1125.
Preston, R. D. (1974). *The Physical Biology of Plant Cell Walls*. London: Chapman &
   Hall.
Raub, T. J. (1984). The microcyst wall of *Didymium iridis*: chemical analyses.
   Canadian Journal of Microbiology, **30**, 162 – 170.
Rogers, H. J. & Perkins, H. R. (1968). *Cell Walls and Mambranes*. London: Spon.
Rosenberger, R. F. (1979). Endogenous lytic enzymes and wall metabolism. In *Fungal
   Walls and Hyphal Growth*, ed. Burnett, J. H. & Trinci, A. P. J., pp.
   265 – 277. Cambridge University Press.
Schopf, J. W. (1970). Precambrian micro-organisms and evolutionary events prior to the
   origin of vascular plants. *Biological Reviews*, **45**, 319 – 352.
Schopf, J. W. & Oehler, D. Z. (1976). How old are the eukaryotes? *Science*, **193**,
   47 – 49.
Schopf, J. W. & Walter, M. R. (1983). Archean microfossils: new evidence of ancient
   microbes. In *Earth's Earliest Biosphere: Its Origin and Evolution*, ed. Schopf,
   J. W., pp. 214 – 239. Princeton, USA: Princeton University Press.
Sing, V. O. & Bartnicki-Garcia, S. (1975). Adhesion of *Phytophthora palmivora*
   zoospores: electron microscopy of cell attachment and cyst wall fibril
   formation. *Journal of Cell Science*, **18**, 123 – 132.
Tokunaga, J. & Bartnicki-Garcia, S. (1971). Structure and differentiation of the cell wall
   of *Phytophthora palmivora*: cysts, hyphae and sporangia. *Archiv für
   Mikrobiologie*, **79**, 293 – 310.
Vogel, H. J. (1964). Distribution of lysine pathways among fungi: evolutionary
   implications. *The American Naturalist*, **98**, 435 – 446.
Wessels, J. G. H. & Sietsma, J. H. (1981). Fungal cell walls: a survey. In *Encyclopedia
   of Plant Physiology: Plant Carbohydrates II*, vol. 13b, ed. Tanner, W. &
   Loewus, F. A., pp. 352 – 394. Heidelberg: Springer – Verlag.
Whittaker, R. H. (1969). New concepts of kingdoms and organisms. *Science*, **163**, 150
   – 160.

27

Oomycete phylogeny: ultrastructural perspectives

GORDON W. BEAKES

Department of Plant Biology, The University, Newcastle upon Tyne NE1 7RU, UK

Introduction

The absence of key information makes the unravelling of phylogenetic relationships amongst oomycete fungi a perplexing business. As Waterhouse (1962) pointed out, parallel and convergent evolution are very difficult to detect in organisms which have comparatively few morphological features and no fossil roots. The advent of electron microscopy has, however, revealed a wealth of new structural information, much of which appears to be evolutionarily relatively conservative (Hibberd, 1979). However, we have yet to assess its full evolutionary significance and to quote Manton (1965): ' . . . It is essential to use information from fine-structure as additions to the pool of knowledge provided by general morphology, biochemistry, life histories etc., so that phylogenetic conclusions when reached are based on all evidence and conflict with none . . . '. The fossil record suggests that filamentous coenocytic Xanthophyte-like organisms were present as early as the late Pre-Cambrian and it is plausible that Oomycetes could also have arisen around this time from an algal ancestor by loss of plastids (alternatively xanthophyte algae could have arisen by symbiotic association between a 'prokaryote' plastid progenitor with some form of heterotrophic Oomycete ancestor!). Since there has, therefore, been ample time for divergent, convergent and parallel evolution and because evidence from extant organisms must be relied upon for phylogenetic deductions, the latter will be necessarily speculative.

What are Oomycetes?

In terms of general characteristics, Oomycetes produce zoospores with a forward directed 'tinsel' flagellum and a backward directed, smooth, 'whiplash' flagellum. Their cell walls consist mostly of amorphous β-1,3 glucans with cellulose as the microfibrillar component, although those of the Leptomitaceae also contain chitin (Lee, Swafford & Aronson, 1976). Sexual reproduction is usually oogamous, although in *Lagenidium* Schnepf, Deichgraber & Drebes (1978b) report the sexual fusion of isogamous cysts derived from zoomeiospores. The thallus is usually coenocytic, forming branched eucarpic filaments but smaller holocarpic thalli are not uncommon. The vegetative thallus is diploid or polyploid with meiosis occurring during gametangium differentiation (see Dick, 1972). Many biochemical parameters separate the Oomycetes from other fungi including the presence of water soluble β-1,3 glucan storage reserves related to leucosin and laminarin (Bartnicki-Garcia & Wang, 1973), indicating close affinity with heterokont xanthophycean algae such as *Vaucheria* (Manton, 1965).

Four major orders have been recognised, the Saprolegniales, Leptomitales, Lagenidiales and Peronosporales for which detailed accounts are given by Dick (1973a & b), Sparrow (1973) and Waterhouse (1973). This chapter will discuss the likely phylogenetic relationships between these orders together with a consideration of two groups of uncertain affinity, the marine Thraustochytridiales and the uniflagellate Hyphochytridiomycetes.

Ultrastructural characters

The vegetative thallus

Although the vegetative thallus has provided relatively few characters of phylogenetic significance, the structure of the hyphal apex appears to be reasonably similar for all species so far examined and quite distinct from other fungi (see Bartnicki-Garcia & Wang, 1983). The hyphae of the Leptomitaceae also contain in their vacuoles refractile cellulin granules, comprising a mixture of amorphous glucans and chitin (Lee *et al.*, 1976).

Mitosis has been intensively studied (particularly for *Saprolegnia*) and the phylogenetic aspects have been evaluated in depth by Heath (1980) and Beakes (1981a). The apparently primitive pattern of intranuclear mitosis is remarkably consistent throughout the Oomycetes and is quite similar to that described in *Vaucheria* although differences, particularly at telophase, are significant (Ott & Brown, 1972). In contrast, mitosis in the Thraustochytriales is of the semi-open type, associated with the partial breakdown of the nuclear envelope (Heath, 1980; Beakes, 1981a).

Oospore differentiation

Although traditionally important in classification (Dick, 1969) oospore differentiation has been relatively little studied ultrastructurally (Beakes, 1981b). The rediscovery by Sansome that Oomycetes undergo gametangial meiosis has been confirmed in all four orders of Oomycetes (Dick, 1972) and descriptions of the rather atypical synaptonemal complexes are given by Beakes (1981a). In spite of light microscopic evidence suggesting variability (Dick, 1969), ultrastructural studies have demonstrated the structural homology of oospores in the Saprolegniales and Peronosporales (Beakes, 1981b). All have a prominent storage vacuole (ooplast) derived from the coalescence of the dense-body/fingerprint vesicles. These are characterised by lamellae associated with the inclusion bodies (Beakes, 1981b; Bartnicki-Garcia & Wang, 1983), forming structures unique to these fungi and similar to the 'lamellate vesicles' in the Eustigmatophyceae (Hibberd & Leedale, 1972). In terms of the structure of the oospore wall, the Saprolegniales and Leptomitaceae (unpublished observations of *Apodachlya*) seem to differ from those of the Peronosporales in the absence of an outer periplasm-derived layer(s), although careful interpretation of homology between layers is necessary (Beakes, 1981b).

Zoosporogenesis

By contrast with chytrids (Lange & Olson, 1979), studies of zoospore and cyst ultrastructure in Oomycetes have been limited (Table 27.1). This may reflect the mistaken view that Oomycete zoospores are all rather similar (Lange & Olson, 1979), which has probably arisen because most of the published accounts concentrate on members belonging to two groups, the Saprolegniaceae and the Pythiales (Table 27.1). Two points need to be stressed. Firstly, Oomycete zoospores contain a wide variety of vesicles and other organelles (Holloway & Heath, 1977a & b) and the multiple zoospore/cyst transformations involve considerable changes both in organelle complement and morphology (Holloway & Heath, 1977a & b; Grove & Bracker, 1978; Beakes, 1983). Secondly, there is no standard terminology for these organelles, so that whilst a few attempts have been made to categorise vesicle types (Holloway & Heath, 1977a; Lunney & Bland, 1976a) not all the names selected are appropriate, e.g. the phospholipid vesicles of Lunney & Bland (1976a) do not contain phospholipid!

General ultrastructural organisation Oomycete zoospores all have

Table 27.1. *Summary of ultrastructural studies of secondary type zoospores and cysts of oomycete fungi*

		Storage reserves			Transitional felix	Flagella base and roots			
		Lipid	DBV	WEV		A1	A2	P1	P2
Saprolegniales									
Saprolegniaceae	*Achlya flagellata*	+ +	+P	+	TH				
	Aphanomyces euteiches	+ +	+r	+					
	A. euteiches	+ +	+r	+	TH	?	?	<12	?
	Brevilegnia spp.	+ +	+r	+					
	Dictyuchus sterilis	+	+ +r	+					
	Leptolegnia caudata	+ +	+r	+					
	Saprolegnia diclina	+ +	+P	+	TH	3	2	8	2
	S. ferax	+ +	+rp	+	TH				
	S. monoica	+ +	+r	+	TH	3	2	8	2
	S. parasitica	+ +	+P	+					
Ectrogellaceae	*Ectrogella perforans* (M)	+	?	–					
Lagenidiales									
Lagenidiaceae	*Lagenidium callinectes* (M)	+	+r	–					
Olpidiopsiaceae	*Olpidiosis saprolegniae*	+	+P	+	TH	3	2	5	2
	Petersenia palmariae (M)	–	?	–					
Sirolpidiaceae	*Lagenisma coscinodisci* (M)	+ +	?	–					
	Haliphthoros milfordensis (M)	–	?	–	TH				
Leptomitales									
Leptomitaceae	*Apodachlya pyrifera*	+ +	+r	+					
	Leptomitus lacteus	+ +	+r	+					
Rhipidiaceae	*Sapromyces elongatus*	+	+r	+	TH				
Peronosporales									
Pythiaceaee	*Phytophthora palmivora var. nicotiane*	+ +	+ +r	+					
	P. palmivora var. sojae	+ +	+ +r/c	+					
	P. parasitica				TH	3	2	8	3
	Pythium aphanidermatum	+ +	+ +c	+	TH				
	P. mamillatum	+ +	+ +c	+					
	P. proliferum	+ +	+ +r	+				5<10	
Peronosporaceae	*Plasmopara viticola*	+ +	+ +r	+	TH				

Explanation of abbreviations used in table:
An entry or + means structure present in spores
A − means structure not observed in spores
No entry means information not available
? − means structures resembling organelle present but identification not unequivocable.
(M): marine organisms
Lipid: relative abundance c. + + 10<30% spore volume, + c.<10%
DBV: Densebody Vesicles. abundance as above. p = largely peripherally distributed, r = randomly distributed, c = largely centrally distributed
Flagella roots: A1, A2 anterior rootlets; P1, P2 posterior rootlets
TH: transitional helix reported or shown

Transitional helix

Fibrillar coat

roughly the same complement of organelles and lack localised regions containing ribosomes and specialised lipid-microbody associations. Differences in the organisation of 'primary' and 'secondary' zoospores of *Saprolegnia* have been summarised by Holloway & Heath (1977a) and subtle variations in the relative quantities and distribution of

Encystment apparatus					K/U body	Cyst Coat Outer electron dense layer	Ornamentation	Fibrillar coat	Citations
EV	PMB	PC	FV/PV	EV					
SP	−	+	+	−	TA	+	−		Beakes unpublished
SP	−	+	+	−	TC	+	−		Powell et al. 1985[+]
SP	−	+	+	−	T−	+	−*		Hoch & Mitchell, 1972
SP−B	−	+	+	−	?	+	−		Powell et al. 1985; Beakes unpublished
Elongate	−	+	+	−	?	+	long spines		Hallett & Dick 1986
SP−B	−	+	+	+	−A	+	−*		Beakes unpublished
B	−	+	+	−	TA	+P	Single boathook*		Beakes 1983; Barr & Allan 1985
B	−	+	+	−	TA	+P	Single boathook*		Heath & Greenwood 1970
B	−	+	+	−	TC	+P	Single boathook*		Holloway & Heath 1977a,b
Elongate	−	+	+	−	TA/C	+	long boathook (clusters)*		Beakes 1983
SP	−	−	−	−	−	+	−		Kumar 1980
SP	−	−	−	−	−		−		Gotelli 1974
−	−	+	+	−	T−		−		Bortnick et al. 1985
SP	−	−	−	−	−	+	granules		Pueschel & van der Meer 1985
SP	−	?	?	−	−	+(−)	−		Schnepf et al. 1978a,b
SP	−	−	+	+	−	+	tapered spines		Overton et al. 1983
SP−B	−	+	+	−	TA	+	tubule bundles*	+	Beakes & Heath unpublished; Powell et al. 1985
SP	−	+	+	−	−A	+	−		Beakes unpublished
−	+	+	+	−	−	−	−		Beakes unpublished
−	+	+	+	−	−	−	−	+	Reichle 1969[+]
−	+	+	+	−	−	−	−		Bimpong & Hickman; 1975; Sing & Bartnicki-Garcia 1975
								+	Barr & Allan 1985
−	+	+	+	−	−	−	−		Grove & Bracker 1978; Hallet & Dick 1986
−	+	+	+	+	−	−	−	+	Beakes unpublished
−	+	+	+	+	−	−	−		Lunney & Bland 1976a,b
−	?	?	+	−	−	−	−		Gay & Dennington unpublished

EV: Encystment Vesicles. SP: spherical in profile; B: capsule or 'bar' shaped
PMB: peripheral microbodies
PC: peripheral cisternae
FV/PV: fibrillar or peripheral vesicles
WV: wall vesicles
K/U body: kinetosome-associated organelles
TA: tubular inclusions, amorphous granular matrix; TC: tubular inclusions, crystalline matrix; T−: tubular inclusions, little or no matrix; −A: no tubules, amorphous matrix
Cyst coat. +P: present in localized plaques
Ornamentation: * small globular structures attached spore coat.

densebody/fingerprint vesicles, lipid and other organelles do exist in 'secondary' spore types (Table 27.1). Although the anteriorly flagellate Hyphochytridiomycete zoospores contain many similar organelles to Oomycetes they differ significantly in the clustering of cytoplasmic ribosomes around the nucleus (Fuller & Reichle, 1965; Lange & Olson,

1979; Cooney, Barr & Barstow, 1985). In contrast, the biflagellate Thraustochytrid zoospores lack many of the vesicular components found in Oomycete spores and have scales on their surface (Darley, Porter & Fuller, 1973; Kazama, 1980).

Fine structure of flagella and flagellar roots The fine ornamentation of *Saprolegnia* flagella was found to be similar to that in heterokont algae and supports the proposed close phylogenetic relationship between these two groups (Manton, 1965). More recently, examination of the basal region of the flagellum in Oomycetes (Table 27.1) and algae in the Chrysophyceae, Xanthophyceae and Eustigmatophyceae (Hibberd, 1979) has revealed a unique 'concertina-like', 'hooped' or 'spiral' transitional helix (Fig. 27.1). Significantly, this structure also occurs in the Hyphochytridiomycete *Rhizidiomyces* (Fig. 27.1; Barr & Allan, 1985) but not in the Thraustochytrids which instead have characteristic electron-dense kinetosome inclusions (Fig. 27.1; Kazama, 1980; Barr & Allan, 1985).

The few detailed serial reconstructions in Oomycetes of the microtubular and microfibril arrays associated with the kinetosomes (Holloway & Heath, 1977b; Barr & Allan, 1985; Bortnick, Powell & Bangert, 1985), have revealed a root system comprised of four distinct segments (Table 27.1 and Fig. 27.1). A broadly similar organisation occurs in *Thraustochytrium* (Barr & Allan, 1985) whereas the uniflagellate Hyphochytrid, *Rhizidiomyces*, has a much reduced posterior system; presumably due to loss of the associated flagellum (Barr & Allan, 1985). The root system in *Saprolegnia* and *Phytophthora* is also associated with a fan-shaped striate body, reminiscent of the much more extensive 'rhizoplast' structures reported in Chrysophyte algae such as *Ochromonas* (Hibberd, 1970)., Brown algal zoospores also possess a four root system, although lacking the typical Oomycete conformation (O'Kelly & Floyd, 1984).

The encystment apparatus This comprises a group of organelles directly or indirectly involved in attachment and the subsequent assembly of the cyst wall (Beakes, 1983). A summary of their characteristics is given in Table 27.1 and Fig. 27.2. There are at least five vesicular components.

(1) Encystment or cyst coat vesicles in *Saprolegnia ferax* were initially termed 'bar bodies' because of their capsule shape (Heath & Greenwood, 1970; Holloway & Heath, 1977a) but in fact, show considerable morphological variation (Beakes, 1983; Table 27.1). They are derived from endoplasmic reticulum and are characterised by their structured peripheral layer which, upon discharge, gives rise to the thin, often

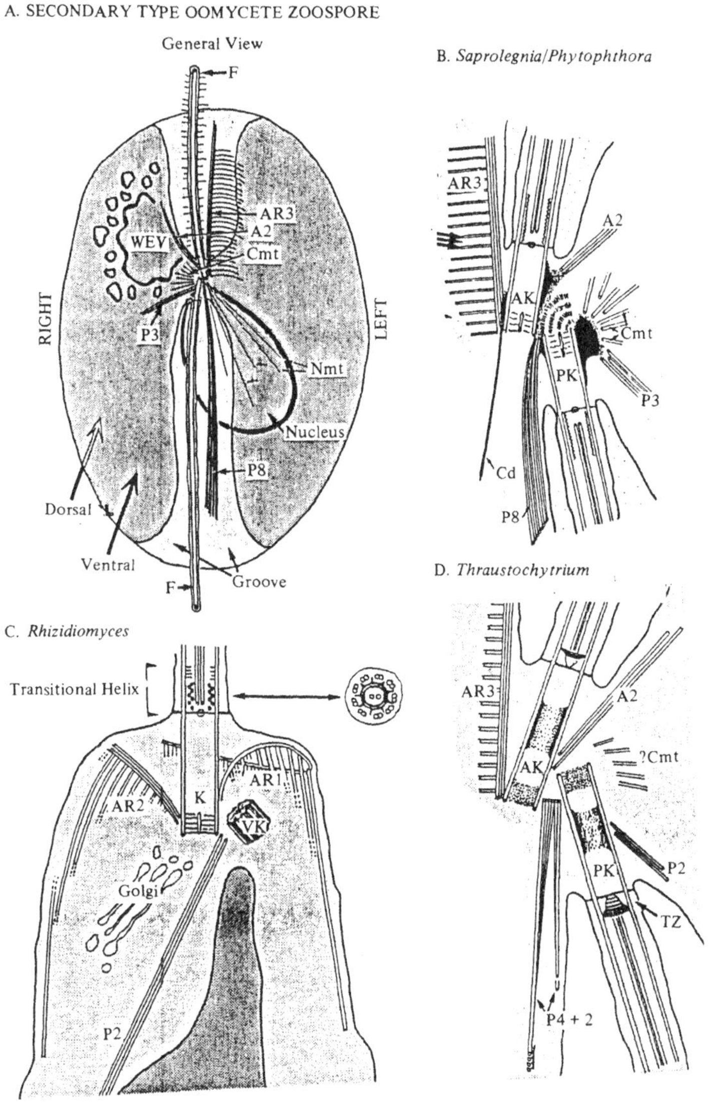

A. SECONDARY TYPE OOMYCETE ZOOSPORE

General View

B. *Saprolegnia/Phytophthora*

C. *Rhizidiomyces*

D. *Thraustochytrium*

Fig. 27.1. Diagrams illustrating the fine structure of the flagellum base and microtubular rootlet systems in Oomycete fungi (A & B), a Hyphochytrid, *Rhizidiomyces* (C) and Thrautochytrid, *Thraustochytrium* (D). Adapted from Barr & Allen (1985). Reproduced with the permission of the *Canadian Journal of Botany*, National Research Council of Canada.

incomplete, electron-dense outer cyst wall layer (Beakes, 1983). Similar vesicles have also been observed in the Xanthophyte algae, *Pseudobumil-leriopsis* (Deason, 1973) and *Vaucheria* (Ott & Brown, 1975). In the Saprolegniales and Leptomitaceae these vesicles contain an assortment of 'spheres', 'tubular hairs', 'boathooks' and 'spines' which, upon release, ornament the cyst walls (Beakes, 1983 & unpublished; Hallett & Dick, 1986; I. B. Heath unpublished; Fig. 27.1; Table 27.1). Comparable structures have not been seen in zoospores of *Pythium*, *Phytophthora* and *Sapromyces* (Table 27.1) and their cysts, correspondingly, lack the outer

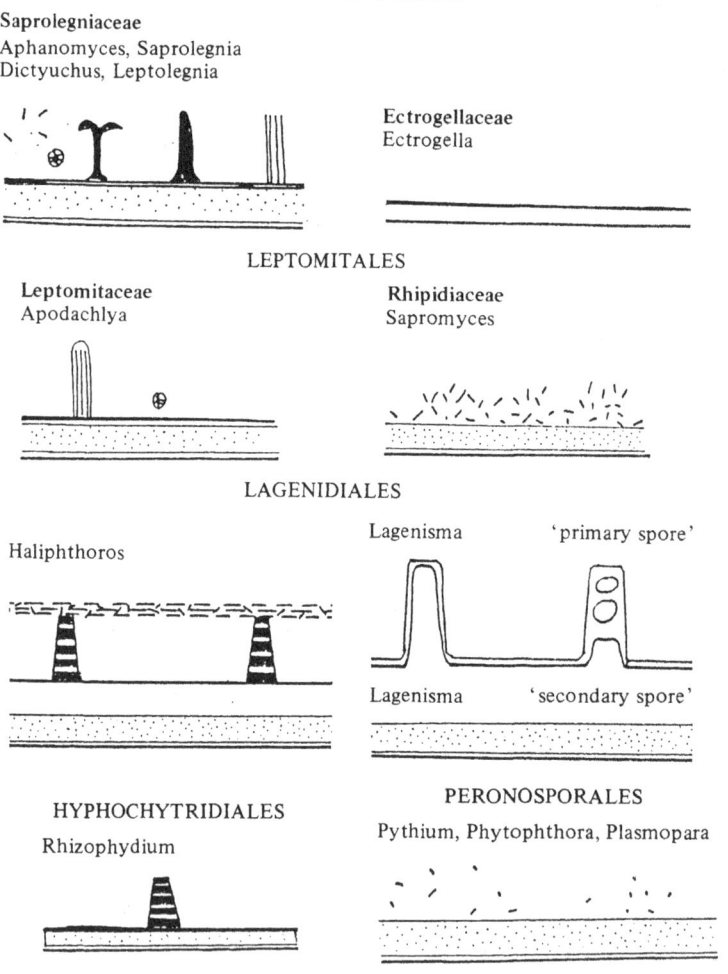

SAPROLEGNIALES

Saprolegniaceae
Aphanomyces, Saprolegnia
Dictyuchus, Leptolegnia

Ectrogellaceae
Ectrogella

LEPTOMITALES

Leptomitaceae
Apodachlya

Rhipidiaceae
Sapromyces

LAGENIDIALES

Haliphthoros

Lagenisma   'primary spore'

Lagenisma   'secondary spore'

HYPHOCHYTRIDIALES

Rhizophydium

PERONOSPORALES

Pythium, Phytophthora, Plasmopara

Fig. 27.2. Diagram summarising cyst coat structure in oomycete fungi and the Hyphochytrid *Rhizophydium*.

electron-dense coat (Fig. 27.2), although they may have fibrillar decorations (Hallett & Dick, 1986). Whilst the zoospores of the lagenidian species, *Lagenisma* (Schnepf *et al.*, 1978b), *Olpidiopsis* (Bortnick *et al.*, 1985) and *Petersenia* (Pueschal & Van der Meer, 1985) all contain abundant 'cyst coat' vesicles and produce cysts with electron-dense coats; the second generation cysts of *Lagenisma* have walls of the peronosporalean type (Schnepf, Deichgraber & Drebes, 1978a; Fig. 27.2). In *Haliphthoros* (Overton, Tharp & Bland, 1983) the cyst coat vesicles give rise to an electron-dense wall layer with tapered and banded appendages which forms below an outer fibrillar layer. These appendages are virtually identical to those on the cyst coats of the Hyphochytrid *Rhizidiomyces* (Fuller & Reichle, 1965).

(2) Peripheral microbodies (also termed U-bodies and parastrasomes) are structures, 300-500 nm in diameter, with partially crystalline matrices, found in zoospores of *Pythium*, *Phytophthora* and *Sapromyces* (Rhipidiaceae). They disappear upon encystment, although their discharge has not been observed (Grove & Bracker, 1978). Whilst the presence of catalase in their core has led their classification as microbodies (Bimpong & Hickman, 1975; Powell, Lehnen & Bortnick, 1985) it, nevertheless, still seems likely that they are homologous to the saprolegnian cyst coat vesicles and may have some role in the encystment processes.

(3) Peripheral/Fibrillar vesicles are Golgi-derived spherical structures (600 to 800 nm) occurring in secondary type zoospores throughout the Oomycetes (Table 27.1). They usually lie in close contact with the plasma membrane and contain fibrillar contents which can aggregate into discrete inclusion bodies, particularly in the Saprolegniales (Beakes, 1983). They contain a glycoprotein adhesive material (Sing & Bartnicki-Garcia, 1975; Beakes, 1983) and in *Phytophthora*, are discharged immediately upon encystment and are responsible for the adhesion of the spores to solid substrata (Sing & Bartnicki-Garcia, 1975). Morphologically and ontogenetically, these vesicles are similar to the so called mucilage vesicles described in certain Chrysophyte algae such as *Olithodiscus* (Leadbeater, 1970). This vesicle fraction probably gives rise, in *Sapromyces*, to an outer reticulate cyst coat (Beakes unpublished) whereas, in *Haliphthoros*, a discrete outer fibrillar layer results (Fig. 27.2).

(4) Wall vesicles are similar to vesicles associated with wall formation in hyphae and other stages. Usually, they do not appear in the peripheral cytoplasm until after settling, although similar structures are present in the zoospores of some *Pythium* and *Leptolegnia* species before encystment (Table 27.1).

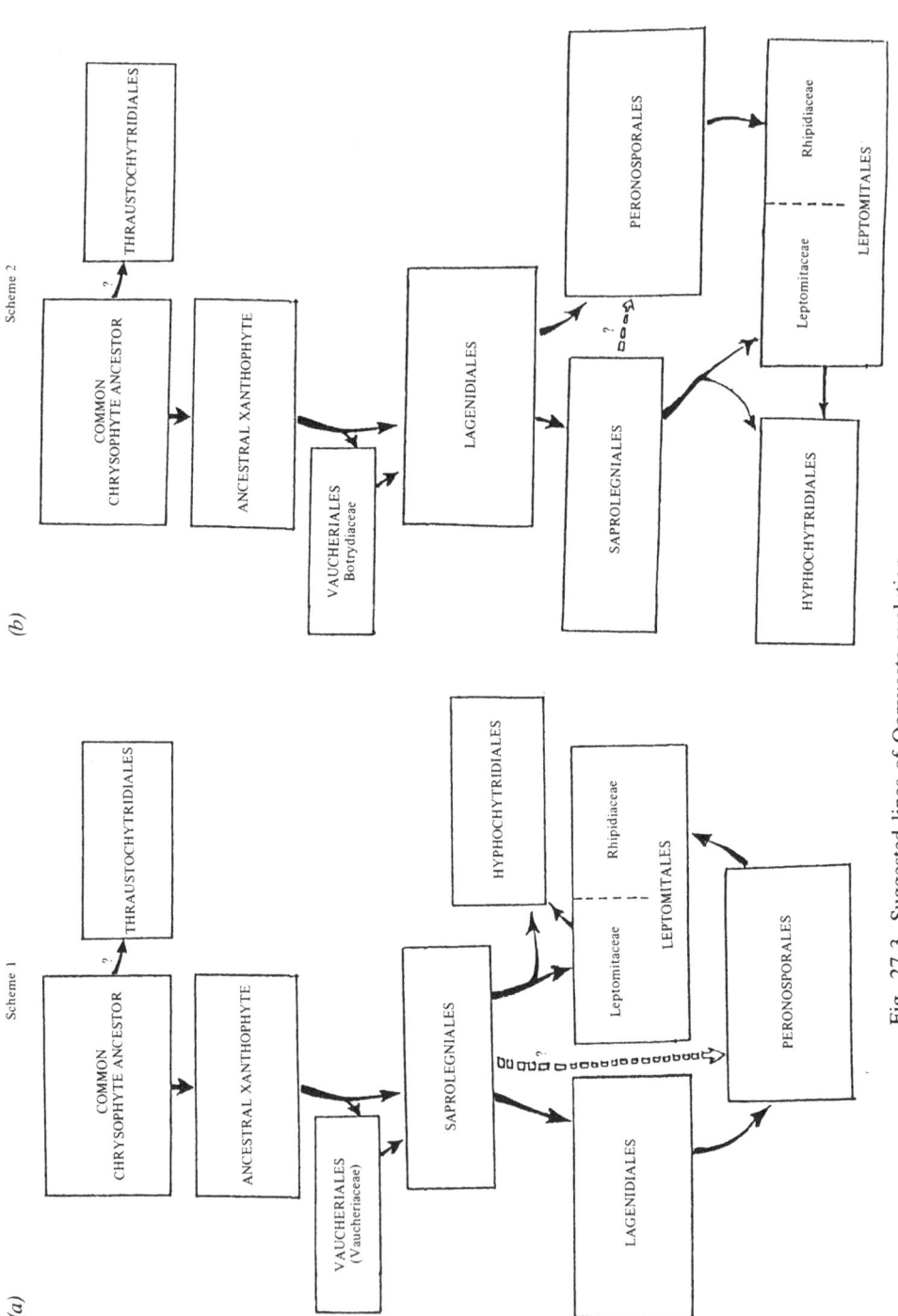

Fig. 27.3. Suggested lines of Oomycete evolution.

(5) Peripheral cisternae of smooth endoplasmic reticulum lie underneath the plasma membrane of secondary type spores of saprolegnian and peronosporalean Oomycetes (Grove & Bracker, 1978, Beakes, 1983). These cisternae are not present in primary spores of *Saprolegnia* (Holloway & Heath, 1977a) or in marine species (Schnepf *et al.*, 1978b; Kumar, 1980; Overton *et al.*, 1983; Pueschel & Van der Meer, 1985), indicating a possible role in osmotic stabilisation. They rapidly disperse upon encystment (Hoch & Mitchell, 1972; Grove & Bracker, 1978; Beakes, 1983).

Finally, some mention must also be made of the organelles called U-bodies (this time standing for 'unidentified', cf. U-bodies mentioned above; Hoch & Mitchell, 1972) or K-bodies (for kinetosome associated; Holloway & Heath, 1977), which are consistently located close to the kinetosomes in saprolegnian zoospores and cysts (Table. 27.1). A detailed account of their comparative fine structure and possible phylogenetic significance has been given by Powell *et al.* (1985). K-bodies also occur in the Leptomitaceae (Powell *et al.*, 1985; Beakes, unpublished) and significantly, in the lagenidian fungus, *Olpidiopsis* (Bortnick *et al.*, 1975), but never in the Peronosporales. They may have some relationship with cyst coat vesicles since intergradation between the two organelle types occurs. They vary considerably in their architecture, but until the full range of intergeneric (Powell *et al.*, 1985) and intraspecific (Beakes, unpublished) variation is ascertained it would be unwise to read too much phylogenetic significance into this.

Phylogenetic conclusions
Problem orders
Hyphochytridiomycetes show many structural affinities with Oomycete fungi as indicated by the presence of a transitional helix, related flagella root structure (Barr & Allan, 1985) and similar encystment apparatus and cyst coat structure (Fuller & Reichle, 1965; Lange & Olson, 1979; Cooney *et al.*, 1985). The presence of both chitin and cellulose in their walls suggests they are an offshoot, either directly of the Leptomitaceae, or of the pathway from the Saprolegniales to this group (Fig. 27.3a & b). However, the occurrence of semi-open mitosis (Heath, 1980) and ribosomal aggregation in the zoospores (Cooney *et al.*, 1985) probably justify the continued placing of these organisms in their own class.

Notwithstanding Barr & Allan's (1985) suggestion that the Thrausto-chytrids should be retained in the Oomycetes, features such as their semi-open mitosis, infurrowing cleavage mechanism, lack of transitional helix, galactose based scales and rhizoid structure are sufficient to merit their separation into a Division of their own (Darley *et al.*, 1974). There are, however, sufficient similarities to the Oomycetes to support their continued inclusion with the fungi (Kazama, 1980), perhaps having evolved from the Chrysophyte line which gave rise to scaly flagellates with pseudopodia, such as *Phaeaster* (Belcher & Swale, 1971).

Possible origins of Oomycete orders

The four main orders have many common ultrastructural features with respect to their flagella, hyphal apices, pattern of mitosis, zoospores and oospores and possession of the characteristic dense-body/fingerprint vesicles. Together with other features, this indicates their common origin possibly from an ancestral Xanthophycean alga which also gave rise to the Vaucheriales (Fig. 27.3).

However, inter-relationships between the orders are less clear. Two main pathways are suggested (Fig. 27.3a & b). One envisages the

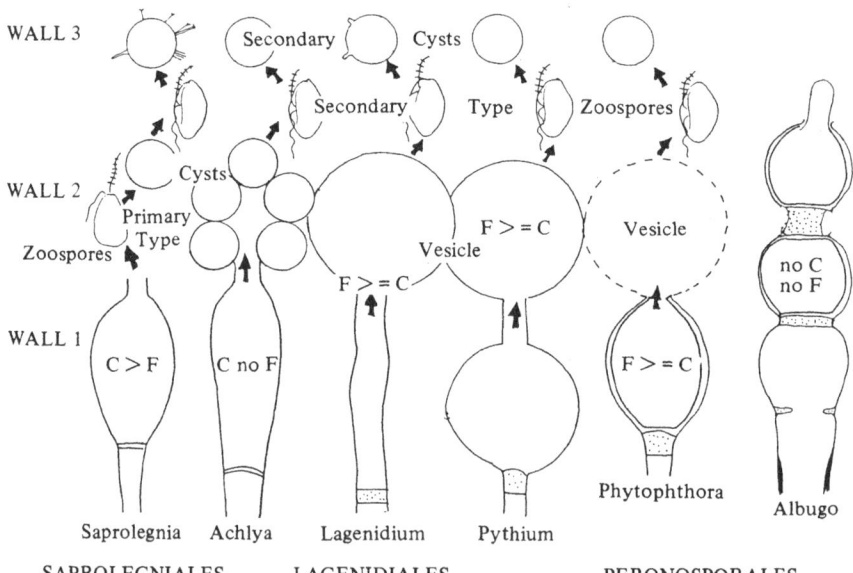

Fig.27.4. Diagram illustrating patterns of sporangium development in Oomycetes. F, flagellum synthesis; C, cytoplasmic cleavage.

Saprolegniales as the main ancestral group, giving rise sequentially, or possibly independently, to the Leptomitales, Lagenidiales and Peronosporales (Fig. 27.3a). This scheme is supported by the morphological and structural similarities between *Saprolegnia* and *Vaucheria* outlined above (Ott & Brown, 1972, 1975). The possibility that the Saprolegniales are primitive is further suggested by their largely saprotrophic life style, the contention that multi-ovulate oogonia without periplasm are more primitive than uniovulate peronosporalean oogonia (Dick, 1969) and that diplanetic (dimorphic) behaviour of zoospores is unlikely to be secondarily derived (but see below). It is certainly possible to propose a logical scheme assuming three phases of wall synthesis during zoosporogenesis; in Saprolegniales represented by the sporangium wall (wall 1, Fig. 27.4); the primary cyst wall (wall 2) and the secondary cyst wall (wall 3). Following suppression of the motile primary zoospore stage (which also occurs within the Saprolegniales in genera such as *Achlya* and *Dictyuchus*), it is suggested that in *Lagenidium* (Gotelli, 1974) and *Pythium* (Lunney & Bland, 1976a) the primary wall phase is represented by the 'sporangial vesicle' and in *Phytophthora* by the inner sporangium wall and evanescent vesicle (Fig. 27.4).

The second scheme envisages the parasitic Lagenidiales as ancestral with the saprolegnian and peronosporalean lines evolving in parallel, rather than sequentially (Fig. 27.3b). This is supported strongly by the isogamous sexual reproduction of *Lagenisma* (Schnepf *et al.*, 1978a). Also, there appears to be morphological similarity between these Oomycetes and the alga *Botrydium*, which might represent another type of ancestral Xanthophyte (Fig. 27.3b). The dimorphic zoospores which characterise the genus *Saprolegnia* could have arisen from monomorphic zoospores in the Lagenidiales as illustrated in Fig. 27.5. The first-formed zoospores of *Lagenidium* (Gotelli, 1974), *Lagenisma* (Schnepf *et al.*, 1978a & b), *Petersenia* (Pueschel & Van der Meer, 1985) and *Haliphthoros* (Overton *et al.*, 1983) retract their flagella axonemes upon settling and illustrate a trend towards pyriform morphology and apical flagellum insertion (Fig. 27.5). By contrast the first formed spores of *Pythium* (Grove & Bracker, 1978) and *Phytophthora palmivora* (Bimpong & Hickman, 1975) and 'secondary type' saprolegnian spores (Hoch & Mitchell, 1972; Holloway & Heath, 1977a & b) all shed their flagella upon settling, although flagellum retraction has been shown in *Phytophthora palmivora* var. *nicotiana* (Reichle, 1969b). If flagella retraction is the prime dividing character between these two spore types, the Lagenidiales could also be considered as diplanetic.

Both the Lagenidiales (Sparrow, 1973, 1976) and Leptomitales (Dick, 1969) have been proposed as intermediate between the Saprolegniales and Peronosporales. The Lagenidiales clearly exhibit both saprolegnian and peronosporalean traits (Table 27.1, Figs. 27.4, 27.5). In contrast, morphological and biochemical evidence suggests the two main families in the Leptomitales fall neatly into two groups, with the Leptomitaceae showing saprolegnian affinities (Table 27.1). The absence of K-bodies and cyst coat vesicles in *Sapromyces*, together with the presence of oospore periplasm in the Rhipidiaceae, support the contention that this group has peronosporalean connections (Sparrow, 1976). Therefore, in the phylogenetic schemes (Fig. 27.3a & b) it is proposed that the Lagenidiales either gave rise to, or are intermediate between, the Saprolegniales and Peronosporales, whilst the Leptomitales are derived from both saprolegnian and peronosporalean ancestors.

Acknowledgements I wish to thank Brent Heath, Donald Barr, John Gay and Curt Pueschel for supplying unpublished information, or permitting me to use their published material.

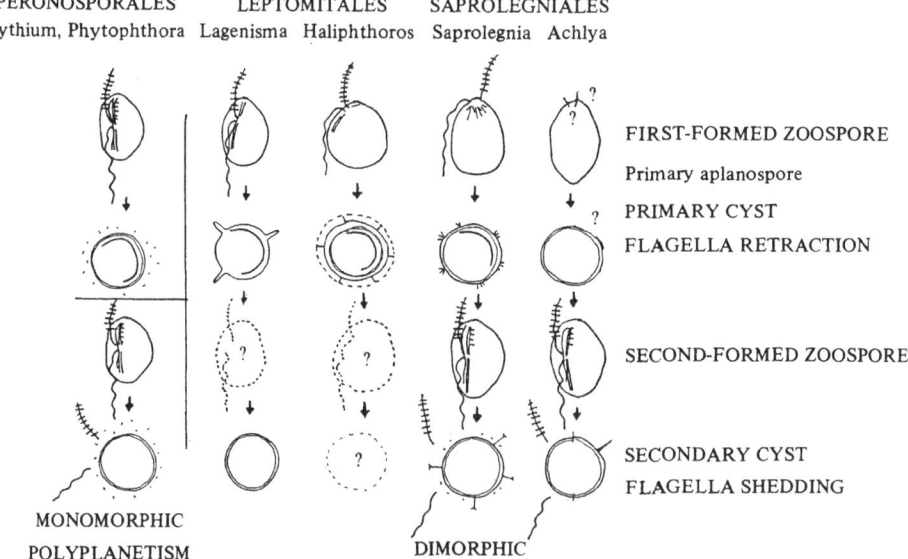

Fig. 27.5. Diagram illustrating the different patterns of zoospore behaviour in Oomycete fungi. (? details not known.)

References

Barr, D. J. S. & Allan, P. M. E. (1985). A comparison of the flagellar apparatus in *Phytophthora, Saprolegnia, Thraustochytrium*, and *Rhizidiomyces. Canadian Journal of Botany*, **63**, 138 − 154.

Bartnicki-Garcia, S. & Wang, M. C. (1983). Biochemical aspects of morphogenesis in *Phytophthora*. In *Phytophthora, its Biology, Ecology, Taxonomy and Pathology*, ed. Erwin, D. C., Bartnicki-Garcia, S. & Tsoa, P. H., pp. 121 − 137. St. Paul, Minnesota: American Phytopathological Society.

Beakes, G. W. (1981a). Ultrastructure of the phycomycete nucleus. In *The Fungal Nucleus*, British Mycological Society Symposium volume 5, ed. Gull, K. & Oliver, S. G., pp. 1 − 35. Cambridge University Press.

Beakes, G. W. (1981b). Ultrastructural aspects of oospore differentiation. In *The Fungal Spore: Morphogenetic Controls*, ed. Hohl, H. R. & Turian, G., pp. 71 − 94. London & New York: Academic Press.

Beakes, G. W. (1983). A comparative account of cyst coat ontogeny in saprophytic and fish-lesion (pathogenic) isolates of the *Saprolegnia diclina − parasitica* complex. *Canadian Journal of Botany*, **61**, 603 − 625.

Belcher, J. H. & Swale, E. M. F. (1971). The microanatomy of *Phaeaster pascheri* Scherffel (Chrysophyceae). *British Phycological Journal*, **6**, 157 − 169.

Bimpong, C. F. & Hickman, C. J. (1975). Ultrastructural and cytochemical studies of zoospores, cysts, and germinating cysts of *Phytophthora palmivora. Canadian Journal of Botany*, **53**, 1310 − 1327.

Bortnick, R. N., Powell, M. J. & Bangert, T. N. (1985). Zoospore fine structure of the parasite *Olpidiopsis saprolegniae* variety *saprolegniae* (Oomycetes, Lagenidiales). *Mycologia*, **77**, 861 − 879.

Cooney, E. W., Barr, D. J. S. & Barstow, W. E. (1985). The ultrastructure of the zoospore of *Hyphochytrium catenoides. Canadian Journal of Botany*, **63**, 497 − 505.

Darley, W. M., Porter, D. & Fuller, M. S. (1973). Cell wall composition and synthesis via Golgi-directed scale formation in the marine eucaryote, *Schizochytrium aggregatum*, with a note on *Thraustochytrium* sp. *Archiv für Mikrobiologie*, **90**, 89 − 106.

Deason, T. R. (1971). The fine-structure of sporogenesis in the Xanthophycean alga *Pseudobummilleriopsis pyrenoidosa. Journal of Phycology*, **7**, 101 − 107.

Dick, M. W. (1969). Morphology and taxonomy of the Oomycetes, with special reference to the Saprolegniaceae, Leptomitaceae and Pythiaceae. I. Sexual reproduction. *New Phytologist*, **68**, 751 − 775.

Dick, M. W. (1972). Morphology and taxonomy of the Oomycetes, with special reference to Saprolegniaceae, Leptomitaceae, and Pythiaceae. II. Cytogenetic systems. *New Phytologist*, **71**, 1151 − 1159.

Dick, M. W. (1973a). Saprolegniales. In *The Fungi, An Advanced Treatise. Vol 4*, ed. Ainsworth, G. C., Sparrow, F. K. & Sussman, A. S., pp. 113 − 144. New York: Academic Press.

Dick, M. W. (1973b). Leptomitales. In *The Fungi, An Advanced Treatise. Vol 4*, ed. Ainsworth, G. C., Sparrow, F. K. & Sussman, A. S., pp. 145 − 158. New York: Academic Press.

Fuller, M. S. & Reichle, R. (1965). The zoospore and early development of *Rhizidiomyces apophysatus. Mycologia*, **57**, 946 − 961.

Gotelli, D. (1974). The morphology of *Lagenidium callinectes*. II. Zoosporogenesis. *Mycologia*, **66**, 846 − 858.

Grove, S. N. & Bracker, C. E. (1978). Protoplasmic changes during zoospore encystment and cyst germination in *Pythium aphanidermatum*. *Experimental Mycology*, **2**, 51 – 98.

Hallett, I. C. & Dick, M. W. (1986). Fine structure of zoospore cyst ornamentation in the Saprolegniaceae and Pythiaceae. *Transactions of the British Mycological Society*, **86**, 457 – 463.

Heath, I. B. (1980). Variant mitoses in lower eukaryotes: indicators of the evolution of mitosis? *International Review of Cytology*, **64**, 1 – 80.

Hoch, H. C. & Mitchell, J. E. (1972). The ultrastructure of zoospores of *Aphanomyces euteiches* and of their encystment and subsequent germination. *Protoplasma*, **75**, 113 – 138.

Holloway, S. A. & Heath, I. B. (1977a). An ultrastructural analysis of the changes in organelle arrangement and structure between the various spore types of *Saprolegnia*. *Canadian Journal of Botany*, **55**, 1328 – 1339.

Holloway, S. A. & Heath, I. B. (1977b). Morphogenesis and the role of microtubules in synchronous populations of *Saprolegnia* zoospores. *Experimental Mycology*, **1**, 9 – 29.

Hibberd, D. J. (1970). Observations on the cytology and ultrastructure of *Ochromonas tuberculatus* sp. nov. (Chrysophyceae), with special reference to the discobolocysts. *British Phycological Journal*, **5**, 119 – 143.

Hibberd, D. J. (1979). The ultrastructure and phylogenetic significance of the flagellar transition region in the chlorophyll c-containing algae. *BioSystems*, **11**, 243 – 261.

Hibberd, D. J. & Leedale, G. F. (1972). Observations on the cytology and ultrastructure of the new algal class, Eustigmatophyceae. *Annals of Botany*, **36**, 49 – 71.

Kazama, F. (1980). The zoospore of *Schizochytrium aggregatum*. *Canadian Journal of Botany*, **58**, 2434 – 2446.

Kumar, C. R. (1980). An ultrastructural study of the marine diatom *Lichomorpha hyalina* and its parasite *Ectrogella perforans*. II. Development of the fungus in its host. *Canadian Journal of Botany*, **58**, 2557 – 2574.

Lange, L. & Olson, L. W. (1979). The uniflagellate phycomycete zoospore. *Dansk Botanisk Arkiv*, **33**, 6 – 95.

Leadbeater, B. S. C. (1970). A fine structural study of *Olisthodiscus luteus* Carter. *British Phycological Journal*, **4**, 3 – 17.

Lee, H. Y., Swafford, J. R. & Aronson, J. M. (1976). Architecture and deposition of cellulin granules in *Apodachlya* sp. *Mycologia*, **68**, 87 – 98.

Lunney, C. Z. & Bland, C. E. (1976a). An ultrastructural study of zoosporogenesis in *Pythium proliferum* de Bary. *Protoplasma*, **88**, 85 – 100.

Lunney, C. Z. & Bland, C. E. (1976b). Ultrastructural observations of mature and encysting zoospores of *Pythium proliferum* de Bary. *Protoplasma*, **90**, 119 – 137.

Manton, I. (1965). Some phyletic implications of flagellar structure in plants. In *Advances in Botanical Research*, ed. Preston, R., pp. 1 – 34. London: Academic Press.

O'Kelly C. J. & Floyd, G. L. (1984). The absolute configuration of the flagellar apparatus in zoospores from two species of Laminariales (Phaeophyceae). *Protoplasma*, **123**, 18 – 25.

Ott, D. W. & Brown, R. M. (1972). Light and electron microscopical observations on mitosis in *Vaucheria littorea* Hofman ex C. Agardh. *British Phycological Journal*, **7**, 361 – 374.

Ott, D. W. & Brown, R. M. (1975). Developmental cytology of the genus *Vaucheria*. III. Emergence, settlement and germination of the mature zoospore of *V. fontinalis* (L.) Christensen. *British Phycological Journal*, **10**, 49 – 56.

Overton, S. V., Tharp, T. P. & Bland, C. E. (1983). Fine structure of swimming, encysting, and germinating spores of *Haliphthoros milfordensis*. *Canadian Journal of Botany*, **61**, 1165 – 1177.

Powell, M. J., Lehnen, L. P. & Bortnick R. N. (1985). Microbody-like organelles as taxonomic markers among Oomycetes. *BioSystems*, **18**, 321 – 334.

Pueschel, C. M. & van der Meer, J. P. (1985). Ultrastructure of the fungus *Petersenia palmariae* (Oomycetes) parasitic on the alga *Palmaria mollis* (Rhodophyceae). *Canadian Journal of Botany*, **63**, 409 – 418.

Reichle, R. E. (1969). Retraction of flagella of *Phytophthora palmivora* var. *nicotiana* zoospores. *Archiv für Mikrobiologie*, **66**, 340 – 347.

Schnepf, E., Deichgraber, G. & Drebes, G. (1978a). Development and ultrastructure of the marine, parasitic Oomycete, *Lagenisma coscinodisci* Drebes (Lagenidiales). The infection. *Archives of Microbiology*, **116**, 133 – 139.

Schnepf, E., Deichgraber, G. & Drebes, G. (1978b). Development of the marine, parasitic Oomycete, *Lagenisma coscinodisci* (Lagenidiales): encystment of primary zoospores. *Canadian Journal of Botany*, **56**, 1309 – 1314.

Sing, V. O. & Bartnicki-Garcia, S. (1975a). Adhesion of *Phytophthora palmivora* zoospores: electron microscopy of cell attachment and cyst wall fibril formation. *Journal of Cell Science*, **18**, 123 – 132.

Sparrow, F. K. (1973). Lagenidiales. In *The Fungi, An Advanced Treatise. Vol. 4*, ed. Ainsworth, G. C., Sparrow, F. K. & Sussman, A. S., pp. 158 – 164. New York: Academic Press.

Sparrow, F. K. (1976). The present status and classification of biflagellate fungi. In *Recent Advances in Aquatic Mycology*, pp. 213 – 222. London: Elek Press.

Waterhouse, G. M. (1962). The zoospore. *Transactions of the British Mycological Society*, **45**, 1 – 20.

Waterhouse, G. M. (1973). Peronosporales. In *The Fungi, An Advanced Treatise. Vol. 4*, ed. Ainsworth, G. C., Sparrow, F. K. & Sussman, A. S., pp. 164 – 185. New York: Academic Press.

28

Xylariaceous fungi: use of secondary metabolites

A . J . S . W H A L L E Y A N D R . L . E D W A R D S *

Biology Department, Liverpool Polytechnic, Liverpool L3 3AF, UK and
*Department of Chemistry, University of Bradford, Bradford BD7 1DP, UK

Introduction

According to Cain (1972), little is to be gained from any sort of phylogenetic tree based on an arrangement of present-day genera and families. In the absence of an adequate fossil record, however, it is only present-day organisms which are available for investigation and sound understanding of the relationships between them is surely a prerequisite for any phylogenetic interpretation. Few families in the Ascomycotina have been the subject of such wide-ranging investigations, resulting in a reasonable understanding of at least some intergeneric relationships, as the Xylariaceae (Rogers, 1979). This family, consequently, provides a suitable framework for investigation of chemical characters and their correlation with putative lines of descent.

The Xylariaceae: current taxonomy

The Xylariaceae is an assemblage of sphaeriaceous genera with obscure but apparently common ancestry, which are worldwide in their distribution. A variety of generic schemes have been proposed, as shown in Table 28.1. As a result of the paucity of recent collections of many of these genera and the consequent lack of detailed or experimental data, it is difficult to be confident about inter-generic relationships except for the 'core' genera recognised by Rogers (1979). These latter have, in addition to the use of traditional taxonomic characters, been subjected to studies on ascospore ornamentation (Rogers, 1977, Rogers & Whalley, 1978), nature of the apical apparatus and ascus tip (Griffiths, 1973, Pouzar, 1985a & b), peridial anatomy (Jensen, 1985), anamorph morphology and development (Martin, 1967; Greenhalgh & Chesters, 1968; Jong & Rogers, 1972), stromal pigmentation (Greenhalgh & Whalley, 1970) and

Table 28.1. *Genera of the Xylariaceae*

Rogers (1979)	Cannon *et al.* (1985)*	Eriksson & Hawksworth (1985)†
Anthostomella	Anthostoma	Anthostomella
Areolospora	Anthostomella	Areolospora
Camillea	Ascotricha	Ascotricha
Daldinia	Camarops	Camillea
Entonaema	Coniochaeta	Daldinia
Graphostroma	Coniochaetidium	Engleromyces
Hypocopra	Daldinia	Entonaema
Hypoxylon	Hypocopra	Helicogermslita
Kretzschmaria	Hypoxylon	Hypocopra
Lopadostoma	Lopadostoma	Hypoxylon
Nummularia	Nummulariella	Kretzschmaria
Penzigia	Podosordaria	Lopadostoma
Podosordaria	Poronia	Nummulariella
Poronia	Rosellinia	Penzigia
Rosellinia	Ustulina	Phylacia
Stromatoneurospora	Wawelia	Podosordaria
Thamnomyces	Xylaria	Poroconiochaeta
Ustulina		Poronia
Xylaria		Pulveria
		Rhopalostroma
		Rosellinia
		Sarcoxylon
		Stilbohypoxylon
		Stromatoneurospora
		Thamnomyces
		Theissenia
		Ustulina
		Wawelia
		Xylaria

* British genera only; † *Astrocystis*, *Batistia* and *Paucithecium* may also be included.

cytology (Rogers, 1964, 1970, 1973, 1975). Based on the resulting data, a scheme depicting relationships between these genera may be drawn up as shown in Fig. 28.1.

Hypoxylon is the most studied genus and is the only substantial one recently to have been monographed, by Miller (1961) who proposed four sections on the basis of traditional morphological features (Table 28.2). Subsequent experimental studies have partly confirmed this scheme but have also resulted in some proposed modifications (Whalley & Green-halgh, 1973; Whalley, 1977; Pouzar, 1979, 1985a & b; Rogers, 1979). It

Table 28.2. *Classification of* Hypoxylon *(Miller, 1961)*

Section Hypoxylon:	Stromata leathery or woody, never carbonaceous, bright coloured, usually some shade of red, purple or brown; ostioles umbilicate
Section Papillata:	Ostioles papillate
subsection Papillata:	Stromata coloured at least when young becoming brown to black at maturity
subsection Primo-cinerea:	Stromata white or light grey when young becoming black and very carbonaceous at maturity
Section Annulata:	Ostioles papillate and surrounded by an annular disc: stromata usually coloured
Section Applanata:	Stromata restricted or indefinitely effused, carbonaceous and applanate; ostioles umbilicate or papillate

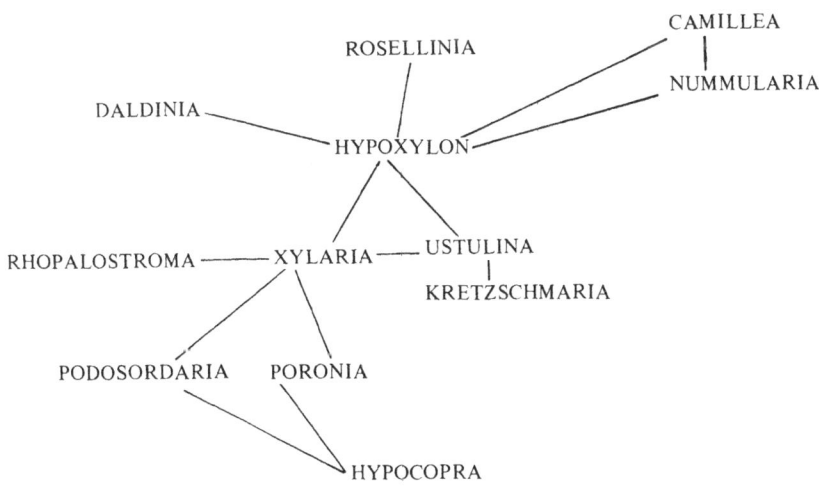

Fig. 28.1. Proposed relationships in the Xylariaceae.

is with respect to the relationships, both between these different sections and between these sections and other genera, that most progress has been made and where a knowledge of secondary metabolites has a major impact.

Secondary metabolites
Nature and distribution
It is still useful to regard secondary metabolites as structurally diverse compounds, frequently exhibiting taxonomic specificity in their production, which typically occur during later or stationary phases of growth (Moss, 1984). In spite of many references to their taxonomic specificity, they have been much neglected in fungal systematics, especially in the Ascomycotina. Only the lichenised orders have been investigated in depth, with over 4500 species examined; in non-lichenised forms only certain groups in the Clavicipitales, Endomycetales and Eurotiales are considered to have received suitable attention (Hawksworth, 1985).

The Xylariaceae are a rich source of secondary metabolites, identified mainly as butyrolactones (Edwards & Whalley, 1979; Anderson, Edwards & Whalley, 1982), dihydroisocoumarins (Anderson, Edwards & Whalley, 1983) and succinic acid derivatives (Anderson, Edwards & Whalley, 1985). The structure and distribution of the more commonly occurring compounds are given in Fig. 28.2 and Tables 28.3 and 28.4. In addition, sesquiterpene alcohols (Anderson, Briant *et al.*, 1984; Anderson, Edwards *et al.*, 1984), pyrenophorin (Anderson, Edwards & Whalley, 1983) and the allenic ether, chestersiene (Edwards, Anderson & Whalley, 1982) have limited occurrence; a range of cytochalasins occur sporadically throughout the family and griseofulvin, 5-hydroxygriseofulvin and dechlorogriseofulvin have been isolated from representatives of two different genera (Edwards & Whalley, unpublished).

Taxonomic significance
The dihydroisocoumarins, with 5-methylmellein as a common constituent, are generally associated with *Hypoxylon* species. Mellein, iso-ochracein and ramulosin are characteristic for a limited number of species belonging to the section Hypoxylon, although mellein also occurs elsewhere. The relationship between applanate species of *Hypoxylon* and certain members of the former genus *Nummularia* (= *Nummulariella* Eckblad & Granmo) is of special interest since there is some controversy surrounding their separation (Whalley & Edwards, 1985). Miller (1932)

separated *Nummularia* from applanate *Hypoxylon* species on the basis of its cup-shaped stroma. Pouzar (1979) regarded this separation as untenable and united some species of *Hypoxylon* with *Nummularia* in the newly erected *Biscogniauxia*. Rogers (1979) believed this might be premature since *Biscogniauxia* now contained species with light coloured, ornamented ascospores together with those possessing smooth, dark spores. The presence of 5-methyl mellein amongst members of *Nummularia*

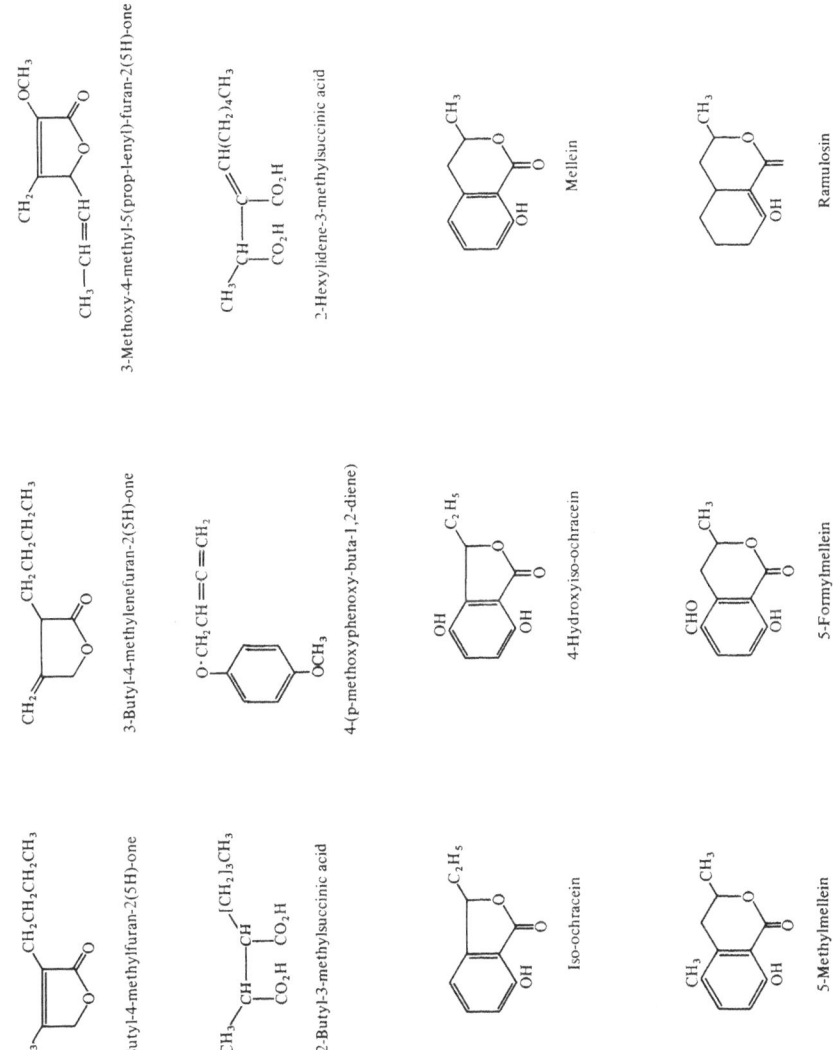

Fig. 28.2. Commonly occurring secondary metabolites.

Table 28.3. *Distribution of dihydroisocoumarins*

Species	Section	Dihydroisocoumarins 1 2 3 4 5 6 7 8 9 10
H. fragiforme		+ +
H. howeianum		+ + + +
H. haematostroma		+ + +
H. venustuissimum	Hypoxylon	+ + +
H. jecorinum		+
H. rubiginosum		+
H. fraxinophilum		+
H. cohaerens		
H. multiforme	Papillata	
H. investiens		
H. rutilum		+
H. confluens		+
H. udum		
H. mammatum		+ + +
H. illitum	Primo-cinerea	+ + +
H. chestersii		+
H. serpens		+
H. truncatum	Annulata	+
H. stygium		+
H. grenadense var. macrospora		+
H. atropuncatum		+ +
H. mediterraneum	Applanata	+
H. tinctor		+
H. punctulatum		
H. microplacum		+
U. deusta	Ustulina	+
N. discreta	Nummularia/	+ + +
N. broomeiana	Nummulariella/	+ +
N. dennisii	Biscogniauxia/	+ +

Key to the identity of the dihydroisocoumarins: 1 = Mellein; 2 = Iso-ochracein; 3 = Ramulosin; 4 = 4-Hydroxyiso-ochracein; 5 = 5-Methylmellein; 6 = 5-Formylmellein; 7 = 5-Methoxycarbonylmellein; 8 = 6-Methoxy-5-methylmellein; 9 = 5-Hydroxymethylmellein; 10 = 5-Carboxymellein.

confirms its close affinity to *Hypoxylon*. However, 5-formylmellein is restricted to *Nummularia* and is absent from the applanata section of *Hypoxylon*, indicating the need for further information before the merger is accepted (Whalley & Edwards, 1985).

Table 28.4. *Distribution of succinic acid derivatives*

Species	Diacid 1	Diacid 2
Ustulina deusta		+
Poronia pileiformis		+
Xylaria polymorpha		+
X. longipes		+
X. hypoxylon		+
X. mali		+
Hypoxylon illitum	+	

Diacid 1 = 2-Butyl-3-methylsuccinic acid;
Diacid 2 = 2-Hexylidene-3-methylsuccinic acid.

Just as the dihydroisocoumarins are characteristic of *Hypoxylon* and *Nummularia*, so the succinic acid derivatives are associated with *Xylaria* and its close relatives. Furthermore the presence of 2-hexylidene 3-methyl succinic acid in *Poronia pileiformis* suggests that it really belongs in *Xylaria* (Paden, 1978), and *Rhopolostroma gracile*, which is considered on other grounds to be related to *Xylaria* (Hawksworth & Whalley, 1985) produces succinic acid as a major metabolite (Anderson, Edwards & Whalley, 1985). *Ustulina deusta* has long been believed to be intermediate between *Hypoxylon* and *Xylaria* (Rogers, 1968) and the presence of the hexylidene methyl succinic acid supports its association with the latter genus. Moreover, the production of 2-butyl-3-methylsuccinic acid (Anderson, Edwards & Whalley, 1985) by *H. illitum* closely links it with another member of the subsection primo-cinerea, *H. serpens*, which produces a butyrolactone bearing the same alkyl substituents (Edwards & Whalley, 1979), as well as to *Xylaria*.

The taxonomic significance of the other chemicals, e.g. cytochalasins, is not yet known but they are, at least, useful markers at the species level.

Phylogenetic considerations

Present day descendants of the Xylariaceae are primarily parasitic and saprotrophic colonisers of angiosperms, there being very few records of pteridophyte or gymnosperm hosts. The earliest evidence of xylariaceous fungi in the fossil record is as ascospores in the early Cenozoic, long after angiosperm evolution and diversification was advanced (Tiffney & Barghoorn, 1974). Axelrod (1970) presented a strong case to support the origin and initial development of angiosperms in open areas which were subjected to regular periods of drought. It is, therefore, of interest that present day xylariaceous fungi exhibit structures and life-history features that almost certainly developed as adaptations to hosts growing under seasonally dry conditions (Rogers, 1979). Thus, the massive sessile stromata of *Daldinia* and many species of *Hypoxylon* may be a development to minimise water loss from the perithecia. In other representatives of the family, e.g. *Hypoxylon udum*, *Anthostomella* and *Lopadostoma*, where the perithecia are much reduced, they are protected from desiccation through immersion in the substratum. In the coprophilous *Hypocopra*, the stromata are more or less rudimentary and it has been suggested that the dung itself may provide the fungus with the advantages of a massive stroma (Rogers, 1979). Similarly, it has been argued that the upright stromata of species of *Xylaria* were probably adaptations by colonisers of buried substrata on dry sites to raise their perithecia above ground level for more effective ascospore dispersal. The recently described *Wawelia octospora* seems to be strongly xerophilic (Minter & Webster, 1983). A special adaptation to dry environments is shown by *Nummularia* and applanate species of *Hypoxylon*, where a bipartite stroma develops within the bark of host trees. The inner and outer regions of the stroma are separated by a gelatinous zone which swells up, forcing the two stromal layers apart and rupturing the bark, which is then shed together with the outer stromal layer. This mechanism ensures that the developing stroma is protected by the bark from excessive water loss. Additionally,

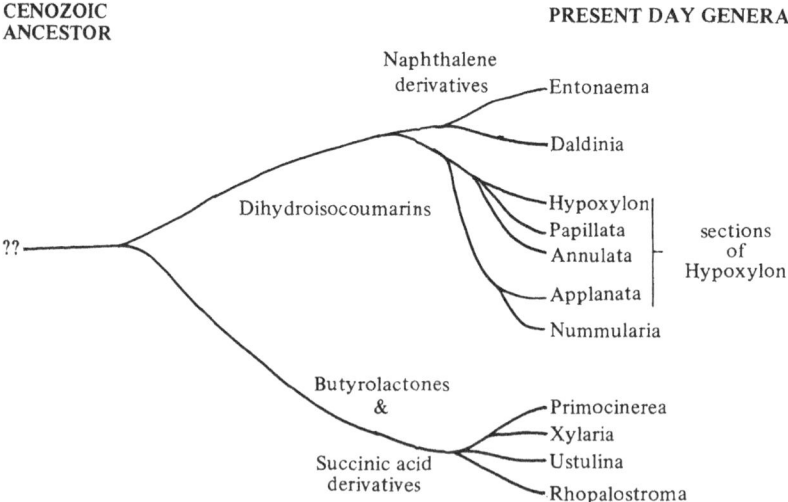

Fig. 28.3. Chemical diversity and possible phylogenetic relationships.

the relatively long period of ascospore maturation and discharge, rapid germination of ascospores in water, and the discharge of ascospores when water is available, are all features which might be considered to reflect adaptation to dry environments (Rogers, 1979).

If one accepts that the Xylariaceae evolved on angiosperms living in dry conditions and that features associated with this have been retained, then it is possible to propose a number of groups based on these adaptive features (Fig. 28.3). There is a high degree of association of other characteristics with these groups and here, chemical considerations are of special interest. Thus, *Daldinia* and the sections Hypoxylon, Annulata and subsection Papillata of *Hypoxylon* are, in general, characterised by well developed, sessile and often massive stromata which in the majority of cases are coloured (Greenhalgh & Whalley, 1970); by the occurrence of the germ slit on the convex side of the ascospore (Martin, 1967; Pouzar, 1979); and by the possession of anamorphs which can be accommodated in the form genus *Nodulisporium* and morphologically allied hypho-mycetes (Jong & Rogers, 1972). This group is also characterised by the production of dihydroisocoumarins. In the section Applanata of *Hypo-xylon* and in *Nummularia*, which have the bipartite stromal structure, the anamorphs generally belong to *Basidiobotrys* but a strong association with the other sections of *Hypoxylon* is shown through the occurrence of dihydroisocoumarins. The, as yet, exclusive occurrence of 5-formyl mellein in *Nummularia* is perhaps an indication of its development independently of *Hypoxylon*. *Daldinia* remains chemically distinct, since *D. concentrica* produces naphthalene derivatives and dihydroxy ketones (Allport & Bu'lock. 1960).

In the remaining section of *Hypoxylon*, the Primo-cinerea, the stromata are in general very much reduced and frequently immersed. Here butyro-lactones, not dihydroisocoumarins, are produced and the close chemical similarity between the butyrolactones and the succinic acid derivatives isolated from *Xylaria* and *Ustulina* suggests a closer association with these genera than exists with other species of *Hypoxylon*. The report of succinic acid or derivatives in *Rhopalostroma gracile* and *Anthostomella avocetta* (Anderson, Edwards & Whalley, 1985), also links these fungi with the *Xylaria/Hypoxylon* Primo-cinerea group.

In linking the different genera by chemical constituents and correlating these with other characters (Fig. 28.3), it is useful to remember that ' . . . it is relatively easy to find characters which appear to be connecting links between various taxa at almost any level including genera, families and orders . . . unfortunately, when all these links are put together you find that any particular taxon can be linked in all directions rather than in an orderly ascending phylogenetic tree . . . ' (Cain, 1972). Nonetheless, even with many gaps in present knowledge, it is tempting to postulate that in the Xylariaceae a major evolutionary division occurred, resulting in the separate development of the dihydroisocoumarin group, with their well developed stromata and the butyrolactone/succinic acid derivative group, which shows either reduced or stalked stromata.

References

Allport, D. C. & Bu'lock, J. D. (1960). Biosynthetic pathways in *Daldinia concentrica*. *Journal of the Chemical Society*, 654 − 662.

Anderson, J. R., Briant, C. E., Edwards, R. L., Mabelis, R. P., Poyser, J. P., Spencer, H. & Whalley, A. J. S. (1984). Punctatin A (Antibiotic M95464): X-Ray crystal structure of a sesquiterpene alcohol with a new carbon skeleton from the fungus, *Poronia punctata. Journal of the Chemical Society. Chemical Communications*, 405 − 406.

Anderson, J. R., Edwards, R. L., Freer, A. A., Mabelis, R. P., Poyser, J. P., Spencer, H. & Whalley, A. J. S. (1984). Punctatins B and C (Antibiotics M95154 and M95155) Further sesquiterpene alcohols from the fungus *Poronia punctata. Journal of the Chemical Society, Chemical Communications*, 917 − 919.

Anderson, J. R., Edwards, R. L. & Whalley, A. J. S. (1982). Metabolites of the higher fungi. Part 19. Serpenone, 3-methoxy-4-methyl-5 prop-1-enylfuran-2 (5H)-one. A new γ-Butyrolactone from the fungus *Hypoxylon serpens* (Barron's strain) (Persoon ex Fries) Kickx. *Journal of the Chemical Society. Perkins Transactions*, **1**, 215 − 221.

Anderson, J. R., Edwards, R. L. & Whalley, A. J. S. (1983). Metabolites of the higher fungi. Part 21. 3-methyl-3, 4-dihydroisocoumarins and related compounds from the ascomycete family Xylariaceae. *Journal of the Chemical Society. Perkins Transactions*, **1**, 2185 − 2192.

Anderson, J. R., Edwards, R. L. & Whalley, A. J. S. (1985). Metabolites of the higher fungi. Part 22. 2-butyl-3-methylsuccinic acid and 2-hexylidine-3-methyl-succinic acid from xylariaceous fungi. *Journal of the Chemical Society, Perkins Transactions*, **1**, 1481 − 1485.

Axelrod, D. I. (1970). Mesozoic paleogeography and early angiosperm history. *Botanical Review*, **36**, 277 − 319.

Cain, R. F. (1972). Evolution of the fungi. *Mycologia*, **64**, 1 − 14.

Cannon, P. F., Hawksworth, D. L. & Sherwood-Pike, M. A. (1985). *The British Ascomycotina: An Annotated Checklist*. Slough: Commonwealth Agricultural Bureau.

Edwards, R. L., Anderson, J. R. & Whalley, A. J. S. (1982). Metabolites of the higher fungi. Part 20. Chestersiene, 4-(4-methoxyphenoxy)buta-1, 2-diene: an allenic ether from the fungus *Hypoxylon chestersii. Phytochemistry*, **21**, 1721 − 1723.

Edwards, R. L. & Whalley, A. J. S. (1979). Metabolites of the higher fungi. Part 18. 3-Butyl-4-methyl-2 (5H) furanone and 3-Butyl-4-methylene-2 (5H) furanone.

New γ-Butyrolactones from the fungus *Hypoxylon serpens* (Persoon ex Fries) Kickx. *Journal of the Chemical Society. Perkins Transactions*, **1**, 803 − 806.

Eriksson, O. & Hawksworth, D. L. (1985). Outline of the ascomycetes − 1985. *Systema Ascomycetum*, **4**, 1-79.

Greenhalgh, G. N. & Chesters, C. G. C. (1968). Conidiophore morphology in some British members of the Xylariaceae. *Transactions of the British Mycological Society*, **51**, 57 − 82.

Greenhalgh, G. N. & Whalley, A. J. S. (1970). Stromal pigments of some species of *Hypoxylon*. *Transactions of the British Mycological Society*, **55**, 89 − 96.

Griffiths, H. B. (1973). Fine structure of 7 unitunicate pyrenomycete asci. *Transactions of the British Mycological Society*, **60**, 261 − 271.

Hawksworth, D. L. (1985). Problems and prospects in the systematics of the Ascomycotina. *Proceedings of the Indian Academy of Sciences (Plant Science)*, **94**, 319 − 339.

Hawksworth, D. L. & Whalley, A. J. S. (1985). A new species of *Rhopalostroma* with a *Nodulisporium* anamorph from Thailand. *Transactions of the British Mycological Society*, **84**, 560 − 562.

Jensen, J. D. (1985). Peridial anatomy and pyrenomycete taxonomy. *Mycologia*, **77**, 688 − 701.

Jong, S. C. & Rogers, J. D. (1972). Illustrations and descriptions of conidial states of some *Hypoxylon* species. *Washington State University Agricultural Experiment Station. Technical Bulletin*, **77**, 1 − 49.

Martin, P. (1967). Studies in the Xylariaceae: 1. New and old concepts. *Journal of South African Botany*, **33**, 205 − 240.

Miller, J. H. (1932). British Xylariaceae. II. *Transactions of the British Mycological Society*, **17**, 125 − 135.

Miller, J. H. (1961). *A monograph of the world species of Hypoxylon*. Athens, Georgia USA: University of Georgia Press.

Minter, D. W. & Webster, J. (1983). *Wawelia octospora* sp. nov., a xerophilous and coprophilous member of the Xylariaceae. *Transactions of the British Mycological Society*, **80**, 370 − 373.

Moss, M. O. (1984). The mycelial habit and secondary metabolite production. In *The Ecology and Physiology of the Fungal Mycelium*, British Mycological Society Symposium volume 8, ed. Jennings, D. H. & Rayner, A. D. M., pp. 127 − 142. Cambridge University Press.

Paden, J. W. (1978). Morphology, growth in culture, and conidium formation in *Poronia pileiformis*. *Canadian Journal of Botany*, **56**, 1774 − 1776.

Pouzar, Z. (1979). Notes on taxonomy and nomenclature of Nummularia (Pyrenomycetes). *Česká mycologie*, **33**, 207 − 219.

Pouzar, Z. (1985a). Reassessment of *Hypoxylon serpens* − complex I. *Česká mycologie*, **39**, 15 − 25.

Pouzar, Z. (1985b). Reassessment of *Hypoxylon serpens* − complex II. *Česká mycologie*, **39**, 129 − 134.

Rogers, J. D. (1964). *Hypoxylon pruinatum*: the chromosome number. *Mycologia*, **56**, 369 − 373.

Rogers, J. D. (1968). *Hypoxylon deustum*: the chromosome number. *Mycopathologia et Mycologia Applicata*, **35**, 249 − 255.

Rogers, J. D. (1970). Cytology of *Poronia oedipus* and *Poronia punctata*. *Canadian Journal of Botany*, **48**, 1665 − 1668.

Rogers, J. D. (1973). Cytology of *Podosordaria leporina*. *Canadian Journal of Botany*, **51**, 791 − 793.

Rogers, J. D. (1975). *Xylaria polymorpha* II. Cytology of a form with typical robust stromata. *Canadian Journal of Botany*, **53**, 1736 – 1743.

Rogers, J. D. (1977). Surface features of the light colored ascospores of some applanate *Hypoxylon* species. *Canadian Journal of Botany*, **55**, 2394 – 2398.

Rogers, J. D. (1979). The Xylariaceae: systematic, biological and evolutionary aspects. *Mycologia*, **71**, 1 – 42.

Rogers, J. D. & Whalley, A. J. S. (1978). A new *Hypoxylon* species from Wales. *Canadian Journal of Botany*, **56**, 1346 – 1348.

Tiffney, B. H. & Barghoorn, E. S. (1974). The fossil record of the fungi. *Occasional Papers of the Farlow Herbarium*, **7**, 1 – 42.

Whalley, A. J. S. (1977). Relationships of coloured papillate species of *Hypoxylon* to other members of the genus. In *Abstracts, Second International Mycological Congress, Tampa, Florida, USA*, ed. Bigelow, H. E. & Simmons, E. G., p. 735.

Whalley, A. J. S. & Edwards, R. L. (1985). *Nummulariella marginata*; its conidial state, secondary metabolites and taxonomic relationships. *Transactions of the British Mycological Society*, **85**, 385 – 390.

Whalley, A. J. S. & Greenhalgh, G. N. (1973). Numerical taxonomy of *Hypoxylon*. 1. Comparison of the cultural and perfect states. *Transactions of the British Mycological Society*, **61**, 435 – 454.

29

Whence cometh the Agarics?
A reappraisal

O. K. MILLER Jr. AND ROY WATLING*

*Department of Biology, Virginia Polytechnic Institute and State University, Blacksburg, Virginia 24061-0794, USA, and *Royal Botanic Garden, Edinburgh EH3 5LR, Scotland*

Introduction

As discussed by Savile (1968), many evolutionary schemes have been proposed for the origin of the basidiomycetes. However, an approach with which we sympathise is that of Cain (1972), even though he himself regarded his ideas as reading like science fiction. Cain suggested that the ancestors of the Basidiomycota were autotrophic, forming a loose, filamentous, chlorophyllous thallus. He hypothesised further that both haplont and dikaryotic phases were present. The former produced a parasitic dikaryotic stage, the sole or main function of which was to produce dikaryotic disseminating propagules which, on germination, produced a thallus supporting, following nuclear fusion and meiotic division, a further haplont phase. Propagules formed by this last stage, in response to moving from marine to land habitats, became forcibly discharged, a feature associated with basidiomycetes today.

Notwithstanding the validity or otherwise of Cain's hypothesis, the approach adopted here will be to consider the origins and relationships of the agarics and extrapolate backwards instead of working forwards from an unknown. Since there is little doubt that the ancestors of the homobasidiomycetes have long since vanished, we can only probe the possibilities by analysis of a spattering of sometimes disjunct end-products of evolution.

The analysis involves identification of parallel morphological series within the major groups of larger fungi and of a combination of additional data to indicate in which direction these series can be read. Although having to rely on the morphology, anatomy and development of present-day basidiomes — which are undoubtedly rather recent in their evolution — some general concepts arise which demand further exploration. As always, there is the problem of discriminating between primitive and reduced characteristics. For example, the gymnocarpic state, being developmentally simple, would appear to be more primitive than the mixangio-, pilangio-, bulbangio-carpic states (Watling, 1978; 1985a), but does this always apply?

Geological considerations

As mentioned by Lewis (Chapter 11), the fossil evidence, which has been strengthened further by recent studies, suggests that basidiomycetes and ascomycetes occurred prior to the Carboniferous. Rothwell (1972) described, in soil from Middle Pennsylvanian petrifications, sclerotia with fungal tissue composed of cells with common walls which appear to be a *textura angularis*. This type of sclerotium, which is common today, was fully developed, therefore, some 300 million years ago. The development of plectenchymatous structures such as this, along with the presence of clamp connections and well-developed hyphal systems in the wood of Middle Carboniferous progymnosperms and ferns, as reported by Dennis (1969), alters perception of the time frame in which the evolution of higher fungi took place (Berch, Miller & Thiers, 1984).

More recently, Stubblefield, Taylor & Beck (1985) presented convincing evidence of a wood decay fungus in *Callixylon newberryi*, a progymnosperm from the Upper Devonian. The extensive proliferation of hyphae in both tracheid and ray cells leaves no doubt that a well-developed, septate, hyphal thallus had evolved in the Eumycota before the Carboniferous period. The late Devonian marked the emergence of the progymnosperms (Beck, 1976) with a secondary meristem and this allowed the development of forests which, of course, created vast new ecosystems. The accompanying development of decomposition processes also occurred against the background of a flora evolving when the super continent, Pangaea, provided a large, contiguous, land mass. The earliest reconstructions for the Late Cambrian show that a majority of the continents were dispersed around the globe in low tropical latitudes (Bambach, Scotese & Ziegler, 1981). Open polar oceans were most likely warmed by oceanic circulation and frigid polar climates have not been detected. Plant communities continued to have every opportunity to develop during the Devonian and Carboniferous periods within two large Paleozoic continents, Laurasia and Gondwana, as they slowly moved together. The land bridges were probably reduced by the Jurassic, hence preventing some gene flow between continents, but most, including Schuster (1976), agree that a cosmopolitan flora was present as late as this period. However, Schuster does conclude from fossil gymnosperms that Northern and Southern floras were incipient. By the Cretaceous the ectomycorrhizal Pinaceae was widespread, as were the angiosperms (Miller, 1977) which, by some estimates, made up half of the flora by biomass and frequently occupied niches formerly taken by conifers.

The most protected place in which to be when an organism emerges from a fluid to a dry environment, is, at least partly, within the cells or tissues of another one. Thus, on moist tropical soils a primitive basidiomycete could exist concurrently with other evolving organisms, later to undergo an explosion of adaptive radiation as the angiosperms developed, offering not only the early soil profile as a home, but all the mass of leafy debris, free of terpenes and related compounds associated with the gymnospermous plants and radiomimetic substances in ferns.

Significance of mycotrophy

The present representatives of the earliest angiosperms appear to occupy three major lines of evolution which started from the ancestral Magnoliidae (Cronquist, 1981): one to the present-day Magnoliidae which are associated with endomycorrhizae; a second primitive line to the Hamamelidae, ranging from the Australian Casuarinales (Warcup, 1980) to the Hamamelidales of the Northern Hemisphere composed of ectomycorrhizal associates; and a third to the Dilleniidae. The Casuarinales and Hamamelidales are currently associated with basidiomycetes and we assume that this relationship has existed throughout time (Malloch, Pirozynski & Raven, 1980). The third line today contains ectomycorrhizal orders, such as the Malvales and Salicales, involving many basidiomycete taxa. However, the ericoid endomycorrhizal habit (Miller, 1983; Read, 1983), is present in both the Ericales and Diapensiales and it is within these that we may have uncovered one of the primitive mutualisms among the Basidiomycotina.

The ericoid endomycorrhizal habit is well established throughout both hemispheres, usually with ascomycetes (see Lewis, Chapter 11). However, Seviour, Willing and Chilvers (1973) described basidiomes from Australia associated with ericoid mycorrhizas of imported Ericaceae. They tentatively identified the fungus as *Clavaria vermicularis*, although one of us (RW) later identified it as *C. argillacea*, which is also associated with *Calluna vulgaris* in Europe. Another ericoid mutualism involving a *Clavaria* occurs with *Rhododendron* in North America (Englander & Hull, 1980; Peterson, Mueller & Englander, 1980). Here, the fungal symbiont is also *C. argillacea* according to R. H. Petersen (personal communication), who identified the same taxon from the Northern forests of North America. The coiled hyphae of *C. argillacea* within the host cells have dolipore septa, known only in the Basidiomycotina, but are accompanied by ascomycetous septate hyphae containing Woronin bodies (Englander & Hull, 1980).

It is appealing to consider the clavarioid basidiome as primitive and the ultimate expression of simplicity. As Savile (1955) points out ' . . . hymenium production follows closely behind the growing point, and a modest total growth ensures some spore production . . . '. The simple sporocarp composed of monomitic tissue, thin-walled, hyaline basidiospores, and narrow hyphal-like stichobasidium combines several primitive features. However, there is little doubt that the group of fungi producing such basidiomes, as it is treated now by Julich (1984), Corner (1950), Petersen (1973) and others, is polyphyletic. For example, mycorrhizal taxa of *Clavaria*, basidiolichens such as *Multiclavula* and parasites with sclerotia, as in *Typhula*, were all at one time grouped with dimitic species in *Lentaria* which are decomposers. Moreover, ontogenetic responses documented by Miller & Stewart (1971) show reversion by lamellate taxa to clavarioid forms. Nevertheless, an hypothesis that accommodation to a terrestrial environment by green plants required the close association of a fungus can still be met, and just as present-day clavarioid taxa are associated with both algae and higher green plants as mutuals, so could the progenitors, some even being parasitic. The parasitic condition is considered primitive by Savile (1955, 1968), but the possibility that mycorrhizas are equally primitive is considerable (Pirozynski & Malloch, 1975).

The close association of mycorrhizal fungi with their respective hosts seems to have resulted in several general trends. Firstly, a relatively simple sporocarp composed of monomitic tissue is formed with comparatively simple, thin-walled basidiospores and a loss or reduction of conidiophores and asexual propagules. Secondly, mutualistic or mycorrhizal fungi, as a group, grow poorly or not at all on synthetic media and require complex substances to promote their growth and survival. Thirdly, they lack the complex laccase enzyme systems so well developed in active decomposers and lack the extension growth rate necessary for rapid colonisation of an available substratum. Finally, the interaction of hormones by mycorrhizas with the host, cytokinin and IAA growth stimulation and the presence of common vitamin deficiencies all reflect the dependency of these fungi on their host. The remarkable ability of a mycorrhizal fungus to mitigate the uptake of phosphorus by the host (Ford *et al.*, 1985), lends strength to the theory that favoured host–fungus relationships can give a particular plant an advantage over others and so affect the course of evolution.

Ecological phenomena

Ecological information must always be considered, and wherever possible, correlated with other data. Thus, Nobles (1965) and Redhead & Ginns (1985), found a relationship between the type of wood-rot and degree of evolution within polypores and pleurotoid fungi. Watling (1985b) has shown the importance of strategies in the understanding of relationships within the Bolbitiaceae. This is why we believe that one theme running through the evolution of agarics is the mycorrhizal strategy. However, Bondartzeva (1963) showed anatomy can reflect ecology, not necessarily relationships, so caution is paramount.

Mycorrhizal formation makes two distinct energy demands on the host; that necessary for the formation of the mycorrhizal mantle, Hartig net, and intercellular hyphal coils, and that involving formation of basidiomes. The latter usually occurs while the host is photosynthetically active and must not jeopardise the host reserves needed for periods of dormancy. This constrains both the size and complexity of the basidiome, as well as controlling the fruiting response (Hacskaylo, 1965).

The selection pressures dictating evolution of mycorrhizal basidiomes have, thus, been the need to protect the developing spore from desiccation and to provide an increased hymenial surface for the production of basidiospores within a short-lived, fragile sporocarp. Neither the perennial habit nor elaborately constructed basidiomes are found, contrasting with vegetatively persistent members of the Aphyllophorales. Similarly, development of complex-walled, pigmented basidiospores is a feature of widely distributed agaric decomposers, but not mycorrhizal formers (Miller, 1983).

The logical extension from the clavarioid condition among epigeous taxa is the cantharelloid basidiome. The widespread development of the upland ectomycorrhizal gymnosperm forests of both the Northern and Southern hemispheres, especially the former, provided the opportunity for the full development of the wrinkled and ridged hymenium of the Cantharellaceae in which the taxa, although still gymnocarpic, are somewhat protected by a pileus. Kühner (1984) enumerates the primitive characters possessed by the chanterelles which parallel those of the Clavariaceae. It would appear that wind borne spores, forcibly discharged, provided the dominant means of early spore dissemination.

However, as the upland gymnosperm forest communities developed, a litter layer would have provided the necessary cover to allow simple, hypogeous or resupinate basidiomes to develop, probably resembling ectomycorrhizal species of *Tomentella* (Danielson, Zak & Parkinson, 1984) and others. Wherever decomposition processes in temperate or boreal biomes are slow, a litter layer develops and this simple type of basidiome can exist and thrive, largely unaltered in form from its ancestral progenitors.

The advent and further colonisation of upland habitats by angiosperms was also accompanied by the evolution and widespread distribution of invertebrates and later, of rodents and small marsupials in both hemispheres. The retention of the basidiospores would be a response to invertebrate dissemination and perhaps a further stage of speciation was in response to mammal and marsupial transportation, leading to the secotioid character. This is exemplified by the high numbers of rodent species correlated with similar high numbers of hypogeous/subhypogeous fungi in the North American Western cordillera. The same may be applicable to the mollusc dispersal of spores of *Gastroboletus* in NW America and the speciation of *Weraroa* in New Zealand. Not all animal transportation has resulted in the hypogeous state, as demonstrated by the sometimes floral, often bizarre, but successful members of the Phallales.

It is unlikely that angiocarpic, hypogeous or epigeous sporocarps evolved significantly prior to the Mesozoic. However, extremes in climatic cycles most likely supplied the selection pressure which resulted in the pileus margin clasping the stipe to conserve humidity around the developing spores and loss of forcible spore discharge in many groups.

The resulting alteration in sporocarp morphology took place to accommodate life above ground in xeric ecotypes and deserts where dispersal is achieved by abrasion and erosion of the outer pileal surface to expose the propagules and convergent evolution is strongly suggested by the extremely variable taxa involved.

The key question of the derivation of the secotioid agaric has long been debated (Heim, 1948; Savile, 1968; Watling, 1978, 1982). Thus, whilst Singer (1951, 1958) envisaged that at least some secotioid Gasteromycetes gave rise to certain elements of the Agaricales and Russulales, the opposite view has been convincingly presented by Thiers (1984). Through work on *Macowanites luteolus*, *Zelleromyces ravenelii*, and *Gymnomyces yubaensis* by S. L. Miller (1985), we have become aware of the presence of statismosporic, heterotropic and orthotropic basidiospores on the tramal plates of a single secotioid specimen. Thus, two types of basidia have been observed many times and it is therefore, not surprising that sporadic spore deposits have been obtained from *Macowanites* as reported by several mycologists (Singer & Smith, 1960; Pegler & Young, 1981) and in certain gasteroid Cortinariaceae and Bolbitiaceae (Watling, 1977).

Of greater importance is that heterotropic and orthotropic basidiospores are developmentally very similar and can be expressed by the same genome. As Miller (1985) indicates ' . . . development of asymmetric sterigmata in orthotropic basidiosporogenesis corresponds closely to establishment of early spore asymmetry in heterotropic basidiospore formation and suggests that heterotropy and orthotropy may be more similar than previously believed . . . '. In terms of evolution, the presence of secotioid members of a family or order may not, therefore, signify a major departure nor does it suggest derivation of an agaricoid line from secotioid ancestors. Rather, these changes imply recent responses to environmental change coupled with alternatives to forcible spore discharge as a means of spore dissemination.

It is also striking that the vast majority of secotioid taxa found in mesic forest communities are related to ectomycorrhizal agaric families (Trappe, 1962) while the epigeous, secotioid taxa in xeric and desert habitats are related to families of dark spored decomposers (Miller, 1982, 1983). In the latter cases, the production of secondarily thickened cell walls provides tough, desiccation-resistant tissues and basidiospores and melanin pigments further protect the spore from damage by UV radiation. The thick-walled spores usually have an apical germ pore and possess an open-pored hilum to accommodate spore germination. In contrast, mycorrhizal, secotioid, and often hypogeous, agaric relatives have developed odours at maturity to attract animals or insects and tough tissues which protect the developing basidiospores. Modifications in spore morphology have occurred which enable these propagules to survive in soil without the vulnerability incurred by a germ pore.

Interrelations between agaric genera

In an earlier Symposium of this Society, in 1979, Watling (unpublished) offered a scheme of interrelationships between the families of agarics (Fig. 29.1) suggesting an early separation of two main streams of agarics; one pale or white-spored agarics and another coloured spored, the former evolving from the latter by simplification of the spore-wall, a phenomenon seen even in present-day taxa, e.g. *Tricholoma cystidiosum*, which is a white spored *Inocybe*.

The Cortinariaceae (i.e. the Agaricales *sensu stricto* of Kühner, 1980)

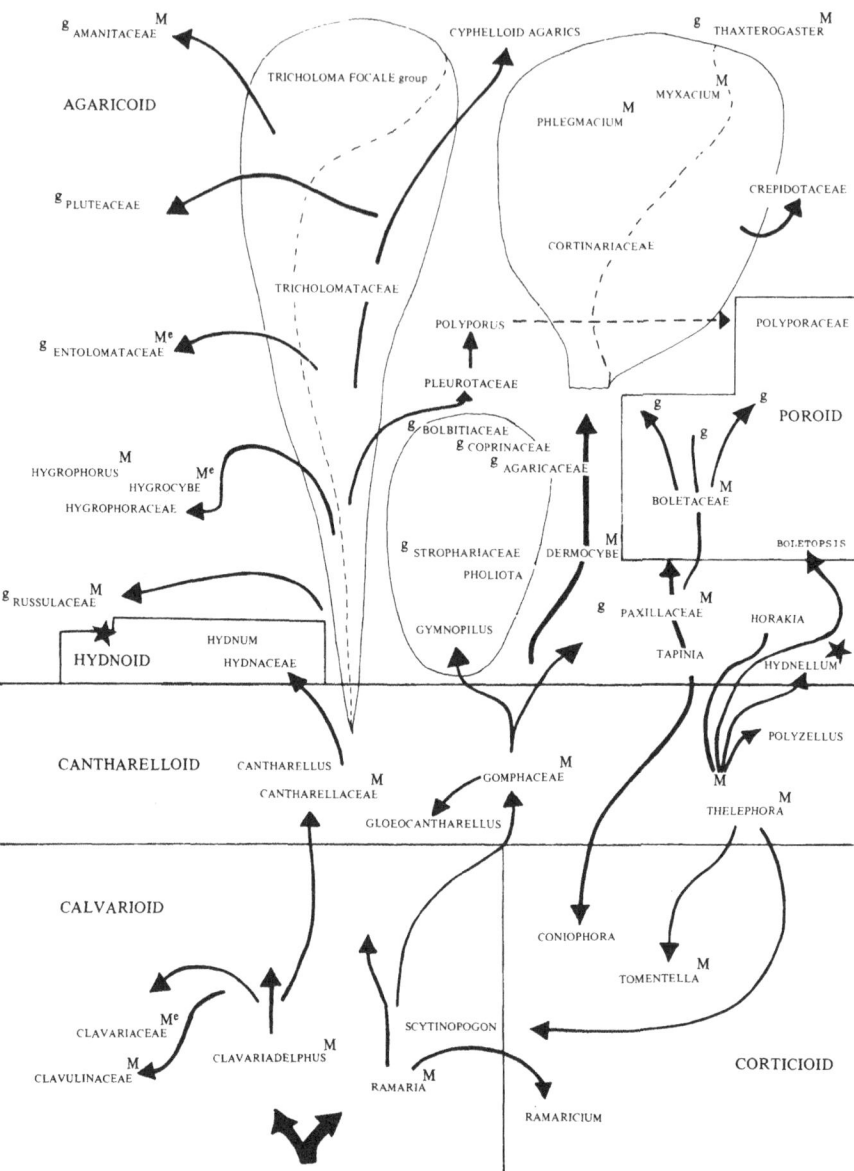

Fig. 29.1. Selected extant genera and major groups of agarics and allies linked by common morphological, anatomical, developmental and chemical characters to offer a feasible origin of agaricoid fungi. Key: m, over 75% mycorrhizal; m^e, possibly endomycorrhizal; g, includes gasteroid members; *, two unrelated hydnoid (toothed) genera. *Myxacium* and *Phlegmacium* are subgenera of *Cortinarius*.

arose from a gomphoid ancestor from which one branch led to *Paxillus* plus the array of bolete genera and another branch led to *Gymnopilus* and thence to the Strophariaceae, Bolbitiaceae, Coprinaceae and Agaricaceae, all of which are saprotrophic. A third line, to *Cortinarius* through subgenus *Dermocybe* as outlined by Høiland (1983), separates further to Inocybeae and differentiated Cortinarieae.

From a *Cantharellus*-like ancestor the tricholomataceous agarics arose through a *Clitocybe/Omphalina* consortium to *Tricholoma*, Amanitaceae, Entolomataceae and Plutaceae. We do not agree with Høiland (1983) in relating *Tricholoma sensu stricto* with *Cortinarius* (subgenus *Phlegmacium*) except for *T. bulbiger* which would now be placed in *Leucocortinarius*.

Cantharellus, *Clavariadelphus*, and *Hydnum* (*H. repandum* group) are linked together, as are *Gomphus*, *Ramaria*, *Ramaricium*, *Kavina*, and *Lentaria*. Thus, the general clavarioid basidioma can apply to all the major lines of Homobasidiomycetes. Far from the amphigenic hymenium being a hindrance to these suggestions, as implied by Savile (1968), it is, rather, an integral stage leading to gymnocarpic agarics, via reduction of the hymenium to a specific limited site, and later to velangiocarpic agarics, via protection of the hymenium, as suggested by Church (1920) and later Corner (1964). The almost amphigenic hymenium of some immature agarics supports this view (Watling, 1985a).

Cyphelloid fungi, although equally worthy of attention, have not been discussed in the past in as much detail as the secotioid agarics. However, it has become abundantly clear that some of these fungi should be integrated into existing agaric families. Thus *Arrhenia* can be related to *Omphalina* (Redhead, 1984), *Hemimycena* to *Mycena* and *Calyptella* to *Marasmius*, all Tricholomataceae. As with secotioid fungi, the cyphelloid character has occurred in different groups at different times. The majority of cyphelloid fungi are saprotrophic, with at least one parasitic, whereas secotioid fungi may be biotrophic or saprotrophic depending on from which family or genus they arose, e.g. *Gastroboletus* (Boletaceae) − mycorrhizal, *Galeropsis* (Strophariaceae) − saprotrophic. The ease by which the secotioid and cyphelloid states can be achieved is evident from the behaviour in culture *Psilocybe merdaria* (Watling, 1971).

Donk (1964) first incorporated a range of macro- and micro-chemical reactions into the familial classification of the Aphyllophorales and by following his concepts for agarics we can link a series of morphotypes generally kept separate using classical approaches.

In the Tricholomataceae there is a line of mycorrhizal taxa from

Table 29.1. *Families of agaricoid fungi with the estimated number of constituent taxa taken from the Dictionary of the Fungi (Hawksworth, Sutton & Ainsworth, 1983) and modified by an expansion factor calculated by RW from the number of species recently described in the Index of Fungi, with an estimate of their suspected mycorrhiza-formers. Based in part on Watling (1982). Gasteroid members of these families, numbering about 500 in all, are not included*

Agaricoid family	Estimated number of taxa	Estimate of suspected mycorrhizal species (% of total)
Agaricales		
Agaricaceae	250	0
Bolbitiaceae	175	0
Coprinaceae	850	0
Cortinariaceae	1700	45, poss. → 75
Lepiotaceae	225	0
Strophariaceae	200	0
Tricholomatales		
Hygrophoraceae	200	50, poss. → 100
Pleurotaceae	150	0
Tricholomataceae	1350	11, poss. → 17
Amanitaceae	250	95
Pluteales		
Pluteaceae	100	0
Entolomataceae	300	? 100
Russulales		
Russulaceae	750	95
Boletales		
Boletaceae (inc. Strobilomycetaceae)	375	90+
Gomphidiaceae	30	100
Paxillaceae	30	poss. 100
Thelephorales		
Thelephoraceae (inc. Bankeraceae)	150	90+
Gomphales		
Gomphaceae and Ramariaceae	150	90+

Cantharellus to the North temperate, annulate *Tricholoma* species, with many large saprotrophic units branching off, e.g. *Collybia, Mycena, Marasmius*, in which several taxa exhibit considerable resource selectivity, e.g. *Marasmius buxi* only on *Buxus* (Rayner, Watling & Frankland,

1985). In the present Cortinariaceae, the mycorrhizal facies is dominant, but an early branch of saprotrophs, *Gymnopilus*, gave rise to a large group of other coloured spored agarics. A second, unrelated, saprotrophic branch would be the Crepidotaceae, many members becoming parallel in field characters to the Pleurotaceae. Although, in the Tricholmataceae the mycorrhizal characteristic is less prominent than the Cortinariaceae, lichenisation is exhibited, a feature only found elsewhere in some aphyllophoraceous fungi, e.g. Dictyonemataceae, but more importantly in *Multiclavula* in the *Clavariadelphus* consortium.

Kühner (1980) has suggested a link between the cantharelloid fungi and *Phyllotopsis* and if this is the case, the tough Pleurotaceae *sensu stricto* and Polyporaceae can be related as a line parallel to that from the fleshly Paxillaceae to the boletes. Undoubtedly, Polyporaceae is a derived group, some becoming reduced to what we now call Corticiaceae. The Russulaceae appear to have few obvious links but the most probable connection is with the Laccariaceae (Kühner, 1980) considered herein as part of the Tricholomataceae.

Conclusion

The simple series from clavarioid to agaricoid basidiomes outlined above is found either entirely or in part in many quite distinct basidiomycete lines. Its beauty is its intrinsic simplicity, and the ease by which a column of tissue can be formed with an amphigenous hymenium producing sexual spores. It is by modification of this column that the agarics, polypores and hydnums have developed in a parallel way to the morels from the ascomycetous disc. In certain basidiomycetous lines some links have been lost, or the lines terminate very early. Mycorrhizas probably also underlie the success of many basidiomycetes and it is estimated that between 3500 − 5000 taxa are involved (Table 29.1). A re-examination of the available data of both fossil and extant taxa shows a reliable picture can be constructed. With the limited fossil data, the authors believe that a logical picture as to the directions of evolution in the agaricoid fungi can still be shown from a study of present-day species.

References

Bambach, R. K., Scotese, C. R. & Ziegler, A. M. (1981). Before Pangea: the geographies of the Paleozoic world. In *Climates Past and Present*, ed. Skinner, B. J., pp. 48 − 60. Los Altos, California: William Kaufmann Inc.

Beck, C. B. (1976). Current status of the Progymnospermopsida. *Review of Palaeobotany and Palynology*, **21**, 5 – 23.

Berch, S. M., Miller, O. K. & Thiers, H. D. (1984). Evolution of mycorrhizae. *Abstracts, 6th North American Conference on Mycorrhizae*, pp. 189 – 192. Corvallis, Oregon: College of Forestry, Oregon State University.

Bondartzeva, M. A. (1963). On the anatomical criterion in the taxonomy of Aphyllophorales. *Botaničeskij žurnal SSSR*, **48**, 362 – 372.

Cain, R. F. (1972). Evolution of the fungi. *Mycologia*, **64**, 1 – 14.

Church, A. H. (1920). Elementary notes on the morphology of fungi. *Botanical Memoirs*, **7**, 1 – 29.

Corner, E. J. H. (1950). A monograph of *Clavaria* and allied genera. *Annals of Botany*, Memoirs no. 1, 1 – 740.

Corner, E. J. H. (1964). *The Life of Plants*. London: Weidenfeld & Nicolson.

Cronquist, A. (1981). *An Integrated System of Classification of Flowering Plants*. Boston: Houghton Mifflin Co.

Danielson, R. M., Zak, J. C. & Parkinson, D. (1984). Mycorrhizal inoculum in a peat deposit formed under a white spruce stand in Alberta. *Canadian Journal of Botany*, **62**, 2557 – 2560.

Dennis, R. L. (1969). Fossil mycelium with clamp connections from the middle Pennsylvanian. *Science*, **163**, 670 – 671.

Donk, M. A. (1964). A conspectus of the families of Aphyllophorales. *Persoonia*, **3**, 199 – 324.

Englander, L. & Hull, R. J. (1977). A possible mycorrhizal association between *Clavaria* species and ericaceous plants. *Proceedings of the American Phytopathological Society*, **4**, 186 – 187.

Englander, L. & Hull, R. J. (1980). Reciprocal transfer of nutrients between ericaceous plants and a *Clavaria* sp. *New Phytologist*, **84**, 661 – 667.

Ford, V. L., Torbert, J. L., Burger, J. A. & Miller, O. K. (1985). Comparative effects of four mycorrhizal fungi on Loblolly pine seedlings growing in a greenhouse in a Piedmont soil. *Plant & Soil*, **83**, 215 – 221.

Hacskaylo, E. (1965). *Thelophora terrestris* and mycorrhizae of Virginia Pine. *Forest Science*, **11**, 401 – 404.

Hawksworth, D. L., Sutton, B. C. & Ainsworth, G. C. (1983). *Dictionary of the Fungi*. Kew, Surrey: Commonwealth Mycological Institute.

Heim, R. (1948). Phylogeny and natural classification of macro-fungi. *Transactions of the British Mycological Society*, **30**, 161 – 178.

Høiland, K. (1983). *Cortinarius* subgenus *Dermocybe* with special regard to the species in the Nordic countries. *Opera Botanica*, **71**, 1 – 112.

Julich, W. (1984). *Die Nichtblatterpilze, Gallertpilze und Bauchpilze*. Band IIb/1, Basidiomyceten, 1 Teil. New York: Gustav Fischer Verlag.

Kühner, R. (1980). Les Hyménomycètes agaricoides. *Bulletin mensuel de la Société Linnéenne de Lyon*, numéro spécial, 1 – 1028.

Kühner, R. (1984). Some mainlines of classification in the gill fungi. *Mycologia*, **76**, 1059 – 1074.

Malloch, D. W., Pirozynski, K. A. & Raven, P. H. (1980). Ecological and evolutionary significance of mycorrhizal symbioses in vascular plants (a review). *Proceedings of the National Academy of Sciences, USA*, **77**, 2112 – 2118.

Miller, C. N. (1977). Mesozoic conifers. *Botanical Review*, **43**, 217 – 280.

Miller, O. K. (1982). Taxonomy of ecto- and ectendomycorrhizal fungi. In *Methods and Principles of Mycorrhizal Research*, ed. Schenk, N. C., pp. 91 – 101. St Paul, Minnesota: American Phytopathological Society.

Miller, O. K. (1983). Ectomycorrhizae in the Agaricales and Gasteromycetes. *Canadian Journal of Botany*, **61**, 909 − 916.

Miller, O. K. & Stewart, L. (1971). The genus *Lentinellus. Mycologia*, **63**, 333 − 369.

Miller, S. L. (1985). *Basidiosporogenesis and Developmental Anatomy of Spore Release in the Russulales - A Systematic Interpretation.* Doctoral Thesis, Virginia Polytechnic Institute and State University, Blacksburg, Virginia, USA.

Nobles, M. K. (1965). Identification of cultures of wood-inhabiting hymenomycetes. *Canadian Journal of Botany*, **43**, 1097 − 1139.

Pegler, D. N. & Young, T. W. K. (1981). World pollen and spore flora, 10. In *Russulaceae Roze*, ed. Nilsson, S., pp. 1 − 35. Stockholm: The Almquist & Weksell Periodical Co.

Petersen, R. H. (1973). Aphyllophorales II: the clavarioid and cantharelloid Basidiomycetes. In *The Fungi, an Advanced Treatise*, vol. IVB, ed. Ainsworth, G. C., Sparrow, F. K. & Sussman, A. S., pp. 351 − 368. New York: Academic Press.

Petersen, T. A., Mueller, W. C. & Englander, L. (1980). Anatomy and ultrastructure of a *Rhododendron* root-fungus association. *Canadian Journal of Botany*, **58**, 2421 − 2433.

Pirozynski, K. A. & Malloch, D. W. (1975). The origins of land plants: a matter of mycotrophism. *BioSystems*, **6**, 153 − 164.

Rayner, A. D. M., Watling, R. & Frankland, J. C. (1985). Resource relationships − an overview. In *Developmental Biology of Higher Fungi*, British Mycological Society Symposium volume 10, ed. Moore, D., Casselton, L. A., Wood, D. A. & Frankland, J. C., pp. 1 − 40. Cambridge University Press.

Read, D. J. (1983). The biology of mycorrhiza in the Ericales. *Canadian Journal of Botany*, **61**, 985 − 1004.

Redhead, S. A. (1984). *Arrhenia* and *Rimbachia*, expanded generic concepts, and a re-evaluation of *Leptoglossum* with emphasis on muscicolous North American Taxa. *Canadian Journal of Botany*, **62**, 865 − 892.

Redhead, S. A. & Ginns, J. H. (1985). A reappraisal of agaric genera associated with brown rots of wood. *Transactions of the Mycological Society of Japan*, **26**, 349 − 381.

Rothwell, G. W. (1972). *Palaeosclerotium pusillum* gen. et sp. nov., a fossil eumycete from the Pennsylvanian of Illinois. *Canadian Journal of Botany*, **50**, 2353 − 2356.

Savile, D. B. O. (1955). A phylogeny of the Basidiomycetes. *Canadian Journal of Botany*, **33**, 60 − 104.

Savile, D. B. O. (1968). Possible interrelations between fungal groups. In *The Fungi, an Advanced Treatise*, vol. 3, ed. Ainsworth, G. C. & Sussman, A. S., pp. 649 − 675. New York: Academic Press.

Schuster, R. (1976). Plate tectonics and its bearing on the geographical origin and dispersal of the Angiosperms. In *Origin and Early Evolution of Angiosperms*, ed. Beck, C. B., pp. 48 − 138. New York: Columbia University Press.

Seviour, R. J., Willing, R. R. & Chilvers, G. A. (1973). Basidiocarps associated with ericoid mycorrhizas. *New Phytologist*, **72**, 381 − 385.

Singer, R. (1951). The Agaricales (mushrooms) in modern taxonomy. *Lilloa*, **22** (1949), 5 − 832.

Singer, R. (1958). The meaning of the affinity of the Secotiaceae with the Agaricales. *Sydowia*, **12**, 1 − 43.

Singer, R. & Smith, A. H. (1960). Studies on Secotiaceous fungi. IX. The Astrogastraceous series. *Memoirs of the Torrey Botanical Club*, **21**, 1 − 122.

Stubblefield, S. P., Taylor, T. N. & Beck, C. B. (1985). Studies of Paleozoic fungi. IV. Wood-decaying fungi in *Callixylon newberryi* from the Upper Devonian. *American Journal of Botany*, **72**, 1765 — 1774.

Thiers, H. D. (1984). The secotioid syndrome. *Mycologia*, **76**, 1 — 8.

Trappe, J. M. (1962). Fungus associates of ectotrophic mycorrhizae. *Botanical Reviews*, **28**, 538 — 606.

Warcup, J. H. (1980). Ectomycorrhizal associations of Australian indigenous plants. *New Phytologist*, **85**, 531 — 535.

Watling, R. (1971). Polymorphism in *Psilocybe merdaria*. *New Phytologist*, **70**, 307 — 326.

Watling, R. (1977). Observations on the Bolbitiaceae 3. A xeromorphic member of the family. *Kew Bulletin*, **31**, 587 — 594.

Watling, R. (1978). From infancy to adolescence: advances in the study of higher fungi. *Transactions of the Botanical Society of Edinburgh*, **42** (supplement), 61 — 74.

Watling, R. (1982). Taxonomic status and ecological identity in the basidiomycetes. In *Decomposer Basidiomycetes: Their Biology and Ecology*, British Mycological Society Symposium volume 7, ed. Frankland, J. C., Hedger, J. N. & Swift, M. J., pp. 1 — 32. Cambridge University Press.

Watling, R. (1985a). Hymenial surfaces in developing agaric primordia. *Botanical Journal of the Linnean Society*, **91**, 273 — 293.

Watling, R. (1985b). Ecological strategies in the Bolbitiaceae. *Abstracts of the IX Congressus Mycologicus Europaeus*, Oslo.

Index